Membrane Dynamics and Domains

Subcellular Biochemistry

Volume 37

SUBCELLULAR BIOCHEMISTRY

Recent Volumes in this Series

Volume 26	***myo*-Inositol Phosphates, Phosphoinositides, and Signal Transduction** Edited by B.B. Biswas and Susweta Biswas
Volume 27	**Biology of the Lyosome** Edited by John B. Lloyd and Robert W. Mason
Volume 28	**Cholesterol: It's Functions and Metabolism in Biology and Medicine** Edited by Robert Bittman
Volume 29	**Plant–Microbe Interactions** Edited by B.B. Biswas and H.K. Das
Volume 30	**Fat-Soluble Vitamins** Edited by Peter J. Quinn and Valerian E. Kagan
Volume 31	**Intermediate Filaments** Edited by Harald Herrmann and J. Robin Harris
Volume 32	**α-Gal and Anti-Gal: α1,3-Galactosyltransferase, α-Gal Epitopes, and the Natural Anti-Gal Antibody** Edited by Uri Galili and José-Luis Avila
Volume 33	**Bacterial Invasion into Eukaryotic Cells** Edited by Tobias A. Oelschlaeger and Jörg H. Hacker
Volume 34	**Fusion of Biological Membranes and Related Problems** Edited by Herwig J. Hilderson and Stephen Fuller
Volume 35	**Enzyme-Catalyzed Electron and Radical Transfer** Edited by Andreas Holzenburg and Nigel S. Scrutton
Volume 36	**Phospholipid Metabolism in Apoptosis** Edited by Peter J. Quinn and Valerian E. Kagan
Volume 37	**Membrane Dynamics and Domains** Edited by Peter J. Quinn

---

A Continuation Order Plan is available for this series. A continuation order will bring delivery of each new volume immediately upon publication. Volumes are billed only upon actual shipment. For further information please contact the publisher.

Membrane Dynamics and Domains

Subcellular Biochemistry Volume 37

Edited by

Peter J. Quinn
Kings College London
London, UK

Kluwer Academic / Plenum Publishers
New York, Boston, Dordrecht, London, Moscow

ISSN 0306-0225

ISBN 0-306-48425-0

This volume is a continuation of the journal *Sub-Cellular Biochemistry*,
Volume 1-4 of which were published quarterly from 1972 to 1975

233 Spring Street, New York, New York 10013

http//www.wkap.nl/

10 9 8 7 6 5 4 3 2 1

A C.I.P. record for this book is available from the Library of Congress.

Printed in the United States of America

Contributors

Ruedi Aebersold, Institute for Systems Biology, 1441 North 34th Street, Seattle, WA 98103-8904, USA

Nicholas Andrews, Advanced Material Laboratory, Materials and Process Center, Sandia National Laboratories, Albuquerque, NM, USA

Yechezkel Barenholz, Laboratory of Membrane and Liposomal Research, Department of Biochemistry, The Hebrew University – Hadassah Medical School, PO Box 12272, Jerusalem 91120, Israel

Timothy J. Boyle, Advanced Material Laboratory, Materials and Process Center, Sandia National Laboratories, Albuquerque, NM, USA

Scott Bunge, Advanced Material Laboratory, Materials and Process Center, Sandia National Laboratories, Albuquerque, NM, USA

Helen Cox, Centre for Neuroscience, New Hunt's House, Guy's Campus, London SE1 1UL, United Kingdom

Samuel Donohoe, Institute for Systems Biology, 1441 North 34th Street, Seattle, WA 98103-8904, USA

Jimmy Eng, Institute for Systems Biology, 1441 North 34th Street, Seattle, WA 98103-8904, USA

HongYou Fan, Advanced Material Laboratory, Materials and Process Center, Sandia National Laboratories, Albuquerque, NM, USA

Paul L. Fox, Department of Cell Biology, The Lerner Research Institute, Cleveland Clinic Foundation, 9500 Euclid Avenue, Cleveland, OH 44195, USA

David R. Goodlett, Department of Medicinal Chemistry, Box 357610, University of Washington, Seattle, WA 98195, USA

Eric K.-W. Hui, Department of Microbiology, Immunology and Molecular Genetics, UCLA School of Medicine, Los Angeles, CA 90095-1747, USA

Andrew Keller, Institute for Systems Biology, 1441 North 34th Street, Seattle, WA 98103-8904, USA

Paul Kotula, Materials Characterization Department, Materials and Process Center, Sandia National Laboratories, Albuquerque, NM, USA

Kamen S. Koumanov, Institute of Biophysics, Bulgarian Academy of Sciences, 1113 Sofia, Bulgaria

Akihiro Kusumi, Kusumi Membrane Organizer Project, ERATO, Japan Science and Technology Corporation, 5-11-33, Nagoya, Japan

Xiao-jun Li, Institute for Systems Biology, 1441 North 34th Street, Seattle, WA 98103-8904, USA

Karin Leiderman, Department of Mathematics and Statistics, University of New Mexico, School of Medicine, Albuquerque, NM 87131, USA

Karina F. Meiri, Department of Anatomy and Cell Biology, Tuft's University School of Medicine, 136 Harrison Avenue, Boston, Massachusetts 02111, USA

Enrico Mombelli, Molecular Neurobiology Group, MRC Centre for Developmental Neurobiology, 4th Fl South New Hunt's House, Guy's Campus, London SE1 1UL, United Kingdom

Roger Morris, Molecular Neurobiology Group, MRC Centre for Developmental Neurobiology, 4th Fl South New Hunt's House, Guy's Campus, London SE1 1UL, United Kingdom

Debi P. Nayak, Department of Microbiology, Immunology and Molecular Genetics, UCLA School of Medicine, Los Angeles, CA 90095-1747, USA

Alexey I. Nesvizhskii, Institute for Systems Biology, 1441 North 34th Street, Seattle, WA 98103-8904, USA

Janet M. Oliver, Department of Pathology, University of New Mexico, School of Medicine, Albuquerque, NM 87131, USA

Marie-Odile Parat, Department of Cell Biology, The Lerner Research Institute, Cleveland Clinic Foundation, 9500 Euclid Avenue, Cleveland, OH 44195, USA

Janet R. Pfeiffer, Department of Pathology, University of New Mexico, School of Medicine, Albuquerque, NM 87131, USA

Fred Possmayer, Departments of Obstetrics & Gynaecology and Biochemistry, CIHR Group in Fetal and Neonatal Health and Development, The University of Western Ontario, London, ON, N6A 5A5 Canada

Peter J. Quinn, Department of Life Sciences, King's College London, 150 Stamford Street, London SE1 9NN, United Kingdom

Galina Radeva, Department of Chemistry and Biochemistry, University of Guelph, Guelph, Ontario, Canada N1G 2W1

Marilyn D. Resh, Cell Biology Program, Memorial Sloan-Kettering Cancer Center, 1275 York Avenue, Box 143, New York, NY 10021

Ken Ritchie, Department of Biological Science, Nagoya University, Chikusa-ku, 464-8602 Nagoya, Japan

Margaret L. Sanders, Department of Mathematics and Statistics, University of New Mexico, School of Medicine, Albuquerque, NM 87131, USA

Michael Schrader, Department of Cell Biology and Cell Pathology, Philipps-University, Robert Koch Strasse 5, 35037 Marburg, Germany

Frances J. Sharom, Department of Chemistry and Biochemistry, University of Guelph, Guelph, Ontario, N1G 2W1 Canada

Stanley L. Steinberg, Department of Mathematics and Statistics, University of New Mexico, School of Medicine, Albuquerque, NM 87131, USA

Zurab Surviladze, Department of Pathology, University of New Mexico, School of Medicine, Albuquerque, NM 87131, USA

Kelly Vaughn, Institute for Systems Biology, 1441 North 34th Street, Seattle, WA 98103-8904, USA

Ruud Veldhuizen, Departments of Medicine and Physiology and Pharmacology, Lawson Health Research Institute, 268 Grosvenor Street, London, ON, N6A 4V2, Canada

Priska D. von Haller, MacroGenics, Inc., 1441 North 34th Street, Seattle, WA 98103, USA

Julian D. Watts, Institute for Systems Biology, 1441 North 34th Street, Seattle, WA 98103-8904, USA

Bridget S. Wilson, Materials Characterization Department, Materials and Process Center, Sandia National Laboratories, Albuquerque, NM, USA

Carla Wofsy, Department of Mathematics and Statistics, University of New Mexico, School of Medicine, Albuquerque, NM 87131, USA

Claude Wolf, Faculte de Medecine Saint Antoine, INSERM U538, 27 Rue Chaligny, 75571 Paris Cedex 12, France

Bernd Wollscheid, Institute for Systems Biology, 1441 North 34th Street, Seattle, WA 98103-8904, USA

Eugene Yi, Institute for Systems Biology, 1441 North 34th Street, Seattle, WA 98103-8904, USA

Jun Zhang, Department of Computer Science, University of New Mexico, School of Medicine, Albuquerque, NM 87131, USA

Preface

The fluid-mosaic model of membrane structure formulated by Singer and Nicolson in the early 1970s has proven to be a durable concept in terms of the principles governing the organization of the constituent lipids and proteins. During the past 30 or so years a great deal of information has accumulated on the composition of various cell membranes and how this is related to the different functions that membranes perform. Nevertheless, the task of explaining particular functions at the molecular level has been hampered by lack of structural detail at the atomic level. The reason for this is primarily the difficulty of crystallizing membrane proteins which require strategies that differ from those used to crystallize soluble proteins. The unique exception is bacteriorhodopsin of the purple membrane of *Halobacterium halobium* which is interpolated into a membrane that is neither fluid nor in a mosaic configuration. To date only 50 or so membrane proteins have been characterised to atomic resolution by diffraction methods, in contrast to the vast data accumulated on soluble proteins. Another factor that has been difficult to explain is the reason why the lipid compliment of membranes is often extremely complex. Many hundreds of different molecular species of lipid can be identified in some membranes. Remarkably, the particular composition of each membrane appears to be maintained within relatively narrow limits and its identity distinguished from other morphologically-distinct membranes. As yet we may only speculate about the biochemical mechanisms responsible for creating such complexity and the homeostatic mechanisms that are in place to preserve it.

While the original model conceded that specific interactions between different proteins and even between proteins and particular membrane lipids might be responsible for non-random distribution of components within the structure it was suggested that "long-range random arrangements are the norm". With benefit of hindsight it is not surprising that what was considered to be the norm most likely will turn out to be the exception. Amongst the most convincing evidence was the realization that the membrane matrix was not a random assortment of fluid lipids arranged in a bilayer configuration. Firstly, it became apparent that the lipids present in each leaflet of the bilayer were different. Biochemical processes for creating and maintaining membrane lipid asymmetry were subsequently characterised. Next, the importance of lateral phase separation of lipids initially via gel phase separation and subsequently via specific lipid–lipid interactions was recognised. Related to this is the need to prevent phase separation of molecular species of lipids present in all cell membranes that, if unconstrained by interaction with other membrane components, would form nonbilayer structures. One concept to emerge from these ideas was the formulation of the raft hypothesis that postulated the existance of lipid domains in the exoplasmic membrane leaflet created by specific interactions between cholesterol and saturated molecular species of sphingolipids. These domains were said to be in a liquid-ordered state, in contrast to the surrounding fluid lipid bilayer.

This shift in conceptualisation of the arrangement of the lipids led to the idea that modulation of the non-random distribution of membrane components represents an important mechanism for regulating the behaviour of cells and their response to external stimuli. For example, the dissipation of membrane lipid asymmetry is believed to trigger apoptosis, initiate blood clotting and is a factor in response of cells to oxidative stress. Likewise, the raft hypothesis envisages that liquid-ordered domains are dynamic in that their size and distribution is not static. The rafts serve to sort proteins associate with the membrane. Certain classes of protein, for example, glycosylphosphatidylinositol-anchored, fatty acylated, prenylated and proteins tethered to cholesterol appear to be attracted to these liquid-ordered domains. This class of membrane proteins is often associated with transmembrane signalling processes and their concentration into rafts is believed to be integral to their signalling functions.

Sorting of membrane proteins can also be assisted by protein–protein interactions. Interactions between proteins are believed to underlie the formation of intercellular junctions of various types such as gap junctions, zona occludens junctions and desmosomes. Likewise, the interactions between the cell and elements of the extracellular matrix are known to involve specific interactions between proteins. Lastly, homotypic associations between specialized proteins such as the caveolins and clathrin are responsible for trafficking membrane constituents about the cell. The formation of specialized structures resulting from short-range specific interactions between integral membrane proteins in creating large regular arrays was indeed envisaged by the fluid-mosaic model but their utility was hardly conceived.

This volume assembles contemporary evidence of domain formation in membranes gathered by leading researchers in fields ranging from membrane biophysics to cell biology. Questions relating to how non-random membrane arrangements are defined, detected and monitored to characterise their dynamic behaviour are addressed. The components associated with membrane rafts are examined and the different mechanisms employed to create membrane microdomains are considered. Evidence relating to the regulation of membrane microdomain formation in the context of membrane biogenesis and homeostasis is discussed. The role of membrane domain formation in the generation and transmission of signals transduced via membranes to evoke a variety of physiological responses is examined. Finally, examples of where membrane domain structures are involved in infection and disease states are provided.

Peter J. Quinn
London

Contents

Part I: Membrane Domain Structure

Part II: Membrane Domain Composition

Part III: Creation of Membrane Microdomains

Part IV: Regulation of Domain Formation

Part V: Signal Transduction Processes

Part VI: Domain Dynamics in Disease

Part I

Membrane Domain Structure

Chapter 1

Membrane Receptor Mapping: The Membrane Topography of FcεRI Signaling

Janet M. Oliver[1], Janet R. Pfeiffer[1], Zurab Surviladze[1], Stanly L. Steinberg[2], Karin Leiderman[2], Margaret L. Sanders[2*], Carla Wofsy[2#], Jun Zhang[3], HongYou Fan[4], Nicholas Andrews[4], Scott Bunge[4], Timothy J. Boyle[4], Paul Kotula[5], and Bridget S. Wilson[1]

[1]Departments of Pathology; [2]Mathematics and Statistics; [3]Computer Science, University of New Mexico; [4]Advanced Material Laboratory; [5]Materials Characterization Department, Materials and Process Center, Sandia National Laboratories, Albuquerque, NM, USA

Abstract

Ligand binding to membrane receptors initiates cascades of biochemical events leading to physiological responses. Hundreds of proteins and lipids are implicated in signaling networks and programs in genomics and proteomics are continuously adding new components to the signaling "parts lists". Here, we generate high resolution maps of signaling networks using cytoplasmic face-up membrane sheets that can be labeled with immunogold probes (3–10 nm) and imaged in the transmission electron microscope. Our model system is the mast cell and we focus on mapping the topography of the high affinity IgE receptor, FcεRI, its associated tyrosine kinases, Lyn and Syk, and the signaling proteins that propagate signals from these kinases. Crosslinked receptors and their

[*] M.L. Sanders is presently in the Department of Family and Community Medicine at the University of New Mexico.

[#] Our colleague, Carla Wofsy, died on August 4, 2003, after a long illness. This chapter is dedicated to her memory as a pioneering mathematical biologist and as a human of enormous grace, integrity and compassion.

Subcellular Biochemistry, Volume 37: Membrane Dynamics and Domains

Edited by Peter J. Quinn. Kluwer Academic / Plenum Publishers, New York, 2004

signaling partners segregate during signaling to multiple, dynamic membrane domains, including a transient FcεRI-Lyn domain and at least two other distinct domains, one characterized by the presence of receptor, Syk and multiple signaling proteins, but not Lyn (primary signaling domains), and one characterized by the presence of LAT and PLCγ1 but not receptor (secondary signaling domains). PI 3-kinase associates with both primary and secondary signaling domains and may help to recruit specific signaling proteins through the local remodeling of inositol phospholipids. The lipid raft markers, GM1 and Thy-1, fail to localize in native membrane sheets either with each other or with signaling domains. We introduce new probes to localize multiple signaling molecules on the same membrane sheet and new computational tools to capture and analyze their topographical relationships. In the future, we expect that high resolution maps of signaling networks will be integrated with chemical kinetic analyses, with cell fractionation data and with a range of real-time fluorescence measurements, into mathematical models with power to predict mechanisms that regulate the efficiency, specificity, amplitude and duration of signaling pathways.

1. INTRODUCTION

Cell biologists have been exploring deviations of membrane protein topography from the predictions of the fluid mosaic model (Singer and Nicolson, 1972) since the early 1970s. The discovery that selected transmembrane proteins are either included in or excluded from the membrane that is removed from the cell surface during phagocytosis was among the first demonstration that proteins can segregate in the plane of the membrane (Berlin *et al.*, 1974; Oliver *et al.*, 1974). Later evidence that coated pits, membrane compartments specialized for endocytosis, are also distributed non-randomly in membranes reinforced the heterogeneity of membrane organization (Pfeiffer *et al.*, 1980; Oliver and Berlin, 1983). Over the years, it has become abundantly clear that the segregation of proteins and lipids within the fluid membrane of animal cells is critical to many cellular processes including cell differentiation, polarization and directional migration (Edidin, 1997, 2001; Simons and Ikonen, 1997; Jacobson and Dietrich, 1999; Anderson and Jacobson, 2002). The current interest in signaling domains in membranes is a natural extension of these concepts of membrane heterogeneity.

Since the mid-1980s, we have used immunogold labeling and electron microscopy to directly observe the topography of the high affinity IgE receptor, FcεRI, of basophils and mast cells during signal transduction. Early studies using immunogold-labeling and scanning electron microscopy (SEM) showed that crosslinked FcεRI are displaced rapidly from regions of membrane curvature (microvilli, lamellae) and accumulate in clusters on planar

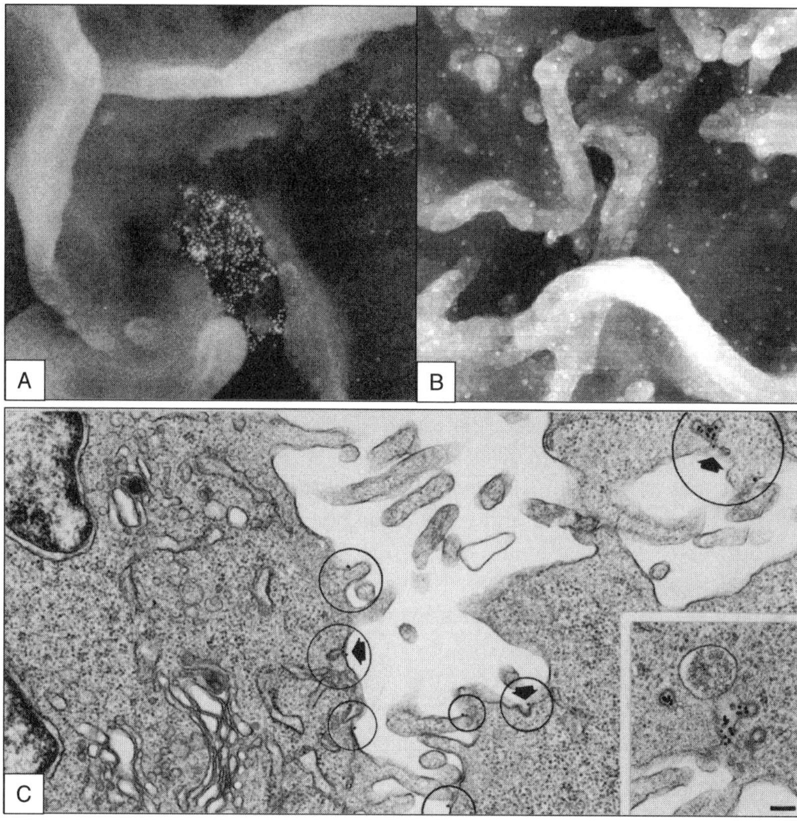

Figure 1. Topography of crosslinked FcεRI and of Thy-1 on mast cells. In A and B, IgE-primed RBL-2H3 mast cells were activated for 5 min with rabbit anti-IgE Ab, then fixed, labeled with mouse anti-Thy-1 mAb, and finally with anti-rabbit-gold (A; to localize crosslinked FcεRI) or with anti-mouse-gold (B; to localize Thy-1). Probes were imaged using scanning electron microscopy. FcεRI crosslinking causes the clustering of receptors, with no accompanying Thy-1 clustering. In C, cells were incubated for 10 min with DNP-BSA-gold, then processed by conventional thin sectioning and transmission electron microscopy. After signaling, crosslinked receptors are internalized through clathrin-coated pits.

regions of membrane (Stump *et al.*, 1989; Seagrave *et al.*, 1991; Figure 1A). Under the same conditions the glycerophospholid-anchored protein, Thy-1, remains distributed over the entire cell surface (Figure 1B). Complementary thin section transmission electron microscopy (TEM) studies showed that receptor redistribution and signaling are followed by endocytosis through clathrin-coated pits (Pfeiffer *et al.*, 1985; Figure 1C).

The possibility of a relationship between the topographically constrained FcεRI clusters observed during signaling and the "lipid rafts" described in other chapters of this book was raised in the mid-1990s, principally through work in the Baird and Holowka laboratory (Field *et al.*, 1995, 1997, 1999) on detergent-resistant membrane fractions (DRMs, "lipid rafts") obtained from mast cells by gradient centrifugation of Triton-X-100-extracted cells. Their work suggested that the interaction of crosslinked FcεRI with Lyn, a dually acylated Src-family tyrosine kinase, in DRMs is the critical event that initiates the FcεRI signaling cascade. Compositional analyses indicated that DRMs are enriched in sphingomyelin, gangliosides, saturated glycerophospholipids and cholesterol and have other properties consistent with a liquid ordered phase that could potentially facilitate their phase segregation from bulk membrane (Fridriksson *et al.*, 1999; Ge *et al.*, 1999).

The interpretation of detergent extraction/gradient centrifugation experiments remains controversial. There is evidence that detergents may force associations between components that are not colocalized in intact cells (Mayor and Maxfield, 1995). It is also known that the biochemical composition of DRMs changes when the Triton-X-100 concentration is changed, when alternative detergents are used, and when detergent is omitted altogether (Montixi *et al.*, 1998; Surviladze *et al.*, 1998, 2001; Field *et al.*, 1999; Parolini *et al.*, 1999; Ilangumaran *et al.*, 1999; Schuck *et al.*, 2003). Methods to observe membrane segregation *in situ* have also generated controversy. Results based on light and fluorescence microscopy of proteins and lipids, including refinements of fluorescence recovery after photobleaching (FRAP) methods and newer single particle tracking (SPT) methods, have led some investigators to conclude that lipid rafts comprise a minor fraction of the cell surface and others to propose that above 50% of the plasma membrane consists of lipid rafts (Saxton and Jacobson, 1997; Kenworthy *et al.*, 2000; Edidin, 2001; Iino and Kusumi, 2001). Several groups estimate lipid rafts to be less than 70 nm in diameter (Friedrichson and Kurzhalia, 1998; Varma and Mayor, 1998; Pralle *et al.*, 2000), and thus below the resolution (>200 nm) of fluorescence microscopy. Estimates of raft stability vary from quite stable to very transient.

It seemed very apparent to us that new methods that would allow direct mapping of signaling complexes on native membranes at molecular resolution were needed to understand membrane organization during signaling

2. MAKING MEMBRANE SHEETS

We implemented a transmission electron microscopy (TEM) method to map the topography of the FcεRI and its associated signaling proteins and lipids at high resolution. The method, described in Wilson *et al.* (2000, 2001, 2002) and Pfeiffer *et al.* (2002), was modified and extended from earlier work by Sanan and Anderson (1991). Cell monolayers adherent to glass coverslips are inverted onto poly-L-lysine-coated nickel EM grids, pressure is applied briefly, and then the coverslips are lifted, leaving cytoplasmic face-up sheets of native membranes. The sheets are fixed lightly, labeled with functionalized immunogold particles (usually 3–10 nm in diameter, but observation of particles as small as 1.4 nm is practical with brief gold enhancement after labeling is complete), then fixed more completely and imaged at high resolution. The TEM images gold probes bound to both the outer and inner leaflet of the plasma membrane. Thus, proteins can be labeled from the outside of the cell before preparing the sheets as well as from the inside on isolated sheets. Cells can be fixed before labeling with probes for extracellular determinants. It is modestly more difficult to produce high quality sheets from prefixed cells that form live cells.

Figure 2 illustrates typical membrane sheets prepared from resting and activated mast cells and labeled with 10 nm gold particles specific for the FcεRI β subunit. The resting cell (Figure 2A) shows a dispersed distribution of small FcεRI clusters over the whole sheet. Monomeric receptors do not associate with coated pits. After 5 minutes of crosslinking (Figure 2B), 10 nm gold particles specific for FcεRI β are concentrated in aggregates on membrane that stains intensely with osmium. Clathrin-coated pits very often occur at the periphery of these osmiophilic patches. After signaling, crosslinked FcεRI become cargo for these endocytic structures.

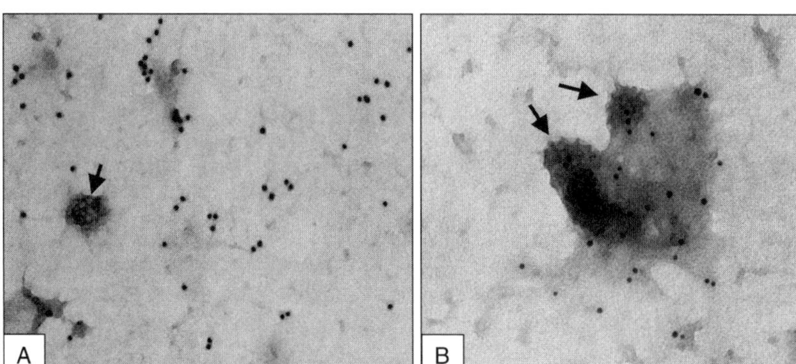

Figure 2. Topography of monomeric and crosslinked FcεRI on membrane sheets. Membrane sheets were prepared from IgE-primed RBL-2H3 cells, without crosslinking (A) and after crosslinking with 1 μg/ml anti-IgE Ab for 5 min (B). Monomeric receptors are distributed as small, dispersed clusters that do not associate with coated pits (arrows). Crosslinked receptors associated with electron-dense membrane regions. After signaling, crosslinked FcεRI are internalized though coated pits that typically bud from the periphery of the patches.

The method is particularly straightforward with cells like RBL-2H3 and fibroblasts that adhere to glass coverslips. For non-adherent cells, we usually plate the cells for 5–10 min onto poly-L-lysine-coated coverslips prior to manipulation and sheet preparation. Some non-adherent cells undergo apoptosis on poly-L-lysine. For these cells, plating onto fibronectin- or silane-coated surfaces generally enables production of satisfactory sheets. Membrane sheets have been used successfully in our laboratory to localize Ras isoforms in fibroblasts (collaborative with the J. Buss laboratory at Iowa State University), the BCR and associated proteins in B cells (collaborative with the B. Vilen laboratory at the University of North Carolina), and the TCR and associated proteins in T cells (collaborative with the M. Davis laboratory at Stanford University). Other groups have used membrane sheets to localize Lck and LAT in T cells (Schade and Levine, 2002) and Ras in COS cells and fibroblasts (Prior *et al.*, 2003).

3. THE BIOCHEMISTRY AND TOPOGRAPHY OF FCεRI SIGNALING

3.1. Signaling proteins in the FcεRI signaling cascade

Our topographical analysis of the FcεRI signaling pathway, below, takes advantage of the detailed "parts list" of signaling proteins discovered in this and many other laboratories, particularly those of H. Metzger, J-P. Kinet, R. Siraganian, M. Beaven, R. Geahlen, J. Bolen, L. Samelson, Y. Kawakami and J. Rivera. The high affinity IgE receptor, FcεRI, of mast cells and basophils is an $\alpha\beta\gamma_2$ tetramer with IgE binding sites in the α subunit and immunoreceptor tyrosine-based activation motifs (ITAMs) in both the β and γ subunit cytoplasmic tails (Blank *et al.*, 1989; Figure 3, inset upper right). FcεRI signaling is initiated by crosslinking two or more adjacent receptors. This crosslinking event activates a palmitylated and myristylated (membrane-associated, putatively lipid raft-seeking) Src-related protein tyrosine kinase, Lyn (Eiseman and Bolen, 1991). Lyn phosphorylates the β and γ subunit ITAMs that in turn provide binding sites for the binding and activation of a separate kinase, that we identified as PTK72 ^{Syk} (Hutchcroft *et al.*, 1992). Syk activation following its recruitment to FcεRIγ leads directly or indirectly to the phosphorylation of multiple cytoplasmic signaling molecules, among them proteins that modify membrane phospholipids (phospholipase Cγ [PLCγ] isoforms, phosphatidylinositol 3-kinase [PI 3-kinase] isoforms and others), proteins that phosphorylate other proteins on serines and threonines as well as on tyrosines (protein kinase C isoforms, Bruton's tyrosine kinase, AKT and others), proteins with GTPase

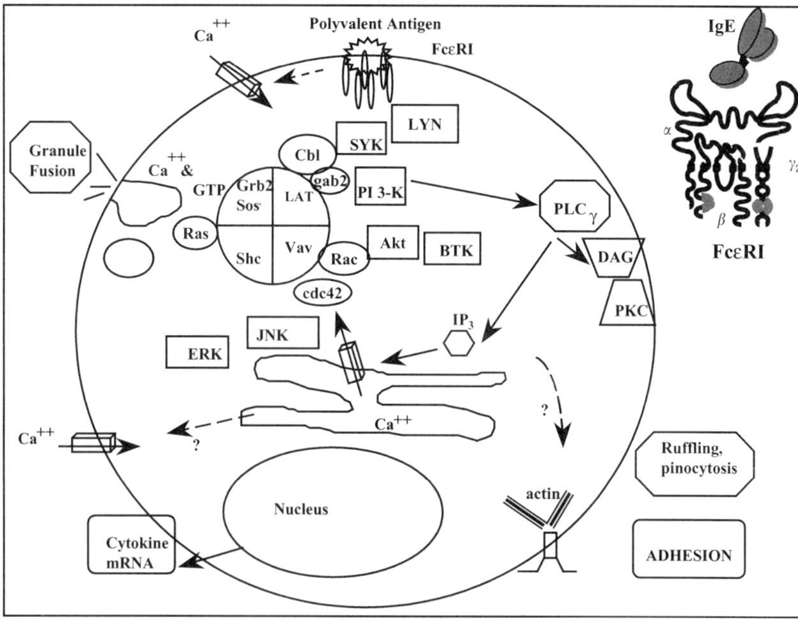

Figure 3. Schematic view of the subunit structure of FcεRI (inset upper right) and of the biochemical and physiological signaling responses initiated by FcεRI crosslinking.

activity (members of the Ras, Rab and Rho protein families implicated respectively in signaling to the nucleus through the MAP kinase cascade, to granule movement and to cytoskeletal reorganization), proteins that activate Ras-related GTPases (Vav, SoS), proteins with ubiquitin ligase as well as adaptor functions (Cbl) and proteins that lack known catalytic activity but contain motifs that allow them to link activated proteins to downstream partners (adaptors; including Gab2, Nck, Grb2, Slp-76, FYB, others) (reviewed in Wilson and Oliver, 2001). Certain transmembrane proteins, including LAT, are also phosphorylated in activated mast cells (Saitoh *et al.*, 2000) and may provide membrane binding sites for signaling proteins with SH2 domains. Figure 3 is a cartoon showing a subset of the signaling proteins whose activities are modulated by FcεRI crosslinking. Figure 3 also notes some of the physiological responses to this crosslinking event.

3.2. The topography of FcεRI signal initiation

Results of experiments where sheets were double-labeled for receptor and the signal initiating kinase, Lyn, are shown in Figure 4. FcεRI co-localizes

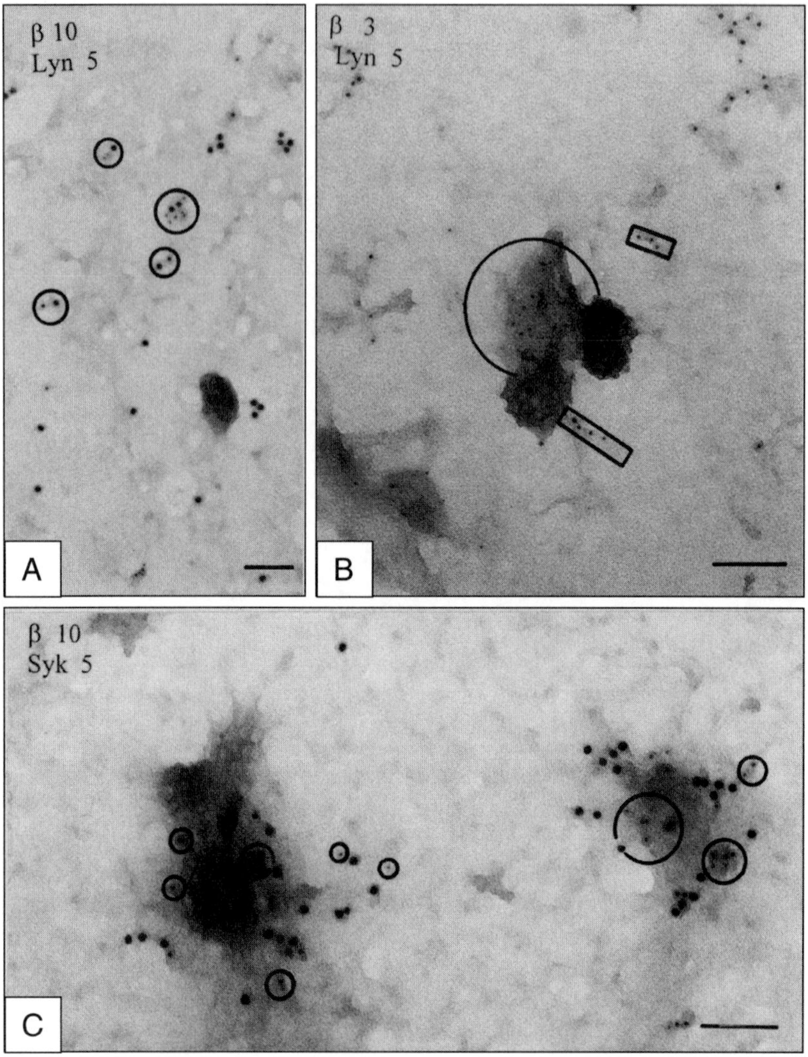

Figure 4. The topography of Lyn and Syk during FcεRI signaling. Membrane sheets were prepared from IgE-primed RBL-2H3 cells, without crosslinking (A) and after crosslinking with 1 μg/ml anti-IgE Ab for 2 min (B,C). Sheets were labeled from the inside with gold specific for FcεRI β, Lyn and Syk. Receptor (10 nm gold) and Lyn (5 nm gold) are loosely colocalized on resting cells (circles in A). Lyn (5 nm gold) is segregated away from the osmiophilic patches that accumulate receptor (3 nm gold) on activated cells (rectangles, B). Conversely, Syk (5 nm gold) is dramatically recruited to the osmiophilic receptor (10 nm gold) patches (C). Bars = 0.1 μm.

loosely with Lyn in small clusters on native membrane sheets prepared from resting cells (circles, Figure 4A). Measurements of association between two sizes of gold particles reveals that around 20% of gold particles marking Lyn lie in close proximity to receptors (Wilson *et al.*, 2000). Remarkably, the redistribution of crosslinked FcεRI into osmiophilic patches is associated with a complete separation of receptor from Lyn (rectangles, Figure 4B). Because Lyn is thought to catalyze receptor subunit phosphorylation, we hypothesized that a transient tightening of the Lyn-receptor clusters might precede this separation. This predicted configuration was captured in cells treated with a particular anti-FcεRIα mAb, H10, that has the unusual properties of supporting the two earliest biochemical events in the FcεRI signaling cascade, Lyn activation and FcεRI subunit tyrosine phosphorylation, but being impaired in Syk activation and secretion (not shown; illustrated in Lara *et al.*, 2001).

If the osmiophilic patches are signaling domains they should recruit Syk, and indeed they do. When membranes from resting cells are labeled for receptor and Syk, no colocalization is seen (not shown). In contrast, Syk is dramatically colocalized with crosslinked FcεRI in the osmiophilic patches (Figure 4C).

From these observations, we hypothesized that signaling is initiated by the rapid phosphorylation of receptor in Lyn microdomains. We proposed that the subsequent separation of crosslinked, phosphorylated receptor from Lyn, and its redistribution to osmiophilic patches that accumulate Syk, may create a primary site of active signaling to downstream responses (Wilson *et al.*, 2000).

3.3. The topography of FcεRI signal propagation

Further studies (Wilson *et al.*, 2001, 2002) showed that the lipid remodeling enzyme, PLCγ2, and the adaptor, Gab2, both co-localize preferentially with Syk in receptor-rich osmiophilic patches in activated cells (illustrated for PLCγ2 in Figure 5A). The other frames in Figure 5 show that the GTP exchange factor, Vav (Figure 5B), the adaptor protein Grb2 (Figure 5C) and the adaptor/ubiquitin ligase, Cbl (Figure 5D) also associate preferentially with the osmiophilic patches in activated cells. Other signaling proteins that associate selectively with FcεRI in primary domains include Gab2 and BTK as well as the endocytosis proteins clathrin, AP-2 and Eps-15 (not shown). The preferential localization of these diverse signaling and endocytic proteins in osmiophilic patches supports the identification of these patches as microdomains specialized for the dual functions of signal propagation and clathrin-mediated endocytosis. Because Cbl may also contribute to the negative regulation of signaling (Rao *et al.*, 2002), these associations provide the first hint that signaling may be both activated and terminated in these domains.

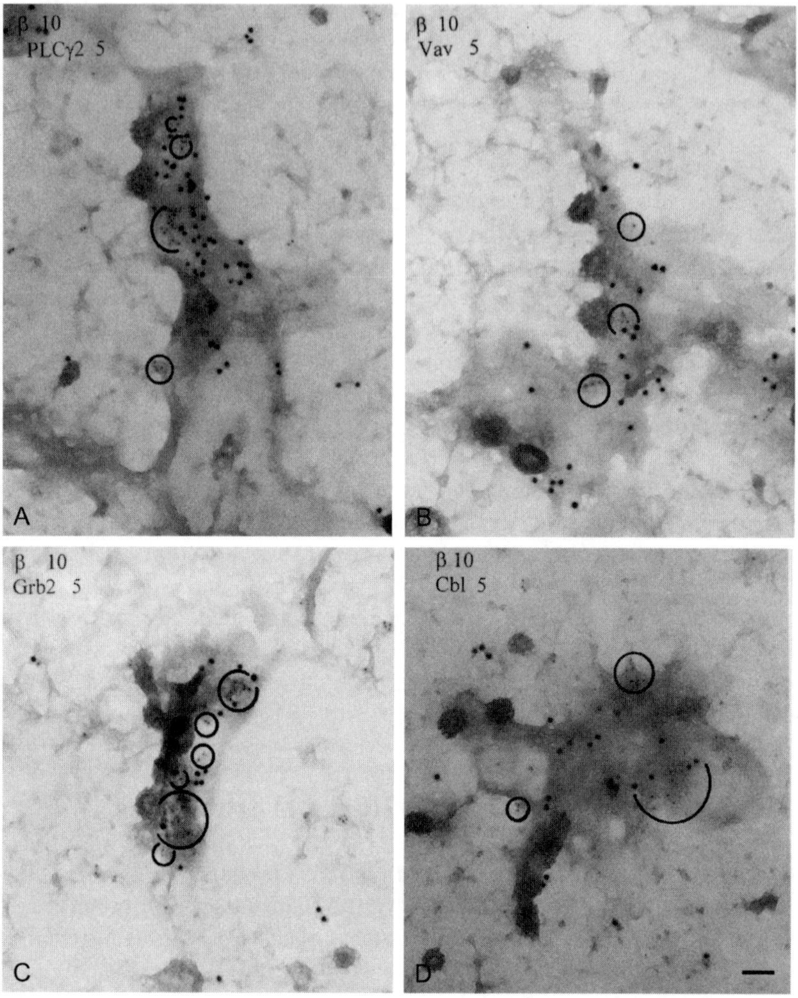

Figure 5. FcεRI signal propagation from the osmiophilic patches. Membrane sheets were prepared from activated RBL-2H3 cells (1 μg/ml anti-IgE Ab, 2 min) and double-labeled for FcεRI β (10 nm gold) and a series of proteins (5 nm gold) that are tyrosine phosphorylated in response to FcεRI crosslinking. The micrographs show that PLCγ2 (A), Vav (B), Grb2 (C) and Cbl (D) are all recruited to the osmiophilic patches. Bar = 0.1 μm.

Remarkably, the dually palmitylated transmembrane protein, LAT, found previously to be essential for FcεRI signaling (Saitoh *et al.*, 2000), does not mix with receptor when observed directly on native membrane sheets. Figure 6A shows that LAT is distributed on resting membranes as clusters usually

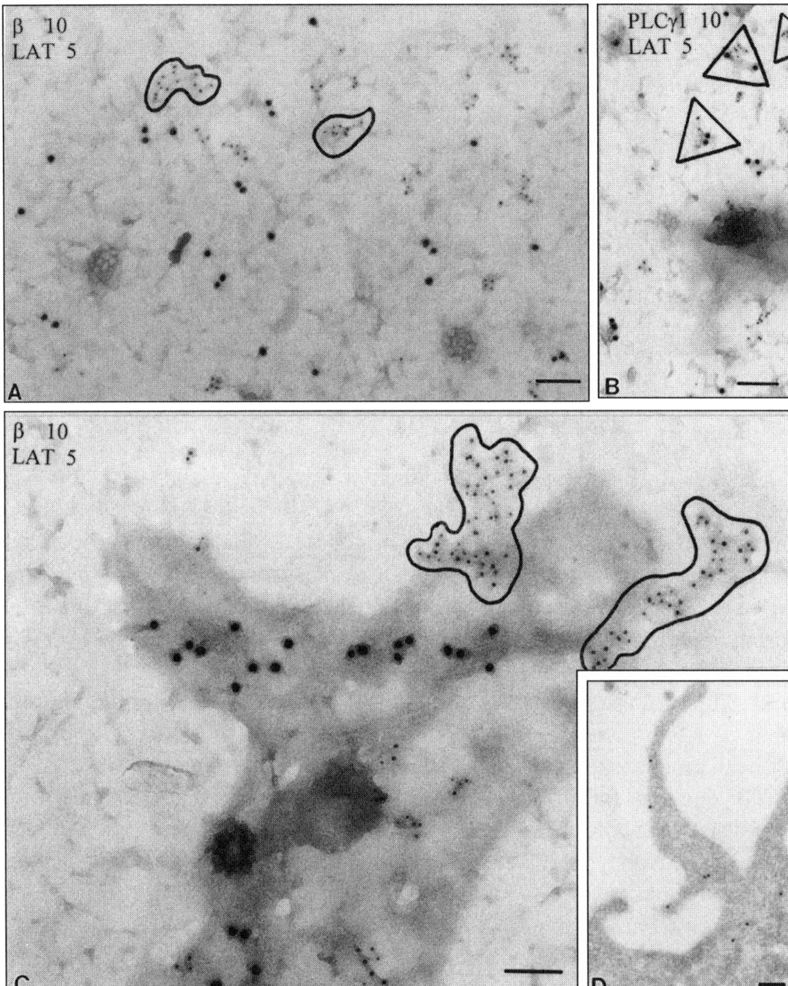

Figure 6. LAT topography during FcεRI signaling. Membrane sheets were prepared from IgE-primed RBL-2H3 cells, without crosslinking (A) and after crosslinking with 1 μg/ml anti-IgE Ab for 2 min (B,C). Sheets were double labeled for FcεRI β or PLCγ1 (10 nm gold) and LAT (5 nm gold). There is little or no colocalization of LAT with FcεRI β in either resting or activated cells (A,B). However LAT clusters intersect osmiophilic patches and become substantially larger in activated cells (B). The colocalization of PLCγ1 with LAT clusters in activated cells is illustrated in (C). The inset (D) shows our earlier localization of PLCγ1 to lamellar membrane by LR white embedding and immunogold labeling on thin sections. Bars = 0.1 μm.

containing fewer than ten 5 nm gold particles. These LAT clusters do not mix with receptor or associate with coated pits. After FcεRI crosslinking, LAT redistributes into much larger, elongated structures (typically 30 or more 5 nm gold particles per cluster) that intersect, but still do not mix with, receptor clusters (Figure 6B).

We hypothesize that LAT rafts function as topographically distinct secondary sites of active signaling. Supporting this, LAT clusters are shown in Figure 6C to colocalize with PLCγ1, identified previously by TEM of immunogold-labeled, LR-White-embedded sections as a lamella-associated protein (Figure 6D; Barker *et al.*, 1998). In contrast, the primary signaling domains are located adjacent to coated pits that are always located on planar regions of membrane.

3.4. The topography of inositol phospholipid synthesis

Although the mapping studies above show clearly that FcεRI signaling may occur in distinct membrane domains, they give little insight into how these domains are formed and stabilized. One hypothesis is that the membrane segregation resulting in the formation of primary and secondary signaling domains may be initiated by the activities of enzymes that remodel membrane inositol phospholipids. Supporting this, the p85 regulatory domain of PI 3-kinase is the only signaling protein found to date associated in activated mast cells with both FcεRI (Figure 7A) and with LAT (Figure 7B). Different PI 3-kinase products attract signaling proteins through interactions between the different phospholipid head groups and specific interaction domains in signaling proteins (FYVE and PX domains for PI(3)P; FERM, TUBBY, PH and ENTH domains for $PI(4,5)P_2$; PH and ENTH domain for $PI(3,4,5)P_3$) (reviewed in Teruel and Meyer, 2000; Hurley and Meyer, 2001; Pawson and Nash, 2003). Thus, topographical restriction of different PI 3-kinase isoforms could provide a direct method to establish multiple, distinct and likely transient, signaling domains. Other lipid remodeling enzymes may similarly contribute to localized lipid remodeling and protein recruitment. For example, diacylglycerol (DAG), a product of PLC activity, binds signaling proteins with C1 domains, while proteins with C2 domains seek membrane regions enriched for phosphatidic acid and phosphatidyl serine. Interestingly, the lipid phosphatase, PTEN, also localizes to primary signaling domains where it may contribute to signal termination through the turnover of phospholipid anchors for recruited proteins.

3.5. Interpreting the topography of FcεRI signaling

Our work to date implicates multiple, dynamic membrane domains, including a transient receptor-Lyn domain and at least two other distinct domains,

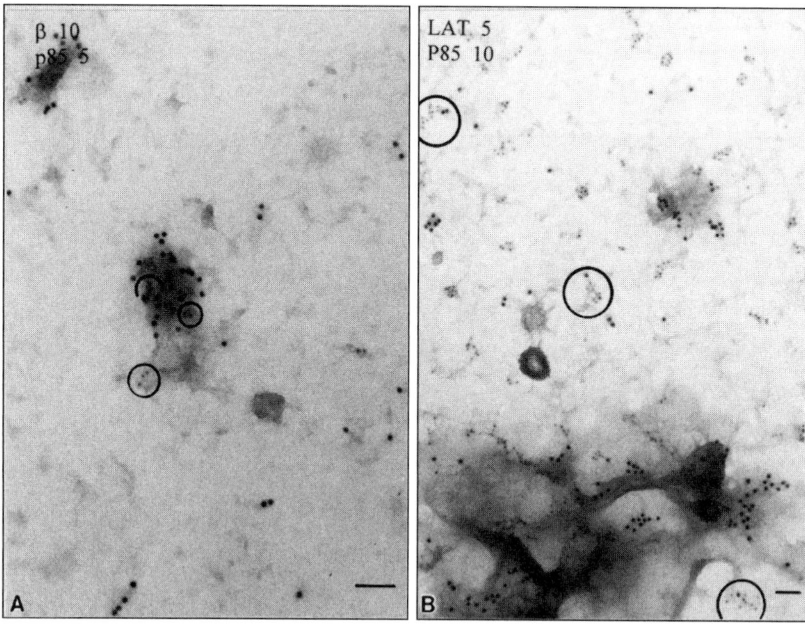

Figure 7. Localizing PI 3-kinase on membrane sheets. Membrane sheets were prepared from activated RBL-2H3 cells (3 μg/ml anti-IgE Ab for 2 min) and double-labeled for the p85 subunit of PI 3-kinase and either FcεRI β or LAT. P85 colocalizes with both FcεRI β (circles in A) and LAT (circles in B).

one characterized by the presence of receptor, Syk and multiple signaling proteins, but not Lyn ("primary signaling domains"), and one characterized by the presence of LAT and PLCγ1 but not receptor ("secondary signaling domains") in FcεRI signaling. The discovery of LAT domains has many implications, including the possibility that different arms of the FcεRI signaling cascade may be propagated from different signaling domains and, of course, that other secondary domains may be present in the native membrane. The p85 regulatory subunit of PI 3-kinase, perhaps in association with unique catalytic domains, is the only protein found so far in maps of both primary and secondary signaling domains. We speculate that PI 3-kinase may contribute to domain formation by catalyzing the selective remodeling of inositol phospholipids, that in turn induce the selective recruitment of proteins with domain motifs recognizing these remodeled lipids. Our work, summarized in cartoon form in Figure 8, clearly adds to the evidence that plasma membranes contain microdomains for signaling and provides new insight into the dynamic and heterogeneous nature of these signaling compartments.

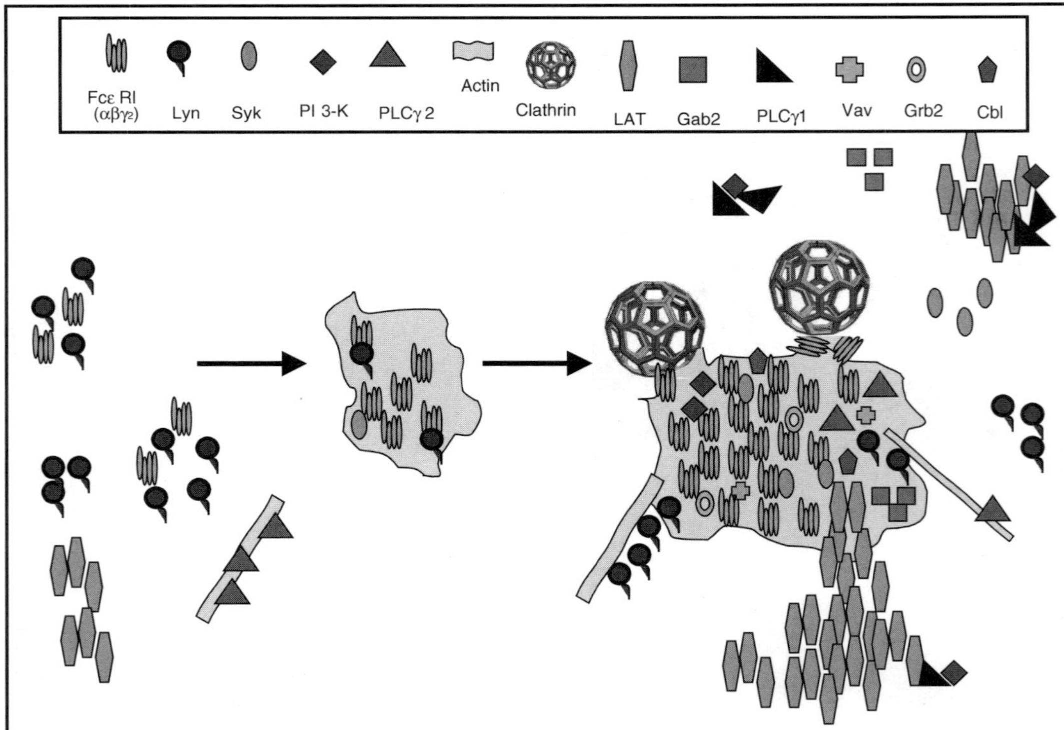

Figure 8. The topography of FcεRI signaling. FcεRI and Lyn are loosely colocalized in resting cells. FcεRI crosslinking induces Lyn-mediated phosphorylation needed for Syk recruitment. Larger FcεRI aggregates exclude Lyn and include Syk to form functional signaling domains. The clathrin-coated vesicles that internalize crosslinked receptors bud from membrane adjacent to the Syk-FcεRI complexes. A subset of downstream signaling molecules associate with the osmiophilic patches ("primary signaling domains"). Secondary signaling domains are organized around LAT. PI 3-kinase may associate with both primary and secondary domains.

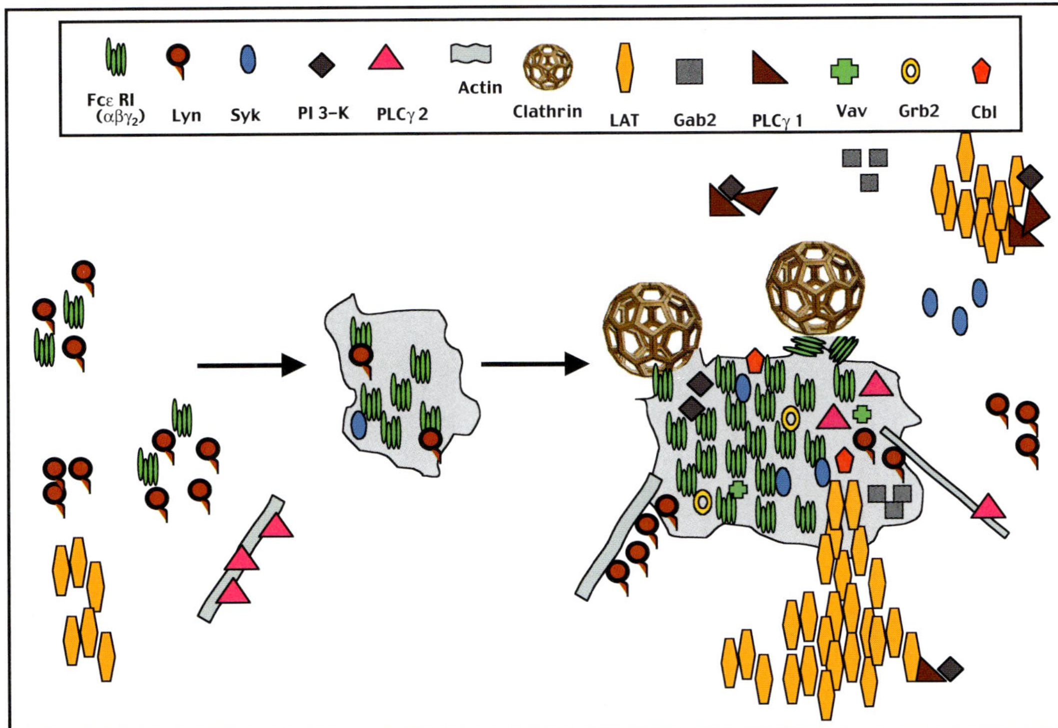

Figure 8. The topography of FcεRI signaling. FcεRI and Lyn are loosely colocalized in resting cells. FcεRI crosslinking induces Lyn-mediated phosphorylation needed for Syk recruitment. Larger FcεRI aggregates exclude Lyn and include Syk to form functional signaling domains. The clathrin-coated vesicles that internalize crosslinked receptors bud from membrane adjacent to the Syk-FcεRI complexes. A subset of downstream signaling molecules associate with the osmiophilic patches ("primary signaling domains"). Secondary signaling domains are organized around LAT. PI 3-kinase may associate with both primary and secondary domains.

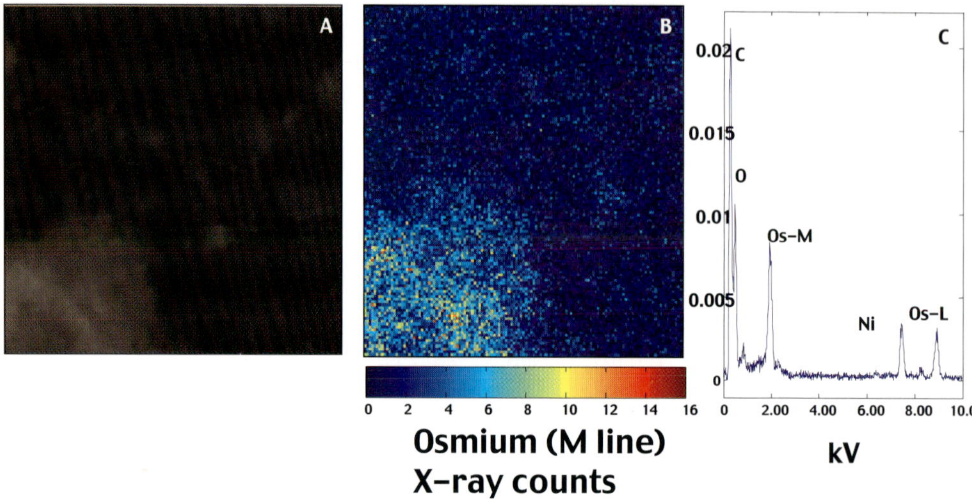

Figure 11. Spectral imaging shows osmium concentration within the dark patches of membrane. (A) Annular dark-field STEM image of an OsO_4-stained membrane sheet from an antigen-stimulated cell, showing an osmiophilic patch. (B) Composite X-ray spectra identifying carbon, oxygen, nickel and osmium present in the same region. (C) Osmium component image calculated by AXSIA.

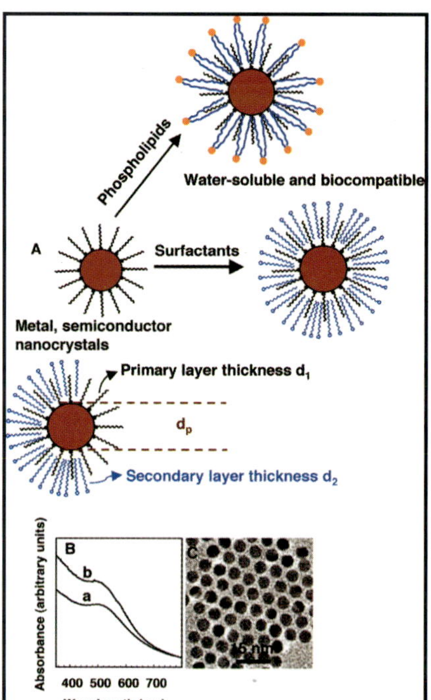

Figure 13. Schematic flow of synthesis of water-soluble nanocrystal-micelles and their characterization by plasmon resonance spectroscopy.

4. RELATIONSHIPS BETWEEN SIGNALING DOMAINS AND LIPID RAFTS

4.1. The topography of membrane markers for "Lipid Rafts"

As already noted, the results of detergent extraction and gradient centrifugation studies suggested that crosslinked FcεRI signal from detergent-resistant membranes enriched for specific lipids (cholesterol, glycerophospholipids, gangliosides) and lipid-modified proteins (acylated Lyn, GPI-anchored Thy-1, palmitylated LAT) (Field *et al.*, 1995, 1997, 1999; Surviladze *et al.*, 1998, 2001). We showed in Figure 6 (above) that crosslinked FcεRI and LAT have distinct distributions when imaged on membrane sheets. To determine if other "raft markers" seggregate with or away from FcεRI, we explored the topography of two additional "raft markers", Thy-1 and GM1. Like Lyn, these membrane components co-localize with crosslinked FcεRI in detergent-resistant membrane fractions from activated cells (Figure 9).

In Figure 10, FcεRI were crosslinked and receptor topography was compared with that of monomeric or crosslinked Thy-1 and GM1. Figures 10A, B show membrane sheets from cells that were activated with DNP-BSA-gold, then fixed and labeled or Thy-1 with anti-Thy-1-gold or for GM-1 with biotinyl-cholera toxin-avidin-gold (CTX-gold). Both micrographs show a primary signaling domains that concentrates crosslinked receptors. Thy-1-gold (A) and CTX-gold (B) are distributed as dispersed clusters. SEM analysis gave the same result (compare the clustered topography of crosslinked FcεRI with the dispersed topography of Thy-1 in Figures 1A, B). Thus, Thy-1 and GM1 monomers do not redistribute with crosslinked FcεRI. Figure 10C shows a membrane sheet from a cell that was incubated simultaneously with DNP-BSA-gold to crosslink FcεRI and CTX-gold to crosslink GM1. GM1 crosslinking, simultaneous with FcεRI crosslinking, causes GMI and FcεRI to accumulate in the same electron dense regions of the plasma membrane. In Figure 10D, FcεRI, Thy-1 and GM1 were all clustered from the outside before fixation and the two "raft" markers were localized on membrane sheets. Crosslinked GM1 is again in electron dense patches. However, crosslinked Thy-1 remain distributed in small clusters across the membrane sheet. Thus, the simultaneous crosslinking of FcεRI and of two recognized "lipid raft" markers yields two quite different behaviors for the raft markers. We conclude that "raft markers" exhibit independent behavior on native membranes, with GM1 showing an inherently dispersed distribution but becoming topographical restricted after crosslinking and Thy-1 distributing as clusters whose topography remains dispersed even after external crosslinking.

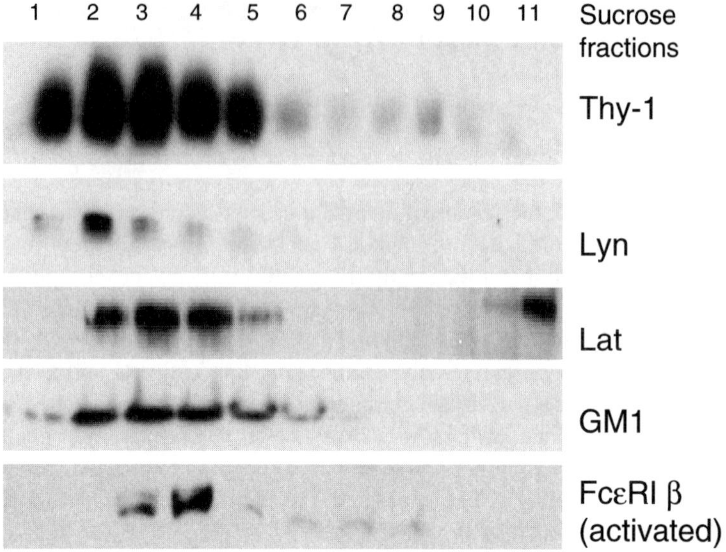

Figure 9. Distributions of "raft markers" on sucrose density gradients. IgE-primed RBL-2H3 cells were activated with DNP-BSA (1 μg/ml; 2 minutes), lysed in buffer containing 0.05% Triton X-100 and lysates overlaid onto 80% sucrose, followed by discontinous layers of 35%, 25%, 20%, 15% and 10%. Following 16 hrs ultracentrifugation at 200,000 × g, fractions were harvested and analysed by SDS-PAGE and western blotting. Thy-1, Lyn and the β subunit of FcεRI were detected on blots using primary antibodies, followed by HRP-conjugated secondary antibodies and fluorography. GM1 was detected using biotinylated cholera toxin (20 μg/ml) and avidin-HRP.

4.2. Analysis of signaling domains by high resolution X-ray spectral imaging

We called the primary signaling domains "osmiophilic patches" because the electron-dense nature of the membrane regions that accumulate crosslinked FcεRI, GM1 and many signaling proteins is markedly reduced when osmium is omitted as a post-fixation stain for TEM analysis (Wilson *et al.*, 2001). In Figure 11, we stained membrane sheets with osmium tetroxide but not uranyl acetate and imaged them using a FEI/Philips Electron Optics Tecnai F30-ST, 300 kV TEM/STEM equipped with an energy dispersive X-ray spectrometer, electron energy-loss spectrometer/imaging filter and Automated eXpert Spectral Image Analysis (AXSIA) software (Kotula *et al.*, 2003). Figure 11A is a STEM annual dark-field image of a membrane sheet prepared

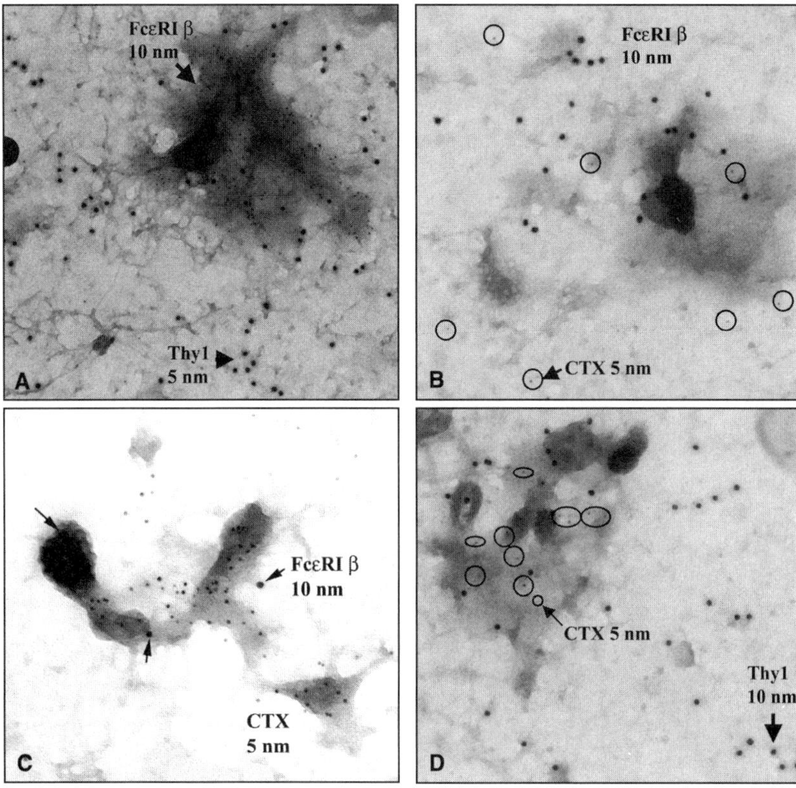

Figure 10. Distributions of GM1 and Thy-1 after FcεRI crosslinking. In A,B, IgE-primed cells were stimulated with DNP-BSA-5 nm colloidal gold for 5 minutes at 37°C, then fixed and stained from the outside with anti-Thy-1–10 nm gold (A) or CTX-gold (B). Aggregated FcεRI is concentrated within osmiophilic patches, whereas both Thy-1 (A) and GM1 (B) maintain dispersed distributions. In C, cells were incubated simultaneously with DNP-BSA-10 nm gold and with biotinyl-cholera toxin-5 nm avidin-gold. Crosslinked FcεRI and crosslinked GM1 colocalize in the osmiophilic patches. In D, FcεRI, Thy-1 and GM1 were crosslinked simultaneously and membranes labeled for the two "raft" markers. Crosslinked GM1 again seeks the osmiophilic patch, whereas crosslinked Thy-1 remains dispersed.

from an antigen-stimulated RBL-2H3 cell. The image shows a typical electron-dense patch (seen here as a "mirror image" with the patch as a lighter region in the bottom left corner). Figure 11B shows the composite spectral analysis of this region, with X-ray peaks corresponding to the grid support composed of nickel, carbon in the sample and two X-ray lines expected for osmium bonded to the membrane. Pixel-by-pixel values for osmium (Figure 11C) show that osmium labeling is markedly enhanced in the patch.

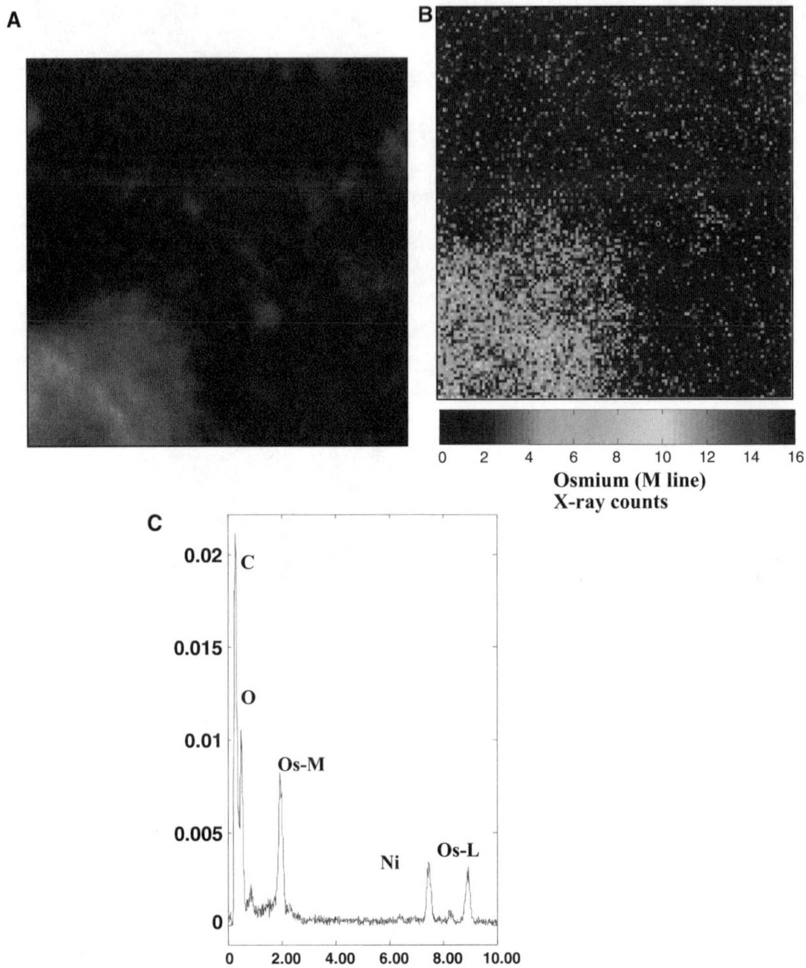

Figure 11. Spectral imaging shows osmium concentration within the dark patches of membrane. (A) Annular dark-field STEM image of an OsO_4-stained membrane sheet from an antigen-stimulated cell, showing an osmiophilic patch. (B) Composite X-ray spectra identifying carbon, oxygen, nickel and osmium present in the same region. (C) Osmium component image calculated by AXSIA.

Osmium is thought to label membranes for TEM analysis principally through its reaction with double bonds in unsaturated lipids (March 1977). Thus, the osmium component image suggests that aggregated FcεRI redistribute to membrane enriched for unsaturated glycerophospholipids. This provides a further contrast with biochemical analyses of DRMs, that are enriched for saturated glycerophospholipids.

4.3. Integrating signaling domains with lipid rafts: the challenge ahead

The relationship between the "signaling microdomains" defined by TEM and "lipid rafts" defined by detergent extraction and gradient centrifugation remain to be clarified. Markers for lipid rafts as defined in biochemical studies do not colocalize either with each other or with crosslinked FcεRI when analyzed in native membrane sheets. This suggests that the topographical relationships between membrane proteins and lipids are not adequately revealed by detergent extraction and gradient centrifugation experiments that disturb the native membrane. Adding to the discordance between TEM and biochemical studies, the osmium staining pattern of primary signaling patches suggests that these domains may be enriched for unsaturated phospholipids, whereas one defining characteristic of lipid rafts is their high content of saturated phospholipids and cholesterol.

The relationship between signaling microdomains defined by TEM and lipid rafts reported from confocal microscopy is also unclear. Stauffer and Meyer (1997) presented the coclustering of fluorescent cholera toxin labeling GM1 and crosslinked fluorescent FcεRI and Syk SH2 domains as evidence that IgE receptors signal from "lipid rafts" in intact mast cells. However cholera toxin is pentameric (Reed *et al.*, 1988; Ribi *et al.*, 1988), so it is likely that the fluorescent toxin used in live cell imaging experiments, like the gold-conjugated toxin used here, induces the redistribution of GM1 to microdomains that independently accumulate receptors and signaling molecules.

It is remarkable that, 30 years from the first evidence of deviations of membrane behavior from the predictions of the fluid mosaic model, there is still no comprehensive model to explain the heterogeneity of membrane topography and membrane function. Electron microscopy clearly reveals a much more complex and dynamic topographical organization of membrane domains than is predicted by biochemical analysis of detergent-resistant membranes or observed to date by fluorescence microscopy. We believe that integrating electron microscopy with complementary chemical kinetics analyses and other fluorescence and cell fractionation data, will ultimately lead to general models with power to understand mechanisms that regulate the efficiency, specificity, amplitude and duration of receptor-mediated signaling.

The rest of the chapter addresses the development of new probes and new computational methods needed to take full advantage of membrane sheet technology for probing the topography of cell signaling.

5. IMPROVING RECEPTOR MAPPING THROUGH TECHNOLOGY AND COMPUTATION

5.1. Developing better probes to label receptors and signaling components

Until recently, TEM analysis of signaling complexes on membrane sheets has been limited by the reduced efficiency of labeling as gold particles increase in size to 10 nm in diameter and by the limited numbers of probes that can be imaged in a single experiment when all the probes are functionalized with antibodies. To overcome this, we are generating and testing new nanoparticles that can be distinguished based on chemical composition, size and shape and can be functionalized with protein domain motifs as well as with antibodies.

The synthesis of monodisperse nanoparticles requires discrete nucleation events, followed by slower growth under specific temperature and reactant addition conditions (Murray *et al.*, 2000). We have successfully controlled particle size and morphology in nanoparticle compositions, including metals, compound semiconductors, and complex oxides, through control of these reaction conditions and the use of coordinating surfactants that "cap" the surface of the nanoparticles (Murray *et al.*, 2000; Bunge *et al.*, 2003a,b; TJ Boyle, NL Andrews, TJ Headley, MA Rodriguez and SD Bunge, unpublished results). The approach uses a novel reaction system composed of a coordinating solvent for the reproducible synthesis of nanoparticles via pyrolysis of a variety of molecular precursors.

Figures 12, A–C, shows a selection of different nanospheres of rigidly controlled diameters. Here, a novel anhydrous route for the synthesis of amine capped Cu°, Ag°, and Au° nanoparticles was applied via the use of polynuclear mesityl (mesityl = 2,4,6-$Me_3C_6H_2$) derivatives (Bunge *et al.*, 2003b). Under an argon atmosphere, crystalline $[Cu(\mu\text{-mesityl})]_5$, $[Ag(\mu\text{-mesityl})]_4$ and $[Au(\mu\text{-mesityl})]_5$ were dissolved in octylamine and subsequently injected into a heated solution generating the corresponding oxide-free metal nanoparticles. The temporal growth for each nanocrystalline material was carefully monitored to obtain particles of various size and dispersity as illustrated by transmission electron microscopy (TEM). Similar success has been obtained using amide instead of mesityl precursors (not shown).

Figures 12, D–F, shows nanoparticles of 3 distinct shapes. The key to the preparation of shape-controlled metal nanoparticles is to specifically retard or promote the plane of growth. This requires control over how supersaturation is achieved plus control over the precursors' decomposition properties (i.e., reduction rate, ligand loss, etc), light, the poisoning/promotion by surfactants,

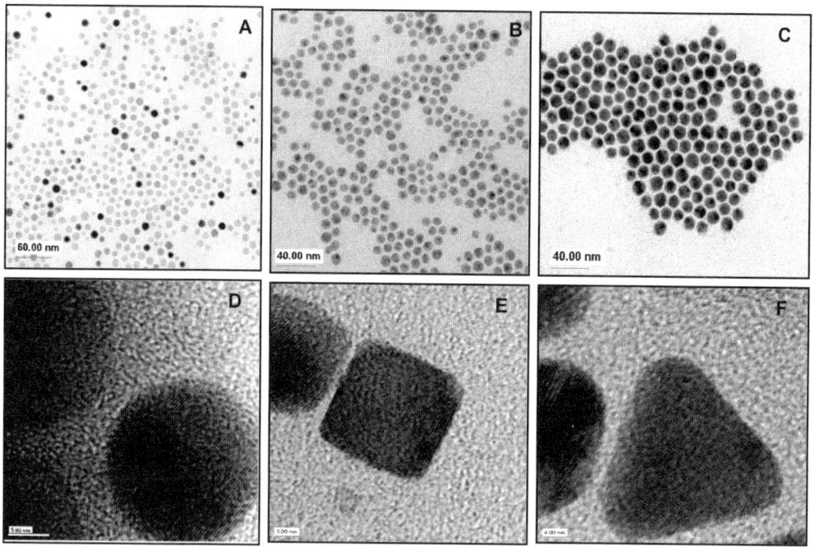

Figure 12. TEM images of different compositions and shapes of nanoparticles. (A,B,C) Cu°, Ag°, and Au° spheres, all approximately 10 nm in diameter. (D,E,F), Cu° spheres, Pt° cubes, and Au° triangles, approx 10 nm in diameter.

temperature, concentration, and a myriad of other variables (Adair and Suvaci, 2000; Murray *et al.*, 2000; Teranishi *et al.*, 2000; Jin *et al.*, 2001; Freemantle, 2001; Fu *et al.*, 2002; Nin *et al.*, 2002).

As well as devising novel routes for nanoparticle synthesis, we have devised new approaches to their functionalization through the generation of water-soluble metal nanocrystal micelles (nanoparticles) (Fan *et al.*, 2003). The new strategy considers the nanoparticle as a hydrophobic molecule that can be enclosed within the hydrophobic core of a micelle composed of a primary layer of alkane thiols of variable chain lengths and a secondary layer made up of a mixture of nonionic polyethylene glycol (PEG), charged surfactants/lipids and functionalized surfactants/lipids. The PEG inhibits nonspecific protein adsorption, the charged head groups (cationic or anionic) are used to prevent the particles from aggregation and the functionalized lipids provide amines, sulfhydryls and other groups for protein binding. Figure 13A illustrates the encapsulation of the nanoparticle inside a hybrid bilayer shell with precisely defined primary and secondary layer thicknesses. During encapsulation, surfactants reduce surface tension between hydrophobic particles and water, enabling the stabilization of hydrophobic particles in water. The hydrophobic interaction between the primary layer (the derivatized alkanethiol

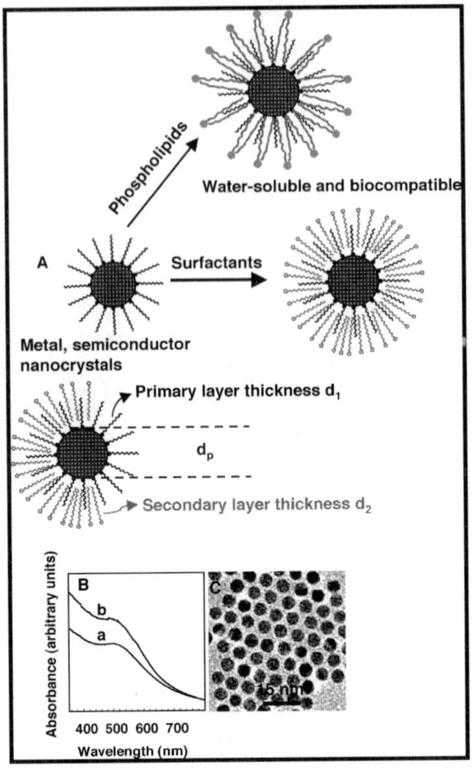

Figure 13. Schematic flow of synthesis of water-soluble nanocrystal-micelles and their characterization by plasmon resonance spectroscopy.

molecules) and the secondary layer (the surfactant molecules) is essential and thermodynamically favors the encapsulation process and the formation of an interdigitated microstructure. Cationic, anionic, and non-ionic surfactants can all form nanoparticle-micelles, allowing ready control of surface charge and functionality. The general method to link bio-macromolecules involves reactions between amines and di-functional couplers such as succinimidyl esters and isothiocyanates. Figure 13B shows the UV spectra for (a) alkanethiol stabilized gold nanoparticles in chloroform and (b) gold nanoparticles after encapsulation in micelles. For both spectra, we observe no differences between the plasmon resonance bands (at ~510 nm) of the alkanethiol stabilized gold nanoparticles in chloroform and the resulting water soluble micelles. Cationic, anionic, and non-ionic surfactants can all form nanoparticle-micelles, allowing ready control of surface charge and functionality. The new probes will be functionalized both with antibodies specific for signaling proteins and lipids and with protein domain motifs with high selectivity for specific signaling pathway components, particularly specific inositol phospholipids. The available repertoire of lipid-targeted motifs includes FYVE and p40phox-PX domains to localize PI(3)P, AKT-PH domains to localize PI(3,4)P_2 and PI(3,4,5)P_3, PLCδ-PH domains for PI(3,4)P_2, BTK-PH domains for PI(3,4,5)P_3 and C1 domain from PKC to localize DAG (Teruel and Meyer, 2000; Hurley and Meyer, 2001; Pawson and Nash, 2003).

5.2. Developing better methods to quantify the topography of receptors and signaling components

Membrane sheets provide an unprecedented view of the topography of signaling. However, computational methods are essential to extract spatial coordinates of probes, to analyze and statistically validate the clustering and co-clustering of these probes and to integrate results between experiments in order to establish the relative spatial distributions of multiple different probes.

Our first quantitative studies of receptor topography during signaling were performed using immunogold-labeled scanning electron micrographs. IgE-primed mast cells were labeled for different times with immunogold particles, typically 15 nm DNP-BSA-gold or anti-IgE-gold, and probe distributions were observed by backscattered electron imaging (BEI) in the SEM. Spatial point coordinates of gold particles in the aggregates were extracted from digitized SEM negatives and mean aggregate size was determined using a variant of Ripley's *K* function to describe distributions and nearest neighbor frequencies. Experimental data were compared with data simulated from a version of the chemical-spatial model of Goldstein and Wofsy (1994). Results of this analysis, elaborated in Wofsy *et al.* (1995) and Sanders (1996), suggested there is more than one scale of FcεRI aggregation during signaling and provided some of the first evidence for the existence of transient receptor subdomains. Several groups subsequently expanded the application of Ripley's *K* function analysis to the analysis of gold particle distributions. Philimonenko *et al.* (2000) developed a computer program called "gold" (available at http://uemweb.biomed.cas.cz/hozak/gold.htm) that uses the Ripley K^{-1} function to test for the presence of clustering and colocalization of gold particles labeling conventional TEM thin section images and the Ripley *K* function for a more detailed statistical evaluation of the data. Prior *et al.* (2003) implemented a related bivariate *K*-function analysis useful to study the co-localization of differently sized gold particles marking H-Ras and K-Ras on fibroblast membrane sheets.

Our current procedures use an ImageJ (http://rsb.info.nih.gov/ij/) plugin that automates the acquisition of gold particle data from micrographs and provides tools for doing cluster analysis, and a MatLab (http://www.mathworks.com/) toolbox that is used for statistical and cluster analysis. To understand clustering, it is important to begin with images containing a maximal area of the cell membrane. Therefore, gold particle distributions are obtained using the lowest magnification (around 30,000x) that still permits identification of the smallest gold particles. The images are recorded on film and then digitized on a high-quality scanner, producing files that are over 80 megabytes in size. Gold particle coordinates are extracted from these files using the plugin. The plugin uses cropping and thresholding to produce a file of gold particle coordinates. This algorithm is fast and over 95% accurate for 10 nm particles, but only about 50% accurate for 5 nm particles. We therefore developed an alternative algorithm based on filtering to identify 5 nm particles. As no algorithm can be expected to be completely accurate, we have also developed a user interface that allows for the easy addition of missed particles and removal of false identifications (Zhang *et al.*, 2003).

There are two main approaches to clustering analysis, one originating in computer science (Jain *et al.*, 1999) and the other in spatial statistics or statistical geometry (Stoyan *et al.*, 1995). Figure 14 shows the outcome of a clustering algorithm based on aggregating into clusters all particles that are pairwise

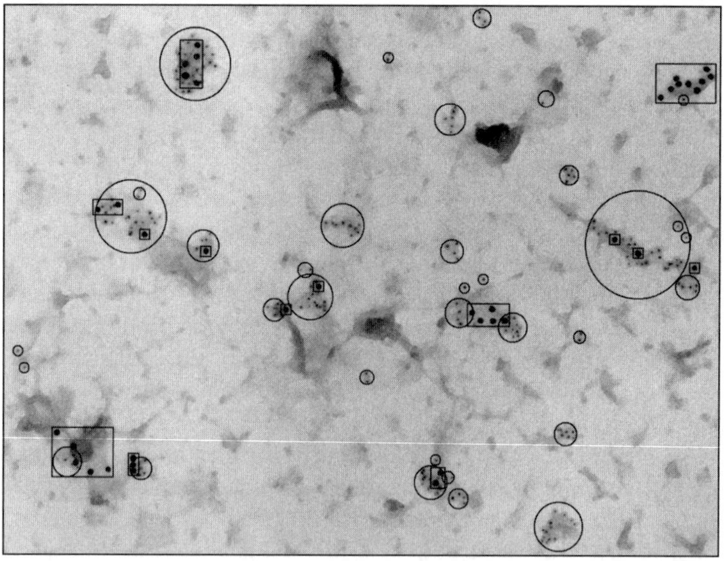

Figure 14. Clusters are identified by putting any two particles that are separated by less than a cutoff distance into the same group. In this micrograph, clusters of 5 nm particles (circles) and 10 nm particles (rectangles) are defined by a cutoff distance = 3d, where d is the diameter of the particles.

within a given small distance of each other. This relatively direct algorithm provides intuitively correct cluster sizes and is not sensitive to modest changes in the cutoff distance. Different sizes of gold particles can be selected for cluster analysis on the same sheet, providing users with a highly intuitive tool to test for the presence or absence of co-clustering. We are currently extending this analysis to produce dendrograms from the data and to quantify the properties of the clusters and co-clusters more precisely.

Because clustering algorithms can produce false positive results, it is important to use statistical analyses to show that the particles are indeed not randomly distributed. The Hopkins statistic, that tests spatial randomness by comparing nearest neighbor distances from random points and randomly chosen gold particles, has proven simple and intuitive for analysis of single-size particle distribution. As described (Jain and Dubes, 1988) the expected distribution for this statistic is known and is used to produce the smooth bell-shaped curves in Figure 15. Figure 15A (left panel) is a image of 300 computer-generated random points. The corresponding histogram of the relative frequency distribution (right panel) shows Hopkins values that have the same shape as the expected distribution and are centered around 0.5 as expected for a random distribution. Figure 15B (left panel) shows a typical plot of gold particle coordinates from experimental data. The plot is not obviously clustered by direct observation. However, the histogram (at right) shows the Hopkins distribution is shifted to the right, indicating that the distribution is somewhat clustered. Figure 15C (left panel) shows another typical plot of gold particle coordinates from experimental data. This plot is obviously clustered by direct observation. The histogram shows a marked shift to the right, confirming a high degree of clustering.

A version of Ripley's *K* analysis, based on the work of Haase (1995) on spatial point pattern analysis in ecology, also gives information on spatial relationships on membrane sheets labeled with one size of gold particle (not shown) and is particularly useful for analyses when there are two sizes of gold particles (Figure 16). The Haase program, SPPA (available at http://home.t-online.de/home/haasep/frm_sppa.htm), uses $L(t) - t$ for each distance t, where $L(t) = (K(t)/\pi)^{1/2}$, and $K(t)$ is the Ripley's *K* function. For a homogenous Poisson process, the expected value of $L(t) - t$ is zero. When more than one type of gold particle is involved, the bivariate Ripley's *K* function is used to test for co-clustering. Results from the Ripley's analysis can be tested for statistical significance using Monte Carlo simulations. Figures 16A and C show experimental data from two different double labeling conditions. In both cases, 5 nm gold particles are represented as triangles and 10 nm particles as squares. Co-clustering is obvious in C but not in A. Figure 16B analyzes the data presented in A. The $L(t) - t$ (solid line) has been plotted against t for each 20 nm interval up to 200 nm. The solid horizontal line at 0 gives the expected

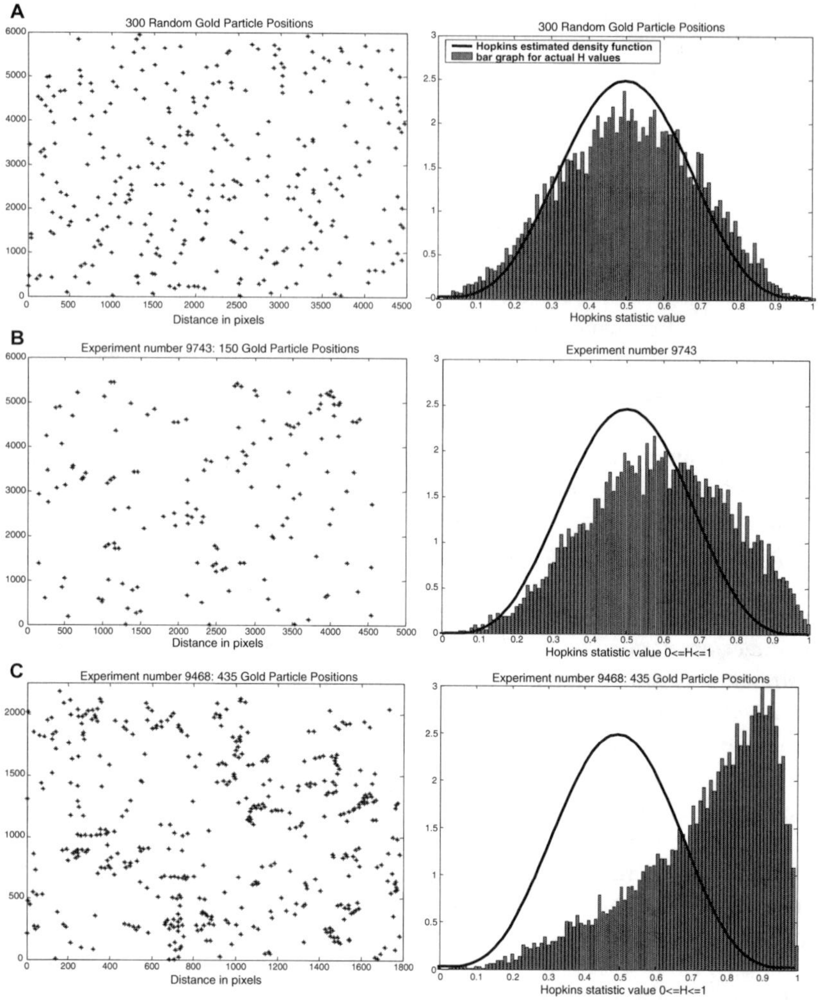

Figure 15. The Hopkins statistic was used on experimental data (B,C) and on randomly generated patterns of particles in a region of the same size as the micrographs (A). For the analysis, we choose $m = 5$, then choose m random sampling points and m unbiased gold particles and compute the value H. We repeat this for $k = 1{,}000$ times and then form a histogram of the data using b bins. The histogram can be viewed as giving the (piecewise constant) function $\mathrm{N}(H)$, $0 \leq H \leq 1$. This histogram is normalized by setting $\mathrm{n(h)} = b\,\mathrm{N}(H)/k$ so that it produces a relative frequency distribution that is compared to the theoretical density function . The calculation is performed using a Matlab tool kit developed for the analysis of gold particle data. Results verify that computer-generated random point patterns are indeed random and show that gold particles are distributed non-randomly in both sets of experimental data.

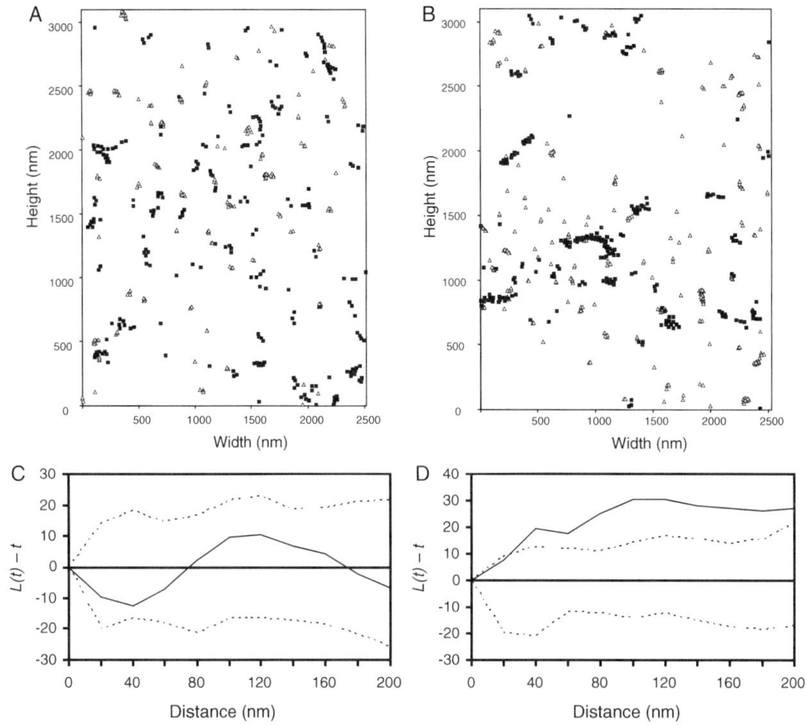

Figure 16. Ripley's analysis. The spatial distribution patterns of 5 nm and 10 nm particles from experimental data are shown in (A,C). Spatial analyses of the bivariate distribution patterns were performed using the SPPA program of P. Haase (B,D).

value for *L(t)* under complete spatial randomness (CSR). The dotted lines give a 99% confidence envelope for CSR estimated from 100 Monte Carlo simulations. The $L(t) - t$ line is within the confidence envelope, indicating that the deviation from CSR is not statistically significant. On the other hand, the analysis on experimental data presented in Figure 16C shows that the deviation from CSR is statistically significant for distance from 30 nm to 200 nm, indicating a high degree of co-clustering between 5 nm and 10 nm gold particles in this distance range (Figure 16D).

Even with these computational tools in place, many challenges remain. One major challenge will be to establish the co-clustering (or lack thereof) of multiple components of signaling pathways when the labeling is limited to a maximum of 3–5 probes per membrane sheet and comparison must be made between experiments.

6. CONCLUSIONS

Clearly, recognition that receptors and their associated signaling partners colocalize in domains that form in specific relationship to membrane features, and that signaling domains are not the unitary structures suggested from detergent extraction/sucrose gradient centrifugation studies adds excitement, complexity and challenge to the study of the role of protein and lipid topography in cell signaling. In the short term, the results of further high resolution mapping studies are likely to reveal additional secondary signaling domains and asymmetries in the topography of membrane lipids as well as proteins. On a longer time scale, we expect that high resolution topographical maps of signaling networks will be aligned with chemical kinetics analyses (as in Faeder *et al.*, 2003), with cell fractionation data (as in Field *et al.*, 1995, 1997, 1999 and Surviladze *et al.*, 1998, 2001) and with a range of real-time fluorescence measurements (as in Stauffer and Meyer, 1997). Mathematical models that integrate data from these multiple sources should have power to predict mechanisms that regulate the efficiency, specificity, amplitude and duration of signaling through the FcεRI and other receptors. Such models should have general value in signal transduction research and specific value in the search for new drug targets useful for the treatment of diseases like cancer, diabetes, allergies and asthma where symptoms can be linked to altered activities of signaling pathways. New drugs like farnesyltransferase inhibitors already target the location rather than the activity of proteins in signaling pathways and provide proof-of-principle for continued research.

ACKNOWLEDGEMENTS

This work was supported in part by NIH grants P20 GM66283, R01 GM49814, R01 AI051575, and by DoD grants to Sandia National Laboratory. Sandia is a multi-program laboratory operated by Sandia Corporation, A Lockheed Martin Company, for the United States Department of Energy (DOE) under contract DC-AC04-94AL85000. Use of the electron microscopy facilities at the UNM-HSC and Sandia National Laboratory is gratefully acknowledged.

REFERENCES

Adair, J.H. and Suvaci, E. 2000. Morphological control of particles. Current Opinion in Colloid and Interface. Science 5: 160-167.

Anderson, R.G.W. and Jacobson, K. 2002. A role for lipid shells in targeting proteins to caveolae, rafts and other lipid domains. Science 296: 1821-1825.

Barker, S.A., Caldwell, K.K., Pfeiffer, J.R., Wilson, B.S. 1998. Wortmannin-sensitive phosphorylation, translocation and activation of PLCγ1, but not PLCγ2, in antigen-stimulated RBL-2H3 mast cells. Mol. Biol. Cell 9: 483-496.

Barker, S.A., Lujan, D., Wilson, B.S. 1999. Multiple roles for PI 3-kinase in the regulation of PLCγ activity and Ca^{2+} mobilization in antigen-stimulated mast cells. J. Leukocyte Biol. 65: 321-329.

Berlin, R.D., Oliver, J.M., Ukena, T.E., Yin, H.H. 1974. Control of cell surface topography. Nature 247: 45-46.

Blank, U., Ra, C., Miller, L., White, K., Metzger H., and Kinet, J-P. 1989. Complete structure and expression in transfected cells of high-affinity IgE receptor. Nature 337: 187-189.

Brown, D.A., London, E. 2000. Structure and function of sphingolipid and cholesterol-rich membrane rafts. J. Biol. Chem. 275: 17221-17224.

Bunge, S.D., Krueger, K.M., Boyle, T.J., Rodriguez, M.A., Headley T.J. and Colvin V.L. 2003a. Synthesis of coinage-metal nanoparticles from mesityl precursors. J. Materials Chem. in press.

Bunge, S.D., Krueger, K.M., Boyle, T.J, Rodriguez, M.A., Headley, T.J. and Colvin, V.L. 2003b. Growth and morphology of cadmium chalcogenides: the synthesis of nanorods, tetrapods and spheres from CdO and $Cd(O_2CCH_3)_2$. J. Materials Chem. 13: 1705-1709.

Cressie, N.A.C. 1993. Statistics for Spatial Data, New York: Wiley.

Edidin, M., 1997. Lipid microdomains in cell-surface membranes. Current Opin. Struct. Biol. 7: 528-532.

Edidin, M. 2001. Shrinking patches and slippery rafts: scales of domains in the plasma membrane. TRENDS in Cell Biology 11: 492-496.

Eiseman, E., Bolen, J. 1992. Engagement of the high affinity IgE receptor activates src protein-related tyrosine kinases. Nature 355: 78-80.

Faeder, J.R., Hlavacek, W.S., Reischl, I., Blinov, M.L., Metzger, H., Redondo, A., Wofsy, C. and Goldstein, B. 2003. Investigation of early events in FcεRI-mediated signaling using a detailed mathematical model J. Immunol. 170: 3769-3781.

Fan, H., Boyle, T.J., Simon, T., Molloy, K. and Brinker, C.J. 2003. Surfactant-assisted self-assembly of ordered gold nanocrystals/silica nanostructures and their electrical properties. Submitted for publication.

Field, K.A., Holowka, D. and Baird, B. 1995. FcεRI-mediated recruitment of $p53/56^{lyn}$ to detergent-resistant membrane domains accompanies cellular signaling. Proc. Natl. Acad. Sci. USA 92: 9201-9205.

Field, K.A., Holowka, D. and Baird, B. 1997. Compartmentalized activation of the high affinity immunoglobulin E receptor within membrane domains. J. Biol. Chem. 272: 4276-4280.

Field, K.A., Holowka, D. and Baird, B. 1999. Structural aspects of the association of FcεRI with detergent-resistant membranes. J. Biol. Chem. 274: 1753-1758.

Freemantle, M. 2001. Nanoparticle shape control. Chemical and Engineering News 79: 10.

Fridriksson, E.K., Shipkova, P.A., Sheets, E.D., Holowka, D., Baird, B. and McLafferty, F.W. 1999. Quantitative analysis of phospholipids in functionally important membrane domains from RBL-2H3 mast cells using tandem high-resolution mass spectrometry. Biochemistry 38: 8056-8063.

Friedrichson, T. and Kurzchalia, T. 1998. Microdomains of GPI-anchored proteins in living cells revealed by crosslinking. Nature 394: 802-805.

Fu, X.Y., Wang, Y., Wu, N.Z., Gui, L.L. and Tang, Y.Q. 2002. Shape-control of platinum nanoparticles stabilized by oxalate. Acta Chimica Sinica 60: 1324-1330.

Ge, M., Field, K.A., Aneja, R., Holowka, D., Baird, B. and Freed, J.H. 1999. Electron spin resonance characterization of liquid ordered phase of detergent-resistant membranes from RBL-2H3 cells. Biophysical Journal 77: 925-933.

Goldstein, B. and Wofsy, C. 1994. Aggregation of cell surface receptors. Lectures on Mathematics in the Life Sciences 24: Cell Biology. Ed. B. Goldstein and C. Wofsy. Providence RI: American Mathematical Society, 109-135.

Haase, P. 1995. Spatial pattern analysis in ecology based on Ripley's *K*-function: Introduction and methods of edge correction. Journal of Vegetation Science 6: 575-582.

Hurley, J.H. and Meyer, T. 2001. Subcellular targeting by membrane lipids. Current Opinion in Cell Biology 13: 146-152.

Hutchcroft, J.E., Geahlen, R.L., Deanin, G.G. and Oliver, J.M. 1992. FcεRI-mediated tyrosine phosphorylation and activation of the 72 kDa protein-tyrosine kinases, PTK72, in RBL-2H3 rat tumor mast cells. Proc. Natl. Acad. Sci. 89: 9107-9111.

Iino, R. and Kusumi, A. 2001. Single-fluorophore dynamic imaging in living cells. J. Fluorescence 11: 187-195.

Ilangumaran, S., Arni, S., van Echten-Deckert, G., Borisch, B. and Hoesli, D.C. 1999. Microdomain-dependent regulation of Lck and Fyn protein-tyrosine kinases in T lymphocyte plasma membranes. Mol. Biol. Cell. 10: 891-905.

Jacobson, K., and Dietrich, C. 1999. Looking at lipid rafts? Trends Cell Biol. 9: 87-91.

Jain, A.J. and Dubes, R.C. 1988. Algorithms for Clustering. Prentice Hall, Englewood Cliffs, New Jersey (1988).

Jain, A.K., Murty, M.N. and Flynn, P.J. 1999. Data clustering: a review. ACM Computing Surveys 31: 264-323.

Jin, R.C., Cao, Y.W., Mirkin, C.A., Kelly, K.L., Schatz, G.C. and Zheng, J.G. 2001. Photoinduced conversion of silver nanospheres to nanoprisms. Science 294: 1901-1903.

Kenworthy, A.K., Petranova, N., Edidin, M. 2000. High-resolution FRET microscopy of cholera toxin B-subunit and GPI-anchored proteins in cell plasma membranes. Mol. Biol. Cell 11: 1645-1655.

Kotula, P.G., Keenan, M.R. and Michael, J.R. 2003. Automated analysis of SEM X-ray spectral images: A powerful new microanalysis tool. Microsc. Microanal. 9: 1-17.

Lara M., Ortega, E., Pecht, I., Pfeiffer, J.R., Martinez, A.M., Lee, R.J., Surviladze, Z., Wilson, B.S. and Oliver, J.M. 2001. Overcoming the signaling defect of Lyn-sequestering, signal-curtailing FcεRI dimers. J. Immunol. 167: 4329-4337.

March, J. 1977. Advanced Organic Chemistry: Reactions, Mechanisms and Structure (2nd ed) McGraw-Hill Book Co (New York) pp 748-749.

Mayor, S. and Maxfield, F.R. 1995. Insolubility and redistribution of GPI-anchored proteins at the cell surface after detergent extraction. Mol. Biol. Cell 6: 929-944.

Montixi, C., Langlet, C., Bernard, A.M., Thimonier, J., Dubois, C., Wurbel, M.A., Chauvin, J.P., Pierres, M. and He, H.T. 1998. Engagement of T cell receptor triggers its recruitment to low-density detergent-insoluble membrane domains. EMBO J. 17: 5334-5348.

Murray, C.B., Kagan, C.R. and Bawendi, M.G. 2000. Synthesis and characterization of monodisperse nanocrystals and close-packed nanocrystal assemblies. Annual Review of Materials Science 30: 545-610.

Niemeyer, C.M. 2001. Nanoparticles, proteins, and nucleic acids: Biotechnology meets materials science. Angew. Chem. Int. Ed. 40: 4128-4158.

Nin, G.L., Chang, Y.F., Liu, Y.L., Teng, F. and Lin, Y.A. 2002. Shape-controlled synthesis of alumina nanoparticles by carboxy-containing organic molecules. Chemical Journal of Chinese Universities-Chinese 23: 345-348.

Oliver, J.M., Ukena, T.E., Berlin, R.D. 1974. Effects of phagocytosis and colchicine on the distribution of lectin-binding sites on cell surfaces. Proc. Nat. Acad. Sci. USA 71: 394-398.

Oliver, J.M. and Berlin, R.D. 1983. Surface and cytoskeleton events regulating leukocyte membrane topography. Sem. Hematol. 20: 282-304.

Parolini, I., Topa, S., Sorice, M., Pace, A., Ceddia, A., Montesoro, E., Pavan, A., Lisanti, M.P., Peschle, C., Sargiacomo, M. 1999. Phorbol ester-induced disruption of the CD4-Lck complex occurs within a detergent-resistant microdomain of the plasma membrane – Involvement of the translocation of activated protein kinase C isoforms. J. Biol. Chem. 274: 14176-14187.

Pawson, T. and Nash, P. 2003. Assembly of cell regulatory systems through protein interaction domains. Science 300: 445-452.

Pfeiffer, J.R., Deanin, G.G., Seagrave, J.C., Davis, B.H., Oliver, J.M. 1985. Membrane and cytoskeletal changes associated with IgE-mediated serotonin release in rat basophilic leukemia cells. J. Cell Biol. 101: 2145-2155.

Pfeiffer, J.R., Oliver, J.M. and Wilson, B.S. 2002. Observing signal transduction, endocytosis and degranulation by immunogold labeling and transmission electron microscopy on membrane sheets. Am. Biotech. Methods. 20: 18-22.

Philimonenko, A.J., Janacek, J. and Hozak, P. 2000. Statistical evaluation of colocalization patterns in immunogold labeling experiments. J Struct. Biol. 132: 201-210.

Pralle, A., Keller, P., Florin, E-L., Simons, K. and Hörber, J.K.H. 2000. Sphingolipid-cholesterol rafts diffuse as small entities in the plasma membrane of mammalian cells. J. Cell Biol. 148: 997-1007.

Prior, I.A., Muncke, C., Parton, R.G. and Hancock, J.F. 2003. Direct visualization of ràs proteins in spatially distinct cell surface microdomains. J Cell Biol 1690: 165-170.

Rao, N., Dodge, I. and Band, H. 2002. The Cbl family of ubiquitin ligases: critical negative regulators of tyrosine kinase signaling in the immune system. J. Leukoc. Biol. 71: 753-763.

Reed, R.A., Mattai, K. and Shipley, G.G. 1987. Interaction of cholera toxin with ganglioside GM1 receptors in supported lipid monolayers. Biochemistry 26: 824-832

Ribi, H.O., Ludwig, D.S., Mercer, K.L., Schoolnik, G.K. and Kornberg, R.D. 1988. 3-Dimensional structure of cholera toxin penetrating a lipid membrane. Science 239: 1272-1276.

Saitoh, S., Arudchandran, R., Manetz, T.S., Zhang, W., Sommers, C.L., Love, P.E., Rivera, J. and Samelson, L.E. 2000. LAT is essential for FcεRI-mediated mast cell activation. Immunity 12: 525-535.

Sanan, D.A., Anderson, R.G.W. 1991. Simultaneous visualization of LDL receptor distribution and clathrin lattices on membranes torn from the upper surface of cultured cells. J. Histochem. Cytochem. 39: 1017-1024.

Sanders, M.L. 1996. Characterizing molecular aggregation on cell surfaces from spatial point patterns. PhD Thesis. Department of Mathematics and Statistics, University of New Mexico.

Schade, A.E. and Levine, A.D. 2002. Lipid raft heterogeneity in human peripheral blood T lymphoblasts: A mechanism for regulating the initiation of TCR signal transduction. J. Immunol. 168: 2233-2239.

Seagrave, J.C., Pfeiffer, J.R., Wofsy, C., Oliver, J.M. 1991. The relationship of IgE receptor topography to secretion in RBL-2H3 mast cells. J. Cell Phys. 148: 139-151.

Schuck S., Honsho, M., Ekroos, K., Shevchenko, A. and Simons, K. 2003. Resistance of cell membranes to different detergents. Proc. Natl. Acad. Sci. USA. 100: 5795-5800.

Simons, K., Ikonen, E. 1997. Functional rafts in cell membranes. Nature 387: 569-572.

Singer, S.J. and Nicolson, G.L. 1972. The fluid mosaic model of the structure of cell membranes. Science 175: 720-731.

Stauffer, T.P. and Meyer, T. 1997. Compartmentalized IgE receptor-mediated signal transduction in living cells. J. Cell Biol. 139: 1447-1454.

Stoyan, D., Kendall, W.S. and Mecke, J. 1995. Stochastic Geometry and its Applications. Wiley Series in Probability and Statistics. Wiley, Chichester, UK.

Stump, R.F., Pfeiffer, J.R., Schneebeck, M.C., Seagrave, J.C. and Oliver, J.M. 1989. Mapping gold-labeled receptors on cell surfaces by backscattered electron imaging and digital image analysis. Studies on the IgE receptor on mast cells. Am. J. Anat. 185:128-141.

Surviladze, Z., Dráberová, L., Kubínová, L., Dráber, P. 1998. Functional heterogeneity of Thy-1 membrane microdomains in rat basophilic leukemia cells. Eur. J. Immunol. 28: 1847-1858.

Surviladze, Z., Dráberová, L., Kovárová, L., Boubelik, M. and Dráber, P. 2001. Differential sensitivity to acute cholesterol lowering of activation mediated via the high-affinity IgE receptor and Thy-1 glycoprotein. Eur. J. Immunol. 31: 1-10.

Teranishi, T., Kurita, R. and Miyake, M. 2000. Shape control of Pt nanoparticles. Journal of Inorganic and Organometallic Polymers 10: 145-156.

Teruel, M.N. and Meyer, T. 2000. Translocation and reversible localization of signaling proteins: a dynamic future for signal transduction. Cell 103:181-184.

Varma, R. and Mayor, S. 1998. GPI-anchored proteins are organized in submicron domains at the cell surface. Nature 394: 798-801.

Wilson, B.S., Pfeiffer, J.R. and Oliver, J.M. 2000. Observing FcεRI signaling from the inside of the mast cell membrane. J. Cell Biol. 149: 1131-1142.

Wilson, B.S., Pfeiffer, J.R., Surviladze, Z., Gaudet, E.A. and Oliver, J.M. 2001. High resolution mapping of mast cell membranes reveals primary and secondary domains of FcεRI and LAT. J. Cell Biol. 15: 645-658.

Wilson, B.S. and Oliver, J.M. 2001. Effector roles of IgE antibodies: Targeting allergen to the high affinity IgE receptor for FcεRI-dependent signaling and antigen presentation. In: B. Zweiman and L.B. Schwartz (Eds), Inflammatory Mechanisms in Allergic Diseases. Marcel Dekker, New York. pp 197-232.

Wilson, B.S., Pfeiffer, J.R. and Oliver, J..M. 2002. FcεRI signaling observed from the inside of the mast cell membrane. Mol. Immunol. 38: 1259-1268.

Wofsy, C., Sanders, M.L., Donahoe, G.W., Pujol, M. and Oliver, J.M. 1995. Quantifying IgE receptor aggregation from SEM-immunocytology. Microscopy and Microanalysis 1: 782-784.

Zhang, J., Leiderman, K., Wilson, B.S., Oliver, J.M. and Steinberg, S.L. 2003. The acquisition and analysis of nanoparticle data. Ms in preparation.

Chapter 2

Rafts, Little Caves and Large Potholes: How Lipid Structure Interacts with Membrane Proteins to Create Functionally Diverse Membrane Environments

Roger Morris[1], Helen Cox[2], Enrico Mombelli[1], and Peter J. Quinn[3]
[1]Molecular Neurobiology Group, MRC Centre for Developmental Neurobiology;
[2]Centre for Neuroscience Research; and [3]School of Life Sciences,
King's College, London, UK

Abstract
This chapter reviews how diverse lipid microdomains form in the membrane and partition proteins into different functional units that regulate cell trafficking, signalling and movement. We will concentrate upon five major issues:

1. the diversity of lipid structure that produces diverse microenvironments into which different subsets of proteins partition;
2. why ordered lipid domains exclude proteins, and the conditions required for select subsets of proteins to enter these domains;
3. the coupling of the inner and outer leaflets within ordered microdomains;
4. the effect of ordered lipid domains upon membrane properties including curvature and hydrophobicity that affect membrane fission, fusion and extension of filopodia;
5. the biological effects of these structural constraints; in particular how the properties of these domains combine to provide a very different signalling, trafficking and membrane fusion environment to that found in disordered (fluid mosaic) membrane.

In addressing these problems, the review draws upon studies ranging from molecular dynamic modelling of lipid interactions, through physical studies of model membrane systems to structural and biological studies of whole cells, examining in the process problems inherent in visualising and purifying these microdomains. While the diversity of structure and function of ordered lipid microdomains is emphasised, some general roles emerge. In particular, the basis for having quite different, non-interacting ordered lipid domains on the same membrane is evident in the diversity of lipid structure and plays a key role in sorting signalling systems. The exclusion of ordered membrane from coated pits, and hence rapid endocytosis, is suggested to underlie the ability of highly ordered domains to establish stable secondary signalling systems required, for instance, in T cell receptor, insulin and neurotrophin signalling.

1. THE AGENDA

Current interest in how specialised lipid microdomains, known as rafts and/or caveolae, are formed by the condensation of cholesterol with lipids on the surface membrane is fuelled by their ability to control a broad range of cellular functions including transmembrane signalling (Brown and London, 1998; Simons and Toomre, 2000), trafficking of vesicles to and from the plasma membrane (Mukherjee and Maxfield, 2000; Simons and Gruenberg, 2000; Ikonen, 2001; Sprong *et al.*, 2001), pathogen entry and exit (Lindwasser and Resh, 2001; Samuel *et al.*, 2001; van der Goot and Harder, 2001; Haldar *et al.*, 2002), cell adhesion (Krauss and Altevogt, 1999; Lacalle *et al.*, 2002) and directed cell movement (Pierini and Maxfield, 2001). To make sense of the deluge of information, at times apparently conflicting, which describes very different biological effects regulated by lipid microdomains, we need to understand how lipids form these domains, and how proteins interact with them.

There is a simple, widely-accepted model for the formation of these domains (Brown and London, 1997; Simons and Ikonen, 1997; Brown and London, 2000; Simons and Toomre, 2000; London, 2002). Cholesterol induces the partitioning of saturated from unsaturated lipids to form a distinct, highly ordered saturated lipid phase (the liquid ordered phase, L_o; Recktenwald and McConnell, 1981; Ipsen *et al.*, 1987). Since sphingolipids are the major saturated lipid species of surface membrane (Holthuis *et al.*, 2001), the common perception is that the interaction of cholesterol with sphingolipids produces a separate L_o domain on the cell surface, the "raft", into which certain proteins partition.

And here one immediately encounters an anomaly. Sphingolipids are present almost exclusively in the outer leaflet of the bilayer (Holthuis *et al.*, 2001). Rafts, however, appear to control transmembrane signalling in part by regulating the entrance into these microdomains of cytoplasmic kinases and adaptor proteins that are anchored to the inner leaflet of the membrane by lipids (myristate and palmitate) that cannot penetrate into the outer leaflet. Thus the catalyst for much current interest arises from selective partitioning of lipid-anchored proteins in the sphingolipid-free, cytoplasmic leaflet of the membrane. Rafts must exist within the inner leaflet; indeed, as we will see, it is likely that the two leaflets are physically coupled in rafts so that the properties of the outer leaflet are mirrored in the inner.

At a broader level, focusing upon a specialist phase produced by interactions between the hydrocarbon chains ignores the contributions of the diverse head groups and linkage to the glycerol backbone that also contribute to the state of lipid phases in membranes. A diversity of ordered lipid environments would be expected purely from the properties of the lipids, to which the enclosed proteins add a reciprocal influence, ordering or disordering the lipids around them. This diversity and interdependence of lipid and protein undoubtedly explains the multiplicity of "raft" environments (Madore *et al.*, 1999; Roper *et al.*, 2000; Drevot *et al.*, 2002; He and Meiri, 2002; Muller, 2002; Prior *et al.*, 2003), including cholesterol-independent rafts (Hansen *et al.*, 2001), that occur in cells.

Our aim in this review is to understand how lipids partition proteins into different environments within the membrane, and what benefit accrues to the proteins as a result. While the impetus for understanding this question is functional, its basis is structural. Unusually for a structural problem, understanding the heterogeneity of lipid microenvironments within membranes involves very diverse elements within both the lipid and protein components that have traditionally been studied in isolated disciplines (lipid biophysics and biochemistry, protein chemistry and cell biology). They now demand to be viewed as an interacting whole. Perhaps not surprisingly there is a degree of confusion and distrust over the methods to be used and the interpretation of results at this rapidly growing interface between different disciplines. Progress so often comes from cross-examining the causes and consequences of such disputes, and this too is the object of this review.

2. WHAT WE MEAN – NAMES OF DOMAINS

Two distinct structures of the plasma membrane, caveolae and "rafts", are the paradigmatic L_o microdomains. Caveolae are readily recognised as non-coated flask-like invaginations of the surface membrane, ~50 nm in diameter (hence their name, "little caves"; Figure 1A). Rafts have no such morphological correlate, and are identified on cells by the clustering of their components on the cell surface seen by immunolabelling in the electron microscope (EM) (Figure 2D–F) or by nm-resolution fluorescent techniques such as fluorescence resonance energy transfer (FRET) (Zacharias *et al.*, 2002). We follow general

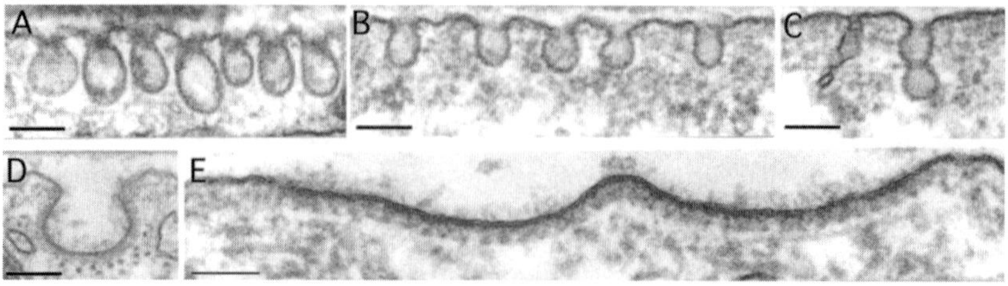

Figure 1. The structure of caveolae and coated pits are disturbed in cells from patients lacking the enzyme required to produced plasmalogens (Thai *et al.*, 2001). (A) caveolae in normal human fibroblasts, showing the distinctive high frequency of flask-like invaginations along the cell surface. Comparable cells from DAPAT-deficient patient show a lower frequency of shallower caveolae (B), some with distorted morphology (C). A typical coated pit is shown in (D) cells from DAPAT-deficient patients have instead quite long (μm) stretches of membrane that are coated but not pitted (E). Scale bars are 100 nm. Photos kindly provided by Dr Wilhelm Just, BZH, Heidelberg.

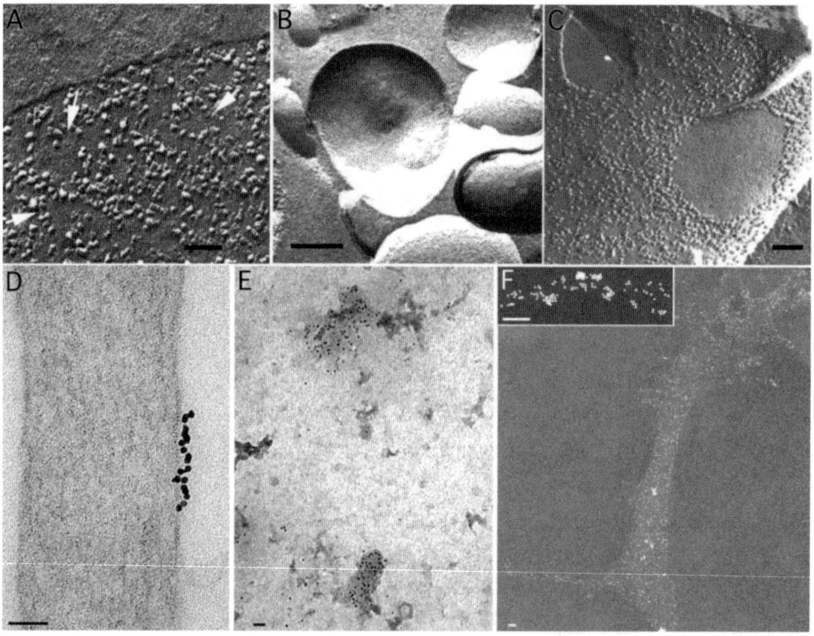

Figure 2. Electron microscopy (EM) of rafts in membranes. A–C: freeze-etch analysis of the surface membrane of an intact human erythrocyte (A), showing nm scale patches (e.g. arrows) devoid of intramembrane particles that match, in size and frequency, rafts seen by immunohistochemical techniques (D–F); (B) an electron micrograph of a freeze-fracture replica obtained from a DRM preparation from human erythrocytes (using 1% Triton X-100 at 4°C) showing the absence of intramembrane particles. It is apparent in the DRM preparation that detergent

practice in referring to caveolae and rafts as the structures that exist in the living membrane. The term we most often use, however, is ordered lipid microdomain, which encompasses rafts, caveolae, and possibly a wider range of microenvironments that selectively admit subsets of membrane proteins as a result of their degree of lipid order.

Caveolae and rafts resist solubilisation in non-ionic detergents (Brown and Rose, 1992; Sargiacomo *et al.*, 1993), forming vesicles, referred to as DRMs (for detergent-resistant membranes),[1] which remain after the intervening

Figure 2. Continued
treatment has resulted in considerable reorganization of the original membrane and often small vesicles can be observed inside larger vesicles. (C) a freeze-fracture replica showing phase separation of gel phase in the plasma membrane of the blue-green alga, *Synechococcus*, in an organism grown at 38°C (at which temperature the distribution of membrane-associated particles was fairly random) and cooled to 15°C prior to thermal quenching to produce domains in the gel phase (assessed by calorimetric and X-ray diffraction evidence (Furtado *et al.*, 1979; Mannock *et al.*, 1985)) corresponding to large regions of smooth fracture plane. The conventional interpretation is that the high melting point lipids separate into a gel phase from which intrinsic membrane proteins are excluded. It is usually inferred that this is a lateral phase separation and this is, in part, true as can be seen by an increase in density of particles around the perifery of the smooth regions of the fracture plane. The additional particles in this high density region, however, cannot account for all the particles originally present in the membrane. This suggests that membrane proteins have either been ejected from the hydrophobic interior of the membrane into the aqueous phase or the constituents of the particles have undergone a reorganization that enables them to accommodate into a lamellar gel phase. From this we may conclude that some intrinsic proteins are arranged in oligomeric complexes and it has been argued that non-bilayer forming membrane lipids not only promote protein complexes but serve to seal the irregular protein–lipid interface to the passage of solutes (Quinn, 1985). It is clear that intrinsic membrane proteins can integrate into gel and liquid-ordered phases but not in the same manner as in a fluid matrix. D–F: different views of immunolabelled GPI-proteins on the surface of primary cultured sensory neurons. (D): a conventional transmission EM of an axon, immunolabelled with Fab anti-Thy-1 directly coupled to 5 nm gold particles; labelling was for 30 min at 4°C on live cells, the cells were then fixed with 1% glutaraldehyde/1% paraformaldehyde and processed conventionally. (E) A view using the "tear-off" technique. Cells were labelled live with Fab anti-prion protein as in D, then an EM grid coated with polylysine was pushed gently against the cells for 1 min; removal of the grid tore off the adhering cell surface, which was fixed as above and viewed in the TEM. (F) Immunogold labelling of an axon near its growth cone, seen by back-scattered electrons in the Field Emission Scanning EM; gold label but not the underlying axon is visible. Cells were fixed at room temperature with 0.5% glutaraldehyde/0.5% paraformaldehyde and labelled at 4°C with 40 nm gold to which anti-prion protein antibodies, produced so as to recognise glutaraldehyde-fixed antigen, were coupled (Madore *et al.*, 1999; Ford *et al.*, 2002). The inset shows part of an axon at higher power. Scale bars are 100 nm in A–C; 50 nm in D; 500 nm in E and F (500 nm is approximately the limit of resolution available with confocal fluorescence microscopy).

[1] DRMs are also known as DIGs (detergent-insoluble glycolipid enriched domains) and GEMs (glycolipid-enriched membranes).

membrane is solubilised and float at low density upon gradient centrifugation. Diverse membrane structures resist solubilisation in mild detergents and float at low density; the assumption that because two components float together, they are within the same structure on the membrane has been a major source of confusion in the field. Where a specific (usually immunoaffinity purified) subtype of either raft or DRM is being discussed, therefore, we will designate its type (e.g. $DRM^{Caveolae}$ or $raft^{Prion\ Protein}$); where its type is not designated, the reader should be aware that the statements made apply to a heterogeneous group of membrane microdomains. Detergent-free isolation methods have also been developed, which rely on a combination of chemical (alkaline) and mechanical (usually sonication) fractionation of the membrane followed by gradient centrifugation to isolate a low density membrane fraction. Since these floating membrane vesicles are put forward as detergent-free equivalents of DRMs, we will refer to them as DfRMs (an intuitively simple acronym despite its inherent illogicality).

3. RAFTS, OR JUST JUNK? – EXAMINING SOME COMMON MISCONCEPTIONS

Before considering how lipid heterogeneity forms microdomains in the membrane, it will be useful to examine some assumptions that dominate both concepts and experimental approaches in this field.

3.1. Rafts cannot be seen by microscopy

The sub-micron size of rafts is below the resolution of conventional fluorescent microscopy, but well within that of electron microscopy (EM). We may well have been looking at rafts for years in the EM without recognising them. Rafts have a very low content of proteins in general, and transmembrane proteins in particular. It has long been evident in freeze fracture that microdomains devoid of intramembrane particles are abundantly embedded within the matrix of intramembrane-particle rich membrane (Figure 2A). These have the size, abundance and lack of transmembrane particles (thought to be lipo-protein complexes)[2] expected of rafts. Caveolae also are devoid of intramembrane particles in freeze-fracture images (Montesano, 1979).

Immunohistochemistry should show whether GPI-anchored proteins and sphingolipids are corralled within specific membrane areas. Initial studies

[2] The evidence is indirect – the diameter of the particles is often more than twice the thickness of a lipid bilayer (Tsvetkova *et al.*, 1994) and the size of the protein complexes associated with membranes are consistent with these dimensions (Williams *et al.*, 1982).

aimed at demonstrating that GPI-anchored proteins occur within caveolae (Ying *et al.*, 1992) illustrated instead the danger of using polyclonal, polyvalent antibodies to investigate, at the ultrastructural level of resolution, the localisation of membrane proteins in poorly fixed membrane. The act of cross-linking by antibodies caused the GPI-anchored proteins to translocate to the caveolae (Mayor *et al.*, 1994; Fujimoto, 1996). These technical problems, which can be simply overcome,[3] are not restricted to EM immunolabelling.[4] However, localisation of proteins in different rafts has been observed by EM immunohistochemistry that correlates with the isolation of these proteins in different DRMs and with functional differences in their properties on cell surface (e.g. Madore *et al.*, 1999; Prior *et al.*, 2003; Brügger *et al.*, 2004); and the general view of raft size and distribution obtained from such studies (Figure 2D–F) is supported by studies of the confinement of fluorescent probes[5]

[3] Multiple monovalent antibody fragments (Fabs) bind to gold, producing multivalent EM probes. However, monovalent immunogold particles can readily be produced using Fab antibody fragments diluted in carrier Fab or other protein (Sunyach *et al.*, 2003). The need to ensure the reagent is monovalent is less if the antigen is immobilised by covalent bonding to stable components in its environment. This is particularly difficult with GPI-anchored proteins, because large carbohyrate chains often cover much of their extracellular domain and they lack a cytoplasmic domain anchored to the cytoskeleton that can be stabilised by cross-linking. These problems are aggravated if paraformaldehyde alone is used as fixative (it takes many hours/days to form stable covalent bonds; Ohba *et al.*, 1979); the inclusion of glutaraldehyde allows instant cross-linking (Habeeb and Hiramoto, 1968; Cheung *et al.*, 1990) at concentrations that preserve ultrastructure, immobilise lipid-anchored proteins and quantitatively retain antigenicity (Morris and Barber, 1983; Madore *et al.*, 1999; Ford *et al.*, 2002; Prior *et al.*, 2003).

[4] Green fluorescent protein and its derivatives, widely used to follow the localisation and trafficking of membrane proteins including raft/caveolae components, have hydrophobic patches that cause them to aggregate when tethered to a membrane (Zacharias *et al.*, 2002). Similarly, the process of adding fluorochromes to antibodies makes them more hydrophobic and prone to aggregate (The and Feltkamp, 1970), an aspect of reagent coupling that must be borne in mind as heavily-conjugated reagents are prepared for single-particle tracking.

[5] Confocal microscopy, even with image resolution improved by computer deconvolution, is limited by the wavelength of UV light to a resolution of hundreds of nm and cannot identify individual rafts. The practice of cross-linking rafts by antibodies in order to see them assumes that this forced convergence is mimicking a physiological mechanism. Emerging ultrastructural evidence for the FceRI receptor (Wilson *et al.*, 2000; Wilson *et al.*, 2001) supports minor changes in raft size following receptor activation but different rafts that are adjacent to each other (co-localised by confocal microscopy) do not merge. The expectation that cross-linking induced raft convergence is associated with transmembrane signalling comes from a host of studies showing that antibody-induced cross-linking of any raft component (GPI-anchored proteins, gangliosides, and artificially engineered GPI proteins) activates raft-associated kinases (Fyn, Lck) (Robinson, 1991). Analysis of Thy-1, the GPI-anchored protein which has frequently featured in such studies (Barboni *et al.*, 1991), shows that its physiological role is to down-regulate, not stimulate, intracellular kinases (Hueber *et al.*, 1997; Hollrigel *et al.*, 1998). Antibody-induced activation may be a consequence of the disruption of a structure that normally down-regulates signalling.

(Varma and Mayor, 1998; Schutz *et al.*, 2000; Dietrich *et al.*, 2002; Zacharias *et al.*, 2002) or of cross-linked proteins (Friedrichson and Kurzchalia, 1998) to nm-scale microdomains on the surface of living cells.

3.2. Rafts float solitarily in a sea of normal membrane

Rafts individually are small but can be very numerous. The total proportion of membrane they occupy can vary from around 20% (e.g. muscle cell surface (Schutz *et al.*, 2000); axonal membrane is of this order (Madore *et al.*, 1999)) through to 30–40% on the luminal Golgi surface (Holthuis *et al.*, 2001) or the cytoplasmic surface of BHK cells (Prior *et al.*, 2003). On endothelial cells, caveolae account for ~30% of the cell surface (Goligorsky *et al.*, 2002). Within specialised compartments such as the apical surface of epithelial cells the proportion of ordered membrane is almost certainly higher. Collectively, rafts are not a minor membrane compartment, and in some membranes may represent the dominant domain.

3.3. A protein that partitions into an ordered lipid microdomain is thereby concentrated

DRMs contain only a few % by weight of membrane proteins, which is why they float at a characteristic light density (e.g. Brown and Rose, 1992; Gafencu *et al.*, 1998; Madore *et al.*, 1999; Pike *et al.*, 2002). It was therefore assumed that rafts occupied only a few % of the cell surface, a view enshrined within their name. Thus when a reasonable proportion (20–80%) of lipidated kinases were also found to float as DRMs, they were said to be concentrated within the rafts. While this clearly need not be the case, the notion that to be present within a raft a protein must be concentrated at high density is as ubiquitous as it is misleading. Raft proteins (e.g. GPI-anchored prion protein) may be clustered at high density, but this is not a necessary property of being within a raft (Tiveron *et al.*, 1994; Madore *et al.*, 1999; Sunyach *et al.*, 2003). For those membranes in which rafts cover >20% of the surface, few endogenously expressed proteins will be at high enough density to appear tightly clustered[6] when assessed by

[6] Thy-1 is expressed at exceptionally high levels on thymocytes (Barclay *et al.*, 1993) and neurons (Morris, 1992); when transgenically expressed at ~8-fold endogenous levels, Thy-1 can be seen by EM immunolabelling to be clustered at high density in its neuronal domains and is totally recovered in DRMs (Madore *et al.*, 1999). Further increasing Thy-1 levels leads to it being increasingly recovered in the detergent-soluble fraction (K. Smith and R. Morris, unpublished). The endogenous level of Thy-1 is therefore ~1/8th that required to saturate its domains. All other surface proteins on thymocytes are expressed at <10% Thy-1's level (Barclay *et al.*, 1993) and are unlikely to be at high density in any surface microdomain.

techniques with nm resolution. The main concentrating effect is likely to be the exact opposite: the >95% of membrane proteins excluded from the rafts will be concentrated within the ~30–50% of the membrane left for them to occupy.

3.4. DRMs are an artefactual fusing of diverse membrane microdomains

The evidence for this oft-repeated and seemingly self-evident judgment is hard to pin down. It would be proven if components seen to be within different microdomains on the cell surface (e.g. by EM immunolabelling) occurred within the same isolated DRMs, or if isolated DRMs were larger than surface rafts, or if the bilayer membrane was detectably perturbed.

To detect mixing of the contents of different solubilised lipid ordered domains, DRMs need to be immunoaffinity purified.[7] Mixing of rafts from totally different membranes within a single DRM was found to occur with Triton X-100 in the solubilisation of brain membranes (Madore *et al.*, 1999) and of model bilayers (Sot *et al.*, 2002). This mixing may result from the selective ability of Triton X-100 to remove inner leaflet lipids (Schuck *et al.*, 2003), presumably transiently producing monolayers that would tend to coalesce. DRMs have been immunoaffinty purified with antibodies to external (Madore *et al.*, 1999; Drevot *et al.*, 2002) and internal (He and Meiri, 2002) raft components, showing both to be exposed on DRMs.[8] In solubilising model membranes, Triton X-100 actually promotes the formation of detergent-insoluble liquid ordered (L_o) domains from liquid disordered (L_d) membrane at 37°C (Heerklotz, 2002). Concern about mixing of components of biological membranes in detergents is well founded.

However, a strong case can equally be made that under controlled conditions detergents do not cause mixing of rafts, even those closely adjacent to each other on the same cell surface. Using the detergent Brij 96, not only did DRMs originating from different membranes not mix, but different GPI-anchored proteins (Thy-1 and prion protein) found in adjacent domains on the neuronal surface were isolated in separate DRMs (Madore *et al.*, 1999) with different lipid compositions (Brügger *et al.*, 2004). Similarly, the T cell receptor and Thy-1,

[7] Immunoaffinity isolation of DRMs is not trivial. The most widely used methods (using large immunolabelled beads (e.g. Sepharose) of ~100–200 μm diameter) work at <5% efficiency with many DRMs; immunolabelled beads of ≤1 μm are much more effective (Madore *et al.*, 1999).

[8] We are not aware of any study that has asked whether both internal and external membrane proteins are exposed on the same surface of a single DRM; nor whether DRMs are sealed vesicles or open membrane fragments or tubes.

both present within the immunological synapse, were recovered in distinct DRMs when Brij 98 was used as detergent (Drevot *et al.*, 2002).

EM analysis of immunolabelled DRMs has found a similar density of label, and size of domain, as on the surface membrane (Madore *et al.*, 1999; Drevot *et al.*, 2002); spin labels show that DRMs contain lipids in the L_o state (Ge *et al.*, 1999). Since sphingolipids are located only in the outer leaflet, the inner leaflet might be assumed to be more susceptible to solubilisation. The evidence, however, indicates this is not so. Thus the "rules" governing which lipidated proteins enter the inner (cytoplasmic) leaflet of rafts studied on the surface of living cells (Stulnig *et al.*, 1998; Liang *et al.*, 2001; Wang *et al.*, 2001; Zacharias *et al.*, 2002) apply identically to DRMs isolated from their membranes (Stulnig *et al.*, 1998; Melkonian *et al.*, 1999; Liang *et al.*, 2001; Wang *et al.*, 2001; Zacharias *et al.*, 2002) and to artificial bilayers in which the L_o phase reproduces the raft environment (Moffett *et al.*, 2000; Wang and Silvius, 2001).

So, although detergents must be used with care, and provide at best a rather blunt instrument with which to dissect the fine structure of the lipid bilayer, they seem surprisingly effective at isolating rafts as membrane fragments. This will undoubtedly improve with the wider use of detergents differing in the saturation of their hydrophobic chains to disrupt different membrane environments (Drevot *et al.*, 2002).[9]

3.5. Detergent-free membrane vesicles better preserve the raft environment

The development of detergent-free methods for isolating rafts (Smart *et al.*, 1995; Song *et al.*, 1996; Prior *et al.*, 2001; Luria *et al.*, 2002; Pike *et al.*, 2002) is undoubtedly useful, although it is not yet possible to compare the same immunoaffinity-isolated vesicle fractions obtained by different methods (e.g. $DRM^{Thy\text{-}1}$ versus $DfRM^{Thy\text{-}1}$, rather than the total low-density fractions). Taking the membrane fraction that collects at the 5/35% sucrose interface, as used in some protocols, is unfortunate since all plasma membrane, most of the Golgi, early endosomes and some endoplasmic reticulum float in this fraction (Holthuis *et al.*, 2001).

Ordered lipid membranes are not intrinsically lighter than disordered membranes – they should have a higher density since cholesterol condenses

[9] A major practical problem with common detergents is that they are not pure chemical reagents, and show considerable batch-to-batch variation. Should pilot studies show a particular detergent to be suitable, it is wise to buy sufficient of a single batch to enable the entire study to be completed.

(i.e. tightly packs) saturated lipids.[10] Membranes float at low density if they have a low protein : lipid ratio, a property expected of rafts because their ordered lipids exclude most membrane proteins. If the membrane has been disrupted along fault lines that match the degree of lipid organisation, vesicles of ordered lipid should float at low density while the rest of the membrane should have a high (protein-rich) density. This has been found for DfRMs containing the different isoforms of Ras, in which the raft dependent form (inactive H-Ras) floated at low density on a continuous sucrose density gradient, with the non-raft forms recovered in high density membrane (Prior and Hancock, 2001; Prior *et al.*, 2001; Prior *et al.*, 2003).

The logic behind the isolation of DfRMs as a vesicle preparation of raft membrane is circuitous. Presumably physical disruption could selectively tear the membrane at the interface between protein-rich and protein-poor areas if each region has strong intrinsic cohesive properties (lipid cohesive properties for rafts, presumably cytoskeletal and other cortical protein networks for disordered membrane). While Ras supports this case, the unproven chain of contingency underlying the isolation of DfRMs needs to be tested and may not hold generally.

The logic underlying the use of detergents to separate these areas is that the tightly-packed ordered lipids within rafts exclude both the majority of membrane proteins, and detergents; the same property leads to the recovery of detergent-resistant membrane vesicles that float at low density. There is a good correlation between detergent insolubility and the degree of lipid order in the membrane (Ahmed *et al.*, 1997; Schroeder *et al.*, 1998; London and Brown, 2000; Li *et al.*, 2001; Wang *et al.*, 2001; Xu *et al.*, 2001) although this too may be an over-simplification (Heerklotz, 2002; Sot *et al.*, 2002). Detergent based methods may not be as bad, nor detergent-free methods as good, as is often supposed.

3.6. Ordered lipid microdomains cannot be isolated at 4°C

The state (phase) of membrane lipids is highly temperature dependent. This is dealt with more fully below, but here it should be noted that the saturated lipids that form ordered microdomains are in an immobile gel phase at 4°C

[10] A cholesterol-condensed bilayer should be ~15% denser than in the absence of cholesterol. Thus the area per molecule of dipalmitoylphosphatidylcholine (MW 733.3 a.u.) ±50% cholesterol (MW 386.7 a.u.) is 44.7 and 62 Å^2, that of cholesterol 32 Å^2, yielding 14.6 a.u./Å^2 with cholesterol, and 11.8 a.u./Å^2 in its absence (values from (Smondyrev and Berkowitz, 1999) figures for dimyristoylphosphatidylcholine are similar (Rog and Pasenkiewicz-Gierula, 2001b)). From our own data (Mombelli *et al.*, in preparation) the corresponding figures for palmitoylsphingomyelin ±33% cholesterol are 14.1 and 12.8 a.u./Å^2.

that would be expected to exclude detergents;[11] this would not however prove that these same domains were present (in a detergent-excluding liquid-ordered rather than gel phase) at 37°C. Initially, non-ionic detergents were found to fully solubilise "raft" components at 37°C (Brown and Rose, 1992), suggesting perhaps that the separation of the different lipid types monitored by detergent insolubility may be an artefact of low temperature. However, it has since been shown that DRMs can be isolated from biological membranes at 37°C (Madore *et al.*, 1999; Drevot *et al.*, 2002). Indeed, in some model membranes condensed cholesterol/sphingolipid complexes are less, not more, soluble at 37°C (Sot *et al.*, 2002). Moreover, there is considerable difference in the detergent insolubility at 4°C of DRMs containing proteins that occupy different microdomains on the cell surface (Madore *et al.*, 1999; Roper *et al.*, 2000), suggesting that ordered lipid microdomains retain differences at 4°C that influence detergent solubility. However, the suspicion remains that the full diversity of ordered lipid microdomains at 37°C is reduced at 4°C.

4. THE PROPERTIES OF LIPIDS ALONE ARE SUFFICIENT TO PRODUCE DIFFERENT MEMBRANE MICRODOMAINS

4.1. Where lipids differ

There are five major sources of heterogeneity in lipids that affect their capacity to form microdomains in membrane (Brown, 1998; Ramstedt and Slotte, 2002). The simple names familiar to cell biologists (sphingomyelin (SM), phosphatidylcholine (PC) etc – see Figure 3) largely hide these differences. Palmitoylsphingomyelin (PSM; Figure 3d), for instance, is very similar to 1,2 dipalmitoyl-*sn*-glycero-3-phosphocholine (DPPC; Figure 3b) in membrane-forming properties (both have phosphocholine headgroups, long saturated hydrocarbon chains and occur primarily in the outer leaflet of the plasma membrane); and both are very dissimilar to 1-oleoyl-2-arachidonyl-*sn*-glycero-3-phosphocholine, a highly unsaturated phosphatidylcholine. The relevant levels of heterogeneity in membrane lipids (Figure 3) are:

1. The backbone, either glycerol or sphingosine, which differ in two important respects. Sphingosine already includes one hydrocarbon chain, equivalent to

[11] This is not true for lipid chains shorter than C16 (Patra *et al.*, 1998; Sot *et al.*, 2002) which might explain the predominance of longer chain lipids in isolated DRMs (Fridriksson *et al.*, 1999; Rietveld *et al.*, 1999).

an *sn-1*-linked C16 saturated fatty acid in a glycerolipid[12] (Figure 3d). The fatty acid is linked by an amide bond, providing an –NH group which together with its –OH group on C3 provides it with two hydrogen bond donors in addition to its hydrogen bond acceptors (phosphate esteric O) to enable sphingolipids to form an extensive hydrogen bonded network at the surface of the membrane (Mombelli *et al.*, 2003). Glycerol has two fatty acids ester-linked to its *sn*-1 and *sn*-2 positions to make a membrane-forming glycerolipid that has only hydrogen bond acceptors. It can only participate in hydrogen-bonded networks via bridging water molecules (Pasenkiewicz-Gierula *et al.*, 1997).

2. The headgroup, conveniently named in the glycerolipids (e.g. phosphatidylcholine (PC)) but not in the more diverse sphingolipids where headgroups include phosphocholine (sphingomyelin), –OH (ceramide), a wide range of sugars including standard hexoses (cerebrosides) and even more complex sugar chains including sialic acid (gangliosides). These sugar residues, and the minimal –OH headgroup of ceramide (Figure 3f) and diacylglycerol (DAG; Figure 3k), greatly change membrane-forming properties of lipids. When assessed for their effect upon the interaction of cholesterol with saturated fatty acid chains, there are relatively small differences between choline, ethanolamine and serine headgroups (Sankaram and Thompson, 1990a; Wang and Silvius, 2003). However, cholesterol has been found not to interact well with PE in the gel phase (even saturated PE; McMullen *et al.*, 1999) and to interact particularly poorly with mono-unsaturated PE (Wang and Quinn, 2002).

3. The length of attached fatty acids (C14–C24 in most mammalian cells); the longer the chain the higher the affinity for ordered lipid domains, with ~2–3-fold increase in affinity for each CH_2CH_2 unit added (Palestini *et al.*, 1995; Wang *et al.*, 2000).

4. The number, position and conformation of double bonds. The number is denoted e.g. C20:4. In mammalian cells the double bonds of the fatty acid residues are invariably *cis*, which introduces a 60° kink in the hydrocarbon chain, preventing the chain becoming linear. The fatty acid in the *sn*-1 position is usually fully saturated and so glycerolipids resemble sphingolipids in this respect; but unlike sphingolipids, the fatty acid in the *sn*-2 position in glycerolipids is normally unsaturated.

5. The chemical bond joining the fatty acid to glycerol, standardly depicted as an ester linkage can also be the more stable vinyl ether linkage (the lipids are then referred to as plasmalogens; Figure 3l) or ether linkage (glyceryl ether phospholipids). Plasmalogens are of particular interest in relationship to rafts (Pike *et al.*, 2002).

[12] The C3 double bond in sphingosine is in *trans*, giving a linear conformation to the chain.

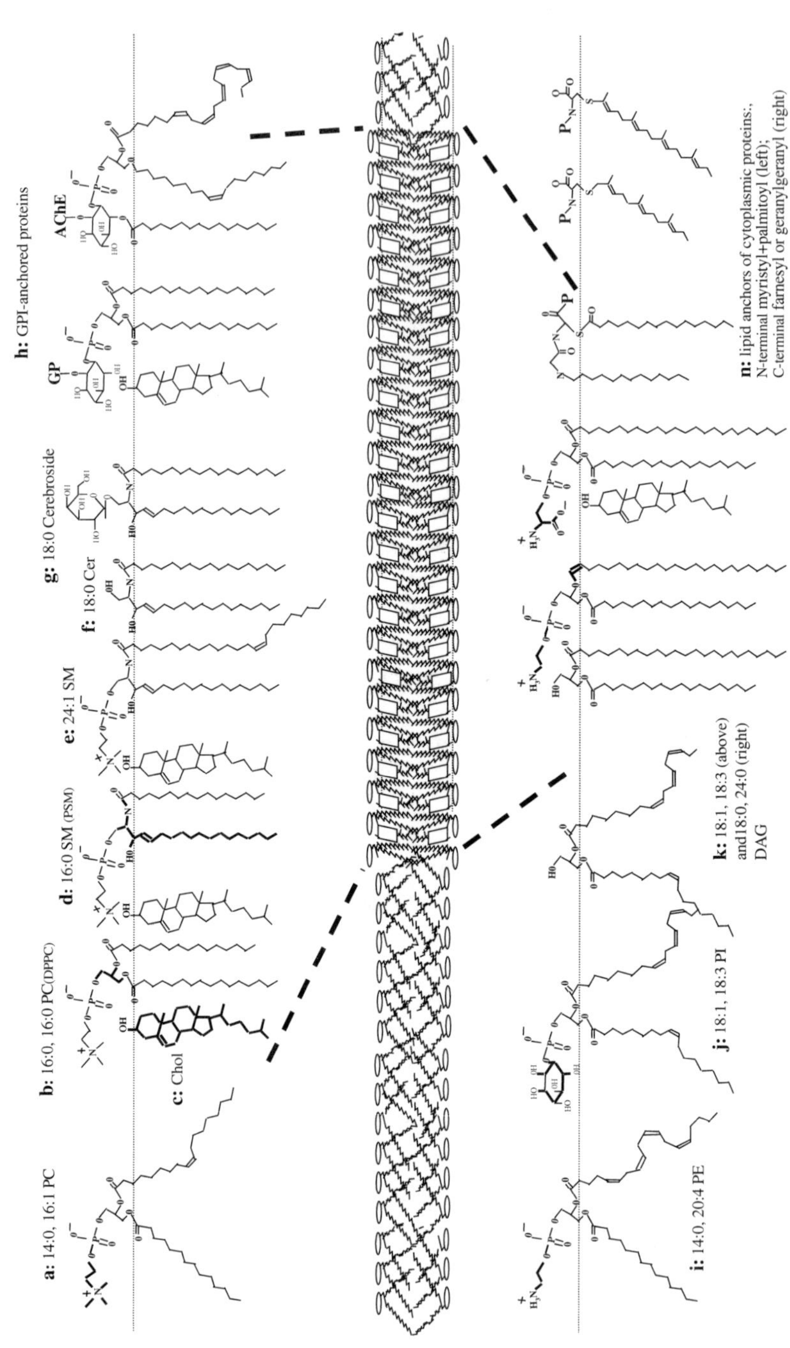
a: 14:0, 16:1 PC
b: 16:0, 16:0 PC(DPPC)
c: Chol
d: 16:0 SM (PSM)
e: 24:1 SM
f: 18:0 Cer
g: 18:0 Cerebroside
h: GPI-anchored proteins
GP
AChE
i: 14:0, 20:4 PE
j: 18:1, 18:3 PI
k: 18:1, 18:3 (above) and18:0, 24:0 (right) DAG
l: plasmanyl PE
m: 18:0, 24:0 PS
n: lipid anchors of cytoplasmic proteins:,
N-terminal myristyl+palmitoyl (left);
C-terminal farnesyl or geranylgeranyl (right)

Figure 3. Structure of membrane lipids found primarily in the outer (a–h) and inner (i–n) leaflets of the bilayer, shown schematically around a bilayer formed from unsaturated lipids flanking an ordered domain of cholesterol (sterol rings represented by a rectangle) and saturated lipids. Dashed lines divide the lipids according to the membrane domains to which they partition. (a) Phosphatidylcholine (PC) with the *sn*-1 fatty acid (myristic) saturated and the *sn*-2 acid unsaturated at the C9 position (palmitoleic). The choline head group is in bold. (b) A saturated (dipalmitoyl) PC; the glycerol backbone is in bold. (c) Cholesterol, in bold; this structure is repeated beside the other lipids that occur within ordered domains. (d) A fully saturated (palmitoyl) sphingomyelin; the sphingosine backbone is in bold. (e) A typical unsaturated SM with the double bond at C15. (f) Stearoyl Ceramide. (g) Galactosyl ceramide (cerebroside). (h) Lipid attachment of GPI-anchored proteins, showing on the left diacyl attachment to inositol-1-phosphate (GP represents the extended *G*lycosyl chain linking to the *P*rotein), and on the right the *sn*-1 alkyl, *sn*-2 acyl linkage to unsaturated fatty acids found with AcetylCholine Esterase (AChE). (i) Phosphatidylethanolamine (PE; ethanolamine group in bold) with arachidonic acid in the *sn*-2 position. (j) Phosphatidylinositol (PI; inositol group in bold), showing unsaturated fatty acids in the *sn*-1 (oleic) and *sn*-2 (α-linolenic) positions as found in certain tissues (e.g. brain). (k) Diacylglycerols (DAGs), unsaturated (left, derived from k) and fully saturated (right, derived from n). (l) Plasmanyl ethanolamine; the vinyl ether linkage is in bold. (m) A saturated phosphatidylserine (PS; serine headgroup in bold). (n) Lipid anchors used by cytoplasmic proteins: the amide-linked myristoyl, thio-acyl linked palmitoyl groups attached to N-terminal Met-Gly-Cys motifs; and the prenoids attached to C-terminal *CAAX* motifs.

In addition to these fatty acid-based lipids, the sterol cholesterol is a major component of certain membranes including the plasma membrane. The four rings of the sterol moiety have a planar conformation and extend into the membrane to a depth equivalent to C10–12 of the fatty acids (Figure 3c), imposing (when tightly packed, i.e. condensed) a linear, relatively immobile conformation on this region of the fatty acid chain. Orientating the sterol rings in the membrane is a minimal headgroup (OH) exposed to the aqueous interface and, buried deep within the bilayer, an extended, flexible hydrocarbon tail. In non-mammalian species, other sterols perform the same function as cholesterol (Miao *et al.*, 2002).

4.2. Phase separation of lipids

The molecular species of membrane lipids separate into different phases as a function of their melting points. At lower temperatures, lipids are quasi-solid, corresponding at the molecular level to a gel or solid-ordered (S_O) phase in which they are virtually immobile (lateral diffusion 10^{-10}–10^{-12} cm^2/sec) with their chains fully extended (with all double bonds in the *trans* conformation) and closely packed in an hexagonal array. At temperatures above their melting temperature (T_M), the acyl chains of the lipids are fluid, corresponding at the molecular level to being loosely packed; the lipids themselves are highly mobile, both in terms of the ability of different lipids to move along the plane of the bilayer (translational mobility), and of their individual fatty acid chains to wave about (conformational order of the fatty acid chains) (Vist and Davis, 1990; Miao *et al.*, 2002). The liquid phase is referred to as Lα (for lamellar liquid-crystalline), or (as we do here) L_d (for lamellar disordered) phase. While all lipids undergo this transition, they do so at very different temperatures: unsaturated lipids remain liquid until near or below 0°C, whereas saturated lipids are in the liquid state only at or above mammalian body temperature (37–45°C; or, for saturated glycolipids, even higher, at 60–70°C (Koynova and Caffrey, 1995). The polar headgroup also influences T_M. For example, for equivalent hydrocarbon chain composition the T_M of phosphatidylethanolamines are more than 20°C higher than the corresponding phosphatidylcholines.

All membranes contain molecular species of lipid that at physiological temperatures do not form bilayer structures. These include phosphatidylethanolamine, monoglycosyldiacylglycerol and cardiolipin. While it is generally assumed that such lipids are constrained into a bilayer arrangement by interaction with other membrane constituents their phase separation to form inverted micelles is an event associated with membrane fusion. Recent NMR evidence, however, has identified lipids undergoing isotropic motion in intact cell membranes (Le Moyec *et al.*, 2000) which are said to be associated with raft domains (Ferretti *et al.*, 2003). Motion of this type may be a feature of the boundary region separating fluid from liquid-ordered domains where packing faults and molecular mismatches are expected to occur.

Since biological membranes are composed of mixtures of saturated and unsaturated lipids, over the normal range of body temperatures (4–40°C) membranes could separate out into two very different phases, the gel and liquid phase. This would probably be disastrous for life – it is difficult to see how the dynamic qualities demanded of living membranes would be possible in the gel phase (Miao *et al.*, 2002). The gel phase is largely prevented from forming by cholesterol (or its homologues in non-vertebrate phyla) which condenses with saturated fatty acid chains to form an intermediate liquid ordered phase (L_o) that has properties of both (Recktenwald and McConnell, 1981; Ipsen *et al.*, 1987; Vist and Davis, 1990; Brown and London, 1997; Figure 4). Like the gel phase, in the L_o phase the fatty acid chains, constrained by the planar ring structure of cholesterol, are more extended and less mobile; like the L_d phase, in the L_o phase the lipids continue to diffuse laterally (hence liquid ordered; L_o lipids cover several μm/sec at 37°C) allowing biological interactions to occur (Vist and Davis, 1990).

In general, the properties of ordered lipid domains in biological membranes match those of the L_o phase characterised in mixtures of lipids and

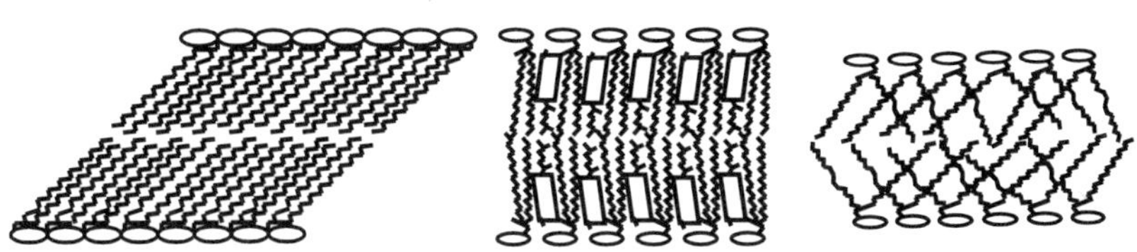

Figure 4. Schematic illustration of saturated lipids in a gel or S_o phase (left) and L_o phase (centre), and unsaturated lipids in a L_d phase (right). In the gel phase the tightly packed, immobile fatty acid chains have maximised van der Waals interactions; their packing is limited by the size of their large headgroups, so their acyl chains tilt at an angle to fill the bilayer. In the L_o phase, the minimal (−OH) headgroup of cholesterol allows its bulky sterol rings (denoted by a single rectangle) to fill in space beneath the lipid headgroups, allowing the fatty acid chains of the lipids to stretch perpendicular to the membrane although their flexibility is constrained by the broad, rigid face of the sterol ring. Although the densely packed L_o bilayer, like the gel phase, excludes mild detergents, the intervening cholesterol molecules associate sufficiently loosely with the phospholipids to enable the latter to freely rotate and diffuse laterally, so this is a fluid rather than gel state. In the prototypical fluid state, L_d, the lipids without intervening cholesterol have nearly the same packing density as they do in the L_o state; of their fatty acid chains, that in the *sn*-1 position is usually staturated, that in the *sn*-2 usually has 1–4 *cis* double bonds that make regular packing impossible. The chains are highly mobile rather than elongated, making the bilayer about 10% thinner in this phase (Vist and Davis, 1990; McIntosh *et al.*, 1992b; McIntosh *et al.*, 1992a; Li *et al.*, 2001).

sterols. On the one hand, lipid rafts/DRMs have the expected composition (high in saturated lipids and cholesterol; Brown and Rose, 1992) and lateral diffusion (Ge *et al.*, 1999; Pralle *et al.*, 2000; Dietrich *et al.*, 2002) typical of the L_0 phase; on the other, L_0 phases made with artificial bilayers are resistant to detergent (Ahmed *et al.*, 1997; Brown and London, 1997; Schroeder *et al.*, 1998; London and Brown, 2000), and GPI-anchored and other diacylated proteins partition into them (Schroeder *et al.*, 1998; London and Brown, 2000; Moffett *et al.*, 2000; Dietrich *et al.*, 2001a; Dietrich *et al.*, 2001b; Li *et al.*, 2001; Radhakrishnan *et al.*, 2001; Wang *et al.*, 2001; Dietrich *et al.*, 2002; Saslowsky *et al.*, 2002).

4.3. Cholesterol condenses with (semi-) saturated lipids

The statement that cholesterol interacts preferentially with sphingolipids rather than glycerolipids is as common and it is misleading. Cholesterol interacts with lipids containing long chain (≥C14) saturated fatty acids, an interaction dominated by van der Waals interactions (Gronberg *et al.*, 1991; Kan *et al.*, 1991; McIntosh *et al.*, 1992b). In order for van der Waals forces to operate, the sterol and lipid must fit closely. This steric constraint requires of the sterol that it have the 3-OH head group, the sterol rings must be planar, the hydrocarbon tail must be of the appropriate length, and it should be the natural enantiomer (Mattjus *et al.*, 1995; Xu and London, 2000; Lalitha *et al.*, 2001; Xu *et al.*, 2001; Miao *et al.*, 2002). On the lipid side, the primary requirement is for long, saturated lipid chains; in particular, *cis* double bonds are poorly tolerated in that part of the fatty acid chain (first ~12 C's) that interacts with the planar sterol rings (Gronberg *et al.*, 1991; Kan *et al.*, 1991; Guo *et al.*, 2002). Glycerolipids, made in the endoplasmic reticulum, often have 1–4 double bonds in this region of their *sn*-2 acyl chain (Barenholz and Thompson, 1980; Leventis and Silvius, 2001), whereas sphingolipids (made in the Golgi) are either fully saturated or have a double bond at the C15 position (Brown, 1998; Holthuis *et al.*, 2001) where it is accommodated by the flexible hydrocarbon tail of cholesterol. However, this requirement for saturation applies strictly only to the fatty acid in the *sn*-1 position for a glycerolipid; a single double bond, even in a relatively superficial position abutting the sterol rings of cholesterol, is tolerated nearly as well as a fully saturated chain in this position (Thewalt and Bloom, 1992; Reyes Mateo *et al.*, 1995; Leventis and Silvius, 2001; Miao *et al.*, 2002). It is the presence of multiple double bonds in one chain, or unsaturation in both chains, that strongly prevents condensation (Leventis and Silvius, 2001).

The range of membrane lipids that can condense with cholesterol and partition into a L_o phase is much greater than is commonly assumed. This includes in particular the glycerolipids of the inner leaflet of the plasma membrane, and of internal membranes within the cell, provided they meet the condition of saturation in the *sn-1* chain and no more than one double bond in the *sn-2* chain (Sankaram and Thompson, 1990a; Wang and Silvius, 2001). The fatty acid attached at *sn-1* to glycerolipids is normally fully saturated except in certain tissues (including brain and sperm; Leventis and Silvius, 2001). Sphingolipids have an inbuilt structural advantage in that they have an invariant linear hydrocarbon chain at this position as part of the sphingosine base (Brown, 1998).

The relative indifference of cholesterol for sphingolipids compared to fatty acid chain-matched glycerolipids runs counter to a number of studies that demonstrate a preference of cholesterol for sphingomyelin rather than chain-matched glycerolipids. SM is similar to PC in that both have phosphocholine headgroups, but differs in having a sphingosine base that forms an amide bond with a fatty acid, rather than the ester links formed by glycerol (Figure 3d). In practice, this means that SM contributes both hydrogen bond acceptors (esteric O) and donors (–OH and –NH) to the membrane surface, whereas PC has only acceptors (esteric O) (Boggs, 1987; Brown, 1998). While SM's –OH group is not involved in interactions with cholesterol (Gronberg *et al.*, 1991; Kan *et al.*, 1991), there is considerable evidence that its –NH group is. This interaction is not sufficiently robust to be detected by solid state NMR studies of hydrated bilayers (Guo *et al.*, 2002), but rather is detected by indirect effects such as the accessibility of cholesterol's 3-OH to cholesterol oxidase (Bittman *et al.*, 1994; Mattjus and Slotte, 1996), the extractability of cholesterol from the membrane (Leventis and Silvius, 2001), or cholesterol's effect on interfacial elasticity (Li *et al.*, 2001). The multiple hydrogen-bonding propensities of each SM molecule leads to the formation of relatively large hydrogen-bonded clusters (4–8 molecules) of sphingomyelin in the membrane (Mombelli *et al.*, 2003). Cholesterol can only form a single hydrogen bond and so acts as a terminator of the hydrogen bonding networks (Mombelli *et al.*, in preparation; Miao *et al.*, 2002). The gain to the membrane of the single hydrogen bond formed with cholesterol is at the expense of larger clusters of hydrogen bonded sphingolipids, and so makes a relatively minor contribution to membrane properties.

4.4. How does cholesterol interact with (semi-) saturated lipids?

The physical methods used to investigate membranes monitor bulk properties, but give only rudimentary information about molecular interactions.

The situation that exists in trying to define the physical basis for the biological role of membrane microdomains is rather like that of genetics before the double helix. The application of molecular modelling to physical data identified the hydrogen bonds that underlie DNA structure; the resulting model has shed light upon diverse areas of genetics and biology ever since. There is a similar need for interaction between model building (now done with computers) and physical methods if we are to understand the molecular interactions that control the formation of membrane domains (Tieleman *et al.*, 1997; Nagle and Tristram-Nagle, 2000; Scott, 2002).

Molecular dynamics (MD) has successfully modelled the condensation of cholesterol with saturated PC. For instance, the restraint upon mobility of the first 12 CH_2 groups of the fatty acid chains caused by the planar sterol rings, long known from NMR studies (Sankaram and Thompson, 1990b), has been obtained in an MD bilayer (Rog and Pasenkiewicz-Gierula, 2001a). Computer modelling, however, has gone beyond experimental limits in showing that cholesterol makes closer contacts with the *sn*-1 rather than *sn*-2 fatty acid chain of glycerolipids (Rog and Pasenkiewicz-Gierula, 2001b), which would explain why unsaturation is allowed in the *sn*-2 chain. More controversially, it shows that the main interaction is not that of cholesterol with the lipid chains, but of the tightly packed lipid chains with each other (Rog and Pasenkiewicz-Gierula, 2001b). Cholesterol acts as a hydrophobic spacer that allows the saturated lipids to pack tightly and thereby interact more with each other along their entire length. Presumably cholesterol's constraint upon the flexibility of the first C12 atoms of the fatty acid chains promotes van der Waals interactions between the chains. The small –OH headgroup of cholesterol is probably of key importance in allowing room for tighter packing of the bulky phosphocholine headgroup (Rog and Pasenkiewicz-Gierula, 2001b; see also Brown (1998) and Li *et al.*, 2001).

If the action of cholesterol upon the condensation of lipids is indirect as just described, it is not surprising that other agents that similarly enhance interactions between fatty acid chains in the bilayer will promote or even sustain ordered lipid domains.

4.5. Stoichiometry of the cholesterol : lipid interaction

Studies of the condensing effect of cholesterol upon sphingomyelins in monolayers suggest a 1 : 2 stoichiometry (although the evidence is rather indirect and this value is indicative only; Radhakrishnan *et al.*, 2001). Cholesterol produces a marked effect on membrane characteristics as its mole% in a monolayer or bilayer of saturated lipids rises above 20% (e.g. Li *et al.*, 2001) to a maximum of 50 mole% after which it forms a separate cholesterol-only phase (e.g. McIntosh *et al.*, 1992b). In DRMs isolated from biological membranes, the mole% of cholesterol is ~30% (Brown and Rose, 1992; Rietveld *et al.*, 1999; Gkantiragas *et al.*, 2001; Pike *et al.*, 2002).

Normally the level of cholesterol on the surface membrane is strictly regulated (Ikonen and Parton, 2000), and can only be manipulated experimentally by removing cholesterol (typically with methyl β-cyclodextrin or cholesterol oxidase). However, red blood cells have only one membrane which is exposed continuously to a rich supply of cholesterol. In DRMs isolated from both human and goat red blood cells (ruminant cells have twice the sphingolipid content of human red cells), there is a strict 1 : 1 stoichiometry between cholesterol and sphingomyelin, although in these cells the cholesterol level is higher in the solubilised (disordered) membrane than in the rafts (see Chapter X). This suggests that lipid ordered microdomains impose an absolute equimolar stoichiometry between cholesterol and the saturated lipid, and as readily exclude excess sterol as they include it when needed.

Although the condensing effect of cholesterol on lipids becomes marked at >20 mole% cholesterol, other effects occur at concentrations too low (≤5%) to form an L_o phase (Sankaram and Thompson, 1990a). In an L_d phase bilayer, molecules of cholesterol in opposed leaflets line up opposite each other with their tails in contact, thereby coupling the two sides of the bilayer (Rogers *et al.*, 1979; Sankaram and Thompson, 1990a; Harris *et al.*, 1995; Mukherjee and Chattopadhyay, 1996). This coupling requires the membrane to be planar and is lost when a high degree of curvature (e.g. in coated pits or caveolae) is introduced (Rukmini *et al.*, 2001).

While cholesterol is the major sterol in mammalian cells, in other species a range of different sterols predominate. It has been argued that the fine tuning of sterol structure during evolution results in (and therefore may be driven by) greater ability to condense lipids into the L_o phase (Miao *et al.*, 2002). However, the fungal sterol ergosterol, and 7-dehydrocholesterol (both of which have a second double bond in the B ring, promoting planarity of the rings) more strongly promote L_o phase formation than native cholesterol (Xu *et al.*, 2001). The latter is the final intermediate in cholesterol synthesis and accumulates in man in Smith-Lemli-Opitz syndrome (Xu *et al.*, 2001). The advantage to the lower phyla of the better condensing sterols may be the need in poikilotherms to maintain membrane fluidity at low temperatures; the advantage of cholesterol for homeotherms may be its enhanced ability to couple the inner and outer leaflets of the bilayer (below).

4.6. In which leaflet(s) of the bilayer is cholesterol found?

Given the asymmetry of biological membranes, this is a critical question. Studies using the fluorescent sterol dehydroergosterol (DHE), which many criteria indicate to be an excellent cholesterol analogue in cells (Hale and Schroeder, 1982; Mukherjee *et al.*, 1998), show it to be primarily (~70%) accessible to quenching agent present on the cytoplasmic, not the external, surface (Kier *et al.*, 1986; Wood *et al.*, 1990; Schroeder *et al.*, 1991; Igbavboa *et al.*, 1997). A totally different approach (cross-linking by a photoactivatable ganglioside homologue) has found cholesterol in the outer leaflet of L_d phase (normal) membrane, but none in the outer leaflet of caveolae, probably reflecting sequestration by cholesterol-binding caveolin in the inner leaflet (Pitto *et al.*, 2000). Ultrastructural labelling of a human B lymphocyte line, using a modified (non-haemolytic) Perfringolysin O toxin which recognises only higher density (~30 mole%) cholesterol in membranes, showed abundant cholesterol in the outer leaflet on the cell surface, particularly concentrated on filopodia (Möbius *et al.*, 2002); comparable labelling could also be found on the cytoplasmic leaflet.[13] Interestingly, by far the highest concentration of cholesterol labelling was found, not on the plasma membrane but on multivesicular bodies (Möbius *et al.*, 2003).

Cholesterol therefore occurs in both leaflets; precise determination of the levels in the inner and outer leaflets of ordered lipid domains would provide a basis for assessing the structure of rafts in each.

4.7. Cholesterol mimetics: vitamin E

Another lipid that phase separates into membrane domains is vitamin E. Vitamin E has a similar structure and polarity as cholesterol and shares many features of its effect on the membrane lipid bilayer matrix. Recent evidence (Porter, 2003) also suggests that vitamin E stimulates cholesterol biosynthesis on the one hand, via activation of a soluble protein factor regulating the conversion of squalene to lanosterol in the downstream pathway for cholesterol biosynthesis, and down regulates expression of scavenger lipo-protein receptors on the other. Clearly there is interdependence which may indicate duality of function in membranes. Furthermore, a variety of functions of vitamin E unrelated to its putative antioxidant role have been identified recently, many of which appear to involve cell signalling processes (Ricciarelli *et al.*, 2002).

In model membrane experiments, the thermal properties of mixed aqueous dispersions of vitamin E (α-tocopherol) with distearoyl-, dipalmitoyl-, dimyristoyl- and dilauroyl-derivatives of phosphatidylcholine (DSPC, DPPC, DMPC,

[13] Penetration of labelled toxin through the matrix of fixed cytoplasmic protein to the cytoplasmic leaflet of the plasma membrane was variable and prevented any quantitative assessment relative to the external leaflet (Möbius *et al.* (2002) and W. Mobius, personal communication).

DLPC) have been examined using differential scanning calorimetry. These showed that increasing proportions of α-tocopherol cause a progressive broadening of the gel to liquid-crystalline phase transition and a decrease of enthalpy change of the transition (Quinn, 1995). This means that vitamin E intercalates between these saturated phospholipids and perturbs the packing of the acyl chains so that they do not form the close alignment required for formation of a typical gel phase. Similar results were obtained by Fourier transform infrared and Raman spectroscopy which monitors the vibrational motion of the methylene groups of the hydrocarbon chains (Lefevre and Picquart, 1996). Proportions of α-tocopherol greater than about 20 mole% resulted in almost complete loss of transition enthalpy indicating that at a 1:5 molar ratio the phospholipid transition was completely removed by the presence of tocopherol. Both thermal and structural evidence have shown that α-tocopherol forms a stoichiometric complex of 1:10 with phosphatidylcholine in both fluid and gel phases (Quinn, 1995).

In contrast to the results obtained with phosphatidylcholines the evidence from studies of binary mixtures of α-tocopherol with phosphatidylethanolamines indicates that the two lipids do not mix under conditions in which the phospholipid is in a fluid lamellar phase (Wang and Quinn, 1999). Translating this result into cell membranes where there are mixtures of bilayer and non-bilayer forming lipids asymmetrically distributed across the membrane, α-tocopherol would not be expected to distribute randomly in the lipid matrix. The approach to investigate this question was to determine how α-tocopherol distributes in mixtures of phospholipids. Calorimetric studies of the effect of α-tocopherol on equimolar mixtures of DPPC/DSPC and DMPC/DSPC have been reported by Ortiz *et al.* (1987). The DPPC/DSPC mixture showed only a single endotherm, which was modified by the presence of α-tocopherol in a similar manner to the individual phosholipids. The DMPC/DSPC mixture, however, showed monotectic behaviour so that increasing proportions of α-tocopherol caused the highest temperature endotherm to broaden and shift towards lower temperatures, while the lowest temperature endotherm was weakened and was absent in the mixture containing 20 mole% α-tocopherol. This was interpreted such that α-tocopherol preferentially affects the lower melting component in the mixture. Increasing proportions of α-tocopherol in an equimolar mixture of 18:0/18:1- and 18:0/22:6-PC also results in a two component transition enthalpy (Stillwell *et al.*, 1996). The lower temperature transition is progressively broadened as the α-tocopherol content of the mixture increases but the higher temperature component is virtually unaffected by the presence of up to 10 mole% α-tocopherol. This was interpreted as a phase separation of α-tocopherol within the mixed phospholipid dispersion with a preference for an association with the more unsaturated molecular species of phospholipid. This type of evidence has often been used to assert that α-tocopherol associates with membrane lipids that are most vulnerable to oxidation. Clearly the result indicates a difference between cholesterol and α-tocopherol, the latter showing a preference for interaction with unsaturated molecular species of phospholipid.

More recent studies have used the inverted hexagonal phase induced by α-tocopherol in PEs as a marker for (PE+α-tocopherol) domain in PC/PE mixtures (Wang and Quinn, 2000a; Wang and Quinn, 2000b). Studies of mixtures of DOPC/DOPE showed that the effect of α-tocopherol on unsaturated phospholipids follows a similar pattern to that observed for their saturated counterparts. However, in mixed aqueous dispersions of α-tocopherol with DOPE/DOPC (1:1) and DOPE/DMPC (1:1), it was found that lamellar gel and liquid-crystalline phases dominated the phase structure and hexagonal phase was largely absent. The detection of nonlamellar phase induced by the presence of α-tocopherol would indicate partition onto PE so the predominance of lamellar structure suggests that α-tocopherol preferentially interacts with PC and not with PE. A similar picture has emerged in studies of α-tocopherol on equimolar mixtures of saturated PE and PC where effects of acyl chain unsaturation are avoided. Direct evidence of acyl chain perturbation in such mixtures showed that α-tocopherol preferentially partitions into PC in PC/PE mixtures irrespective of whether it is in fluid or gel phase. These model studies support the idea that α-tocopherol forms complexes with phosphatidylcholine and tends to be excluded from bilayers of phosphatidylethanolamine. In biological membranes this means that α-tocopherol would locate within the exoplasmic leaflet of membranes where this lipid is largely located and tend to be excluded from the cytoplasmic leaflet where phosphatidylethanolamine predominates.

4.8. Cholesterol-independent microdomains: glycosphingolipids

If the role of cholesterol is simply to promote tighter packing of lipids to enhance van der Waals-bonding of their fatty acid chains, this could be achieved in other ways. In fact, detergent-resistant domains in brush border membranes can form or persist without cholesterol (Hansen *et al.*, 2001; Milhiet *et al.*, 2002). Sphingolipids differ from glycerolipids in their hydrogen bonding propensity (Mombelli *et al.*, 2003), a force in itself not sufficient to condense sphingolipids nor normally to cause them to phase separate from glycerolipids in a bilayer. However, for differences in associative properties to drive phase separation between components, the real driving force is not the associative properties of either component in isolation, but the difference in these properties. In a study in which sphingomyelin separated into microdomains in the presence of PC in a cholesterol-independent fashion, the PC was unsaturated at both the *sn*-1 and *sn*-2 positions (Saslowsky *et al.*, 2002). In those mammalian tissues in which both acyl chains are often unsaturated (Leventis and Silvius, 2001) the associative properties of sphingolipids could well be sufficient to drive phase separation from the highly unsaturated glycerolipids.

One factor that strongly influences hydrogen bonding between sphingolipids is glycosylation. On the cell surface, sphingolipids can be glycosylated. Although normally only a few % of sphingolipids are glycosylated, in the brain (Derry and Wolfe, 1967; Calderon *et al.*, 1995) and on cells such as melanocytes (Ostermeyer *et al.*, 1999) the proportion can reach 30–50% of surface lipids.

The hydrogen bonding capacity of sphingolipids is transformed by sugar residues that extend into the hydrated space above the membrane, reflected in the extraordinarily high melting temperature for transition from the gel to L_d phase for non-charged glycosphingolipids. The simplest glycosylated sphingolipids, galacto- or glucocereboside (Figure 3g), have T_m values of 84–87°C compared to 37–45°C for their parent ceramides (Koynova and Caffrey, 1995) indicating extensive inter-lipid hydrogen bonding conferred by the single sugar residue. These glycosphingolipids form microdomains that are minimally stabilised by sterols (Ahmed *et al.*, 1997; Xu *et al.*, 2001), presumably because the extra-membrane hydrogen bonding network established by the sugar residues, and the absence of a phosphocholine headgroup lying on the membrane, allows these lipids to condense without the assistance of a sterol. The strength of these effects is demonstrated by the remarkable stability (up to 70°C) of the L_o phase formed by monounsaturated C16:0/C18:1 phosphatidylcholine with glucocerebroside (Feng *et al.*, 2004).

Addition of another sugar residue to give lactocerebroside decreases T_M by 10°C (Koynova and Caffrey, 1995), suggesting weaker inter-lipid hydrogen bonding networks are established by the disaccharide. Lactocerebroside (but not galacto- or glucocerebroside) was found to dramatically increase the insolubility of a GPI-anchored protein reconstituted in vesicles formed from saturated (di-C14:0) phosphatidylcholine (Parkin *et al.*, 2001); even trace levels of lactocerebroside produced a 10-fold decrease in solubility of the protein, compared to relatively mild (<2-fold) effects of cholesterol. The sensitivity and specificity of this effect suggests a direct interaction of the lactose chain with the GPI-anchored protein rather than an effect upon the membrane as a whole. It may be that the disaccharide, in forming weaker inter-lipid bonds, is able to interact more strongly with other glycosylated chains such as within the GPI anchor. In any event, it is clear that small differences in structure of the sugar chains can produce markedly different functional effects.

Gangliosides, with more complex sugar chains including negatively charged sialic acid, are reported to have a tendency to self-aggregate that allows them to separate out as distinct domains even within an L_o domain formed with other sphingolipids and cholesterol (Yuan and Johnston, 2000;

Vyas *et al.*, 2001; Yuan and Johnston, 2001; Yuan *et al.*, 2002). This is possibly due to the gangliosides introducing local disorder in both the fatty acid chains and headgroups of surrounding lipids (Roy and Mukhopadhyay, 2002). Gangliosides not only separate out from other lipids, different species of ganglioside separate out from each other (Peters *et al.*, 1984b; Vyas *et al.*, 2001), a process that occurs over a wide range of lipid and sterol compositions (Peters *et al.*, 1984a).

However, other investigators using comparable methods have not found ganglioside-specific effects (Wang and Silvius (2003) and references therein). The variables involved (properties of the probes used; composition of the lipid bilayer; for charged molecules like gangliosides, the salt composition of the aqueous phase) are considerable so that results and interpretations are revised as methods are continuously refined (cf. studies on ceramide; Wang and Silvius, 2003). It is clear, however, that the current preoccupation with the lipid tails of sphingolipids addresses only one aspect of their microdomain-forming properties; their carbohydrate headgroups must also contribute to organisation of the surface membrane, interacting not only with other glycolipids but also with proteins and their carbohydrate chains.

4.9. Plasmalogens: a wild card

Linkage of the *sn-1* fatty acid to glycerol by an alkenyl bond (Figure 31) produces a plasmalogen, primarily plasmenyl phosphatidylethanolamine with some plasmenyl phosphatidylcholine formed by head-group exchange (Gurr *et al.*, 2002). Expectation that the main function of the vinylalkyl rather than acyl linkage was to protect membranes in highly degradative environments (notably the peroxisome, on the luminal face of which the alkyl linkage is formed; Thai *et al.*, 1997) has been challenged by analysis of cells from patients defective in the critical enzyme in plasmalogen synthesis. In these, the size and number of caveolae are reduced, coated pits are absent, and biosynthetic as well as endocytic trafficking perturbed (Thai *et al.*, 2001; Figure 1). Plasmalogens appear to have major structural effects in cell membranes, possibly in promoting concave membrane curvature to allow vesicle formation. Plasmenyl phosphatidylethanolamine, containing arachidonic (C20:4) acid, has been found to be highly enriched in DfRMs, whereas none of this highly unsaturated fatty acid has been found acyl-linked to phosphatidylethanolamine in the same preparation (Pike *et al.*, 2002). How the vinylether linkage affects membrane properties is largely unexplored, but it clearly has very different effects to the standard acyl linkage.

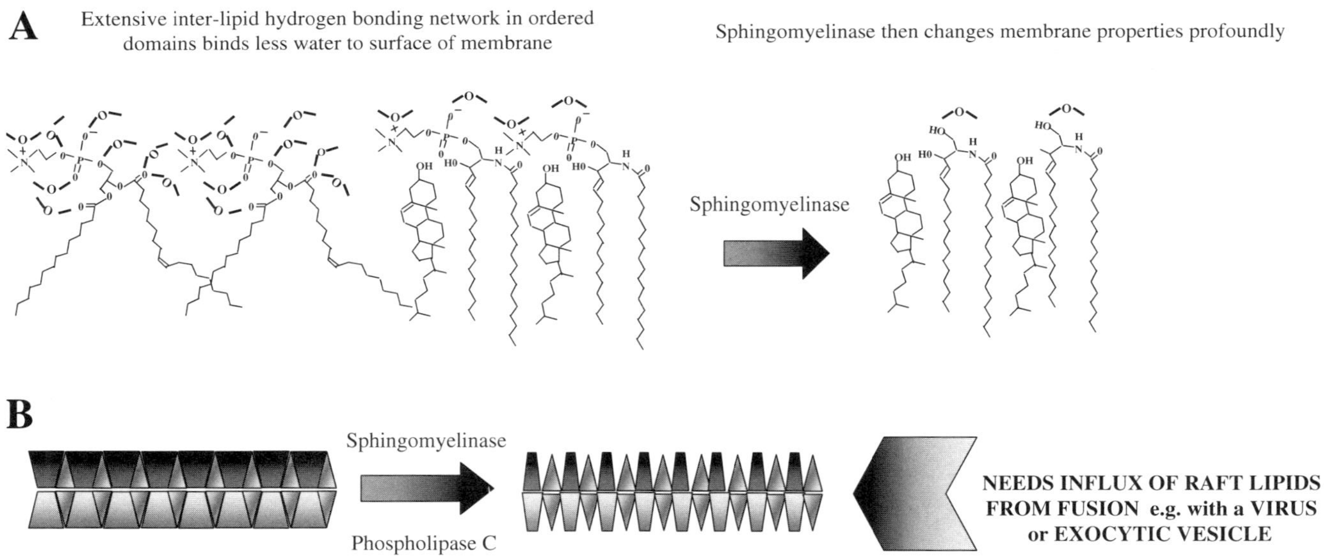

Figure 5. Schematic illustration of some membrane properties of disordered and ordered lipids and their response to signalling. (A) the lack of direct hydrogen-bonding between glycerolipids within L_d domains (Pasenkiewicz-Gierula *et al.*, 1997) allows a higher degree of interaction with water, and deeper penetration of water into the bilayer, than in cholesterol-induced L_o domains (Subczynski *et al.*, 1994). With L_o domains formed with sphingomyelin, signalling-induced removal of the zwitterionic phosphocholine groups further reduces the ability of the membrane surface to hydrogen bond with water. (B) Sphingomyelinase, working on the outer leaflet to produce ceramide, or phospholipase C similarly producing DAG on the inner leaflet, changes the geometry of the membrane lipids by greatly reducing the size of the headgroups to a simple −OH. This alteration, restricted to specific patches of membrane activated by signalling, changes both the permeability and intrinsic curvature of the membrane unless countered by a concomitant influx of large-headgroup containing membrane lipids, for instance by fusion with a virus or vesicle.

4.10. Hydrophobicity at the surface of ordered domains: vesicle fusion and pathogen entry/exit

Water is excluded from ordered membrane phases, including having a diminished hydration shell around the tightly-packed lipid headgroups (Prats *et al.*, 1987; Figure 5A), allowing membranes to come together more readily for fusion (Tatulian and Tamm, 2000). In the fusion of glucose transporter (Glut4) vesicles with the plasma membrane of adipocytes, both target and vesicle SNARE proteins, on the separate fusing membranes, are within DRMs (Chamberlain and Gould, 2002). For infection of cells with a range of viruses, fusion with the cell membrane requires both sphingolipids and cholesterol (Nieva *et al.*, 1994; Keller and Simons, 1998; Nguyen and Hildreth, 2000). However, it seems unlikely that the relative hydrophobicity of ordered sphingolipid domains by themselves is able to explain the preference of pathogens as diverse as viruses and malarial parasites for these microdomains (Lauer *et al.*, 2000; Samuel *et al.*, 2001; van der Goot and Harder, 2001). Viral binding to the cell surface activates signalling pathways within rafts and caveolae (Manes *et al.*, 2000; Pelkmans and Helenius, 2002). Activation of surface sphingomyelinase to produce ceramide (Figure 5) would initiate two synergistic effects promoting membrane fusion: a marked increase in membrane hydrophobicity and curvature induced by the diminutive –OH headgroup of ceramide (Cremesti *et al.*, 2002; Samsonov *et al.*, 2002). Diacylglycerol (DAG) is the comparable lipid on the inner surface; it also is produced in response to signalling and promotes fusion when present in membranes at physiological levels (Siegel *et al.*, 1989). Since cellular genesis is vitally dependent upon membrane fusion/fission, the receptor-controlled production of ceramide and DAG may have evolved first to facilitate membrane trafficking, with the use of these compounds as second messengers for signalling being a later development.

5. INDUCIBLE ALTERATIONS OF MEMBRANE DOMAIN STRUCTURE

In addition to domains formed by complexes between sphingomyelin and cholesterol there is evidence that complexes form between different lipid constituents of membranes and such complexes are often associated with signal transduction. Products of membrane lipid hydrolysis triggered by external factors have long been recognised as a mechanism of transmembrane signal transduction. Two hydrolytic products are produced one of which is more soluble in the cytoplasm than the other. The latter is more likely to remain associated with the membrane since it is less polar than the parent membrane lipid from which it was derived. While the release of free fatty acids to provide substrates for eicosenoid biosynthesis falls into this category this pathway is considered as an indirect signalling process. Dispersal of the hydrolytic products in the membrane in amounts which are formed under physiological conditions would be expected to have negligible impact on the overall structure and stability of the lipid bilayer matrix. There is mounting evidence, however, that lipophilic products form complexes with membrane lipids and partition into domains of specific composition. Such complexes are likely to function as transient lipid domains that are responsible for mediating the signalling processes.

The two major lipophilic products are diacylglycerols (DAGs) and ceramides. Both are very minor components of membrane, being formed and removed rapidly at specific locations in response to signalling. These lipids share a structural and polar similarity but they are produced at different locations by different enzymes responding to the different modulation processes. Their effect on the phase behaviour of the membrane bilayer matrix is also different largely because ceramide can hydrogen bond with neighbouring groups.

There is still some conjecture about whether these metabolites exert their action by directly binding to target effector molecules, rather than indirectly through pertubing membrane structure. Analogues of DAG and ceramide that differ markedly in their amphipathic balance can potentiate biochemical functions, supporting a model of direct binding of the lipid metabolite to target. However, analogues that have a similar amphipathic balance, but quite different structure, to the natural metabolites can also potentiate funtion, suggesting a mode of action dependent on modulation of the phase behaviour of the membrane lipid matrix (Venkataraman and Futerman, 2000; Brose and Rosenmund, 2002). Indeed, there is evidence that subtle differences in amphipathic balance arising from the location of the substrate from which the product is formed may fulfill different signalling functions (Hunt *et al.*, 2002). What then is the evidence that diacylglycerol and ceramide can form complexes that phase separate into distinct domains?

5.1. Diacylglycerol

DAG accumulates transiently in cell membranes as a consequence of phospholipase C-type enzymes activated by a variety of hormones, growth factors and neurotransmitters. The amount of DAG formed can reach 2 mole% in some physiological situations (Preiss *et al.*, 1986). The resulting change in membrane lipid composition represents a molecular signal transducing mechanism responding to the interaction of agonists with their respective cell surface receptors (Kikkawa and Nishizuka, 1986; Berridge, 1987). DAGs are known to induce structural changes in membranes (Das and Rand, 1986).

These physical perturbations of the lipid bilayer may contribute to the ability of DAG to activate protein kinase C (Nishizuka, 1992) as well as to enhance membrane fusion processes (Siegel *et al.*, 1989).

DAGs when codispersed with disaturated phosphatidylcholines in molar ratios of up to 30 mole% create lateral phase separated domains of a stoichiometric complex of DAG and phosphatidylcholine within bilayers of pure phospholipid (Heimberg *et al.*, 1992; Lopez-Garcia *et al.*, 1994). It has been suggested that the domains created by the demixing process destabilizes the phospholipid bilayer (Dibble *et al.*, 1996). It may be expected that physiological proportions of DAG (up to 2 mole%) may result in lateral phase separations in small domains within membranes. Much higher and unphysiological proportions of DAGs (greater than 30 mole%) induce nonbilayer phases such as hexagonal-II (H_{II}) and cubic phases in disaturated phosphatidylcholines (Das and Rand, 1984; Oradd *et al.*, 1995; Takahashi *et al.*, 1996).

The transient nature of lamellar phase complexes between DAG and membrane phospholipids is necessary in the process of switching off the signal generated by the formation of DAG. This is accomplished by DAG kinase which converts the substrate into phosphatidic acid.[14] Control is presumably exercised by the molecular species of DAG that are generated from the parent phospholipid. Studies on the effect of acyl chain length of diacyl glycerol on activation of protein kinase C has shown that DAGs with shorter hydrocarbon chains (C6–C10) are more effective than molecular species containing longer acyl chains (Mori *et al.*, 1982; Sekiguchi *et al.*, 1988). It has been argued that the miscibility of the short-chain DAGs with the phospholipid in liquid-crystalline bilayers is the explanation for the high activity of the short-chain DAGs. By contrast, longer chain DAGs form complexes with the phospholipid which exist in gel phase at temperatures higher than the fluid phase transition of the phospholipid and are consequently unable to activate protein kinase C. Complex formation has been characterised in binary mixtures of dipalmitoylphosphatidylcholine/dipalmitoylglycerol (Lopez-Garcia *et al.*, 1994) and dimyristoylphosphatidylcholine/dimyristoylglycerol (Heimberg *et al.*, 1992). Time-resolved synchrotron X-ray scattering experiments have been used to show that phase separation of a DPPC/DPG complex from domains of pure DPPC occurs in both the gel and the liquid-crystalline phases of the phospholipid (Quinn, 1995). When the chain length of the DAG is shortened to C-12 there is evidence of interdigitation of the molecules in the bilayer comprised of equimolar DPPC/DLG in a complex (Semmler *et al.*, 2000).

[14] That is, adds phosphate to the glycerol –OH of DAG, producing negatively charged phospholipids that would mutually repel each other.

In terms of the effect of DAG on domain formation in membranes it is clear that this is highly dependent on the molecular species of DAG and the degree of unsaturation of the polar lipids comprising the membrane lipid bilayer matrix. Interaction and complex formation between saturated molecular species may generate liquid-ordered domains whereas interaction between unsaturated molecular species tends to destabilize the bilayer in favour of the formation of inverted phases which result in repackaging of intrinsic proteins or membrane fusion (Siegel *et al.*, 1989).

DAG generation from proteolipid complexes containing polyphosphoinositol lipids by intranuclear phospholipase C is a recognised signalling pathway in the control of gene expression (Martelli *et al.*, 2002; Martelli *et al.*, 2003). These complexes have been localized to interchromatin granule clusters by immunostaining with monoclonal antibodies in preparations that have been extracted with cold Triton X-100. On this basis it is said that the complexes do not originate from the membrane bilayer lipid matrix of the nuclear envelope; however, the possibility that that they represent DRM domains of the envelope cannot be discounted on the present evidence.

5.2. Ceramide

The consequences of the presence of ceramide in the membrane are dominated by its tendency to self assemble into rigid bilayer structures (Holopainen *et al.*, 1998) especially in the presence of sphingomyelin and cholesterol (Dobrowsky, 2000; Kolesnick *et al.*, 2000; Massey, 2001). This property results in their recovery in detergent-resistant membrane fractions even from cells in an unstimulated state (Grassme *et al.*, 2001; Veldman *et al.*, 2001). Ceramide domains are more hydrophobic than the parent sphingolipid domains, promoting membrane fusion (Cremesti *et al.*, 2002). In trace levels (3–6 mole%), ceramide synergises with cholesterol in stabilising ordered lipid domains (Xu *et al.*, 2001). Ceramide alone is able to promote order in the lipid chains of phosphatidylcholine, even with monounsaturation in the *sn-2* chain (Hsueh *et al.*, 2002); it partitions into L_0 domains with an affinity markedly higher than that of other (fatty acid-matched) sphingolipids (Wang and Silvius, 2003).

The generation of ceramide in membranes takes place at different subcellular sites. *De novo* biosynthesis occurs at the endoplasmic reticulum, whereas at the plasma membrane ceramide is derived by hydrolysis of sphingomyelin and complex sphingoglycolipids. The hydrolytic enzymes responsible for production of ceramide at the plasma membrane are characterised by their pH optima into acid and neutral sphingomyelinases. Cellular modulators of ceramide production include DAG and other protein kinase C activators and serine proteases (Jaffrezou *et al.*, 2002). Degradation of ceramide by ceramidases is enhanced by a variety of agents including cytokines, cell differentiating agents, death receptor ligands, cancer chemotherapeutic agents and ionizing radiation (Santana *et al.*, 1996; Levade and Jaffrezou, 1999).

It has been proposed that the cellular response to ceramide elevation or depletion depends directly on where the ceramide is located (Van Blitterswijk *et al.*, 2003). According to this view ceramide in the Golgi influences biosynthetic pathways in formation of sphingolipids and their transport in vesicles to the plasma membrane where they promote domain formation. Clustering of receptors on the cell surface, promoted by rafts, induces endocytosis in response to transient ceramide formation. Finally, changes in membrane permeability associated with the action of ceramidases on mitochondrial and plasma membrane are said to mediate apoptotic events.

Apoptosis is one of the main cellular processes regulated by ceramides. Increasing the levels of ceramide by inhibition of ceramidases, which convert ceramide to sphingosine and free fatty acid, results in cell death (Bielawska *et al.*, 1996). Likewise, inhibition of ceramide production by blocking the *de novo* synthesis pathway or inhibiting neutral sphingomyelinase activity slows down apoptosis in response to chemotherapeutic agents, tumor necrosis factor-β, angiotensin-II and B-cell activation (Mathias *et al.*, 1998; Hannun and Luberto, 2000; Kroesen *et al.*, 2001; el Bawab *et al.*, 2002). Finally, depletion of ceramide by overexpression of glucosylceramide synthase has been shown to attenuate apoptosis induced by chemotherapeutic agents and tumor necrosis factor-β (Lavie *et al.*, 1997; Liu *et al.*, 1999).

6. COUPLING OF OUTER AND INNER LEAFLETS

While ordered lipid microdomains can readily exist in both the external and internal leaflets of the membrane, there is no *a priori* reason why rafts in the two leaflets should overlap so they are coupled functionally and structurally. Yet it is generally assumed they are, largely because extracellular and intracellular components of rafts can be co-immunoaffinity purified (e.g. Garnett *et al.*, 1993), and because antibody-mediated patching of extracellular components co-patch intracellular partners (e.g. Harder *et al.*, 1998; Janes *et al.*, 1999; Holowka *et al.*, 2000; Pyeta *et al.*, 2001).

Co-isolation in DRMs of outer and inner leaflet components is not in itself strong evidence of leaflet coupling in rafts; outer and inner leaflet rafts may exist independently and come together only when their apposed (non-raft) leaflet is stripped off by detergent. This is becoming less likely as evidence mounts for associations between specific subsets of intracellular and extracellular raft proteins in different immunoaffinity-isolated DRMs (Drevot *et al.*, 2002; He and Meiri, 2002). In the case of the T cell receptor, components isolated in different

DRMs co-exist within the immunological synapse (T cell receptor with the activated cytoplasmic kinase ZAP-70; GPI-anchored Thy-1 with the transmembrane protein LAT; Drevot *et al.*, 2002). It is difficult to see how specific co-segregation of different, but neighbouring, inner and outer leaflet components could occur unless the two bilayers were physically coupled in rafts.

Microscopic evidence using techniques with nm resolution could provide more direct indication of the extent to which domains on the external and internal leaflets interact. Application of the "tear-off" technique (Sanan and Anderson, 1991), which allows simultaneous EM immunolabelling of proteins on both sides of the bilayer, combined with statistical analysis of co-clustering, is a powerful development pointing the way ahead in the analysis of morphologically featureless microdomains (Prior *et al.*, 2003; see also Wilson *et al.*, 2000; Wilson *et al.*, 2001).[15] Fluorescence microscopy is widely used to address this particular question; it will be effective if the enhanced transmission of FRET across a membrane (Estep and Thompson, 1979; Szollosi *et al.*, 1987; Herman, 1989) enables the assessment of transmembrane relationships in living cells, extending the analysis begun with lipidated proteins on the cytoplasmic surface (Zacharias *et al.*, 2002).

A functional indication of coupling of inner and outer leaflets of rafts comes from the study of the important group of viruses (that include HIV-1, influenza (Scheiffele *et al.*, 1999) and measle viruses (Vincent *et al.*, 2000)) that enter and leave cells via raft membranes. HIV-1 virus, when budding from the surface, acquires the external coat of rafts (cholesterol, sphingolipids, GPI-anchored proteins; Nguyen and Hildreth, 2000), organised in a detergent-resistant L_0 phase (Aloia *et al.*, 1988; 1993). The Gag protein of human HIV-1 is targeted to the cytoplasmic DRMs by multiple myristoylation; addition of a better raft signal, the myristoylation + palmitoylation signal from Fyn, improves Gag partitioning into DRMs and gives a 2–4-fold increase in viral particle production (Lindwasser and Resh, 2001). Similarly dual palmitoylation of HIV-1 envelope protein gp160 targets it to rafts; mutation of the Cys residues involved decreases gp160 incorporation into viral particles, and strongly (160-fold) lowers viral infectivity (Rousso *et al.*, 2000). The targeting of viruses to the inner leaflet of rafts results, upon budding, in the inclusion of outer leaflet raft components in their capsid, a result most readily explained if the inner and outer leaflets are coupled.

Although the coupling of rafts in inner and outer leaflets of biological membranes is more assumed than proven, in model membranes in which

[15] Certain GPI-anchored proteins (Tiveron *et al.*, 1994; Madore *et al.*, 1999) and cholesterol (Möbius *et al.*, 2002) are concentrated on filopodia, organelles that seem highly unlikely to be displayed with the tear-off technique which, for all its ease of use, may selectively discard regions of interest in a cell.

L_o and L_d domains co-exist in both leaflets, the L_o domains are found in register, coupled across the bilayer (Bagatolli and Gratton, 2000a; Bagatolli and Gratton, 2000b; Dietrich *et al.*, 2001a; Rinia *et al.*, 2001; Samsonov *et al.*, 2001; Saslowsky *et al.*, 2002). Some property intrinsic to ordered membrane lipids couples the leaflets.

This coupling could depend upon the properties of either the L_d or L_o phase (coupling of one would physically juxtapose the other). In cellular membranes the concentration of transmembrane proteins in the L_d phase may well be a factor coupling disordered regions of both leaflets.

Within the protein-poor lipid ordered phase, however, coupling may depend more upon lipid than protein interactions. In fact, the properties that promote lipid order also promote coupling of the membrane leaflets. The key factor is likely to be interdigitation across the bilayer, of both cholesterol and fatty acid chains. The modest hydrophilic headgroup (–OH) of cholesterol is not as rigorously anchored to the aqueous interface as large zwitterionic groups such as phosphocholine; instead, cholesterol vibrates continuously in the Z-axis (perpendicular to the membrane surface), penetrating into the other bilayer by 5–11 Å (i.e. into about 30% of the opposed bilayer; Gliss *et al.*, 1999; Endress *et al.*, 2002).[16] In addition, asymmetry in the length of fatty acids attached to lipids, common with sphingolipids (Brown, 1998), can lead to the longer chain crossing into the opposed bilayer (Schmidt *et al.*, 1978; Grant *et al.*, 1987; Morrow *et al.*, 1995). Intuitively, one would suspect that cholesterol, in straightening up and so lengthening the fatty acid chains, would increase the extent of interdigitation between the leaflets of the bilayer. However, cholesterol has been found to prevent interdigitation when the lipid chains are matched (McIntosh *et al.*, 1992b; Cunningham *et al.*, 1995) and only becomes effective when the fatty acid lengths are asymmetric (Maulik *et al.*, 1986; Nabet *et al.*, 1996). Cholesterol, present at 30 mole% and penetrating a full third of the apposed bilayer, is likely to be the dominant factor coupling the leaflets together (Figure 6A).

To the substantial, if transient, degree of penetration of the opposed bilayer by cholesterol and the longer chain fatty acids of the lipids, must be added an additional coupling factor: the penetrating elements, coming from ordered lipid domains, are patterned (Figure 6B). Interdigitation will be much more effective if the partner monolayer is also patterned with a complementary spacing that both accepts and provides *trans*-leaflet incursions. There is a marked difference between the spacing of ordered, condensed lipids (area per residue around 45 $Å^2$) and disordered, unsaturated lipids (area per residue around 65–75 $Å^2$).

[16] This motion is three times that exhibited by ergosterol (Endress *et al.*, 2002) which may be a major reason why, despite the superiority of the fungal sterol in promoting L_o phase formation (Xu *et al.*, 2001), cholesterol is the major membrane sterol in homeotherms.

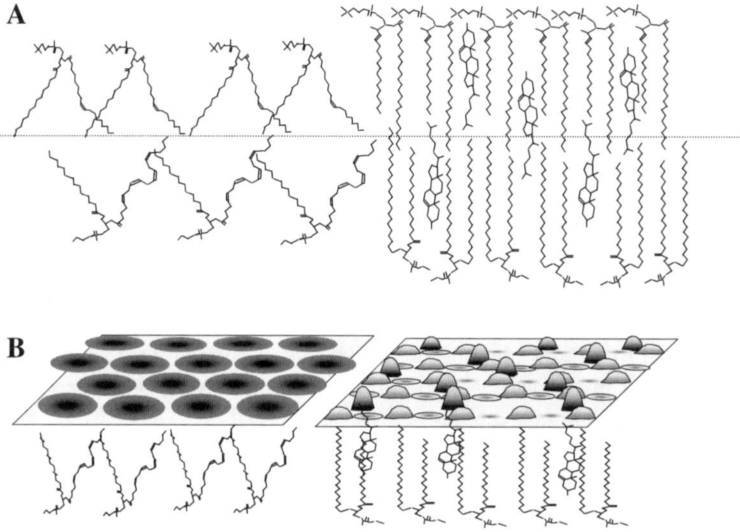

Figure 6. Coupling of ordered domains across the bilayer by interdigitation of fatty acid chains and cholesterol. The profiles of the highly mobile, disordered unsaturated lipids (A, left) project onto the interface between the bilayers (B, left) as flat discs. Interdigitation by the asymmetric fatty acid chains of the ordered phase combined with the movement of cholesterol in the Z axis (A, right) contributes a patterned projection of residues from one bilayer into the other (B, right) that would tend to impose a complementary order on the apposed bilayer.

For L_d phases of the membrane, the mobility of the unsaturated lipid chains is so high that the problem of assembling the two leaflets is that of laying two flat oily surfaces over each other. For the L_o phase, the problem is akin to packing egg boxes: some patterns will intermesh much more stably than others, enabling stably interdigitated microdomains to be established in the membrane.

One of the more remarkable aspects of this field, given the importance of the question and the range of indirect evidence that bears upon it, is how few studies have investigated whether, and how, rafts are coupled within living cell membranes.

7. SELECTIVE PARTITIONING OF LIPID-ANCHORED PROTEINS INTO ORDERED LIPID MICRODOMAINS

The property of floating at low density is increasingly becoming the key diagnostic feature of isolated raft membrane, particularly as DfRMs are characterised. As discussed above, the low density of raft membrane stems from its exclusion of the bulk of membrane proteins, a consequence of the ordered lipid structure of raft membrane. So what proteins do rafts admit, and why?

7.1. Lipid-anchored proteins that partition into the outer, sphingolipid-rich leaflet

The L_0 model of lipid organisation within membrane microdomains suggests a simple rule that should determine whether lipidated proteins enter or are excluded from these domains: proteins with saturated fatty acids attached (palmitoylated or myristoylated) should enter, those with bulky isoprene or prenyl chains should be excluded.

Given that the partitioning of proteins into DRMs was first observed for GPI-anchored proteins that insert into the outer, sphingolipid-containing leaflet of the membrane via a phosphoinositol glycerolipid (Brown and Rose, 1992), one might expect that the rules for entering ordered lipid microdomains were most thoroughly investigated for these proteins. It is widely stated that long chain saturated lipids dominate in mammalian GPI anchors. Benting *et al.* (1999) indeed found the GPI anchors of MDCK and FRT cells to be primarily C16:0/C18:0 and C18:0/C18:0. These authors moreover found that a Trypanosome GPI-anchored protein with a C14:0/C14:0 anchor was only poorly incorporated into a detergent-resistant phase in cholesterol-containing lipid vesicles, suggesting that C14 lipid chains are too short to fit within an ordered lipid domain in the external leaflet.

The lipid anchors of remarkably few individual mammalian GPI-anchored proteins have been characterised. In all four (Thy-1 (Tse *et al.*, 1985); erythrocyte acetylcholinesterase, AChE (Roberts *et al.*, 1988); decay accelerating factor, DAF (Luhrs and Slomiany, 1989); and renal membrane dipeptidase (Brewis *et al.*, 1995)) the lipid chains are long (C16–C24), although in other respects they are diverse. In two, the lipids are bound to glycerol via an *sn-1* alkyl, *sn-2* acyl linkage rather than the acyl linkage of both fatty acids found with conventional glycerolipids. Alkyl/acyl linkage was also dominant in the GPI anchors analysed by Benting *et al.* (1999).

In addition, the fatty acids of GPI anchors can be unsaturated (Figure 3h). In AChE the alkyl-linked *sn-1* lipid is a 2 : 1 mixture of C18:0 and C18:1, the acyl-linked *sn-2* lipid a mixture of C22:4, 5 and 6 chains (Roberts *et al.*, 1988), yet this protein is fully localised within DRMs (Salzer and Prohaska, 2001). In DAF, a C22:0 lipid is acyl-linked at the *sn*-2 position, but predominantly unsaturated lipids (65% C20:1 and C18:1, with 22% C16:0) are alkyl linked to the *sn-1* position. Furthermore, both AChE and DAF have a third lipid ester-linked to the inositol ring. In DAF this too is substantially unsaturated (29% C20:4 with 24% C18:1; the remainder are saturated C16:0–C22:0 fatty acids; Luhrs and Slomiany, 1989), although in AChE it is mainly C16:0 (Roberts *et al.*, 1988). Thy-1 and renal dipeptidase have simple diacyl-linked C18:0/C18:0 or C18:0/C16:0 anchors (Tse *et al.*, 1985; Brewis *et al.*, 1995).

The diversity of GPI anchors shown within this group of four mammalian proteins is typical of such anchors in lower organisms (McConville and Ferguson, 1993). Although there are undoubtedly exceptions (e.g. the C14:0-linked Trypanosome variant surface glycoprotein; Benting *et al.*, 1999) the majority of GPI-anchored proteins, in those species where it has been tested, can be isolated in DRMs (Rietveld *et al.*, 1999; Bagnat *et al.*, 2000; Nolan *et al.*, 2000; Denny *et al.*, 2001) and therefore probably occupy distinctive lipid ordered domains in the membrane. Since cholesterol condenses with mono-unsaturated, as well as fully saturated, lipids (above), it is understandable that C18:1 lipids could be interchangeable with C18:0 lipids in the anchor. It is, however, difficult to see how the multi-unsaturated C22 chain of AChE could fail to disrupt an ordered lipid domain. How the third, inositol-linked chain fits within ordered lipid domains is also unknown; single fatty acids (Chen *et al.*, 2001) do not condense with cholesterol.

The environment of GPI-anchored proteins, even on the same cells, can differ markedly in their detergent resistance (Madore *et al.*, 1999); indeed, prion protein occupies both raft-, and non-raft, environments on the neuronal surface (Sunyach *et al.*, 2003). The structural diversity observed in lipid anchors could contribute to the different microenvironments occupied by GPI-anchored proteins on the membrane.

7.2. Lipid-anchored proteins that partition into the inner, PE-rich leaflet

The factors that determine entry of lipidated cytoplasmic proteins into rafts has been examined in more detail, reflecting interest in the role of these proteins in signalling.

Many cytoplasmic proteins have stably attached to them either of two very different lipid anchors (Figure 3n): a simple C14:0 lipid (myristate) that is amide (N-) linked to their N-terminal Gly residue;[17] or a bulky, polyunsaturated

[17] Only 10 of the 14 myristate carbon atoms penetrate the hydrocarbon core of the bilayer, providing very weak (10^{-4}M K_D) membrane association (Murray *et al.*, 1998).

isoprenyl group thioether linked to a Cys near their C-terminal (Resh, 1999).[18] While addition of these lipids is generally for the lifetime of the protein, neither in itself is sufficient to permanently anchor a protein to membrane; both normally require an additional attachment (Hancock *et al.*, 1991; Shahinian and Silvius, 1995). The additional attachment can be a polybasic patch (e.g. six consecutive Lys residues in Ras (Hancock *et al.*, 1990) that can bind to acidic headgroups of phospholipids (Roy *et al.*, 2000)), or palmitoylation. On the cytoplasmic surface, palmitoylation is via a labile thioester (C-) linkage. This lability is critical, since it enables the state of palmitoylation to be modified in response to receptor-mediated signalling, inducibly changing the membrane-associative properties of the proteins concerned (Robbins *et al.*, 1995; Wedegaertner *et al.*, 1995; Loisel *et al.*, 1999; Resh, 1999; Chen and Manning, 2000).

Unlike the addition of the myristyl or prenoid groups, palmitoylation has a relaxed substrate specifity, so that in normal mammalian tissue, C14:0, C18:0, C18:1, C18:2 and C20:4 (arachidonate) are also found in addition to C16:0 (palmitate) (Resh, 1999; Liang *et al.*, 2002). Moreover, certain cytoplasmic and transmembrane proteins are palmitoylated (usually at two or more sites) without prior myristylation or prenylation (Resh, 1999). This degree of natural variation aids the determination of the structural requirements for incorporation into ordered lipid microdomains.

In cells, most raft proteins are palmitoylated; most prenylated proteins are excluded from rafts (Melkonian *et al.*, 1999). The trimeric G proteins provide a useful test for raft association, since their α subunit is myristylated, palmitoylated or both (Resh, 1999) whereas their γ subunit is prenylated (Wedegaertner *et al.*, 1995). In the resting state, the single $\alpha\beta\gamma$ complex is excluded from DRMs; upon activation, the α subunit dissociates from $\beta\gamma$ and is largely included in DRMs, whereas the $\beta\gamma$ subunits remain excluded (Moffett *et al.*, 2000). Thus to the "rule" that "myristate+palmitate causes partitioning of a protein into rafts, whereas prenylation excludes from rafts", can be added the rider: "where all three modifications occur to a single protein complex, exclusion by prenylation is dominant". Fluorescence studies in living cells using nm-scale FRET analysis confirms that myristate+palmitate or dipalmitate directs a reporter protein to rafts, whereas prenylation excludes it (Zacharias *et al.*, 2002). Similarly, a wide body of evidence shows that inhibition of myristylation and/or palmitoylation of

[18] The N-terminal motif for myristylation is Met-Gly; the Met is removed co-translationally while the polypeptide is still attached to the ribosome and myristoyl transferred from myristoyl-CoA to form an amide linkage with Gly-2 (Resh, 1999). The C-terminal motif encoding prenylation is C*AAX*, where C is Cys, A aliphatic, and X any amino acid; X determines which prenoid is added – when Ser or Met (as in the Ras proteins) the Cys is farnesylated, when Leu (in Rab family proteins) it is geranylgeranylated (Pereira-Leal *et al.*, 2001). As with N-terminal myristylation, *AAX* is removed proteolytically so that the prenylated Cys becomes the C-terminal peptide (Hancock *et al.*, 1989).

Src-family kinases, using either inhibitors (Stulnig *et al.*, 1998; Webb *et al.*, 2000; Liang *et al.*, 2001) or mutation (e.g. Robbins *et al.*, 1995; Kabouridis *et al.*, 1997; Zlatkine *et al.*, 1997; Sheets *et al.*, 1999; van't Hof and Resh, 1999), prevents inclusion of the (mis)anchored protein in DRMs, causes mislocalisation of the kinases at the surface membrane, and inhibits cellular signalling processes in which they are involved. It should also be noted that substitution of palmitate (C18:0) with singly unsaturated palmitoleate (C18:1) allowed DRM entry and signal transduction (Liang *et al.*, 2001).

A more detailed analysis of the structural requirements for inclusion of synthetic lipidated peptides within the L_0 phase of lipid bilayers has been performed using fluorescence quenching (Wang *et al.*, 2000, 2001). These studies confirmed the expected efficacy of long chain saturated lipids (with a single monounsaturated species tolerated) and not of prenoids for directing a peptide into the detergent-resistant L_0 phase. Further, they examined the efficacy of different chemical linkage of the fatty acid upon affinity for the L_0 phase. The GlyCys motif, leading to N-myristylation C-palmitoylation of certain proteins (Resh, 1999), partitions avidly into the L_0 domain, and this affinity is doubled if the C14:0 myristyl group is replaced by C16:0 to give two adjacent palmitates as in Gαs (Wang *et al.*, 2001). However, two adjacent palmitates have 10-fold less (although still significant) affinity for the L_0 phase if both are C-linked (e.g. in the cytoskeletal adaptor GAP43; Resh, 1999); and there is little benefit for additional C-linked palmitoylation of proteins (e.g. Fyn, Lck) if they already contain a dually acylated GlyCys motif (Wang *et al.*, 2001).[19]

These studies demonstrate two important points: first, that the inner leaflet of the plasma membrane includes/excludes lipidated proteins at least as stringently as does the outer, sphingolipid-containing leaflet; second, the rules for inclusion within specific functional domains in living cells match those for inclusion within DRMs and within L_0 domains in model membranes. There can be no doubt that lipid ordered microdomains exist on the inner membrane every bit as much as they do on the outer.

7.3. An exception that improves the rule: Ras

All Ras proteins are farnesylated (farnesyl, the highly unsaturated C15 prenoid; Figure 3n) at their C-terminal *CAAX* box and have palmitate C-linked

[19] The spacing between the fatty acid chains is a neglected, but almost certainly important, variable affecting inclusion of lipidated proteins in ordered lipid domains. Dilipidation mostly, but not invariably, occurs on consecutive residues (Resh, 1999). A Leishmania protein with two intervening Ser residues between myristate N-linked to Gly-2 and palmitate C-linked to Cys-5 (Denny *et al.*, 2000) is not incorporated into DRMs when expressed in mammalian cells (Denny *et al.*, 2001).

to Cys at one (N-Ras and K-Ras4A, at −5/−6 from the prenylated Cys) or two (H-Ras at positions −2 and −5) nearby residues (Prior and Hancock, 2001). Peptide linkage to a bilayer via C-linked farnesyl plus a single palmitate group produces minimal partitioning into an L_o phase (Wang *et al.*, 2001), whereas the dually palmitoylated H-Ras partitions into lipid rafts in the resting state, but leaves them upon activation (Prior *et al.*, 2001). The rafts (~44 nm average diameter) into which non-activated H-Ras partitions were identified by co-localisation (across the plane of the membrane) with a "mildly cross-linked" GPI-anchored protein; and by the cholesterol-dependency of their clustering into microdomains (Roy *et al.*, 1999; Prior *et al.*, 2003). The differential partitioning of H-Ras, and the non-raft localisation of K-Ras, seen by EM immunolabelling, is matched by the flotation of DfRMs to low (inactive H-Ras) and high (activated H-Ras, K-Ras) density fractions in the absence of detergents. The rafts occupied by inactivated H-Ras, however, are only marginally detergent-resistant: they are ~50% solubilised in 0.125% Triton X-100, and completely soluble in 1% detergent (Prior *et al.*, 2001).

The rafts occupied by inactive H-Ras are unconventional in that they have too little lipid order to effectively exclude Triton X-100. Clearly, the farnesyl group prevents inclusion of this protein in highly ordered rafts, in keeping with other studies (Melkonian *et al.*, 1999; Moffett *et al.*, 2000; Wang *et al.*, 2001). However, the rafts are cholesterol dependent, float at low density, and above all are critically involved in directing signalling along the different pathways provided by the different spatial partitioning of the Ras family of proteins on the cyoplasmic surface of the plasma membrane (Prior and Hancock, 2001). Therefore, cholesterol is able to impose a degree of lipid order upon a membrane microdomain sufficient to control entry of particular functional states of a protein, and exclude most other proteins, even when present at a level that confers a low level of detergent resistance to the membrane.

8. TRANSMEMBRANE PROTEINS AND ORDERED LIPID MICRODOMAINS

8.1. Why most transmembrane proteins are excluded

Fitting a transmembrane (TM) domain into disordered membrane is relatively easy – the high mobility and diversity of the acyl chains can accommodate the non-lipid intruder, making room and filling in gaps as required. However, highly ordered gel and L_o phases very poorly accommodate most TM helices (Figure 7). The exclusion of most transmembrane proteins from the ordered gel phase has been identified as a major problem in freezing membranes (Quinn, 1985; Figure 2C), and would be expected to also apply to the L_o phase (London, 2002).

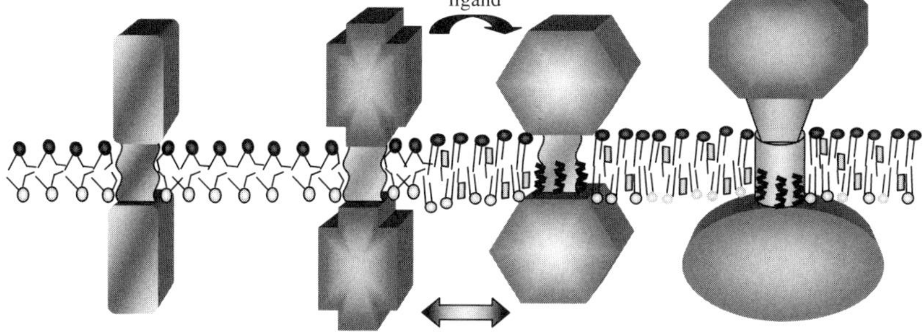

Figure 7. Fitting transmembrane proteins into disordered and ordered membrane environments. The helical transmembrane domain of a typical transmembrane protein (left) is readily accommodated within the fluid mix of unsaturated lipids in disordered membrane. A raft-resident transmembrane protein (right; typified by 'flu' haemagglutinin) has multiple palmitate residues covalently attached to its juxtamembrane cytoplasmic residues, providing a shield of long chain saturated lipids in the inner leaflet of the membrane; the residues in the outer leaflet are also highly specified to ensure they fit within an ordered lipid environment. Between these two extremes stand a large range of proteins, often receptors, that migrate between disordered and ordered lipid environments, accompanied by changes in palmitoylation/depalmitoylation.

In addition to this packing problem, a second steric problem almost certainly arises because cholesterol, in ordering (straightening out) the hydrocarbon chains of lipids, tends to increase the thickness of the bilayer (Nezil and Bloom, 1992; Saslowsky *et al.*, 2002). Since the different thickness of membrane between the Golgi and plasma membrane is thought to retain Golgi resident proteins (with ~18 TM amino acids) while exporting surface proteins (~23–25 TM amino acids; Bretscher and Munro, 1993; Munro, 1995), a TM domain that is resident in rafts should be slightly longer again (Ren *et al.*, 1997; Ren *et al.*, 1999).

The third property of L_0 microdomains that will tend to exclude TM domains is electrical. The high packing density of a cholesterol-condensed phase makes the membrane more impermeable to ions and water (Subczynski *et al.*, 1994), changing the dielectric constant of the membrane. As cholesterol approaches 30 mole%, the electrical potential in the centre of the membrane relative to the surface rises by ~50 mV (Mombelli *et al.*, in preparation). This would inevitably alter the conformation of TM polypeptide domains, contributing to the general exclusion of TM proteins from rafts. However, should the conformational change elongate a protein's TM domain, it could act as part of the switch allowing entry/egress to ordered domains.

8.2. Why certain transmembrane proteins are selectively corralled in ordered domains

We deal in a later section with membrane proteins that we suspect play a structural role in curving ordered lipid membrane. A handful of TM proteins are known to constitutively occupy rafts (without having a major structural role) and can be substantially recovered in DRMs using 1% Triton X-100 or similar detergent extraction. Most if not all of such proteins overcome part of the problem of fitting within L_o domains by having multiple palmitates, covalently attached in the juxtamembrane region of their cytoplasmic domain, inserting into the inner leaflet and presumably shielding the TM helix within this leaflet. However there appears to be no enzymatic mechanism to palmitoylate residues that could insert into the outer leaflet of a raft and shield the TM domain there (with the exception of the Hedgehog protein; Porter *et al.*, 1996).

One solution to this problem is demonstrated by the influenza haemagglutinin (HA) protein. This has 3 palmitates shielding its TM domain in the cytoplasmic leaflet, but what of the outer, unshielded leaflet? Mutational analysis shows that TM residues in the outer (but not inner) leaflet strongly determine the ability of the protein to reside within DRMs (Scheiffele *et al.*, 1997). Moreover, HA's TM domain increases the degree of lipid order when inserted into model bilayers – it positively contributes to the ordering of its lipid environment (Tatulian and Tamm, 2000). Finally, the TM domain of this protein is 27, not 25, amino acids long (Scheiffele *et al.*, 1997). Although the relative importance for raft occupancy of the length, lipid-ordering properties and palmitoylation of the TM domain has not been assessed, all three could play a role.

The adaptor proteins LAT (*L*inker for *A*ctivation of *T* cells; Zhang *et al.*, 1998a) and Cbp (*C*sk *b*inding *p*rotein, Kawabuchi *et al.*, 2000; also known as PAG, Brdicka *et al.*, 2000) are the best characterised mammalian signalling proteins that are resident in ordered lipid microdomains. Structurally, they are very similar: both have minimal extracellular domains of 4 (LAT) and 16 (Cbp) residues, a short TM region of 21 amino acids[20] followed by two

[20] The assigned transmembrane domain for both proteins is not only curiously short; the basic motif (K/R, sometimes with marginally basic H) that normally terminates the transmembrane domain at the cytoplasmic face (van Klompenburg *et al.*, 1997) occurs about six residues more distally. The assigned transmembrane domain (in bold) with juxtamembrane residues including palmitoylated Cys (italics), for Cbp, LAT and HA, are:

Cbp	QMQM**VLWGSLAAVAMFFLITFLILL***C*SS*C*DRDKKPRQH
LAT	MEEA**ILVPCVLGLLLLPILAMLMAL***C*VH*C*HRLPGSYDS
HA	GVYQ**ILAIYATVAGSLSLAIMMAGISFWMCS**NGSLQ*C*RI*C*I

Should the transmembrane helix terminate at the basic RDKK in Cbp and HR in LAT to give a 26 residue transmembrane helix, the deep-lying palmitates would probably interdigitate with the outer leaflet, helping to couple the bilayer.

palmitoylated juxtamembrane Cys residues (Brdicka *et al.*, 2000; Kawabuchi *et al.*, 2000), a modification which has been shown to be required for incorporation of LAT into ordered lipid domains (Zhang *et al.*, 1998b).

The long intracellular domains of both LAT and Cbp contain multiple binding sites for effector proteins. On Cbp these sites mediate two functions: inactivation of Src family kinases,[21] and binding to the actin cytoskeleton (which in turn inhibits raft mobility and T cell activation; Brdicka *et al.*, 2000; Itoh *et al.*, 2002). LAT binds a battery of signalling molecules, some directly (PLCγ1, p85 subunit of PI3K, and Grb2), others indirectly by their binding the adaptor Grb2 (e.g. Cbl, SOS, Vav and SLP-76). LAT acts as an essential bridge between the immediate activation of the T cell receptor by antigen, and downstream effector mechanisms leading to cell proliferation (Zhang *et al.*, 1998a). The involvement of LAT in raft function will be discussed below. While it seems reasonable to assume that the palmitoylation of these proteins accounts for their inclusion within lipid rafts, it will become clear for both neurotrophin receptors and caveolin-1 that such assumptions can be misleading. It remains to be shown why these proteins are in lipid rafts.

8.3. Proteins that move between disordered and ordered domains

An increasingly large number of transmembrane receptors and adhesion molecules are reported to move between ordered and disordered membrane when they are activated by ligand. It is probable that in many cases the TM domain is not optimally adapted to inhabit an ordered lipid environment, since the raft occupied is in some cases fully or mostly soluble in 1% Triton X-100, but insoluble in more selective detergents such as Brij 98 or Lubrol WX, or in very low concentrations of Triton X-100 (e.g. FcεR1 (Field *et al.*, 1995; Wilson *et al.*, 2001) and the T cell receptor (Drevot *et al.*, 2002)). We consider here the interactions of two types of receptor with ordered lipid membrane, both of which illuminate the biological role of rafts and underline our ignorance of the mechanisms involved.

A number of TM proteins act as part of a transmembrane signalling complex that includes a GPI-anchored ligand-binding subunit. These include

[21] Cbp binds and activates Csk, the kinase that inhibits Src-family kinases including Lck, the kinase that initiates T cell receptor signalling (Brdicka *et al.*, 2000; Takeuchi *et al.*, 2000); binding of Csk is controlled by phosphorylation of Cbp by another Src family kinase, Fyn (Brdicka *et al.*, 1998; Yasuda *et al.*, 2002). Very early in T cell signalling, an unknown phosphatase dephosphorylates Cbp, releasing Csk to switch off activation of Lck and thereby limit the receptor-associated component of T cell signalling (Torgersen *et al.*, 2001).

(GPI-protein : TM signal transducer) the adhesion controlled contactin:Caspr complex (Peles *et al.*, 1997; Faivre-Sarrailh *et al.*, 2000; Charles *et al.*, 2002), neurotrophin receptors of the GDNF family such as GFRα1 : Ret (Cacalano *et al.*, 1998; Tansey *et al.*, 2000; Paratcha *et al.*, 2001) and receptors for the STATS cytokines (Sehgal *et al.*, 2002) such as CNTF:gp130 (Stahl and Yancopoulos, 1994). In each case, the transmembrane component only becomes DRM-associated when bound to the GPI-anchored subunit. However, it would be unwise to conclude that entry into ordered lipid membrane is effected because the acyl chains of the GPI-anchor shield the TM helix in the outer leaflet of the plasma membrane, similar to the shielding of TM helix in the inner leaflet by palmitates covalently bound to juxtamembrane cytoplasmic residues. Soluble (non-GPI-anchored) GFRα1 subunit, on binding to Ret, still causes the complex to translocate to DRMs (Paratcha *et al.*, 2001). Presumably binding of the (normally GPI-anchored) ligand-activated subunit causes a conformational shift in Ret, either within its TM helix so that it better fits an ordered lipid domain, and/or within a non-membrane domain to activate binding to a raft-associated protein (e.g. binding to Grb2 would link it to LAT and hence DRMs in lymphoreticular cells). Thus the GPI anchor of GFRα1 is not an adaptation to enable this ligand-binding subunit to conduct the TM component into ordered lipid membrane, but rather an adaptation to enable it to accompany the latter into a raft. This conclusion raises uncomfortable and largely unanswered questions about the role of palmitoylation of transmembrane proteins in determining raft occupancy.

A large number of receptors are reported to traffic through lipid domains as part of their signalling cycle, exemplified by members of the family of seven-transmembrane G-protein-coupled receptors. Not only are many of these receptors palmitoylated to varying extents (e.g. 5 palmitates on Endothelin A receptor, 1 on both the α_{2A} and β_2 adrenoceptors; Resh, 1999), so also are their immediate membrane-bound effectors (G proteins discussed above; RGS proteins (Regulator of G protein Signalling); Tu *et al.*, 1999) and desensitisers (G protein-coupled receptor kinases; Stoffel *et al.*, 1994; Premont *et al.*, 1996; Stoffel *et al.*, 1998). Palmitoylation is responsive to agonist activation (as already discussed for G proteins) so that it is easy to see how such dynamic palmitoylation-controlled exit/entry to ordered lipid domains could be a major factor controlling the signalling and desensitisation of these receptors (Figure 7) (Qanbar and Bouvier, 2003).

Differential use of the lipid environment in signalling is exemplified by β-receptors in cardiomyocytes (Figure 8), where stimulation by catecholamines (activating the predominant β_1-, but also the β_2-adrenoceptor) increases heart rate and cardiac output. Activation of adenylyl cyclase by the β_1 receptor subtype is attenuated by M_2 muscarinic receptor stimulation (by virtue of their co-localisation in non-caveolar disordered domains).

The β_2-adrenoceptor is not M_2 sensitive in neonatal cardiomyocytes, probably because being palmitoylated in its basal state it is confined within the ordered lipids of caveolae (Rybin *et al.*, 2000; Xiang *et al.*, 2002) where it is coupled to Gαs proteins (O'Dowd *et al.*, 1989; Degtyarev *et al.*, 1993). β_2 receptor signalling is dependent upon the integrity of these domains, while β_1 protein kinase A-mediated myocyte signalling and contraction is not (Xiang *et al.*, 2002). Following β_2 activation, the β_2-adenoceptor is depalmitoylated (Loisel *et al.*, 1999) and down-regulated by protein kinase A and a β-adrenoceptor kinase (Moffett *et al.*, 2001) that resides in disordered membrane by virtue of farnesyl-linkage to the membrane (Inglese *et al.*, 1992); β-arrestin then binds and recruits the receptor to clathrin coated pits (Goodman *et al.*, 1996) where it is endocytosed and then recycled to the cell surface (Moore *et al.*, 1995; Seachrist *et al.*, 2000). The β_2-receptor signalling complex is thus assembled within an ordered lipid environment which, following activation, it leaves to be down-regulated.

In contrast, in its basal state the M_2 muscarinic receptor occupies disordered membrane (Figure 8); upon ligand activation, the receptor is palmitoylated (Hayashi and Haga, 1997) and translocates to ordered lipid microdomains (again, caveolae in cardiomyocytes; Feron *et al.*, 1997) where it is coupled to G proteins (Hayashi and Haga, 1997) and one of several signal cascades can be activated. This can include the endothelial subtype of nitric oxide synthetase (eNOS; Feron *et al.*, 1998), held in an inactive state within caveolae but released and activated in response to M_2 signalling. NO produced by the released eNOS may at least in part mediate the negative chronotropic effect of the muscarinic receptor on heart rate, and may also trigger the depalmitoylation of the β_2-adrenoceptor (in stably transfected cells; Adam *et al.*, 1999), driving it from the protection of the caveolae to be desensitised (although this remains controversial; see Vandecasteele *et al.*, 1999; Crystal *et al.*, 2001). The M_2 receptor is itself then down-regulated by sequestration within the caveolae[22] (Dessy *et al.*, 2000).

[22] Down-regulation of the M_2 receptor is dependent on dynamin (Dessy *et al.*, 2000), part of the complex that restricts the neck of coated pits to enable fission of the coated vesicle from the membrane. However, despite earlier expectations that caveolae mediated an alternative, clathrin-independent constitutive endocytic pathway, caveolae appear to be remarkably stable (i.e. not internalised) on the membrane (Le *et al.*, 2002; Thomsen *et al.*, 2002). The strongest evidence that they are involved in endocytosis is that they are the port of entry of certain viruses (e.g. SV40) that are taken up by the cell and enter the endoplasmic reticulum (Pelkmans and Helenius, 2002). It is thought that dynamin-dependent receptor desensitisation of caveolae-resident receptors occurs by occlusion of the caveolae, possibly involving full internalisation. Caveolae (reviewed by Parton, (2003)) have possibly swallowed as many assumptions as extracellular components in the half-century since their discovery.

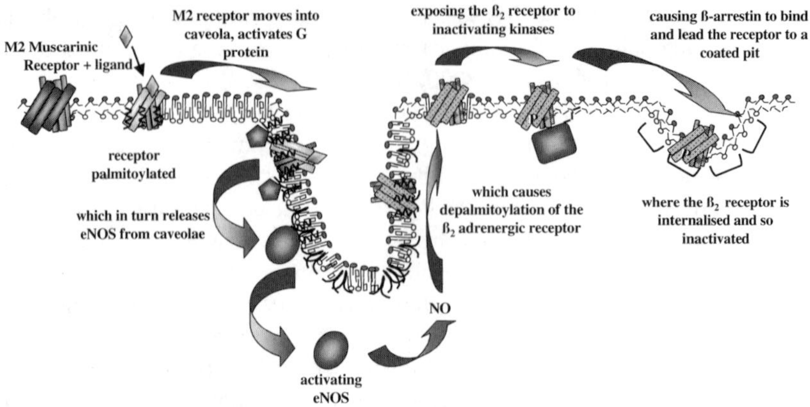

Figure 8. Activation-dependent movement of receptors into and out of ordered lipid domains, exemplified by the M_2 muscarinic receptor and β_2 adrenoceptors in caveolae of embryonic cardiac myocytes.

This example illustrates the potential of ordered lipid microdomains in balancing antagonistic signals, coupling the down-regulation of an accelerator (β_2-adrenoceptor) to application of the brake (M_2 muscarinic receptor activation). It should also be stressed how little we really know about the mechanisms involved. We have implied at a number of steps that palmitoylation/ depalmitoylation is the force driving proteins into/out of the ordered domains, but this has not been proven. Given the example of HA, it seems probable that palmitoylation is necessary, but not sufficient, for entry into an ordered lipid microdomain. The sequences and orientation of 7TM receptor domains are presumably adapted to interact optimally with ordered lipids.

A major problem is that many labs working on molecular aspects of receptor function fail to investigate whether the parameters they study influence the membrane domain occupied by the receptor, due in no small measure to scepticism about the physiological relevance of rafts and the methods used to investigate them. Equally, modifications such as palmitoylation can specifically and differentially affect functions other than (or in addition to) membrane localisation (see for example, Eason *et al.*, 1994; Hawtin *et al.*, 2001), misleading interpretation if too narrow a range of parameters is investigated. Until these problems are addressed by detailed studies of the control of protein entry into ordered lipid microdomains, this rapidly expanding field will remain phenomenological.

9. PLANARITY AND CURVATURE OF ORDERED LIPID MICRODOMAINS: DO INTRAMEMBRANE PROTEINS CURVE ORDERED LIPID BILAYER?

9.1. Planarity/curvature of lipid rafts

Rafts are often associated with tightly curved membranes. Indeed, it is commonly assumed that ordered lipid microdomains are either caveolae or "caveolae-like" domains. However, physical studies suggest that ordered lipid domains should be inherently resistant to forming tightly-curved membrane. We ask here how this physical inflexibility can be reconciled with the apparent biological elasticity of ordered lipid domains. In dealing with curvature, for simplicity we will look at the membrane from the external or luminal surface and refer to it as convex (e.g. a filopodium, or a secreted vesicle such as an exosome) or concave (caveolae or coated pits; intracellular trafficking vesicles) (Figure 9). Convex curvature of the outer leaflet of course requires concave curvature of the inner, and vice versa, a structural asymmetry that possibly contributes to the compositional asymmetry of biological membranes.

Lipids, even in a simple monolayer, have intrinsic curvature if there is a mismatch between the cross-sectional area occupied by their headgroup and their fatty acids; if the headgroup is too small they pack into the membrane as a cone, if too large as an inverted cone, and if just right, as a cylinder (Zimmerberg, 2000). Obviously, in a bilayer composed of identical numbers of lipids, curvature of each leaflet will oppose the other so that where the mismatch is too great (e.g. with unsaturated molecular species of PE) the lipid cannot form a bilayer. The diversity of lipids in biological membranes enables them to take all forms and accommodate proteins of many shapes.

These considerations apply to equilibrium situations and can take hours (or even weeks; Meyer *et al.*, 1998) to unfold with model membranes. Biology is concerned with massive changes in curvature (e.g. to form a filopodium or coated pit) that occur over the space of seconds/minutes. The essential parameters are the elasticity of the bilayer (how much it can bend without breaking); and how readily lipids can be pumped from one leaflet to the other to curve the membrane by simple numerical asymmetry (Farge and Devaux, 1992; Farge *et al.*, 1999; Rauch and Farge, 2000; Huttner and Zimmerberg, 2001).

Cholesterol makes sphingomyelin domains less elastic (Li *et al.*, 2001) and more difficult to deform (Huttner and Zimmerberg, 2001); moreover, it is difficult to see how lipids could be pumped from one leaflet to the other within ordered lipid domains without severing the coupling between the leaflets.

These considerations suggest that ordered lipid domains should naturally be relatively planar and resist deformation into tightly curved structures.

The archetypal example of how disordered, elastic membrane is tightly curved is the coated pit. These rapidly forming membrane invaginations are bent from the outside, by curved scaffolds of protein adherent to the cytoplasmic surface (Brodsky *et al.*, 2001; Figure 9A, B). Ordered lipid membrane is excluded from coated pits. Those GPI-anchored proteins, for instance, that are internalised via coated pits leave their raft environment to do so (Rijnboutt *et al.*, 1996; Vilhardt *et al.*, 1999; Sunyach *et al.*, 2003).[23] Conversely, proteins that remain in rafts as they are internalised, enter via alternative routes (Vilhardt *et al.*, 1999; Lamaze *et al.*, 2001). The morphological apparatus of the alternative mechanism(s) of internalisation (Chatterjee *et al.*, 2001; Lamaze *et al.*, 2001; Fivaz *et al.*, 2002; Sabharanjak *et al.*, 2002) is unknown.

Where rafts exist in tightly curved membranes, one or more proteins are inserted at high abundance into the membrane, acting, we suggest, as a wedge to allow intervening small linear stretches of ordered membrane to remain coupled in both leaflets. Prototypical candidates are caveolin for concave curvature, and prominin or stomatins for convex curvature (Figure 9). The common feature of such proteins is that they insert into the membrane, a property that we suggest is required in order to curve ordered lipid microdomains due to the inflexibility of ordered membrane, both in the direct physical sense (lack of deformability) and in the sense that coupling of rafts in both leaflets prevents simple transfer of components to expand/contract the opposed leaflets.

In proposing these proteins as structural units that curve the naturally planar bilayer of rafts, it must be clear from the outset that the strongest evidence for such a function (for caveolins) is not conclusive; for others it remains speculative. However, considering the possible structural role of these proteins provides a useful framework from which to view a rather complex field. Many (possibly all) of these proteins also perform other functions: they regulate the protein and lipid content of their domains; act as molecular addresses, directing their rafts to particular cellular compartments; and can directly bind certain proteins, thereby restraining them within rafts. Analysis of the role of proteins that exert such pleiotrophic effects within whole cells is fraught with difficulty; dissecting their structural effect upon ordered membrane in simpler systems may clarify a central aspect of their control of cellular function.

[23] This is not to imply that lipid raft components are not internalised via coated pits; indeed, cholesterol is required for clathrin-mediated endocytosis (Rodal *et al.*, 1999; Subtil *et al.*, 1999). DRMs have not been isolated from coated pits/early endosomes, so cholesterol and saturated lipids would appear not to form an L_o phase in these tightly curved organelles.

9.2. Caveolin, the canonical concave curver?

Caveolins[24] are necessary for formation of caveolae, uncoated membrane invaginations of around 50–10 nm diameter (Drab *et al.*, 2001; Park *et al.*, 2002; Woodman *et al.*, 2002a; Figure 1A). In some cells, expression of caveolin-1 is also sufficient to induce the formation of caveolae on the plasma membrane (Fra *et al.*, 1995b). However, the caveolins are present at substantial levels in caveolae-free intracellular membranes (Kurzchalia *et al.*, 1992; Breuza *et al.*, 2002) and can be expressed at the surface of certain cells at high levels without caveolae being formed (Sowa *et al.*, 2001; Breuza *et al.*, 2002). The additional accessory factor required to form caveolae is unknown; it has been suggested to be an ordered lipid environment (Breuza *et al.*, 2002) but caveolin can be recovered in DRMs from a human prostate cell line devoid of caveolae (Sowa *et al.*, 2001).

Caveolins bind cholesterol (Murata *et al.*, 1995) and are closely associated with sphingolipids (Fra *et al.*, 1995a). These proteins insert into the cytoplasmic surface of the membrane, and so control lipid composition from the inner leaflet of the bilayer. They are palmitoylated on three Cys residues (at positions 133, 143 and 156), modifications that are involved in the binding of cholesterol and oligomerisation of the protein (Monier *et al.*, 1996; Uittenbogaard and Smart, 2000). Caveolin's interaction with caveolar membrane is not, however, determined by tri-palmitoylation but rather by a more proximal span of amino acids (residues 82–101) which include a short hydrophobic sequence flanked by basic residues (KYWFYR) that is essential for membrane attachment (Woodman *et al.*, 2002b). It is thought that either this, or an adjacent hydrophobic region (residues 102–134) forms a hairpin loop projecting into the membrane without spanning it (Monier *et al.*, 1996; Woodman *et al.*, 2002b). Caveolins appear therefore to act as a wedge inserted into the inner leaflet of the bilayer (Figure 9C); this fundamental asymmetry is augmented by their concentration of cholesterol in the inner leaflet (Pitto *et al.*, 2000). The lipid composition of DRMs and DfRMs prepared from cells ± caveolae shows no difference between the fractions from parental and caveolin-expressing cells (Pike *et al.*, 2002), implying that rafts and caveolae may have very similar lipid composition.

However, functionally there is a critical difference between caveolin-1 present (at approximately comparable levels) within lipid rafts in caveolae-free cells, and within caveolae. Endothelial nitric oxide synthetase (eNOS), which

[24] Caveolins are a family of 21kD proteins (Okamoto *et al.*, 1998); they are also known as VIP21). Caveolin-1 is required for expression of caveolin-2 (Drab *et al.*, 2001). The latter is present on caveolae-free intracellular membranes (Golgi and lipid droplets; Fujimoto *et al.*, 2002; Breuza *et al.*, 2002), whereas caveolin-1 is also expressed within caveolae on the cell surface. Caveolin-3 is the only caveolin subtype expressed in striated muscle (Woodman *et al.*, 2002a).

produces NO, a potent regulator of diverse vascular functions (Goligorsky *et al.*, 2002; described above in relation to β_2-adrenoreceptor depalmitoylation), is down-regulated when present in caveolae, but not when the caveolin-1 is present at similar levels on the plasma membrane in non-caveolae-producing cells[25] (Sowa *et al.*, 2001). Maintained in its suppressed state within caveolae, eNOS can be activated by a range of physiological stimuli including G-protein coupled receptors that cause the enzyme to translocate away from caveolae (Feron and Kelly, 2001) (Figure 8).

The importance of caveolae for regulating eNOS activity has been validated by analysis of caveolin-1 knockout mice (Drab *et al.*, 2001), that have also shown the caveolins are critical for the regulation of systemic lipid homeostasis (Razani *et al.*, 2002) as well as pulmonary and cardiac development (Drab *et al.*, 2001; Park *et al.*, 2002). Despite intense investigation, the key question as to whether caveolin controls eNOS directly, suppressing its activity by binding to a negative regulatory site, or indirectly by assembling a distinctive organelle into which eNOS is sequestered, has yet to be resolved. A wide range of imaginative experiments have been interpreted as showing that residues 82–101 of caveolin-1 interact with eNOS (see Sowa *et al.*, 2001) and numerous other proteins. Now that this sequence is known to direct caveolin to caveolar membrane (Woodman *et al.*, 2002b), the possibility remains that eNOS binding may primarily be to caveolae rather than to a caveolin, and that suppression of its activity may be mediated by agents other than, or additional to, caveolin.[26] Indeed, the lack of eNOS inhibition by caveolin-1 in caveolae-free cells argues strongly for this result (Sowa *et al.*, 2001).[27]

The oligomerisation, cholesterol- and sphingolipid-binding properties of caveolins pose obvious problems for biochemical analysis that are aggravated by the inclusion of caveolins in detergent-resistant membrane complexes.

[25] Both the caveolin-1 and eNOS were recovered in DRMs isolated from these cells (and both co-localised by immunofluorescence), although it was not shown that both occurred within the same microdomain.

[26] The "scaffolding domain" of caveolin (residues 82–101) has been reported to bind directly to a wide range of proteins that have a "scaffold-binding" motif (Okamoto *et al.*, 1998). The questions raised here about eNOS interaction with this domain apply generally, particularly as analysis of caveolin knockout mice has demonstrated more specific effects than would be expected from the wide specificity described in binding studies (Drab *et al.*, 2001; Razani *et al.*, 2001; Park *et al.*, 2002; Razani *et al.*, 2002; Woodman *et al.*, 2002a).

[27] Direct interaction with another region of caveolin, the palmitate chain on Cys-156, appears to explain the anomalous presence in caveolae of Src kinase (which has a single palmitate rather than the myristate/palmitate motif found in those members of this kinase family that normally enter lipid ordered domains). It appears that caveolin itself contributes the second acyl chain to the ordered packing of the bilayer lipids around Src's palmitate (Lee *et al.*, 2001). A similar mechanism may explain the selective affinity of Gq for caveolae rather than rafts (Oh and Schnitzer, 2001).

Figure 9. The different involvement of proteins in producing tight curvature of disordered and ordered membranes. (A) coated pits, formed in elastic disordered membranes, are curved by the action of the clathrin complex (represented by brackets) attached to, but not inserted into, the cytoplasmic face of the surface membrane. When the pit closes to form a vesicle, the clathrin complex dissociates leaving a vesicle (B) that is stable without a further contribution from structural proteins. In contrast, caveolae (C) require caveolins (shown as loops), whose hairpin insertion into the inner leaflet of the membrane acts, we suggest, as a wedge to enable curvature of the less flexible ordered lipid domains. Intracellular transport vesicles with ordered lipid domains contain similar proteins (e.g. flotillins) inserted into the inner leaflet (D), whereas the opposite curvature is accompanied by insertion of asymmetrically-organised proteins such as prominin into the membrane (E).

Their biosynthetic functions, and in particular their delivery of cholesterol to, and concentration within, the plasma membrane (Roy *et al.*, 1999; Ikonen and Parton, 2000) affects many cellular properties including the organisation of other proteins within cholesterol-sensitive membrane. GPI-anchored proteins are obvious residents of cholesterol-sensitive non-caveolar membrane (and are influenced by caveolin expression; Abrami *et al.*, 2001; Carozzi *et al.*, 2002; Sotgia *et al.*, 2002). However, disordered, detergent-soluble membrane such as that in coated pits is also cholesterol dependent (Rodal *et al.*, 1999; Subtil *et al.*, 1999). Distinguishing primary from secondary influences in the analysis of caveolin function within cells is difficult.

9.3. Convex and concave curves: Flotillins and stomatins

A large family of proteins, of which the flotillins (or reggies) and stomatins define the two major subtypes (Morrow *et al.*, 2002), when expressed at endogenous levels bind to the cytoplasmic surface of rafts rather than caveolae (Lang *et al.*, 1998; Stuermer *et al.*, 2001; Morrow *et al.*, 2002). Their expression is not linked to the caveolins (Schulte *et al.*, 1997; Lang *et al.*, 1998; Volonte *et al.*, 1999); in particular they are major protein components of rafts in cells that do not express caveolins (erythocytes (Salzer and Prohaska, 2001); platelets (Mairhofer *et al.*, 2002); lymphocytes (Solomon *et al.*, 2002); and neurons (Rajaram *et al.*, 1999; Stuermer *et al.*, 2001). How these proteins insert into the membrane is currently unclear, although palmitoylation, oligomerisation and hydrophobic, possibly TM or intra-membrane hair-pin loop sequences, are thought to be involved (Bickel *et al.*, 1997; Schulte *et al.*, 1997; Snyers *et al.*, 1998; Snyers *et al.*, 1999b; Morrow *et al.*, 2002; Figure 9D).

Stomatins and other family members differ from the flotillins/reggie proteins in having an additional 25 amino acids at their immediate N-terminus (Tavernarakis *et al.*, 1999; Morrow *et al.*, 2002). This hydrophobic stretch appears to be a classical type 1 TM domain, ending with an invariant basic Lys at position 26 as required on the cytoplasmic side of the membrane (van Klompenburg *et al.*, 1997); it probably also acts as an uncleaved signal sequence directing these proteins into the membrane of the endoplasmic reticulum (with only 2 amino acids (CG or CE) protruding into the lumen), eventually to be exposed on the cell surface (Morrow *et al.*, 2002).

The flotillins have two hydrophobic stretches which have been suggested to function as TM domains (Bickel, 2002). The second of these is only 18 amino acids long, typical of TM domains in the thinner Golgi membrane (Munro, 1995); a 14 kD N-terminal fragment is protected against proteolysis from the cytoplasmic side of Golgi vesicles, suggesting that in Golgi membranes this domain is inserted into the membrane and functions to retain flotillin-1 within this organelle (in whose lipid rafts it is highly enriched; Gkantiragas *et al.*, 2001). On plasma membrane vesicles, however, flotillin-1 is not protected from cytoplasmic proteolysis, and attachment to the membrane is primarily dependent upon palmitoylation (Morrow *et al.*, 2002). This pool of flotillin-1 is not routed through the Golgi during biosynthesis (Morrow *et al.*, 2002), suggesting that this protein follows at least two different routes during biosynthesis, with different effects upon its targeting in the cell.

The interactions of this complex family with membrane lipids will take some time to unravel. Flotillins are not associated with all lipid rafts (Gkantiragas *et al.*, 2001), and seem particularly to be recruited when raft components are cross-linked (Stuermer *et al.*, 2001), or following receptor activation (Baumann *et al.*, 2000). Expression of flotillin-1 on insect cells produces concave curvature of the membrane, not as marked as with caveolin expression but concave nonetheless (Volonte *et al.*, 1999). It is notable how often flotillins are expressed on concave membrane (e.g. phagosomes (Dermine *et al.*, 2001); late endosomes/lysosomes (Gagescu *et al.*, 2000); biosynthetic vesicles of the export pathway (Marchand *et al.*, 2002), but not on convex membrane (of secreted nanovesicles and microvesicles of erythrocytes (Salzer *et al.*, 2002); and the secreted platelet α granules (Mairhofer *et al.*, 2002)). Conversely, stomatin is often selectively enriched within convex membrane, such as filopodia (Snyers *et al.*, 1999a) and the secreted vesicles mentioned previously (Mairhofer *et al.*, 2002; Salzer *et al.*, 2002).

9.4. Convex curvature: prominin

Also located on organelles (e.g. filopodia) with high membrane curvature, prominin has 5 TM domains holding two large (>250 amino acids in each) glycosylated extracellular domains joined by small intracellular loops (Corbeil *et al.*, 2001; Figure 9E). Its presence on microvilli is independent of the cytoskeleton but very sensitive to cholesterol, which can be cross-linked to prominin in the membrane (Roper *et al.*, 2000). It can be isolated in DRMs using the non-ionic detergent Lubrol WX, but is solubilised by Triton X-100 (Roper *et al.*, 2000). Prominin appears to be enclosed on filopodia by rafts that have a comparatively lower degree of lipid order (and so presumably higher elasticity) than found on planar surfaces of the same cells (Roper *et al.*, 2000). A missense mutation in man leads to retinal degeneration, presumably triggered by the inability to form photoreceptor disc membrane in the absence of prominin (Corbeil *et al.*, 2001).

9.5. A chance association? the MAL family of tetra-span proteins

MAL (VIP 17) is a small, very hydrophobic 17 kD protein with four presumed TM domains that neither binds cholesterol nor requires this sterol to associate with membranes (Puertollano *et al.*, 1999b). MAL is isolated within non-caveolar DRMs (Millan *et al.*, 1997; Frank *et al.*, 1998). It is expressed in a variety of tissues at locations (myelin, microvilli on apical surfaces of kidney epithelia etc) with high convex curvature (Frank *et al.*, 1998). When ectopically expressed in insect cells, very large (200–700 nm diameter) vesicles are formed that are clearly different from the much smaller vesicles induced by caveolin-1 expression in these cells (Puertollano *et al.*, 1997). These properties support the suggestion that MAL contributes to the curvature of convex raft membranes (Frank *et al.*, 1998). However, MAL recycles rapidly between surface and trans-Golgi compartments (Puertollano and Alonso, 1999a), and is necessary for apical sorting of a range of proteins (Puertollano and Alonso, 1999b; Puertollano *et al.*, 1999a; Martin-Belmonte *et al.*, 2000; Martin-Belmonte *et al.*, 2001), both roles that involve vesicles with the opposite (concave) curvature. Other members of this family are necessary for transcytosis (MAL2; de Marco *et al.*, 2002) or co-localise on intracellular vesicles but not plasma membrane with caveolins 1 and 2 (BENE; del Marco *et al.*, 2001). This family of proteins may primarily be involved in intracellular trafficking of raft-containing vesicles rather than influencing membrane curvature.

10. WHAT IS A "RAFT", AND WHAT DOES IT DO?

To the current generation of biomedical scientists raised on a genome-centred culture, in which a noun describes a single entity (gene or its product) which can be identified and isolated following well defined protocols (or, even better, a commercial kit), "rafts" stand out as the epitome of post-genomic trauma. There is no general agreement on what they are, how they can be identified, or how they should be isolated. Worse, even though they are involved in "everything", it is far from clear what anything gets out of its association with a raft. So has our discussion of the problem clarified this situation?

First, what are we talking about? The word "raft" was initially coined to describe specialised lipid microdomains that concentrated and conveyed glucosylceramide to the apical rather than basolateral surface of polarised epithelial cells (Simons and van Meer, 1988). The properties conjured up by the term "raft" (highly selected cargo, small size and directed movement) apply admirably to this sorting and trafficking mechanism for glycosphingolipids. However, "raft" also conjures up expectations of a defined entity which is of itself misleading when applied to the diverse lipid microdomains found within cellular membranes.

What we are really talking about are microdomains of ordered lipids that selectively partition functional groups of proteins within the membrane. They are distinct from the disordered fluid mosaic membrane (Singer and Nicolson, 1972) that acts as the relatively inert solvent for most membrane proteins. In themselves, however, they are a very diverse set of microenvironments. Their selectivity filter for membrane proteins arises from the inability of ordered lipids to accommodate most TM α-helices, a property shared by gel and liquid ordered phases of lipids. Some ordered microdomains in biological membranes may be in the gel (S_O) phase – they would equally sequester that subset of proteins that can be accommodated within ordered lipid phases. The higher lateral mobility of the L_O phase, however, provides a more suitable environment for molecular interactions, and undoubtedly better describes the phase in which most ordered lipid microdomains occur.

10.1. Ordered lipid microdomains come in many different forms

We have outlined how the diversity of lipid structure enables them to form very different L_O microdomains of cell membranes. This underlying lipid-based diversity interacts with the proteins that partition into each microdomain. This is well illustrated by influenza haemagglutinin (Tatulian and Tamm, 2000), but even lipid anchored proteins such as Thy-1 order the lipids around their point of insertion (Reid-Taylor *et al.*, 1999). Diversity within the ordered lipid microdomain is further compounded by the degree of unsaturation of lipids in the disordered membrane, which differs in different cells and must affect the ease with which certain microdomains form. For instance, while lipids with monounsaturation in the *sn-2* fatty acid can condense with cholesterol, they will do so more readily in a membrane with a high content of lipids in which both fatty acid chains are unsaturated; in membranes with a high content of fully saturated lipids these monounsaturated lipids may be excluded from ordered microdomains by competition. While the formation of a distinct phase (e.g. L_O) is an all-or-nothing event, whether such ordered microdomains form, and their individual characteristics, will differ in different membranes.

Ordered lipid microdomains exist within cellular membranes, but are diverse, and each is a negotiated settlement between the particular lipids and proteins within each individual membrane. Although cholesterol and sphingolipids are normal, and important components, neither is necessary. Ordered domains (e.g. in the inner leaflet of the membrane) exist without one or other of them. Given there is no absolute composition, there can be no canonical isolation method, no "Raft-U-Like" isolation kit.

10.2. What do ordered lipid microdomains do for the cell?

Given their structural diversity, are ordered lipid domains better defined functionally?

We suggest there are two key functional consequences of the ordering of lipids. The first is their exclusion of most proteins, so that the few proteins that are admitted to the ordered microenvironment interact with enhanced efficiency, and are restricted to a particular pathway of action. This is well illustrated by the receptors already discussed. Simons and colleagues (Simons and Ikonen, 1997; Simons and Toomre, 2000) have suggested that the very small size of these domains (they are really nano-, rather than micro-, domains) would prevent any one domain containing the variety of proteins required to assemble a signalling complex. In their scheme, the dispersion of interacting subunits in different rafts would suppress signalling until ligand-activation of the receptor itself triggered clustering of rafts to assemble a complete signalling complex within the larger domain. Although there is insufficient data to assess whether this type of mechanism is activated by physiological stimuli, the evidence beginning to emerge from the related family of T cell antigen/Fc receptors suggests that the dispersion of elements of signalling complexes in different rafts is preserved during signalling, with each distinct type of raft sequestering the proteins involved in different phases of signalling.

The second key element, we believe, is the different trafficking of proteins within highly ordered, and disordered, membrane. This difference could have important implications for those signalling systems that require signalling to be sustained well beyond the initial activation of the receptor.

10.3. A general "raft" function: resistance to endocytosis enables sustained cell signalling

When cognate peptide is presented to a responding T lymphocyte, the full molecular machinery required to initiate and sustain signalling from the T cell receptor, including rafts with their associated cytoplasmic kinases (Lck) and adaptors (LAT), is confined within an integrin-sealed patch (the immunological synapse) in contact with the antigen presenting cell (Monks *et al.*, 1998; Dupré *et al.*, 2002; Werlen and Palmer, 2002). The clustering of rafts within the central region of this synapse is often considered to exemplify the mixing of signalling components, although analysis to date has been restricted to immunofluorescence. Higher resolution analysis of the topographical ordering of signalling components has been made by EM immunohistochemistry with the related FcεR1 signalling system (Wilson *et al.*, 2000; Wilson *et al.*, 2001); the general picture obtained agrees with a recent analysis of early events in T cell signalling using the highly selective detergent Brij 98 (Drevot *et al.*, 2002). Each shows that "conventional" raft components (e.g. LAT), present in membrane resistant to 1% Triton X-100, do not mix with activated receptor that is present in different, much more soluble (but still recognisably "raft") domains (Wilson *et al.*, 2000; Wilson *et al.*, 2001; Drevot *et al.*, 2002). The LAT domains can touch or even cross the domains occupied by activated receptor, but their components do not intermix (Wilson *et al.*, 2001). Since the two types of domain differ markedly in their detergent-resistance, the ordering of their lipids is presumably sufficiently different to prevent such mixing.

A very simple role for the more highly ordered (less soluble) microdomains containing the adaptor protein LAT suggests itself, a role moreover that could apply very generally in signalling. The T cell antigen receptor (Lee *et al.*, 2002) and the FcεR1 receptor (Wilson *et al.*, 2000) are rapidly endocytosed after stimulation. It is particularly well established for the T cell receptor that signalling needs to persist for at least 2 hours to drive the cell into proliferation (Lee *et al.*, 2002). Signalling initiated in the small receptor microdomain appears to transfer (via the soluble kinase ZAP70) to the adjacent but separate, more highly ordered adaptor microdomain (Drevot *et al.*, 2002). The FcεR1 receptor accumulates within the clathrin lattice and is rapidly endocytosed via coated pits (Wilson *et al.*, 2000). Highly ordered lipid microdomains cannot enter coated pits, and are internalised by unknown mechanisms that appear to be about ten times slower than clathrin-dependent uptake (Lamaze *et al.*, 2001; Sunyach *et al.*, 2003). In FcεR1 receptor signalling, the more ordered LAT domain does not enter the clathrin lattice and persists on the cell surface (Wilson *et al.*, 2001) where it assembles a secondary stable signalling centre, continuing to activate the cell long after the original receptor had been endocytosed (Figure 10).

This suggested role is consistent with the known requirement for raft-associated LAT to link the T cell receptor to cellular activation (Zhang *et al.*, 1998a; Zhang *et al.*, 1998b), for the different location and trafficking of the T cell receptor and rafts within the immunological synapse (Lee *et al.*, 2002), and for raft stabilisation of transient FcεR1 receptor signalling on mast cells (Holowka *et al.*, 2000). More importantly, the problem it addresses – how to sustain signalling when the receptor itself is often rapidly endocytosed – is a general one. The activated insulin receptor, for instance, is rapidly endocytosed (Pilch *et al.*, 1983) and requires activation of a second signalling pathway in

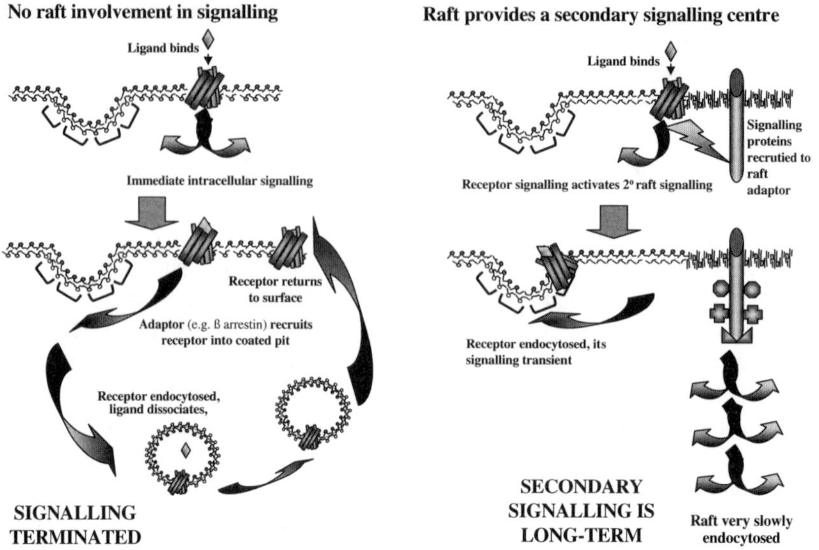

Figure 10. Signalling via surface receptors is transient (left), usually terminated by endocytosis of the receptor followed by dissociation of the ligand. When signalling is transferred from the primary receptor a stable raft-associated protein (prototypically LAT, for T cell and FcεR1 receptors), the latter organises a secondary signalling centre (right) that continues to activate the cell long after the primary receptor has been endocytosed.

order to upregulate surface expression of the glucose transporter GLUT4 (Czech, 2000). This is provided by the flotillin-mediated translocation of its specific adaptor protein, CAP, to ordered lipid microdomains (Baumann *et al.*, 2000). This may again provide persistence of signalling (cf. Vainio *et al.*, 2002), and/or help direct fusion of GLUT4 containing vesicles to the more hydrophobic surface of ordered membrane (Chamberlain and Gould, 2002b). For neurotrophic receptors also, persistence of signalling, leading to neuronal differentiation and survival, is dependent upon retention of the signalling complex within rafts (Peiro *et al.*, 2000; Paratcha *et al.*, 2001).

Resistance to endocytosis undoubtedly interacts with other properties of highly ordered lipid domains to enable them to function as platforms for prolonged membrane signalling. These include their exclusion of inactivating enzymes (such as the phosphatase CD45 from LAT-containing rafts in T cell activation; Edmonds and Ostergaard, 2002) and stabilisation of signalling complexes once formed, by the lower translational diffusion of rafts (Johnson *et al.*, 1996; Pralle *et al.*, 2000; Dietrich *et al.*, 2002).

While current experimental work is concentrated almost exclusively upon nm-scale interactions within rafts, larger scale effects in channelling signalling to particular domains should not be overlooked. In the dendritic compartment of neurons, for instance, a multitude of incoming synaptic signals are integrated to produce a single all-or-nothing response at the axon initial segment. Along the axon the signal is propagated in an energy-driven, self-regenerating spike whose efficacy and speed is augmented by sealing much of the membrane in high-resistance myelin. In contrast, in the dendrite the signal from each synapse decays from its source, and interacts with other synaptic signals received at that time on any of the thousands of other synapses present on the dendritic tree. The enhanced hydrophobicity (greater resistance) of ordered lipid domains, coupled with the exclusion from them of most voltage-gated ion channels,[28] could serve to channel a post-synaptic response along particular pathways of the dendritic surface (Figure 11). The nm scale organisation of ordered lipid domains, and their dynamic, signalling-controlled instability, make them ideally suited to play a role in directing the interaction of signals on the dendritic surface to effect the sophisticated signal integration that underlies the function of the central nervous system. Rafts are involved in the topographical organisation of receptors and adaptor proteins in neural synapses

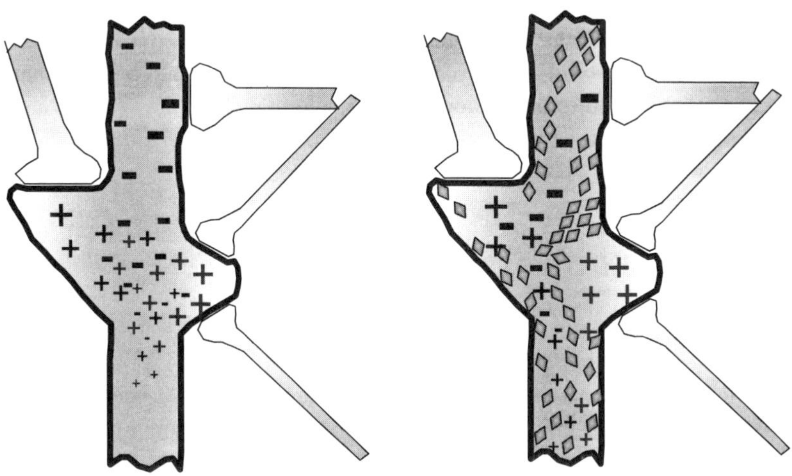

Figure 11. A scheme of four adjacent synapses on a dendritic tree, three excitatory (E1–3, signal indicated by +) and one inhibitory (I1; signal indicated by −). In the absence of rafts (left), the signal simply diminishes with distance; its integration in the dendritic shaft is determined by the relative location of the synapses, duration of signal and timing of the synaptic input. The presence of rafts (right), excluding signal from their area and directing it along the channels of disordered membrane, will enhance the spread of signal from an activated synapse, allowing integration of signal over a wider area of membrane. If the rafts are clustered within specific regions (right) they could actively direct the spread and interaction of the signal, in the example shown making I1 more effective at inhibiting E1, but less effective at inhibiting E2/3, and enhancing the summation of E1–3.

[28] The Shaker-like postassium channel appears to date to be the only raft-associated voltage-gated ion channel (Martens *et al.*, 2001).

(Perez and Bredt, 1998; El-Husseini *et al.*, 2000; Becher *et al.*, 2001; Bruses *et al.*, 2001; El-Husseini *et al.*, 2002), although this is an area in which investigation has hardly commenced.

11. CONCLUSION

The diversity of lipid ordered microdomains in biological membranes poses real challenges, both methodological and conceptual, to the investigator, but is also central to their role in assembling functionally different protein complexes at different sites within the cell. Key areas that need to be elucidated before we can understand the mechanism by which lipids partition proteins include: whether and how the inner and outer leaflets of the membrane are coupled in ordered microdomains; what are the structural requirements for proteins to enter these domains, and to traffic between ordered and disordered domains; and how does the cellular trafficking of proteins differ depending upon their lipid environment? At the methodological level, studies on the composition and function of immunoaffinity-isolated subsets of domains, rather than upon an entire floating fraction, are needed. These, complemented by imaging techniques with nm resolution, will enable us to relate composition of microdomains with organisation in the membrane. When we have this information, we will be in a much stronger position to be able to identify how individual domains control specific cell functions.

ACKNOWLEDGEMENTS

The authors would like to thank Wiebke Mobius and Rob Parton for communicating results prior to publication, and Wilhelm Just for generously providing prints of his published work, reproduced here. EM is supported by EU HPRN-CT-2000-00077, the Training Network on Sphingolipid synthesis and organization, co-ordinated by Gerrit van Meer; it is a pleasure to acknowledge the extent to which this Network has contributed substantially to the development of our ideas.

REFERENCES

Abrami, L., Fivaz, M., Kobayashi, T., Kinoshita, T., Parton, R.G. and van der Goot, F.G., 2001, Cross-talk between caveolae and glycosylphosphatidylinositol-rich domains. *J Biol Chem*, **276**: 30729-36.

Adam, L., Bouvier, M. and Jones, T.L., 1999, Nitric oxide modulates β_2-adrenergic receptor palmitoylation and signaling. *J Biol Chem*, **274**: 26337-43.

Ahmed, S.N., Brown, D.A. and London, E., 1997, On the origin of sphingolipid/cholesterol-rich detergent-insoluble cell membranes: physiological concentrations of cholesterol and sphingolipid induce formation of a detergent-insoluble, liquid-ordered lipid phase in model membranes. *Biochemistry*, **36**: 10944-53.

Aloia, R.C., Jensen, F.C., Curtain, C.C., Mobley, P.W. and Gordon, L.M., 1988, Lipid composition and fluidity of the human immunodeficiency virus. *Proc Natl Acad Sci USA*, **85**: 900-904.

Aloia, R.C., Tian, H. and Jensen, F.C., 1993, Lipid composition and fluidity of the human immunodeficiency virus envelope and host cell plasma membranes. *Proc Natl Acad Sci USA*, **90**: 5181-85.

Bagatolli, L.A. and Gratton, E., 2000a, A correlation between lipid domain shape and binary phospholipid mixture composition in free standing bilayers: A two-photon fluorescence microscopy study. *Biophys J*, **79**: 434-447.

Bagatolli, L.A. and Gratton, E., 2000b, Two photon fluorescence microscopy of coexisting lipid domains in giant unilamellar vesicles of binary phospholipid mixtures. *Biophys J*, **78**: 290-305.

Bagnat, M., Keranen, S., Shevchenko, A. and Simons, K., 2000, Lipid rafts function in biosynthetic delivery of proteins to the cell surface in yeast. *Proc Natl Acad Sci U S A*, **97**: 3254-59.

Barboni, E., Gormley, A.M., Pliego Rivero, F.B., Vidal, M. and Morris, R.J., 1991, Activation of T lymphocytes by cross-linking of glycophospholipid-anchored Thy-1 mobilizes separate pools of intracellular second messengers to those induced by the antigen-receptor/CD3 complex. *Immunology*, **72**: 457-463.

Barclay, A.N., Birkeland, M.L., Brown, M.H., Beyers, A.D., Davis, S.J., Somoza, C. and Williams, A.F., 1993, *The Leucocyte Antigen FactsBook*. Academic Press, London.

Barenholz, Y. and Thompson, T.E., 1980, Sphingomyelins in bilayers and biological membranes. *Biochim Biophys Acta*, **604**: 129-158.

Baumann, C.A., Ribon, V., Kanzaki, M., Thurmond, D.C., Mora, S., Shigematsu, S., Bickel, P.E., Pessin, J.E. and Saltiel, A.R., 2000, CAP defines a second signalling pathway required for insulin-stimulated glucose transport. *Nature*, **407**: 202-207.

Becher, A., White, J.H. and McIlhinney, R.A., 2001, The γ-aminobutyric acid receptor B, but not the metabotropic glutamate receptor type-1, associates with lipid rafts in the rat cerebellum. *J Neurochem*, **79**: 787-795.

Benting, J., Rietveld, A., Ansorge, I. and Simons, K., 1999, Acyl and alkyl chain length of GPI-anchors is critical for raft association in vitro. *FEBS Lett*, **462**: 47-50.

Berridge, M.J., 1987, Inositol-triphosphate and diacylglycerol: two interacting second messengers. *Annu Rev Biochem*, **56**: 159-193.

Bickel, P.E., 2002, Lipid rafts and insulin signaling. *Am J Physiol Endocrinol Metab*, **282**: E1-E10.

Bickel, P.E., Scherer, P.E., Schnitzer, J.E., Oh, P., Lisanti, M.P. and Lodish, H.F., 1997, Flotillin and epidermal surface antigen define a new family of caveolae- associated integral membrane proteins. *J Biol Chem*, **272**: 13793-802.

Bielawska, A., Greenberg, M.S., Perry, D., Jayadev, S., Shayman, J.A., McKay, C. and Hannun, Y.A., 1996, 1(S,2R)-D-erythro-2-(N-myristoylamino)-1-phenyl-1-propanol as an inhibitor of ceramidase. *J Biol Chem*, **271**: 12646-54.

Bittman, R., Kasireddy, C.R., Mattjus, P. and Slotte, J.P., 1994, Interaction of cholesterol with sphingomyelin in monolayers and vesicles. *Biochemistry*, **33**: 11776-81.

Boggs, J.M., 1987, Lipid intermolecular hydrogen bonding: influence on structural organization and membrane function. *Biochim Biophys Acta*, **906**: 353-404.

Brdicka, T., Cerny, J. and Horejsi, V., 1998, T cell receptor signalling results in rapid tyrosine phosphorylation of the linker protein LAT present in detergent-resistant membrane microdomains. *Biochem Biophys Res Commun*, **248**: 356-360.

Brdicka, T., Pavlistova, D., Leo, A., Bruyns, E., Korinek, V., Angelisova, P., Scherer, J., Shevchenko, A., Hilgert, I., Cerny, J., Drbal, K., Kuramitsu, Y., Kornacker, B., Horejsi, V. and Schraven, B., 2000, Phosphoprotein associated with glycosphingolipid-enriched microdomains, (PAG), a novel ubiquitously expressed transmembrane adaptor protein, binds the protein tyrosine kinase csk and is involved in regulation of T cell activation. *J Exp Med*, **191**: 1591-1604.

Bretscher, M.S. and Munro, S., 1993, Cholesterol and the Golgi apparatus. *Science*, **261**: 1280-81.

Breuza, L., Corby, S., Arsanto, J.P., Delgrossi, M.H., Scheiffele, P. and Le Bivic, A., 2002, The scaffolding domain of caveolin 2 is responsible for its Golgi localization in Caco-2 cells. *J Cell Sci*, **115**: 4457-67.

Brewis, I.A., Ferguson, M.A., Mehlert, A., Turner, A.J. and Hooper, N.M., 1995, Structures of the glycosyl-phosphatidylinositol anchors of porcine and human renal membrane dipeptidase. Comprehensive structural studies on the porcine anchor and interspecies comparison of the glycan core structures. *J Biol Chem*, **270**: 22946-56.

Brodsky, F.M., Chen, C.Y., Knuehl, C., Towler, M.C. and Wakeham, D.E., 2001, Biological basket weaving: formation and function of clathrin-coated vesicles. *Annu Rev Cell Dev Biol*, **17**: 517-568.

Brose, N. and Rosenmund, C., 2002, Move over protein kinase C, you've got company: alternative cellular effectors of diacylglycerol and phorbol esters. *J Cell Sci*, **115**: 4399-4411.

Brown, D.A. and London, E., 1997, Structure of detergent-resistant membrane domains: does phase separation occur in biological membranes? *Biochem Biophys Res Commun*, **240**: 1-7.

Brown, D.A. and London, E., 1998, Functions of lipid rafts in biological membranes. *Annu Rev Cell Dev Biol*, **14**: 111-136.

Brown, D.A. and London, E., 2000, Structure and function of sphingolipid- and cholesterol-rich membrane rafts. *J. Biol Chem*, **275**: 17221-24.

Brown, D.A. and Rose, J.K., 1992, Sorting of GPI-anchored proteins to glycolipid-enriched membrane subdomains during transport to the apical cell surface. *Cell*, **68**: 533-544.

Brown, R.E., 1998, Sphingolipid organization in biomembranes: what physical studies of model membranes reveal. *J. Cell Sci*, **111**: 1-9.

Brügger, B., Graham, C.H., Leibrecht, I., Mombelli, E., Jen, A., Wieland, F.T. and Morris, R., 2004, The membrane domains occupied by GPI-anchored prion protein and Thy-1 differ in lipid composition. *J. Biol. Chem.*, in press.

Bruses, J.L., Chauvet, N. and Rutishauser, U., 2001, Membrane lipid rafts are necessary for the maintenance of the $\alpha 7$ nicotinic acetylcholine receptor in somatic spines of ciliary neurons. *J Neurosci*, **21**: 504-512.

Cacalano, G., Farinas, I., Wang, L.C., Hagler, K., Forgie, A., Moore, M., Armanini, M., Phillips, H., Ryan, A.M., Reichardt, L.F., Hynes, M., Davies, A. and Rosenthal, A., 1998, GFRalpha1 is an essential receptor component for GDNF in the developing nervous system and kidney. *Neuron*, **21**: 53-62.

Calderon, R.O., Attema, B. and DeVries, G.H., 1995, Lipid composition of neuronal cell bodies and neurites from cultured dorsal root ganglia. *J. Neurochem*, **64**: 424-429.

Carozzi, A.J., Roy, S., Morrow, I.C., Pol, A., Wyse, B., Clyde-Smith, J., Prior, I.A., Nixon, S.J., Hancock, J.F. and Parton, R.G., 2002, Inhibition of lipid raft-dependent signaling by a dystrophy-associated mutant of caveolin-3. *J Biol Chem*, **277**: 17944-17949.

Chamberlain, L.H. and Gould, G.W., 2002, The v-and t-SNARE proteins that mediate Glut4 vesicle fusion are localised in detergent-insoluble lipid rafts present on distinct intracellular membranes. *J Biol Chem*, **277**: 49750-54.

Charles, P., Tait, S., Faivre-Sarrailh, C., Barbin, G., Gunn-Moore, F., Denisenko-Nehrbass, N., Guennoc, A.M., Girault, J.A., Brophy, P.J. and Lubetzki, C., 2002, Neurofascin is a glial receptor for the paranodin/Caspr-contactin axonal complex at the axoglial junction. *Curr Biol*, **12**: 217-220.

Chatterjee, S., Smith, E.R., Hanada, K., Stevens, V.L. and Mayor, S., 2001, GPI anchoring leads to sphingolipid-dependent retention of endocytosed proteins in the recycling endosomal compartment. *EMBO J*, **20**: 1583-92.

Chen, C.A. and Manning, D.R., 2000, Regulation of galpha i palmitoylation by activation of the 5-hydroxytryptamine-1A receptor. *J Biol Chem*, **275**: 23516-22.

Chen, H., Mendelsohn, R., Rerek, M.E. and Moore, D.J., 2001, Effect of cholesterol on miscibility and phase behavior in binary mixtures with synthetic ceramide 2 and octadecanoic acid. Infrared studies. *Biochim Biophys Acta*, **1512**: 345-356.

Cheung, D.T., Tong, D., Perelman, N., Ertl, D. and Nimni, M.E., 1990, Mechanism of crosslinking of proteins by glutaraldehyde. IV. *In vitro* and *in vivo* stability of a crosslinked collagen matrix. *Connective Tissue Res.*, **25**: 27-34.

Corbeil, D., Roper, K., Fargeas, C.A., Joester, A. and Huttner, W.B., 2001, Prominin: a story of cholesterol, plasma membrane protrusions and human pathology. *Traffic*, **2**: 82-91.

Cremesti, A.E., Goni, F.M. and Kolesnick, R., 2002, Role of sphingomyelinase and ceramide in modulating rafts: do biophysical properties determine biologic outcome? *FEBS Lett*, **531**: 47-53.

Crystal, G.J., Zhou, X., Alam, S., Piotrowski, A. and Hu, G., 2001, Lack of role for nitric oxide in cholinergic modulation of myocardial contractility in vivo. *Am J Physiol Heart Circ Physiol*, **281**: H198-206.

Cunningham, B.A., Midmore, L., Kucuk, O., Lis, L.J., Westerman, M.P., Bras, W., Wolfe, D.H., Quinn, P.J. and Qadri, S.B., 1995, Sterols stabilize the ripple phase structure in dihexadecylphosphatidylcholine. *Biochim Biophys Acta*, **1233**: 75-83.

Czech, M.P., 2000, Lipid rafts and insulin action. *Nature*, **407**: 147-148.

Das, S. and Rand, R.P., 1984, Diacylglycerol causes major structural transitions in phospholipid bilayer membranes. *Biochem Biophys Res Commun*, **124**: 491-496.

Das, S. and Rand, R.P., 1986, Modification by diacylglycerol of the structure and interaction of various phospholipid bilayer membranes. *Biochemistry*, **25**: 2882-89.

de Marco, M.C., Martin-Belmonte, F., Kremer, L., Albar, J.P., Correas, I., Vaerman, J.P., Marazuela, M., Byrne, J.A. and Alonso, M.A., 2002, MAL2, a novel raft protein of the MAL family, is an essential component of the machinery for transcytosis in hepatoma HepG2 cells. *J Cell Biol*, **159**: 37-44.

Degtyarev, M.Y., Spiegel, A.M. and Jones, T.L., 1993, Increased palmitoylation of the Gs protein alpha subunit after activation by the beta-adrenergic receptor or cholera toxin. *J Biol Chem*, **268**: 23769-72.

del Marco, M.C., Kremer, L., Albar, J.P., Martinez-Menarguez, J.A., Ballesta, J., Garcia-Lopez, M.A., Marazuela, M., Puertollano, R. and Alonso, M.A., 2001, BENE, a novel raft-associated protein of the MAL proteolipid family, interacts with caveolin-1 in human endothelial-like ECV304 cells. *J Biol Chem*, **276**: 23009-17.

Denny, P.W., Field, M.C. and Smith, D.F., 2001, GPI-anchored proteins and glycoconjugates segregate into lipid rafts in Kinetoplastida. *FEBS Lett*, **491**: 148-153.

Denny, P.W., Gokool, S., Russell, D.G., Field, M.C. and Smith, D.F., 2000, Acylation-dependent protein export in Leishmania. *J Biol Chem*, **275**: 11017-25.

Dermine, J.F., Duclos, S., Garin, J., St-Louis, F., Rea, S., Parton, R.G. and Desjardins, M., 2001, Flotillin-1-enriched lipid raft domains accumulate on maturing phagosomes. *J Biol Chem*, **276**: 18507-12.

Derry, D.M. and Wolfe, L.S., 1967, Gangliosides in isolated neurons and glial cells. *Science*, **158**: 1450-52.

Dessy, C., Kelly, R.A., Balligand, J.L. and Feron, O., 2000, Dynamin mediates caveolar sequestration of muscarinic cholinergic receptors and alteration in NO signaling. *EMBO J*, **19**: 4272-80.

Dibble, A.R., Hinderliter, A.K., Sando, J.J. and Biltonen, R.L., 1996, Lipid lateral heterogeneity in phosphatidylcholine/phosphatidylserine/diacylglycerol vesicles and its influence on protein kinase C activation. *Biophys J*, **71**: 1877-90.

Dietrich, C., Bagatolli, L.A., Volovyk, Z.N., Thompson, N.L., Levi, M., Jacobson, K. and Gratton, E., 2001a, Lipid rafts reconstituted in model membranes. *Biophys J*, **80**: 1417-28.

Dietrich, C., Volovyk, Z.N., Levi, M., Thompson, N.L. and Jacobson, K., 2001b, Partitioning of Thy-1, GM1, and cross-linked phospholipid analogs into lipid rafts reconstituted in supported model membrane monolayers. *Proc Natl Acad Sci U S A*, **98**: 10642-47.

Dietrich, C., Yang, B., Fujiwara, T., Kusumi, A. and Jacobson, K., 2002, Relationship of lipid rafts to transient confinement zones detected by single particle tracking. *Biophys J*, **82**: 274-284.

Dobrowsky, R.T., 2000, Sphingolipid signalling domains floating on rafts or buried in caves? *Cell Signal*, **12**: 81-90.

Drab, M., Verkade, P., Elger, M., Kasper, M., Lohn, M., Lauterbach, B., Menne, J., Lindschau, C., Mende, F., Luft, F.C., Schedl, A., Haller, H. and Kurzchalia, T.V., 2001, Loss of caveolae, vascular dysfunction, and pulmonary defects in caveolin-1 gene-disrupted mice. *Science*, **293**: 2449-52.

Drevot, P., Langlet, C., Guo, X.J., Bernard, A.M., Colard, O., Chauvin, J.P., Lasserre, R. and He, H.T., 2002, TCR signal initiation machinery is pre-assembled and activated in a subset of membrane rafts. *EMBO J.*, **21**: 1899-1908.

Dupré, L., Aiuti, A., Trifari, S., Martino, S., Saracco, P., Bordignon, C. and Roncarolo, M.-G., 2002, Wiskott-Aldrich syndrome protein regulates lipdi raft dynamics during immunological synapse formation. *Immunity*, **17**: 157-166.

Eason, M.G., Jacinto, M.T., Theiss, C.T. and Liggett, S.B., 1994, The palmitoylated cysteine of the cytoplasmic tail of alpha 2A-adrenergic receptors confers subtype-specific agonist-promoted downregulation. *Proc Natl Acad Sci USA*, **91**: 11178-11182.

Edmonds, S.D. and Ostergaard, H.L., 2002, Dynamic association of CD45 with detergent-insoluble microdomains in T lymphocytes. *J Immunol*, **169**, : 5036-42.

el Bawab, S., Mao, C., Obeid, L.M. and Hannun, Y.A., 2002, Ceramidases in the regulation of ceramide levels and function. *Subcell Biochem*, **36**: 187-205.

El-Husseini, A.E., Craven, S.E., Chetkovich, D.M., Firestein, B.L., Schnell, E., Aoki, C. and Bredt, D.S., 2000, Dual palmitoylation of PSD-95 mediates its vesiculotubular sorting, postsynaptic targeting, and ion channel clustering. *J Cell Biol*, **148**: 159-172.

El-Husseini A. E, D., Schnell, E., Dakoji, S., Sweeney, N., Zhou, Q., Prange, O., Gauthier-Campbell, C., Aguilera-Moreno, A., Nicoll, R.A. and Bredt, D.S., 2002, Synaptic strength regulated by palmitate cycling on PSD-95. *Cell*, **108**: 849-863.

Endress, E., Heller, H., Casalta, H., Brown, M.F. and Bayerl, T.M., 2002, Anisotropic motion and molecular dynamics of cholesterol, lanosterol, and ergosterol in lecithin bilayers studied by quasi-elastic neutron scattering. *Biochemistry*, **41**: 13078-86.

Estep, T.N. and Thompson, T.E., 1979, Energy transfer in lipid bilayers. *Biophys. J*, **26**: 195-208.

Faivre-Sarrailh, C., Gauthier, F., Denisenko-Nehrbass, N., Le Bivic, A., Rougon, G. and Girault, J.A., 2000, The glycosylphosphatidyl inositol-anchored adhesion molecule

F3/contactin is required for surface transport of paranodin/contactin-associated protein, caspr,. *J Cell Biol*, **149**: 491-502.
Farge, E. and Devaux, P.F., 1992, Shape changes of giant liposomes induced by an asymmetric transmembrane distribution of phospholipids. *Biophys J*, **61**: 347-357.
Farge, E., Ojcius, D.M., Subtil, A. and Dautry-Varsat, A., 1999, Enhancement of endocytosis due to aminophospholipid transport across the plasma membrane of living cells. *Am J Physiol*, **276**: C725-733.
Feng, Y., Rainteau, D., Chachaty, C., Yu, Z-W., Wolf, C. and Quinn, P.J., 2004, Characterisation of a quasi-crystalline phase in codispersions of phosphatidylethanolamine and glucocerebroside. *Biophys. J.*, **86**: in press.
Feron, O., Dessy, C., Opel, D.J., Arstall, M.A., Kelly, R.A. and Michel, T., 1998, Modulation of the endothelial nitric-oxide synthase-caveolin interaction in cardiac myocytes. Implications for the autonomic regulation of heart rate. *J Biol Chem*, **273**: 30249-54.
Feron, O. and Kelly, R.A., 2001, The caveolar paradox: suppressing, inducing, and terminating eNOS signaling. *Circ Res*, **88**: 129-131.
Feron, O., Smith, T.W., Michel, T. and Kelly, R.A., 1997, Dynamic targeting of the agonist-stimulated m2 muscarinic acetylcholine receptor to caveolae in cardiac myocytes. *J Biol Chem*, **272**: 17744-48.
Ferretti, A., Knijn, A., Raggi, C. and Sargiacomo, M., 2003, High resolution proton NMR measures mobile lipids associated with Triton-resistant membrane domains in haematopoietic K562 cells lacking or expressing caveolin-1. *Eur Biophys J*, **32**: 83-95.
Field, K.A., Holowka, D. and Baird, B., 1995, FcεRI-mediated recruitment of p53/56^{lyn} to detergent-resistant membrane domains accompanies cellular signaling. *Proc. Natl. Acad. Sci. USA*, **92**: 9201-9205.
Fivaz, M., Vilbois, F., Thurnheer, S., Passquali, C., Abrami, L., Bickel, P.E., Parton, R.G. and van der Goot, F.G., 2002, Differential sorting and fate of endocytosed GPI-anchored proteins. *EMBO J*, **21**: 3989-4000.
Ford, M.L., Burton, L.J., Li, H., Graham, C.H., Frobert, Y., Grassi, J., Hall, S.M. and Morris, R.J., 2002, A marked disparity between the expression of prion protein and its message by neurons of the central nervous system. *Neuroscience*, **111**: 533-551.
Fra, A.M., Masserini, M., Palestini, P., Sonnino, S. and Simons, K., 1995a, A photo-reactive derivative of ganglioside GM1 specifically cross-links VIP21-caveolin on the cell surface. *FEBS Letters*, **375**: 11-14.
Fra, A.M., Williamson, E., Simons, K. and Parton, R.G., 1995b, De novo formation of caveolae in lymphocytes by expression of VIP21-caveolin. *Proceedings of the National Academy of Sciences of the United States of America*, **92**: 8655-59.
Frank, M., van der Haar, M.E., Schaeren-Wiemers, N. and Schwab, M.E., 1998, rMAL is a glycosphingolipid-associated protein of myelin and apical membranes of epithelial cells in kidney and stomach. *J Neurosci*, **18**: 4901-4913.
Fridriksson, E.K., Shipkova, P.A., Sheets, E.D., Holowka, D., Baird, B. and McLafferty, F.W., 1999, Quantitative analysis of phospholipids in functionally important membrane domains from RBL-2H3 mast cells using tandem high-resolution mass spectrometry. *Biochemistry*, **38**: 8056-63.
Friedrichson, T. and Kurzchalia, T.V., 1998, Microdomains of GPI-anchored proteins in living cells revealed by crosslinking. *Nature*, **394**: 802-805.
Fujimoto, T., 1996, GPI-anchored proteins, glycosphingolipids, and sphingomyelin are sequestered to caveolae only after crosslinking. *J. Histochem. Cytochem* : **44**, 929-941.
Fujimoto, T., Kogo, H., Ishiguro, K., Tauchi, K. and Nomura, R.., 2001, Caveolin-2 is targeted to lipid droplets, a new "membrane domain" in the cell. *J Cell Biol*, **152**: 1079-1085.

Furtado, D., Williams, W.P., Brain, A.P.R. and Quinn, P.J., 1979, Phase separation in membranes of Anacystis nidulans grown at different temperatures. *Biochim Biophys Acta*, **555**: 352-357.

Gafencu, A., Stanescu, M., Toderici, A.M., Heltianu, C. and Simionescu, M., 1998, Protein and fatty acid composition of caveolae from apical plasmalemma of aortic endothelial cells. *Cell Tissue Res*, **293**: 101-110.

Gagescu, R., Demaurex, N., Parton, R.G., Hunziker, W., Huber, L.A. and Gruenberg, J., 2000, The recycling endosome of madin-darby canine kidney cells is a mildly acidic compartment rich in raft components. *Mol Biol Cell*, **11**: 2775-91.

Garnett, D., Barclay, A.N., Carmo, A.M. and Beyers, A.D., 1993, The association of the protein tyrosine kinases p56lck and p60fyn with the glycosyl phosphatidylinositol-anchored proteins Thy-1 and CD48 in rat thymocytes is dependent on the state of cellular activation. *Eur J Immunol*, **23**: 2540-2544.

Ge, M., Field, K.A., Aneja, R., Holowka, D., Baird, B. and Freed, J.H., 1999, Electron spin resonance characterization of liquid ordered phase of detergent-resistant membranes from RBL-2H3 cells. *Biophys J*, **77**: 925-933.

Gkantiragas, I., Brugger, B., Stuven, E., Kaloyanova, D., Li, X.Y., Lohr, K., Lottspeich, F., Wieland, F.T. and Helms, J.B., 2001, Sphingomyelin-enriched microdomains at the Golgi complex. *Mol Biol Cell*, **12**: 1819-1833.

Gliss, C., Randel, O., Casalta, H., Sackmann, E., Zorn, R. and Bayer, T., 1999, Anisotrophic motion of cholesterol in orientated DPPC bilayers studied by quasielastic neutron scattering: the liquid-ordered phase. *Biophysical J*, **77**: 331-340.

Goligorsky, M.S., Li, H., Brodsky, S. and Chen, J., 2002, Relationships between caveolae and eNOS: everything in proximity and the proximity of everything. *Am J Physiol Renal Physiol*, **283**: F1-10.

Goodman, O.B., Krupnick, J.G., Santini, F., Gurevich, V.V., Penn, R.B., Gagnon, A.W., Keen, J.H. and Benovic, J.L., 1996, β-Arrestin acts as a clathrin adaptor in endocytosis of the β_2-adrenergic receptor. *Nature*, **383**: 447-450.

Grant, C.W., Mehlhorn, I.E., Florio, E. and Barber, K.R., 1987, A long chain spin label for glycosphingolipid studies: transbilayer fatty acid interdigitation of lactosyl ceramide. *Biochim Biophys Acta*, **902**: 169-177.

Grassme, H., Jekle, A., Riehle, A., Schwarz, H., Berger, J., Sandhoff, K., Kolesnick, R. and Gulbins, E., 2001, CD95 signaling via ceramide-rich membrane rafts. *J Biol Chem*, **276**: 20589-96.

Gronberg, L., Ruan, Z.S., Bittman, R. and Slotte, J.P., 1991, Interaction of cholesterol with synthetic sphingomyelin derivatives in mixed monolayers. *Biochemistry*, **30**: 10746-54.

Guo, W., Kurze, V., Huber, T., Afdhal, N.H., Beyer, K. and Hamilton, J.A., 2002, A solid-state NMR study of phospholipid-cholesterol interactions: sphingomyelin-cholesterol binary systems. *Biophys J*, **83**: 1465-1478.

Gurr, M.I., Harwood, J.L. and Frayn, K.N., 2002, *Lipid Biochemistry*. Blackwell Science Ltd, Oxford.

Habeeb, A.F.S.A. and Hiramoto, R., 1968, Reaction of proteins with glutaraldehyde. *Arch. Biochem Biophys*, **126**: 16-26.

Haldar, K., Mohandas, N., Samuel, B.U., Harrison, T., Hiller, N.L., Akompong, T. and Cheresh, P., 2002, Protein and lipid trafficking induced in erythrocytes infected by malaria parasites. *Cell Microbiol*, **4**: 383-395.

Hale, J.E. and Schroeder, F., 1982, Asymmetric transbilayer distribution of sterol across plasma membranes determined by fluorescence quenching of dehydroergosterol. *Eur J Biochem*, **122**: 649-661.

Hancock, J.F., Cadwallader, K., Paterson, H. and Marshall, C.J., 1991, A CAAX or a CAAL motif and a second signal are sufficient for plasma membrane targeting of ras proteins. *EMBO J*, **10**: 4033-4039.

Hancock, J.F., Magee, A.I., Childs, J.E. and Marshall, C.J., 1989, All ras proteins are polyisoprenylated but only some are palmitoylated. *Cell*, **57**: 1167-1177.

Hancock, J.F., Paterson, H. and Marshall, C.J., 1990, A polybasic domain or palmitoylation is required in addition to the CAAX motif to localize p21ras to the plasma membrane. *Cell*, **63**: 133-139.

Hannun, Y.A. and Luberto, C., 2000, Ceramide in the eukaryotic stress response. *Trends Cell Biol*, **10**: 73-80.

Hansen, G.H., Immerdal, L., Thorsen, E., Niels-Christiansen, L.L., Nystrom, B.T., Demant, E.J. and Danielsen, E.M., 2001, Lipid rafts exist as stable cholesterol-independent microdomains in the brush border membrane of enterocytes. *J Biol Chem*, **276**: 32338-44.

Harder, T., Scheiffele, P., Verkade, P. and Simons, K., 1998, Lipid domain structure of the plasma membrane revealed by patching of membrane components. *J. Cell Biol*, **141**: 929-942.

Harris, J.S., Epps, D.E., Davio, S.R. and Kezdy, F.J., 1995, Evidence for transbilayer, tail-to-tail cholesterol dimers in dipalmitoylglycerophosphocholine liposomes. *Biochemistry*, **34**: 3851-57.

Hawtin, S.R., Tobin, A.B., Patel, S. and Wheatley, M., 2001, Palmitoylation of the vasopressin V1a receptor reveals different conformational requirements for signaling, agonist-induced receptor phosphorylation, and sequestration. *J Biol Chem*, **276**: 38139-46.

Hayashi, M.K. and Haga, T., 1997, Palmitoylation of muscarinic acetylcholine receptor m2 subtypes: reduction in their ability to activate G proteins by mutation of a putative palmitoylation site, cysteine 457, in the carboxyl-terminal tail. *Arch Biochem Biophys*, **340**: 376-382.

He, Q. and Meiri, K.F., 2002, Isolation and characterization of detergent-resistant microdomains responsive to NCAM-mediated signaling from growth cones. *Mol Cell Neurosci*, **19**: 18-31.

Heerklotz, H., 2002, Triton promotes domain formation in lipid raft mixtures. *Biophys J*, **83**: 2693-2701.

Heimberg, T., Würtz, U. and Marsh, D., 1992, Binary phase diagram of hydrated dimyristoylglycerol-dimyristoylphosphatidylcholine mixtures. *Biophys. J.*, **63**: 1369-1378.

Herman, B., 1989, Resonance energy transfer microscopy. *Methods Cell Biol*, **30**: 219-243.

Hollrigel, G.S., Morris, R.J. and Soltensz, I., 1998, Enhanced inhibitory charge transfer during bursts of IPSCs in dentate granule cells in mice with regionally inhibited LTP. *Proc. Roy Soc Lond B*, **265**: 63-69.

Holopainen, J.M., Subramanian, M. and Kinnunen, P.K., 1998, Sphingomyelinase induces lipid microdomain formation in a fluid phosphatidylcholine/sphingomyelin membrane. *Biochemistry*, **37**: 17562-70.

Holowka, D., Sheets, E.D. and Baird, B., 2000, Interactions between FcεRI and lipid raft components are regulated by the actin cytoskeleton. *J Cell Sci*, **113**: 1009-19.

Holthuis, J.C., Pomorski, T., Raggers, R.J., Sprong, H. and Van Meer, G., 2001, The organizing potential of sphingolipids in intracellular membrane transport. *Physiol Rev*, **81**: 1689-1723.

Hsueh, Y.W., Giles, R., Kitson, N. and Thewalt, J., 2002, The effect of ceramide on phosphatidylcholine membranes: a deuterium NMR study. *Biophys J*, **82**: 3089-95.

Hueber, A.O., Bernard, A.M., Battari, C.L., Marguet, D., Massol, P., Foa, C., Brun, N., Garcia, S., Stewart, C., Pierres, M. and He, H.T., 1997, Thymocytes in Thy-1 −/− mice show augmented TCR signaling and impaired differentiation. *Curr Biol*, **7**: 705-708.

Hunt, A.N., Clark, G.T., Neale, J.R. and Postle, A.D., 2002, A comparison of the molecular specificities of whole cell and endonuclear phosphatidylcholine synthesis. *FEBS Lett*, **530**: 89-93.

Huttner, W.B. and Zimmerberg, J., 2001, Implications of lipid microdomains for membrane curvature, budding and fission. *Curr Opin Cell Biol*, **13**, 478-484.

Igbavboa, U., Avdulov, N.A., Chochina, S.V. and Wood, W.G., 1997, Transbilayer distribution of cholesterol is modified in brain synaptic plasma membranes of knockout mice deficient in the low-density lipoprotein receptor, apolipoprotein E, or both proteins. *J Neurochem*, **69**: 1661-1667.

Ikonen, E., 2001, Roles of lipid rafts in membrane transport. *Curr Opin Cell Biol*, **13**: 470-477.

Ikonen, E. and Parton, R.G., 2000, Caveolins and cellular cholesterol balance. *Traffic*, **1**: 212-217.

Inglese, J., Koch, W.J., Caron, M.G. and Lefkowitz, R.J., 1992, Isoprenylation in regulation of signal transduction by G-protein- coupled receptor kinases. *Nature*, **359**: 147-150.

Ipsen, J.H., Karlstrom, G., Mouritsen, O.G., Wennerstrom, H. and Zuckermann, M.J., 1987, Phase equilibria in the phosphatidylcholine-cholesterol system. *Biochim Biophys Acta*, **905**: 162-172.

Itoh, K., Sakakibara, M., Yamasaki, S., Takeuchi, A., Arase, H., Miyazaki, M., Nakajima, N., Okada, M. and Saito, T., 2002, Cutting edge: negative regulation of immune synapse formation by anchoring lipid raft to cytoskeleton through Cbp-EBP50-ERM assembly. *J Immunol*, **168**: 541-544.

Jaffrezou, J.P., Laurent, G. and Levade, T., 2002, Ceramide in regulation of apoptosis. Implication in multitoxicant resistance. *Subcell Biochem*, **36**: 269-284.

Janes, P.W., Ley, S.C. and Magee, A.I., 1999, Aggregation of lipid rafts accompanies signaling via the T cell antigen receptor. *J Cell Biol*, **147**: 447-461.

Johnson, M.E., Berk, D.A., Blankschtein, D., Golan, D.E., Jain, R.K. and Langer, R.S., 1996, Lateral diffusion of small compounds in human stratum corneum and model lipid bilayer systems. *Biophys J*, **71**: 2656-2668.

Kabouridis, P.S., Magee, A.I. and Ley, S.C., 1997, S-acylation of LCK protein tyrosine kinase is essential for its signalling function in T lymphocytes. *EMBO J*, **16**: 4983-4998.

Kan, C.C., Ruan, Z.S. and Bittman, R., 1991, Interaction of cholesterol with sphingomyelin in bilayer membranes: evidence that the hydroxy group of sphingomyelin does not modulate the rate of cholesterol exchange between vesicles. *Biochemistry*, **30**: 7759-7766.

Kawabuchi, M., Satomi, Y., Takao, T., Shimonishi, Y., Nada, S., Nagai, K., Tarakhovsky, A. and Okada, M., 2000, Transmembrane phosphoprotein Cbp regulates the activities of Src-family tyrosine kinases. *Nature*, **404**: 999-1003.

Keller, P. and Simons, K., 1998, Cholesterol is required for surface transport of influenza virus hemagglutinin. *J Cell Biol*, **140**: 1357-1367.

Kier, A.B., Sweet, W.D., Cowlen, M.S. and Schroeder, F., 1986, Regulation of transbilayer distribution of a fluorescent sterol in tumor cell plasma membranes. *Biochim Biophys Acta*, **861**: 287-301.

Kikkawa, U. and Nishizuka, Y., 1986, The role of protein kinase C in transmembrane signalling. *Annu Rev Cell Biol*, **2**: 149-178.

Kolesnick, R.N., Goni, F.M. and Alonso, A., 2000, Compartmentalization of ceramide signaling: physical foundations and biological effects. *J Cell Physiol*, **184**: 285-300.

Koynova, R. and Caffrey, M., 1995, Phases and phase transitions of the sphingolipids. *Biochim Biophys Acta*, **1255**: 213-236.

Krauss, K. and Altevogt, P., 1999, Integrin leukocyte function-associated antigen-1-mediated cell binding can be activated by clustering of membrane rafts. *J Biol Chem*, **274**: 36921-27.

Kroesen, B.J., Pettus, B., Luberto, C., Busman, M., Sietsma, H., de Leij, L. and Hannun, Y.A., 2001, Induction of apoptosis through B-cell receptor cross-linking occurs via de novo generated C16-ceramide and involves mitochondria. *J Biol Chem*, **276**: 13606-14.

Kurzchalia, T.V., Dupree, P., Parton, R.G., Kellner, R., Virta, H., Lehnert, M. and Simons, K., 1992, VIP21, a 21-kD membrane protein is an integral component of trans-Golgi-network-derived transport vesicles. *J. Cell Biol.*, **118**: 1003-1014.

Lacalle, R.A., Mira, E., Gomez-Mouton, C., Jimenez-Baranda, S., Martinez-A., C. and Manes, S., 2002, Specific SHP-2 partitioning in raft domains triggers integrin-mediated signaling via Rho activation. *J Cell Biol*, **157**: 277-289.

Lalitha, S., Kumar, A.S., Stine, K.J. and Covey, D.F., 2001, Enantiospecificity of sterol-lipid interactions: first evidence that the absolute configuration of cholesterol affects the physical properties of cholesterol-sphingomyelin membranes. *Chem. Commun.*, **2001**: 1192-1193.

Lamaze, C., Dujeancourt, A., Baba, T., Lo, C.G., Benmerah, A. and Dautry-Varsat, A., 2001, Interleukin 2 receptors and detergent-resistant membrane domains define a clathrin-independent endocytic pathway. *Mol Cell*, **7**: 661-671.

Lang, D.M., Lommel, S., Jung, M., Ankerhold, R., Petrausch, B., Laessing, U., Wiechers, M.F., Plattner, H. and Stuermer, C.A., 1998, Identification of reggie-1 and reggie-2 as plasmamembrane-associated proteins which cocluster with activated GPI-anchored cell adhesion molecules in non-caveolar micropatches in neurons. *J Neurobiol*, **37**: 502-523.

Lauer, S., VanWye, J., Harrison, T., McManus, H., Samuel, B.U., Hiller, N.L., Mohandas, N. and Haldar, K., 2000, Vacuolar uptake of host components, and a role for cholesterol and sphingomyelin in malarial infection. *EMBO J*, **19**: 3556-3564.

Lavie, Y., Cao, H., Volner, A., Lucci, A., Han, T.Y., Geffen, V., Giuliano, A.E. and Cabot, M.C., 1997, Agents that reverse multidrug resistance, tamoxifen, verapamil, and cyclosporin A, block glycosphingolipid metabolism by inhibiting ceramide glycosylation in human cancer cells. *J Biol Chem*, **272**, :1682-7.

Le Moyec, L., Legrand, O., Larue, V., Kawakami, M., Marie, J.P., Calvo, F., Hantz, E. and Taillandier, E., 2000, Magnetic resonance spectroscopy of cellular lipid extracts from sensitive, resistant and reverting K562 cells and flow cytometry for investigating the P-glycoprotein function in resistance reversion. *NMR Biomed.*, **13**: 92-101.

Le, P.U., Guay, G., Altschuler, Y. and Nabi, I.R., 2002, Caveolin-1 is a negative regulator of caveolae-mediated endocytosis to the endoplasmic reticulum. *J Biol Chem*, **277**: 3371-3379.

Lee, H., Woodman, S.E., Engelman, J.A., Volonte, D., Galbiati, F., Kaufman, H.L., Lublin, D.M. and Lisanti, M.P., 2001, Palmitoylation of caveolin-1 at a single site, Cys-156, controls its coupling to the c-Src tyrosine kinase: targeting of dually acylated molecules (GPI-linked, transmembrane, or cytoplasmic) to caveolae effectively uncouples c-Src and caveolin-1 (TYR-14). *J Biol Chem*, **276**: 35150-58.

Lee, K.H., Holdorf, A.D., Dustin, M.L., Chan, A.C., Allen, P.M. and Shaw, A.S., 2002, T cell receptor signaling precedes immunological synapse formation. *Science*, **295**: 1539-1542.

Lefevre, T. and Picquart, M., 1996, Vitamin E-phospholipid interactions in model multilayer membranes: a spectroscopic study. *Biospectroscopy*, **2**, 391-403.

Levade, T. and Jaffrezou, J.P., 1999, Signalling sphingomyelinases: which, where, how and why? *Biochim Biophys Acta*, **1438**: 1-17.

Leventis, R. and Silvius, J.R., 2001, Use of cyclodextrins to monitor transbilayer movement and differential lipid affinities of cholesterol. *Biophys J*, **81**: 2257-67.

Li, X.M., Momsen, M.M., Smaby, J.M., Brockman, H.L. and Brown, R.E., 2001, Cholesterol decreases the interfacial elasticity and detergent solubility of sphingomyelins. *Biochemistry*, **40**: 5954-63.

Liang, X., Lu, Y., Neubert, T.A. and Resh, M.D., 2002, Mass spectrometric analysis of GAP-43/Neuromodulin reveals the presence of a variety of fatty acylated species. *J Biol Chem*, **277**: 33032-40.

Liang, X., Nazarian, A., Erdjument-Bromage, H., Bornmann, W., Tempst, P. and Resh, M.D., 2001, Heterogeneous fatty acylation of Src family kinases with polyunsaturated fatty acids regulates raft localization and signal transduction. *J Biol Chem*, **276**: 30987-94.

Lindwasser, O.W. and Resh, M.D., 2001, Multimerization of human immunodeficiency virus type 1 Gag promotes its localization to barges, raft-like membrane microdomains. *J Virol*, **75**: 7913-7924.

Liu, Y.Y., Han, T.Y., Giuliano, A.E. and Cabot, M.C., 1999, Expression of glucosylceramide synthase, converting ceramide to glucosylceramide, confers adriamycin resistance in human breast cancer cells. *J Biol Chem*, **274**: 1140-1146.

Loisel, T.P., Ansanay, H., Adam, L., Marullo, S., Seifert, R., Lagace, M. and Bouvier, M., 1999, Activation of the β_2-adrenergic receptor-Gαs complex leads to rapid depalmitoylation and inhibition of repalmitoylation of both the receptor and Gαs. *J Biol Chem*, **274**: 31014-19.

London, E., 2002, Insights into lipid raft structure and formation from experiments in model membranes. *Curr Opin Struct Biol*, **12**: 480-486.

London, E. and Brown, D.A., 2000, Insolubility of lipids in triton X-100: physical origin and relationship to sphingolipid/cholesterol membrane domains, (rafts). *Biochim Biophys Acta*, **1508**: 182-195.

Lopez-Garcia, F., Villalain, J., Gomez-Fernandez, J.C. and Quinn, P.J., 1994, The phase behavior of mixed aqueous dispersions of dipalmitoyl derivatives of phosphatidylcholine and diacylglycerol. *Biophys J*, **66**: 1991-2004.

Luhrs, C.A. and Slomiany, B.L., 1989, A human membrane-associated folate binding protein is anchored by a glycosyl-phosphatidylinositol tail. *J Biol Chem*, **264**: 21446-49.

Luria, A., Vegelyte-Avery, V., Stith, B., Tsvetkova, N.M., Wolkers, W.F., Crowe, J.H., Tablin, F. and Nuccitelli, R., 2002, Detergent-free domain isolated from Xenopus egg plasma membrane with properties similar to those of detergent-resistant membranes. *Biochemistry*, **41**: 13189-97.

Madore, N., Smith, K.L., Graham, C.H., Jen, A., Brady, K., Hall, S. and Morris, R., 1999, Functionally different GPI proteins are organised in different domains on the neuronal surface. *EMBO J*, **18**: 6917-6926.

Mairhofer, M., Steiner, M., Mosgoeller, W., Prohaska, R. and Salzer, U., 2002, Stomatin is a major lipid-raft component of platelet alpha granules. *Blood*, **100**: 897-904.

Manes, S., del Real, G., Lacalle, R.A., Lucas, P., Gomez-Mouton, C., Sanchez-Palomino, S., Delgado, R., Alcami, J., Mira, E. and Martinez-A.C., 2000, Membrane raft microdomains mediate lateral assemblies required for HIV- 1 infection. *EMBO Rep*, **1**: 190-196.

Mannock, D.A., Brain, A.P.R. and Williams, W.P., 1985, Phase behaviour of the membrane lipids of the thermophilic blue-green alga Anacystis nidulans. *Biochim. Biophys. Acta*, **821**: 153-164.

Marchand, S., Devillers-Thiery, A., Pons, S., Changeux, J.P. and Cartaud, J., 2002, Rapsyn escorts the nicotinic acetylcholine receptor along the exocytic pathway via association with lipid rafts. *J Neurosci*, **22**: 8891-8901.

Martelli, A.M., Manzoli, L., Faenza, I., Bortul, R., Billi, A. and Cocco, L., 2002, Nuclear inositol lipid signaling and its potential involvement in malignant transformation. *Biochim Biophys Acta*, **1603**: 11-17.

Martelli, A.M., Tabellini, G., Borgatti, P., Bortul, R., Capitani, S. and Neri, L.M., 2003, Nuclear lipids: New functions for old molecules? *J Cell Biochem*, **88**: 455-461.

Martens, J.R., Sakamoto, N., Sullivan, S.A., Grobaski, T.D. and Tamkun, M.M., 2001, Isoform-specific localization of voltage-gated K+ channels to distinct lipid raft populations. Targeting of Kv1.5 to caveolae. *J Biol Chem*, **276**: 8409-14.

Martin-Belmonte, F., Arvan, P. and Alonso, M.A., 2001, MAL mediates apical transport of secretory proteins in polarized epithelial Madin-Darby canine kidney cells. *J Biol Chem*, **276**: 49337-42.

Martin-Belmonte, F., Puertollano, R., Millan, J. and Alonso, M.A., 2000, The MAL proteolipid is necessary for the overall apical delivery of membrane proteins in the polarized epithelial Madin-Darby canine kidney and fischer rat thyroid cell lines. *Mol Biol Cell*, **11**: 2033-2045.

Massey, J.B., 2001, Interaction of ceramides with phosphatidylcholine, sphingomyelin and sphingomyelin/cholesterol bilayers. *Biochim Biophys Acta*, **1510**: 167-184.

Mathias, S., Pena, L.A. and Kolesnick, R.N., 1998, Signal transduction of stress via ceramide. *Biochem J*, **335**: 465-480.

Mattjus, P., Bittman, R., Vilcheze, C. and Slotte, J.P., 1995, Lateral domain formation in cholesterol/phospholipid monolayers as affected by the sterol side chain conformation. *Biochim Biophys Acta*, **1240**: 237-247.

Mattjus, P. and Slotte, J.P., 1996, Does cholesterol discriminate between sphingomyelin and phosphatidylcholine in mixed monolayers containing both phospholipids? *Chem Phys Lipids*, **81**: 69-80.

Maulik, P.R., Atkinson, D. and Shipley, G.G., 1986, X-ray scattering of vesicles of N-acyl sphingomyelins. Determination of bilayer thickness. *Biophys J*, **50**: 1071-1077.

Mayor, S., Rothberg, K.G. and Maxfield, F.R., 1994, Sequestration of GPI-anchored proteins in caveolae triggered by cross-linking. *Science*, **264**: 1948-1951.

McConville, M.J. and Ferguson, M.A.J., 1993, The structure, biosynthesis and function of glycosylated phosphatidylinositols in the parasitic protozoa and higher eukaryotes. *Biochem. J*, **294**: 305-324.

McIntosh, T.J., Simon, S.A., Needham, D. and Huang, C.H., 1992a, Interbilayer interactions between sphingomyelin and sphingomyelin/cholesterol bilayers. *Biochemistry*, **31**: 2020-2024.

McIntosh, T.J., Simon, S.A., Needham, D. and Huang, C.H., 1992b, Structure and cohesive properties of sphingomyelin/cholesterol bilayers. *Biochemistry*, **31**: 2012-2020.

McMullen, T.P., Lewis, R.N. and McElhaney, R.N., 1999, Calorimetric and spectroscopic studies of the effects of cholesterol on the thermotropic phase behavior and organization of a homologous series of linear saturated phosphatidylethanolamine bilayers. *Biochim Biophys Acta*, **1416**: 119-134.

Melkonian, K.A., Ostermeyer, A.G., Chen, J.Z., Roth, M.G. and Brown, D.A., 1999, Role of lipid modifications in targeting proteins to detergent- resistant membrane rafts. Many raft proteins are acylated, while few are prenylated. *J Biol Chem*, **274**: 3910-17.

Meyer, H.W., Westermann, M., Stumpf, M., Richter, W., Ulrich, A.S. and Hoischen, C., 1998, Minimal radius of curvature of lipid bilayers in the gel phase state corresponds to the dimension of biomembrane structures "caveolae". *J Struct Biol*, **124**: 77-87.

Miao, L., Nielsen, M., Thewalt, J., Ipsen, J.H., Bloom, M., Zuckermann, M.J. and Mouritsen, O.G., 2002, From lanosterol to cholesterol: structural evolution and differential effects on lipid bilayers. *Biophys J*, **82**: 1429-1444.

Milhiet, P.E., Giocondi, M.C. and Le Grimellec, C., 2002, Cholesterol is not crucial for the existence of microdomains in kidney brush-border membrane models. *J Biol Chem*, **277**: 875-878.

Millan, J., Puertollano, R., Fan, L. and Alonso, M.A., 1997, Caveolin and MAL, two protein components of internal detergent-insoluble membranes, are in distinct lipid microenvironments in MDCK cells. *Biochem Biophys Res Commun*, **233**: 707-712.

Möbius, W., Ohno-Iwashita, Y., van Donselaar, E.G., Oorschot, V.M., Shimada, Y., Fujimoto, T., Heijnen, H.F., Geuze, H.J. and Slot, J.W., 2002, Immunoelectron microscopic localization of cholesterol using biotinylated and non-cytolytic perfringolysin O. *J Histochem Cytochem*, **50**: 43-55.

Möbius, W., van Donselaar, E., Ohno-Iwashita, Y., Shimada, Y., Heijnen, H.F.G., Slot, J.W. and Geuze, H.J., 2003, Recycling compartments and the internal vesicles of multivesicular bodies harbor most of the cholesterol found in the endocytic pathway. *Traffic*, **4**: 222-31.

Moffett, S., Brown, D.A. and Linder, M.E., 2000, Lipid-dependent targeting of G proteins into rafts. *J Biol Chem*, **275**: 2191-2198.

Moffett, S., Rousseau, G., Lagace, M. and Bouvier, M., 2001, The palmitoylation state of the β_2-adrenergic receptor regulates the synergistic action of cyclic AMP-dependent protein kinase and beta-adrenergic receptor kinase involved in its phosphorylation and desensitization. *J Neurochem*, **76**: 269-279.

Mombelli, E., Morris, R.J., Taylor, W. and Fraternali, F., 2003, Hydrogen bonding propensities of sphingomyelin in solution and in a bilayer assembly: a molecular dynamics study. *Biophys J*, **84**: 1507-1517.

Monier, S., Dietzen, D.J., Hastings, W.R., Lublin, D.M. and Kurzchalia, T.V., 1996, Oligomerization of VIP21-caveolin in vitro is stabilized by long chain fatty acylation or cholesterol. *FEBS Lett*, **388**: 143-149.

Monks, C.R.F., Freiberg, B.A., Kupfer, H., Sciaky, N. and Kupfer, A., 1998, Three-dimensional segregation of supramolecular activation clusters in T cells. *Nature*, **395**: 82-86.

Montesano, R., 1979, Inhomogeneous distribution of filipin-sterol complexes in smooth muscle cell plasma membrane. *Nature*, **280**: 328-329.

Moore, R.H., Sadovnikoff, N., Hoffenberg, S., Liu, S., Woodford, P., Angelides, K., Trial, J.A., Carsrud, N.D., Dickey, B.F. and Knoll, B.J., 1995, Ligand-stimulated beta 2-adrenergic receptor internalization via the constitutive endocytic pathway into rab5-containing endosomes. *J Cell Sci*, **108**: 2983-2991.

Mori, T., Takai, Y., Yu, B., Takahashi, J., Nishizuka, Y. and Fujikura, T., 1982, Specificity of the fatty acyl moieties of diacylglycerol for the activation of calcium-activated, phospholipid-dependent protein kinase. *J Biochem (Tokyo)*, **91**: 427-431.

Morris, R., 1992, Thy-1, the enigmatic extrovert on the neuronal surface. *BioEssays*, **14**: 715-722.

Morris, R.J. and Barber, P.C., 1983, Fixation of Thy-1 in nervous tissue for immunohistochemistry: a quantitative assessment of the effect of different fixation conditions upon retention of antigenicity and the cross-linking of Thy-1. *J Histochem Cytochem*, **31**: 263-274.

Morrow, I.C., Rea, S., Martin, S., Prior, I.A., Prohaska, R., Hancock, J.F., James, D.E. and Parton, R.G., 2002, Flotillin-1/Reggie-2 traffics to surface raft domains via a Golgi-independent pathway; Identification of a novel membrane targeting domain and a role for palmitoylation. *J Biol Chem*, **277**: 48834-41.

Morrow, M.R., Singh, D., Lu, D. and Grant, C.W.M., 1995, Glycosphingolipid fatty acid arrangement in phospholipid bilayers: cholesterol effects. *Biophys J*, **68**: 179-186.

Mukherjee, S. and Chattopadhyay, A., 1996, Membrane organization at low cholesterol concentrations: a study using 7-nitrobenz-2-oxa-1,3-diazol-4-yl-labeled cholesterol. *Biochemistry*, **35**: 1311-1322.

Mukherjee, S. and Maxfield, F.R., 2000, Role of membrane organization and membrane domains in endocytic lipid trafficking. *Traffic*, **1**: 203-211.

Mukherjee, S., Zha, X., Tabas, I. and Maxfield, F.R., 1998, Cholesterol distribution in living cells: fluorescence imaging using dehydroergosterol as a fluorescent cholesterol analog. *Biophys J*, **75**: 1915-1925.

Muller, G., 2002, Dynamics of plasma membrane microdomains and cross-talk to the insulin signalling cascade. *FEBS Lett*, **531**: 81-87.

Munro, S., 1995, An investigation of the role of transmembrane domains in Golgi protein retention. *EMBO J*, **14**: 4695-4704.

Murata, M., Peranen, J., Schreiner, R., Wieland, F., Kurzchalia, T.V. and Simons, K., 1995, VIP21/caveolin is a cholesterol-binding protein. *Proc. Natl. Acad. Sci. USA*, **92**: 10339-43.

Murray, D., Hermida-Matsumoto, L., Buser, C.A., Tsang, J., Sigal, C.T., Ben-Tal, N., Honig, B., Resh, M.D. and McLaughlin, S., 1998, Electrostatics and the membrane association of Src: theory and experiment. *Biochemistry*, **37**: 2145-2159.

Nabet, A., Boggs, J.M. and Pezolet, M., 1996, Study by infrared spectroscopy of the interdigitation of C26:0 cerebroside sulfate into phosphatidylcholine bilayers. *Biochemistry*, **35**: 6674-6683.

Nagle, J.F. and Tristram-Nagle, S., 2000, Structure of lipid bilayers. *Biochim Biophys Acta*, **1469**: 159-195.

Nezil, F.A. and Bloom, M., 1992, Combined influence of cholesterol and synthetic amphiphillic peptides upon bilayer thickness in model membranes. *Biophys J*, **61**: 1176-1183.

Nguyen, D.H. and Hildreth, J.E., 2000, Evidence for budding of human immunodeficiency virus type 1 selectively from glycolipid-enriched membrane lipid rafts. *J Virol*, **74**: 3264-72.

Nieva, J.L., Bron, R., Corver, J. and Wilschut, J., 1994, Membrane fusion of Semliki Forest virus requires sphingolipids in the target membrane. *EMBO J*, **13**: 2797-2804.

Nishizuka, Y., 1992, Intracellular signaling by hydrolysis of phospholipids and activation of protein kinase C. *Science*, **258**: 607-614.

Nolan, D.P., Jackson, D.G., Biggs, M.J., Brabazon, E.D., Pays, A., Van Laethem, F., Paturiaux-Hanocq, F., Elliot, J.F., Voorheis, H.P. and Pays, E., 2000, Characterization of a novel alanine-rich protein located in surface microdomains in Trypanosoma brucei. *J Biol Chem*, **275**: 4072-80.

O'Dowd, B.F., Hnatowich, M., Caron, M.G., Lefkowitz, R.J. and Bouvier, M., 1989, Palmitoylation of the human beta 2-adrenergic receptor. Mutation of Cys341 in the carboxyl tail leads to an uncoupled nonpalmitoylated form of the receptor. *J Biol Chem*, **264**: 7564-7569.

Oh, P. and Schnitzer, J.E., 2001, Segregation of heterotrimeric G proteins in cell surface microdomains. G(q) binds caveolin to concentrate in caveolae, whereas G(i) and G(s) target lipid rafts by default. *Mol Biol Cell*, **12**: 685-698.

Ohba, Y., Morimitsu, Y. and Watarai, A., 1979, Reaction of formaldehyde with calf-thymus nucleohistone. *Eur J Biochem.*, **100**: 285-293.

Okamoto, T., Schlegel, A., Scherer, P.E. and Lisanti, M.P., 1998, Caveolins, a family of scaffolding proteins for organizing "preassembled signaling complexes" at the plasma membrane. *J Biol Chem*, **273**: 5419-5422.

Oradd, G., Lindblom, G., Fontell, K. and Ljusberg-Wahren, H., 1995, Phase diagram of soybean phosphatidylcholine-diacylglycerol-water studied by x-ray diffraction and 31P- and pulsed field gradient 1H-NMR: evidence for reversed micelles in the cubic phase. *Biophys J*, **68**: 1856-1863.

Ortiz, A., Aranda, F.J. and Gomez-Fernandez, J.C., 1987, A differential scanning calorimetry study of the interaction of alpha-tocopherol with mixtures of phospholipids. *Biochim Biophys Acta*, **898**: 214-222.

Ostermeyer, A.G., Beckrich, B.T., Ivarson, K.A., Grove, K.E. and Brown, D.A., 1999, Glycosphingolipids are not essential for formation of detergent- resistant membrane rafts in melanoma cells. Methyl-beta-cyclodextrin does not affect cell surface transport of a GPI-anchored protein. *J Biol Chem*, **274**: 34459-66.

Palestini, P., Allietta, M., Sonnino, S., Tettamanti, G., Thompson, T.E. and Tillack, T.W., 1995, Gel phase preference of ganglioside GM1 at low concentration in two-component, two-phase phosphatidylcholine bilayers depends upon the ceramide moiety. *Biochim Biophys Acta*, **1235**: 221-230.

Paratcha, G., Ledda, F., Baars, L., Coulpier, M., Besset, V., Anders, J., Scott, R. and Ibanez, C., 2001, Released GFRα1 potentiates downstream signaling, neuronal survival, and differentiation via a novel recruitment of c-ret to lipid rafts. *Neuron*, **29**: 171-184.

Park, D.S., Woodman, S.E., Schubert, W., Cohen, A.W., Frank, P.G., Chandra, M., Shirani, J., Razani, B., Tang, B., Jelicks, L.A., Factor, S.M., Weiss, L.M., Tanowitz, H.B. and Lisanti, M.P., 2002, Caveolin-1/3 double-knockout mice are viable, but lack both muscle and non-muscle caveolae, and develop a severe cardiomyopathic phenotype. *Am J Pathol*, **160**: 2207-2217.

Parkin, E.T., Turner, A.J. and Hooper, N.M., 2001, Differential effects of glycosphingolipids on the detergent-insolubility of the glycosylphosphatidylinositol-anchored membrane dipeptidase. *Biochem J*, **358**: 209-216.

Parton, R.G., 2003, Timeline: Caveolae – from ultrastructure to molecular mechanisms. *Nat Rev Mol Cell Biol*, **4**: 162-167.

Pasenkiewicz-Gierula, M., Takaoka, Y., Miyagawa, H., Kitamura, K. and Kusumi, A., 1997, Hydrogen bonding of water to phosphatidylcholine in the membrane as studied by a molecular dynamics simulation: location, geometry, and lipid-lipid bridging vi hydrogen-bonded water. *J Phys Chem*, **101**: 3677-3691.

Patra, S.K., Alonso, A. and Goni, F.M., 1998, Detergent solubilisation of phospholipid bilayers in the gel state: the role of polar and hydrophobic forces. *Biochim Biophys Acta*, **1373**: 112-118.

Peiro, S., Comella, J.X., Enrich, C., Martin-Zanca, D. and Rocamora, N., 2000, PC12 cells have caveolae that contain TrkA. Caveolae-disrupting drugs inhibit nerve growth factor-induced, but not epidermal growth factor-induced, MAPK phosphorylation. *J Biol Chem*, **275**: 37846-52.

Peles, E., Nativ, M., Lustig, M., Grumet, M., Schilling, J., Martinez, R., Plowman, G.D. and Schlessinger, J., 1997, Identification of a novel contactin-associated transmembrane receptor with multiple domains implicated in protein-protein interactions. *EMBO J*, **16**: 978-988.

Pelkmans, L. and Helenius, A., 2002, Endocytosis via caveolae. *Traffic*, **3**, 311-320.

Pereira-Leal, J.B., Hume, A.N. and Seabra, M.C., 2001, Prenylation of Rab GTPases: molecular mechanisms and involvement in genetic disease. *FEBS Lett*, **498**: 197-200.

Perez, A.S. and Bredt, D.S., 1998, The N-terminal PDZ-containing region of postsynaptic density-95 mediates association with caveolar-like lipid domains. *Neurosci Lett*, **258**: 121-123.

Peters, M.W., Barber, K.R. and Grant, C.W., 1984a, Lateral distribution of gangliosides in bilayer membranes: lipid and ionic effects. *J Neurosci Res*, **12**: 343-353.

Peters, M.W., Mehlhorn, I.E., Barber, K.R. and Grant, C.W., 1984b, Evidence of a distribution difference between two gangliosides in bilayer membranes. *Biochim Biophys Acta*, **778**: 419-428.

Pierini, L.M. and Maxfield, F.R., 2001, Flotillas of lipid rafts fore and aft. *Proc Natl Acad Sci U S A*, **98**: 9471-3.

Pike, L.J., Han, X., Chung, K.N. and Gross, R.W., 2002, Lipid rafts are enriched in arachidonic acid and plasmenylethanolamine and their composition is independent of caveolin-1 expression: a quantitative electrospray ionization/mass spectrometric analysis. *Biochemistry*, **41**: 2075-2088.

Pilch, P.F., Shia, M.A., Benson, R.J. and Fine, R.E., 1983, Coated vesicles participate in the receptor-mediated endocytosis of insulin. *J Cell Biol*, **96**: 133-138.

Pitto, M., Brunner, J., Ferraretto, A., Ravasi, D., Palestini, P. and Masserini, M., 2000, Use of a photoactivable GM1 ganglioside analogue to assess lipid distribution in caveolae bilayer. *Glycoconj J*, **17**: 215-222.

Porter, J.A., Young, K.E. and Beachy, P.A., 1996, Cholesterol modification of Hedgehog signaling porteins in animal development. *Science*, **274**: 255-259.

Porter, T.D., 2003, Supernatant protein factor and tocopherol-associated protein: an unexpected link between cholesterol synthesis and vitamin E (review). *J Nutr Biochem*, **14**: 3-6.

Pralle, A., Keller, P., Florin, E.L., Simons, K. and Horber, J.K., 2000, Sphingolipid-cholesterol rafts diffuse as small entities in the plasma membrane of mammalian cells. *J Cell Biol*, **148**: 997-1008.

Prats, M., Tocanne, J.F. and Teissie, J., 1987, Lateral proton conduction at a lipid/water interface. Effect of lipid nature and ionic content of the aqueous phase. *Eur J Biochem*, **162**: 379-385.

Preiss, J., Loomis, C.R., Bishop, W.R., Stein, R., Niedel, J.E. and Bell, R.M., 1986, Quantitative measurement of sn-1,2-diacylglycerols present in platelets, hepatocytes, and ras- and sis-transformed normal rat kidney cells. *J Biol Chem*, **261**: 8597-8600.

Premont, R.T., Macrae, A.D., Stoffel, R.H., Chung, N., Pitcher, J.A., Ambrose, C., Inglese, J., MacDonald, M.E. and Lefkowitz, R.J., 1996, Characterization of the G protein-coupled receptor kinase GRK4. Identification of four splice variants. *J Biol Chem*, **271**: 6403-6410.

Prior, I.A. and Hancock, J.F., 2001, Compartmentalization of Ras proteins. *J Cell Sci*, **114**: 1603-1608.

Prior, I.A., Harding, A., Yan, J., Sluimer, J., Parton, R.G. and Hancock, J.F., 2001, GTP-dependent segregation of H-ras from lipid rafts is required for biological activity. *Nat Cell Biol*, **3**: 368-375.

Prior, I.A., Muncke, C., Parton, R.G. and Hancock, J.F., 2003, Direct visualization of Ras proteins in spatially distinct cell surface microdomains. *J. Cell Biol.*, **160**: 165-170.

Puertollano, R. and Alonso, M.A., 1999a, MAL, an integral element of the apical sorting machinery, is an itinerant protein that cycles between the trans-Golgi network and the plasma membrane. *Mol Biol Cell*, **10**: 3435-3447.

Puertollano, R. and Alonso, M.A., 1999b, Targeting of MAL, a putative element of the apical sorting machinery, to glycolipid-enriched membranes requires a pre-golgi sorting event. *Biochem Biophys Res Commun*, **254**: 689-692.

Puertollano, R., Li, S., Lisanti, M.P. and Alonso, M.A., 1997, Recombinant expression of the MAL proteolipid, a component of glycolipid-enriched membrane microdomains, induces the formation of vesicular structures in insect cells. *J Biol Chem*, **272**: 18311-15.

Puertollano, R., Martin-Belmonte, F., Millan, J., de Marco, M.C., Albar, J.P., Kremer, L. and Alonso, M.A., 1999a, The MAL proteolipid is necessary for normal apical transport and accurate sorting of the influenza virus hemagglutinin in Madin-Darby canine kidney cells. *J Cell Biol*, **145**: 141-151.

Puertollano, R., Menendez, M. and Alonso, M.A., 1999b, Incorporation of MAL, an integral protein element of the machinery for the glycolipid and cholesterol-mediated apical pathway of transport, into artificial membranes requires neither of these lipid species. *Biochem Biophys Res Commun*, **266**: 330-333.

Pyeta, P.S., Holowka, D. and Baird, B., 2001, Cross-correlation analysis of inner-leaflet-anchored green fluorescent protein co-redistributed with IgE receptors and outer leaflet lipid raft components. *Biophys J*, **80**: 2120-32.

Qanbar, R. and Bouvier, M., 2003, Role of palmitoylation/depalmitoylation reactions in G-protein-coupled receptor function. *Pharmacol Ther*, **97**: 1-33.

Quinn, P.J., 1985, A lipid-phase separation model of low-temperature damage to biological membranes. *Cryobiology*, **22**: 28-46.

Quinn, P.J., 1995, Characterisation of clusters of alpha-tocopherol in gel and fluid phases of dipalmitoylglycerophosphocholine. *Eur J Biochem*, **233**: 916-925.

Radhakrishnan, A., Li, X.M., Brown, R.E. and McConnell, H.M., 2001, Stoichiometry of cholesterol-sphingomyelin condensed complexes in monolayers. *Biochim Biophys Acta*, **1511**: 1-6.

Rajaram, S., Spangler, T.L., Sedensky, M.M. and Morgan, P.G., 1999, A stomatin and a degenerin interact to control anesthetic sensitivity in Caenorhabditis elegans. *Genetics*, **153**: 1673-1682.

Ramstedt, B. and Slotte, J.P., 2002, Membrane properties of sphingomyelins. *FEBS Lett*, **531**: 33-37.

Rauch, C. and Farge, E., 2000, Endocytosis switch controlled by transmembrane osmotic pressure and phospholipid number asymmetry. *Biophys J*, **78**: 3036-3047.

Razani, B., Combs, T.P., Wang, X.B., Frank, P.G., Park, D.S., Russell, R.G., Li, M., Tang, B., Jelicks, L.A., Scherer, P.E. and Lisanti, M.P., 2002, Caveolin-1-deficient mice are lean, resistant to diet-induced obesity, and show hypertriglyceridemia with adipocyte abnormalities. *J Biol Chem*, **277**: 8635-47.

Razani, B., Engelman, J.A., Wang, X.B., Schubert, W., Zhang, X.L., Marks, C.B., Macaluso, F., Russell, R.G., Li, M., Pestell, R.G., Di Vizio, D., Hou, H., Jr., Kneitz, B., Lagaud, G., Christ, G.J., Edelmann, W. and Lisanti, M.P., 2001, Caveolin-1 null mice are viable but show evidence of hyperproliferative and vascular abnormalities. *J Biol Chem*, **276**: 38121-38.

Recktenwald, D.J. and McConnell, H.M., 1981, Phase equilibria in binary mixtures of phosphatidylcholine and cholesterol. *Biochemistry*, **20**: 4505-10.

Reid-Taylor, K.L., Chu, J.W. and Sharom, F.J., 1999, Reconstitution of the glycosylphosphatidylinositol-anchored protein Thy-1: interaction with membrane phospholipids and galactosylceramide. *Biochem Cell Biol*, **77**: 189-200.

Ren, J., Lew, S., Wang, J. and London, E., 1999, Control of the transmembrane orientation and interhelical interactions within membranes by hydrophobic helix length. *Biochemistry*, **38**: 5905-12.

Ren, J., Lew, S., Wang, Z. and London, E., 1997, Transmembrane orientation of hydrophobic alpha-helices is regulated both by the relationship of helix length to bilayer thickness and by the cholesterol concentration. *Biochemistry*, **36**: 10213-20.

Resh, M.D., 1999, Fatty acylation of proteins: new insights into membrane targetting of myristolyated and palmitoylated proteins. *Biochim Biophys Acta*, **1451**: 1-16.

Reyes Mateo, C., Ulises Acuna, A. and Brochon, J.C., 1995, Liquid-crystalline phases of cholesterol/lipid bilayers as revealed by the fluorescence of trans-parinaric acid. *Biophys J*, **68**: 978-987.

Ricciarelli, R., Zingg, J.M. and Azzi, A., 2002, The 80th anniversary of vitamin E: beyond its antioxidant properties. *Biol Chem*, **383**: 457-465.

Rietveld, A., Neutz, S., Simons, K. and Eaton, S., 1999, Association of sterol- and glycosylphosphatidylinositol-linked proteins with *Drosophila* raft lipid microdomains. *J Biol Chem*, **274**: 12049-54.

Rijnboutt, S., Jansen, G., Posthuma, G., Hynes, J.B., Schornagel, J.H. and Strous, G.J., 1996, Endocytosis of GPI-linked membrane folate receptor-α. *J Cell Biol*, **132**: 35-47.

Rinia, H.A., Snel, M.M., van der Eerden, J.P. and de Kruijff, B., 2001, Visualizing detergent resistant domains in model membranes with atomic force microscopy. *FEBS Lett*, **501**: 92-96.

Robbins, S.M., Quintrell, N.A. and Bishop, J.M., 1995, Myristoylation and differential palmitoylation of the HCK protein- tyrosine kinases govern their attachment to membranes and association with caveolae. *Mol Cell Biol*, **15**: 3507-3515.

Roberts, W.L., Myher, J.J., Kuksis, A., Low, M.G. and Rosenberry, T.L., 1988, Lipid analysis of the glycoinositol phospholipid membrane anchor of human erythrocyte acetylcholinesterase. Palmitoylation of inositol results in resistance to phosphatidylinositol-specific phospholipase C. *J Biol Chem*, **263**: 18766-75.

Robinson, P.J., 1991, Phosphatidylinositol membrane anchors and T-cell activation. *Immunol Today*, **12**: 35-41.

Rodal, S.K., Skretting, G., Garred, O., Vilhardt, F., van Deurs, B. and Sandvig, K., 1999, Extraction of cholesterol with methyl-beta-cyclodextrin perturbs formation of clathrin-coated endocytic vesicles. *Mol Biol Cell*, **10**: 961-974.

Rog, T. and Pasenkiewicz-Gierula, M., 2001a, Cholesterol effects on the phosphatidylcholine bilayer nonpolar region: a molecular simulation study. *Biophys J*, **81**: 2190-2202.

Rog, T. and Pasenkiewicz-Gierula, M., 2001b, Cholesterol effects on the phospholipid condensation and packing in the bilayer: a molecular simulation study. *FEBS Lett*, **502**: 68-71.

Rogers, J., Lee, A.G. and Wilton, D.C., 1979, The organisation of cholesterol and ergosterol in lipid bilayers based on studies using non-perturbing fluorescent sterol probes. *Biochim Biophys Acta*, **552**: 23-37.

Roper, K., Corbeil, D. and Huttner, W.B., 2000, Retention of prominin in microvilli reveals distinct cholesterol-based lipid micro-domains in the apical plasma membrane. *Nat Cell Biol*, **2**: 582-592.

Rousso, I., Mixon, M.B., Chen, B.K. and Kim, P.S., 2000, Palmitoylation of the HIV-1 envelope glycoprotein is critical for viral infectivity. *Proc Natl Acad Sci U S A*, **97**: 13523-25.

Roy, D. and Mukhopadhyay, C., 2002, Molecular dynamics simulation of GM1 in phospholipid bilayer. *J Biomol Struct Dyn*, **19**: 1121-1132.

Roy, M.O., Leventis, R. and Silvius, J.R., 2000, Mutational and biochemical analysis of plasma membrane targeting mediated by the farnesylated, polybasic carboxy terminus of K-ras4B. *Biochemistry*, **39**, 8298-8307.

Roy, S., Luetterforst, R., Harding, A., Apolloni, A., Etheridge, M., Stang, E., Rolls, B., Hancock, J.F. and Parton, R.G., 1999, Dominant-negative caveolin inhibits H-Ras function by disrupting cholesterol-rich plasma membrane domains. *Nat Cell Biol*, **1**: 98-105.

Rukmini, R., Rawat, S.S., Biswas, S.C. and Chattopadhyay, A., 2001, Cholesterol organization in membranes at low concentrations: effects of curvature stress and membrane thickness. *Biophys J*, **81**: 2122-34.

Rybin, V.O., Xu, X., Lisanti, M.P. and Steinberg, S.F., 2000, Differential targeting of beta -adrenergic receptor subtypes and adenylyl cyclase to cardiomyocyte caveolae. A mechanism to functionally regulate the cAMP signaling pathway. *J Biol Chem*, **275**: 41447-57.

Sabharanjak, S., Sharma, P., Parton, R.G. and Mayor, S., 2002, GPI-anchored proteins are delivered to recycling endosomes via a distinct cdc42-regulated, clathrin-independent pinocytic pathway. *Dev Cell*, **2**: 411-423.

Salzer, U., Hinterdorfer, P., Hunger, U., Borken, C. and Prohaska, R., 2002, Ca^{2+}-dependent vesicle release from erythrocytes involves stomatin-specific lipid rafts, synexin, annexin VII, and sorcin. *Blood*, **99**: 2569-2577.

Salzer, U. and Prohaska, R., 2001, Stomatin, flotillin-1, and flotillin-2 are major integral proteins of erythrocyte lipid rafts. *Blood*, **97**: 1141-1143.

Samsonov, A.V., Chatterjee, P.K., Razinkov, V.I., Eng, C.H., Kielian, M. and Cohen, F.S., 2002, Effects of membrane potential and sphingolipid structures on fusion of Semliki Forest virus. *J Virol*, **76**, 12691-702.

Samsonov, A.V., Mihalyov, I. and Cohen, F.S., 2001, Characterization of cholesterol-sphingomyelin domains and their dynamics in bilayer membranes. *Biophys J*, **81**: 1486-1500.

Samuel, B.U., Mohandas, N., Harrison, T., McManus, H., Rosse, W., Reid, M. and Haldar, K., 2001, The role of cholesterol and glycosylphosphatidylinositol-anchored proteins of erythrocyte rafts in regulating raft protein content and malarial infection. *J Biol Chem*, **276**: 29319-29.

Sanan, D.A. and Anderson, R.G., 1991, Simultaneous visualization of LDL receptor distribution and clathrin lattices on membranes torn from the upper surface of cultured cells. *J Histochem Cytochem*, **39**: 1017-1024.

Sankaram, M.B. and Thompson, T.E., 1990a, Interaction of cholesterol with various glycerophospholipids and sphingomyelin. *Biochemistry*, **29**: 10670-75.

Sankaram, M.B. and Thompson, T.E., 1990b, Modulation of phospholipid acyl chain order by cholesterol. A solid-state 2H nuclear magnetic resonance study. *Biochemistry*, **29**: 10676-84.

Santana, P., Pena, L.A., Haimovitz-Friedman, A., Martin, S., Green, D., McLoughlin, M., Cordon-Cardo, C., Schuchman, E.H., Fuks, Z. and Kolesnick, R., 1996, Acid sphingomyelinase-deficient human lymphoblasts and mice are defective in radiation-induced apoptosis. *Cell*, **86**: 189-199.

Sargiacomo, M., Sudol, M., Tang, Z. and Lisanti, M.P., 1993, Signal transducing molecules and glycosyl-phosphatidylinositol-linked proteins form a caveolin-rich insoluble complex in MDCK cells. *J. Cell Biol.*, **122**: 789-807.

Saslowsky, D.E., Lawrence, J., Ren, X., Brown, D.A., Henderson, R.M. and Edwardson, J.M., 2002, Placental alkaline phosphatase is efficiently targeted to rafts in supported lipid bilayers. *J Biol Chem*, **277**: 26966-70.

Scheiffele, P., Rietveld, A., Wilk, T. and Simons, K., 1999, Influenza viruses select ordered lipid domains during budding from the plasma membrane. *J Biol Chem*, **274**: 2038-44.

Scheiffele, P., Roth, M.G. and Simons, K., 1997, Interaction of influenza virus haemagglutinin with sphingolipid-cholesterol membrane domains via its transmembrane domain. *EMBO J*, **16**: 5501-08.

Schmidt, C.F., Barenholz, Y., Huang, C. and Thompson, T.E., 1978, Monolayer coupling in sphingomyelin bilayer systems. *Nature*, **271**: 775-777.

Schroeder, F., Nemecz, G., Wood, W.G., Joiner, C., Morrot, G., Ayraut-Jarrier, M. and Devaux, P.F., 1991, Transmembrane distribution of sterol in the human erythrocyte. *Biochim Biophys Acta*, **1066**: 183-192.

Schroeder, R.J., Ahmed, S.N., Zhu, Y., London, E. and Brown, D.A., 1998, Cholesterol and sphingolipid enhance the Triton X-100 insolubility of glycosylphosphatidylinositol-anchored proteins by promoting the formation of detergent-insoluble ordered membrane domains. *J Biol Chem*, **273**: 1150-1157.

Schuck, S., Honsho, M., Ekroos, K., Shevchenko, A. and Simons, K., 2003, Resistance of cell membranes to different detergents. *Proc Natl Acad Sci USA*, **100**: 5795-5800.

Schulte, T., Paschke, K.A., Laessing, U., Lottspeich, F. and Stuermer, C.A., 1997, Reggie-1 and reggie-2, two cell surface proteins expressed by retinal ganglion cells during axon regeneration. *Development*, **124**: 577-587.

Schutz, G.J., Kada, G., Pastushenko, V.P. and Schindler, H., 2000, Properties of lipid microdomains in a muscle cell membrane visualized by single molecule microscopy. *EMBO J*, **19**: 892-901.

Scott, H.L., 2002, Modeling the lipid component of membranes. *Curr Opin Struct Biol*, **12**: 495-502.

Seachrist, J.L., Anborgh, P.H. and Ferguson, S.S., 2000, β_2- adrenergic receptor internalization, endosomal sorting and plasma membrane recycling are regulated by Rab GTPases. *J Biol Chem*, **275**: 27221-28.

Sehgal, P.B., Guo, G.G., Shah, M., Kumar, V. and Patel, K., 2002, Cytokine signaling: STATS in plasma membrane rafts. *J Biol Chem*, **277**: 12067-74.

Sekiguchi, K., Tsukuda, M., Ase, K., Kikkawa, U. and Nishizuka, Y., 1988, Mode of activation and kinetic properties of three distinct forms of protein kinase C from rat brain. *J Biochem (Tokyo)*, **103**: 759-765.

Semmler, K., Meyer, H.W. and Quinn, P.J., 2000, The structure and thermotropic phase behaviour of dipalmitoylphosphatidylcholine codispersed with a branched-chain phosphatidylcholine. *Biochim Biophys Acta*, **1509**: 385-396.

Shahinian, S. and Silvius, J.R., 1995, Doubly-lipid-modified protein sequence motifs exhibit long-lived anchorage to lipid bilayer membranes. *Biochemistry*, **34**: 3813-22.

Sheets, E.D., Holowka, D. and Baird, B., 1999, Critical role for cholesterol in Lyn-mediated tyrosine phosphorylation of FcεRI and their association with detergent-resistant membranes. *J Cell Biol*, **145**: 877-887.

Siegel, D.P., Banschbach, J., Alford, D., Ellens, H., Lis, L.J., Quinn, P.J., Yeagle, P.L. and Bentz, J., 1989, Physiological levels of diacylglycerols in phospholipid membranes induce membrane fusion and stabilize inverted phases. *Biochemistry*, **28**: 3703-09.

Simons, K. and Gruenberg, J., 2000, Jamming the endosomal system: lipid rafts and lysosomal storage diseases. *Trends Cell Biol*, **10**: 459-462.

Simons, K. and Ikonen, E., 1997, Functional rafts in cell membranes. *Nature*, **387**, 569-572.

Simons, K. and Toomre, D., 2000, Lipid rafts and signal transduction. *Nature Reviews: Molecular Cell Biology*, **1**: 31-39.

Simons, K. and van Meer, G., 1988, Lipid sorting in epithelial cells. *Biochemistry*, **27**: 6197-6202.

Singer, S.J. and Nicolson, G.L., 1972, The fluid mosaic model of the structure of cell membranes. *Science*, **175**: 720-731.

Smart, E.J., Ying, Y.S., Mineo, C. and Anderson, R.G., 1995, A detergent-free method for purifying caveolae membrane from tissue culture cells. *Proc Natl Acad Sci USA*, **92**: 10104-8.

Smondyrev, A.M. and Berkowitz, M.L., 1999, Structure of dipalmitoylphosphatidylcholine/cholesterol bilayer at low and high cholesterol concentrations: molecular dynamics simulation. *Biophys J*, **77**: 2075-89.

Snyers, L., Umlauf, E. and Prohaska, R., 1998, Oligomeric nature of the integral membrane protein stomatin. *J Biol Chem*, **273**: 17221-26.

Snyers, L., Umlauf, E. and Prohaska, R., 1999a, Association of stomatin with lipid-protein complexes in the plasma membrane and the endocytic compartment. *Eur J Cell Biol*, **78**: 802-812.

Snyers, L., Umlauf, E. and Prohaska, R., 1999b, Cysteine 29 is the major palmitoylation site on stomatin. *FEBS Lett*, **449**: 101-104.

Solomon, S., Masilamani, M., Rajendran, L., Bastmeyer, M., Stuermer, C.A. and Illges, H., 2002, The lipid raft microdomain-associated protein reggie-1/flotillin-2 is expressed in human B cells and localized at the plasma membrane and centrosome in PBMCs. *Immunobiology*, **205**: 108-119.

Song, S.K., Li, S., Okamoto, T., Quilliam, L.A., Sargiacomo, M. and Lisanti, M.P., 1996, Co-purification and direct interaction of Ras with caveolin, an integral membrane protein of caveolae microdomains. Detergent-free purification of caveolae microdomains. *J Biol Chem*, **271**: 9690-9697.

Sot, J., Collado, M.I., Arrondo, J.R.L., Alonso, A. and Goñi, F.M., 2002, Triton X-100-resistant bilayers: effect of lipid composition and relevance to the raft phenomenon. *Langmuir*, **18**: 2828-2835.

Sotgia, F., Razani, B., Bonuccelli, G., Schubert, W., Battista, M., Lee, H., Capozza, F., Schubert, A.L., Minetti, C., Buckley, J.T. and Lisanti, M.P., 2002, Intracellular retention of glycosylphosphatidyl inositol-linked proteins in caveolin-deficient cells. *Mol Cell Biol*, **22**: 3905-26.

Sowa, G., Pypaert, M. and Sessa, W.C., 2001, Distinction between signaling mechanisms in lipid rafts vs. caveolae. *Proc Natl Acad Sci U S A*, **98**: 14072-77.

Sprong, H., van der Sluijs, P. and van Meer, G., 2001, How proteins move lipids and lipids move proteins. *Nat Rev Mol Cell Biol*, **2**: 504-513.

Stahl, N. and Yancopoulos, G.D., 1994, The tripartite CNTF receptor complex: activation and signaling involves components shared with other cytokines. *J Neurobiol*, **25**: 1454-1466.

Stillwell, W., Dallman, T., Dumaual, A.C., Crump, F.T. and Jenski, L.J., 1996, Cholesterol versus alpha-tocopherol: effects on properties of bilayers made from heteroacid phosphatidylcholines. *Biochemistry*, **35**: 13353-13362.

Stoffel, R.H., Inglese, J., Macrae, A.D., Lefkowitz, R.J. and Premont, R.T., 1998, Palmitoylation increases the kinase activity of the G protein-coupled receptor kinase, GRK6. *Biochemistry*, **37**: 16053-16059.

Stoffel, R.H., Randall, R.R., Premont, R.T., Lefkowitz, R.J. and Inglese, J., 1994, Palmitoylation of G protein-coupled receptor kinase, GRK6. Lipid modification diversity in the GRK family. *J Biol Chem*, **269**: 27791-94.

Stuermer, C.A., Lang, D.M., Kirsch, F., Wiechers, M., Deininger, S.O. and Plattner, H., 2001, Glycosylphosphatidyl inositol-anchored proteins and fyn kinase assemble in non-caveolar plasma membrane microdomains defined by reggie-1 and -2. *Mol Biol Cell*, **12**: 3031-3045.

Stulnig, T.M., Berger, M., Sigmund, T., Raederstorff, D., Stockinger, H. and Waldhäusl, W., 1998, Polyunsaturated fatty acids inhibit T cell signal transduction by modification of detergent-insoluble membrane domains. *J Cell Biol*, **143**, 637-644.

Subczynski, W.K., Wisniewska, A., Yin, J.J., Hyde, J.S. and Kusumi, A., 1994, Hydrophobic barriers of lipid bilayer membranes formed by reduction of water penetration by alkyl chain unsaturation and cholesterol. *Biochemistry*, **33**: 7670-7681.

Subtil, A., Gaidarov, I., Kobylarz, K., Lampson, M.A., Keen, J.H. and McGraw, T.E., 1999, Acute cholesterol depletion inhibits clathrin-coated pit budding. *Proc Natl Acad Sci USA*, **96**: 6775-6780.

Sunyach, C., Jen, A., Deng, J., Fitzgerald, K., Frobert, Y., McCaffrey, M. and Morris, R.J., 2003, The mechanism of internalisation of GPI anchored prion protein. *EMBO J*, **22**: 2591-2601.

Szollosi, J., Damjanovich, S., Mulhern, S.A. and Tron, L., 1987, Fluorescence energy transfer and membrane potential measurements monitor dynamic properties of cell membranes: a critical review. *Prog. Biophys. molec. Biol.*, **49**: 65-87.

Takahashi, H., Hatta, I. and Quinn, P.J., 1996, Cubic phases in hydrated 1:1 and 1:2 dipalmitoylphosphatidylcholine- dipalmitoylglycerol mixtures. *Biophys J*, **70**: 1407-1411.

Takeuchi, S., Takayama, Y., Ogawa, A., Tamura, K. and Okada, M., 2000, Transmembrane phosphoprotein cbp positively regulates the activity of the carboxyl-terminal src kinase, Csk. *J Biol Chem*, **275**: 29183-86.

Tansey, M.G., Baloh, R.H., Milbrandt, J. and Johnson, E.M., Jr., 2000, GFRalpha-mediated localization of RET to lipid rafts is required for effective downstream signaling, differentiation, and neuronal survival. *Neuron*, **25**: 611-623.

Tatulian, S.A. and Tamm, L.K., 2000, Secondary structure, orientation, oligomerization, and lipid interactions of the transmembrane domain of influenza hemagglutinin. *Biochemistry*, **39**: 496-507.

Tavernarakis, N., Driscoll, M. and Kyrpides, N.C., 1999, The SPFH domain: implicated in regulating targeted protein turnover in stomatins and other membrane-associated proteins. *Trends Biochem Sci*, **24**: 425-427.

Thai, T.P., Heid, H., Rackwitz, H.R., Hunziker, A., Gorgas, K. and Just, W.W., 1997, Ether lipid biosynthesis: isolation and molecular characterization of human dihydroxyacetonephosphate acyltransferase. *FEBS Lett*, **420**: 205-211.

Thai, T.P., Rodemer, C., Jauch, A., Hunziker, A., Moser, A., Gorgas, K. and Just, W.W., 2001, Impaired membrane traffic in defective ether lipid biosynthesis. *Hum Mol Genet*, **10**: 127-136.

The, T.H. and Feltkamp, T.E.W., 1970, Conjugation of fluorescein isothiocyanate to antibodies. I. Experiments on the conditions of conjugation. *Immunology*, **18**: 865-873.

Thewalt, J.L. and Bloom, M., 1992, Phosphatidylcholine: cholesterol phase diagrams. *Biophys J*, **63**, 1176-1181.

Thomsen, P., Roepstorff, K., Stahlhut, M. and van Deurs, B., 2002, Caveolae are highly immobile plasma membrane microdomains, which are not involved in constitutive endocytic trafficking. *Mol Biol Cell*, **13**: 238-250.

Tieleman, D.P., Marrink, S.J. and Berendsen, H.J., 1997, A computer perspective of membranes: molecular dynamics studies of lipid bilayer systems. *Biochim Biophys Acta*, **1331**: 235-270.

Tiveron, M.C., Nosten-Bertrand, M., Jani, H., Garnett, D., Hirst, E.M.A., Grosveld, F. and Morris, R.J., 1994, The mode of anchorage to the cell surface determines both the function and membrane location of Thy-1 glycoprotein. *J Cell Sci*, **107**: 1783-96.

Torgersen, K.M., Vang, T., Abrahamsen, H., Yaqub, S., Horejsi, V., Schraven, B., Rolstad, B., Mustelin, T. and Tasken, K., 2001, Release from tonic inhibition of T cell activation through transient displacement of C-terminal Src kinase, Csk, from lipid rafts. *J Biol Chem*, **276**: 29313-18.

Tse, A.G.D., Barclay, A.N., Watts, A. and Williams, A.F., 1985, A glycophospholipid tail at the carboxyl terminus fo the Thy-1 glycoprotein of neurons and thymocytes. *Science*, **230**: 1003-1008.

Tsvetkova, N.M., Brain, A.P.R. and Quinn, P.J., 1994, Structural characteristics of thylakoid membranes of Arabidopsis mutants deficient in lipid fatty acid desaturation. *Biochim Biophys. Acta*, **1192**: 263-271.

Tu, Y., Popov, S., Slaughter, C. and Ross, E.M., 1999, Palmitoylation of a conserved cysteine in the regulator of G protein signaling, RGS, domain modulates the GTPase-activating activity of RGS4 and RGS10. *J Biol Chem*, **274**, 38260-67.

Uittenbogaard, A. and Smart, E.J., 2000, Palmitoylation of caveolin-1 is required for cholesterol binding, chaperone complex formation, and rapid transport of cholesterol to caveolae. *J Biol Chem*, **275**: 25595-99.

Vainio, S., Heino, S., Mansson, J.E., Fredman, P., Kuismanen, E., Vaarala, O. and Ikonen, E., 2002, Dynamic association of human insulin receptor with lipid rafts in cells lacking caveolae. *EMBO Rep*, **3**: 95-100.

Van Blitterswijk, W.J., Van Der Luit, A.H., Veldman, R.J., Verheij, M. and Borst, J., 2003, Ceramide: second messenger or modulator of membrane structure and dynamics? *Biochem J*, **369**: 199-211.

van der Goot, F.G. and Harder, T., 2001, Raft membrane domains: from a liquid-ordered membrane phase to a site of pathogen attack. *Semin Immunol*, **13**: 89-97.

van Klompenburg, W., Nilsson, I., von Heijne, G. and de Kruijff, B., 1997, Anionic phospholipids are determinants of membrane protein topology. *EMBO J*, **16**: 4261-6.

van't Hof, W. and Resh, M.D., 1999, Dual fatty acylation of p59(Fyn), is required for association with the T cell receptor zeta chain through phosphotyrosine-Src homology domain-2 interactions. *J Cell Biol*, **145**: 377-389.

Vandecasteele, G., Eschenhagen, T., Scholz, H., Stein, B., Verde, I. and Fischmeister, R., 1999, Muscarinic and beta-adrenergic regulation of heart rate, force of contraction and calcium current is preserved in mice lacking endothelial nitric oxide synthase. *Nat Med*, **5**: 331-334.

Varma, R. and Mayor, S., 1998, GPI-anchored proteins are organized in submicron domains at the cell surface. *Nature*, **394**: 798-801.

Veldman, R.J., Maestre, N., Aduib, O.M., Medin, J.A., Salvayre, R. and Levade, T., 2001, A neutral sphingomyelinase resides in sphingolipid-enriched microdomains and is inhibited by the caveolin-scaffolding domain: potential implications in tumour necrosis factor signalling. *Biochem J*, **355**: 859-868.

Venkataraman, K. and Futerman, A.H., 2000, Ceramide as a second messenger: sticky solutions to sticky problems. *Trends Cell Biol*, **10**: 408-412.

Vilhardt, F., Nielsen, M., Sandvig, K. and van Deurs, B., 1999, Urokinase-type plasminogen activator receptor is internalized by different mechanisms in polarized and nonpolarized Madin-Darby canine kidney epithelial cells. *Mol Biol Cell*, **10**: 179-195.

Vincent, S., Gerlier, D. and Manie, S.N., 2000, Measles virus assembly within membrane rafts. *J Virol*, **74**: 9911-5.

Vist, M.R. and Davis, J.H., 1990, Phase equilibria of cholesterol/dipalmitoylphosphatidylcholine mixtures: 2H nuclear magnetic resonance and differential scanning calorimetry. *Biochemistry*, **29**: 451-464.

Volonte, D., Galbiati, F., Li, S., Nishiyama, K., Okamoto, T. and Lisanti, M.P., 1999, Flotillins/cavatellins are differentially expressed in cells and tissues and form a hetero-oligomeric complex with caveolins in vivo. Characterization and epitope-mapping of a novel flotillin-1 monoclonal antibody probe. *J Biol Chem*, **274**: 12702-09.

Vyas, K.A., Patel, H.V., Vyas, A.A. and Schnaar, R.L., 2001, Segregation of gangliosides GM1 and GD3 on cell membranes, isolated membrane rafts, and defined supported lipid monolayers. *Biol Chem*, **382**: 241-250.

Wang, T.Y., Leventis, R. and Silvius, J.R., 2000, Fluorescence-based evaluation of the partitioning of lipids and lipidated peptides into liquid-ordered lipid microdomains: a model for molecular partitioning into "lipid rafts". *Biophys J*, **79**: 919-933.

Wang, T.Y., Leventis, R. and Silvius, J.R., 2001, Partitioning of lipidated peptide sequences into liquid-ordered lipid domains in model and biological membranes. *Biochemistry*, **40**: 13031-40.

Wang, T.Y. and Silvius, J.R., 2001, Cholesterol does not induce segregation of liquid-ordered domains in bilayers modeling the inner leaflet of the plasma membrane. *Biophys J*, **81**: 2762-73.

Wang, T.Y. and Silvius, J.R., 2003, Sphingolipid partitioning into ordered domains in cholesterol-free and cholesterol-containing lipid bilayers. *Biophys J*, **84**: 367-378.

Wang, X. and Quinn, P.J., 1999, The effect of alpha-tocopherol on the thermotropic phase behaviour of dipalmitoylphosphatidylethanolamine. A synchrotron X-ray diffraction study. *Eur J Biochem*, **264**: 1-8.

Wang, X. and Quinn, P.J., 2000a, The distribution of alpha-tocopherol in mixed aqueous dispersions of phosphatidylcholine and phosphatidylethanolamine. *Biochim Biophys Acta*, **1509**: 361-372.

Wang, X. and Quinn, P.J., 2000b, Preferential interaction of alpha-tocopherol with phosphatidylcholines in mixed aqueous dispersions of phosphatidylcholine and phosphatidylethanolamine. *Eur J Biochem*, **267**: 6362-6368.

Wang, X. and Quinn, P.J., 2002, Cubic phase is induced by cholesterol in the dispersion of 1-palmitoyl-2-oleoyl-phosphatidylethanolamine. *Biochim Biophys Acta*, **1564**: 66-72.

Webb, Y., Hermida-Matsumoto, L. and Resh, M.D., 2000, Inhibition of protein palmitoylation, raft localization, and T cell signaling by 2-bromopalmitate and polyunsaturated fatty acids. *J Biol Chem*, **275**: 261-270.

Wedegaertner, P.B., Wilson, P.T. and Bourne, H.R., 1995, Lipid modifications of trimeric G proteins. *J Biol Chem*, **270**: 503-506.

Werlen, G. and Palmer, E., 2002, The T-cell receptor signalosome: a dynamic structure with expanding complexity. *Curr Opin Immunol*, **14**: 299-305.

Williams, W.P., Sen, A. and Quinn, P.J., 1982, Protein-lipid interactions in the photosynthetic membrane. *Biochem Soc Trans*, **10**: 335-338.

Wilson, B.S., Pfeiffer, J.R. and Oliver, J.M., 2000, Observing FcepsilonRI signaling from the inside of the mast cell membrane. *J Cell Biol*, **149**: 1131-1142.

Wilson, B.S., Pfeiffer, J.R., Surviladze, Z., Gaudet, E.A. and Oliver, J.M., 2001, High resolution mapping of mast cell membranes reveals primary and secondary domains of FcεRI and LAT. *J Cell Biol*, **154**: 645-658.

Wood, W.G., Schroeder, F., Hogy, L., Rao, A.M. and Nemecz, G., 1990, Asymmetric distribution of a fluorescent sterol in synaptic plasma membranes: effects of chronic ethanol consumption. *Biochim Biophys Acta*, **1025**: 243-246.

Woodman, S.E., Park, D.S., Cohen, A.W., Cheung, M.W., Chandra, M., Shirani, J., Tang, B., Jelicks, L.A., Kitsis, R.N., Christ, G.J., Factor, S.M., Tanowitz, H.B. and Lisanti, M.P., 2002a, Caveolin-3 knock-out mice develop a progressive cardiomyopathy and show hyperactivation of the p42/44 MAPK cascade. *J Biol Chem*, **277**: 38988-97.

Woodman, S.E., Schlegel, A., Cohen, A.W. and Lisanti, M.P., 2002b, Mutational analysis identifies a short atypical membrane attachment sequence (KYWFYR), within caveolin-1. *Biochemistry*, **41**: 3790-95.

Xiang, Y., Rybin, V.O., Steinberg, S.F. and Kobilka, B., 2002, Caveolar localization dictates physiologic signaling of beta 2-adrenoceptors in neonatal cardiac myocytes. *J Biol Chem*, **277**: 34280-86.

Xu, X., Bittman, R., Duportail, G., Heissler, D., Vilcheze, C. and London, E., 2001, Effect of the structure of natural sterols and sphingolipids on the formation of ordered sphingolipid/sterol domains, (rafts). Comparison of cholesterol to plant, fungal, and disease-associated sterols and comparison of sphingomyelin, cerebrosides, and ceramide. *J Biol Chem*, **276**: 33540-46.

Xu, X. and London, E., 2000, The effect of sterol structure on membrane lipid domains reveals how cholesterol can induce lipid domain formation. *Biochemistry*, **39**: 843-849.

Yasuda, K., Nagafuku, M., Shima, T., Okada, M., Yagi, T., Yamada, T., Minaki, Y., Kato, A., Tani-Ichi, S., Hamaoka, T. and Kosugi, A., 2002, Cutting edge: Fyn is essential for tyrosine phosphorylation of Csk-binding protein/phosphoprotein associated with glycolipid-enriched microdomains in lipid rafts in resting T cells. *J Immunol*, **169**: 2813-17.

Ying, Y.S., Anderson, R.G. and Rothberg, K.G., 1992, Each caveola contains multiple glycosylphosphatidylinositol-anchored membrane proteins. *Cold Spring Harbor Symp Quant Biol*, **57**: 593-604.

Yuan, C., Furlong, J., Burgos, P. and Johnston, L.J., 2002, The size of lipid rafts: an atomic force microscopy study of ganglioside GM1 domains in sphingomyelin/DOPC/cholesterol membranes. *Biophys J*, **82**: 2526-35.

Yuan, C. and Johnston, L.J., 2000, Distribution of ganglioside GM1 in L-alpha-dipalmitoylphosphatidylcholine/cholesterol monolayers: a model for lipid rafts. *Biophys J*, **79**: 2768-81.

Yuan, C. and Johnston, L.J., 2001, Atomic force microscopy studies of ganglioside GM1 domains in phosphatidylcholine and phosphatidylcholine/cholesterol bilayers. *Biophys J*, **81**: 1059-1069.

Zacharias, D.A., Violin, J.D., Newton, A.C. and Tsien, R.Y., 2002, Partitioning of lipid-modified monomeric GFPs into membrane microdomains of live cells. *Science*, **296**: 913-916.

Zhang, W., Sloan-Lancaster, J., Kitchen, J., Trible, R.P. and Samelson, L.E., 1998a, LAT: the ZAP-70 tyrosine kinase substrate that links T cell receptor to cellular activation. *Cell*, **92**: 83-92.

Zhang, W., Trible, R.P. and Samelson, L.E., 1998b, LAT palmitoylation: its essential role in membrane microdomain targeting and tyrosine phosphorylation during T cell activation. *Immunity*, **9**: 239-246.

Zimmerberg, J., 2000, Are the curves in all the right places? *Traffic*, **1**: 366-368.

Zlatkine, P., Mehul, B. and Magee, A.I., 1997, Retargeting of cytosolic proteins to the plasma membrane by the Lck protein tyrosine kinase dual acylation motif. *J Cell Sci*, **110**: 673-679.

Part II

Membrane Domain Composition

Chapter 3

Lipid Raft Proteins and Their Identification in T Lymphocytes

Bernd Wollscheid, Priska D. von Haller*, Eugene Yi, Samuel Donohoe, Kelly Vaughn, Andrew Keller, Alexey I. Nesvizhskii, Jimmy Eng, Xiao-jun Li, David R. Goodlett#, Ruedi Aebersold, and Julian D. Watts
Institute for Systems Biology, 1441 North 34th Street, Seattle WA 98103, USA

Abstract
This review focuses on how membrane lipid rafts have been detected and isolated, mostly from lymphocytes, and their associated proteins identified. These proteins include transmembrane antigens/receptors, GPI-anchored proteins, cytoskeletal proteins, Src-family protein kinases, G-proteins, and other proteins involved in signal transduction. To further understand the biology of lipid rafts, new methodological approaches are needed to help characterize the raft protein component, and changes that occur in this component as a result of cell perturbation. We describe the application of new proteomic approaches to the identification and quantification of raft proteins in T-lymphocytes. Similar approaches, applied to other model cell systems, will provide valuable new insights into both cellular signal transduction and lipid raft biology.

1. INTRODUCTION

The molecular composition and dynamic organization of the plasma membrane determines how a cell can interact with its environment at any given

* *Priska D. von Haller is presently at MacroGenics Inc., 1441 North 34th Street, Seattle, WA 98103, USA.*
# *David R. Goodlett is presently in the Department of Medicinal Chemistry, Box 357610, University of Washington, Seattle, WA 98195, USA.*

moment in time. It is the lipids and proteins that reside in the outer and inner leaflets of the plasma membrane that are responsible for the transfer, coordination and restriction of signals entering and leaving the cell. To confer specificity and efficiency in signaling, it is now known that the plasma membrane is organized into a variety of functional microdomains. The molecular composition of such domains differs from that of the surrounding membrane. By either excluding or including lipids and proteins with distinct properties, these microdomains compartmentalize and organize the various membrane functions. One such type of microdomain is known as the 'lipid raft'. Lipid rafts are known to play a vital role in cell signaling, a relationship that has recently become the focus of much study in lymphocytes in particular. Thus, in order to elucidate the specific functions of membrane microdomains such as lipid rafts, they must first be defined, and the proteins that dynamically partition within them need to be identified.

2. DETECTION OF LIPID RAFTS, RAFT DISTRIBUTION AND RAFT TRAFFICKING

Membrane lipid domains were first detected in human and hamster fibroblasts as a detergent-insoluble glycoprotein matrix rich in glycosphingolipids (Carter and Hakomori, 1981). Since then, microdomains and their composition have generally defined by the methods used to detect them. Depending on the spatial, temporal and chemical resolution of the method used, several different domains have been defined (Edidin, 2001). Such microdomains found in the plasma membrane include coated pits (Anderson and Jacobson, 2002), so-called for their appearance in an electron micrograph, and caveolae (Brodsky *et al.*, 2001; Harris *et al.*, 2002; Razani *et al.*, 2002), named for their high enrichment for the protein caveolin. More recently, the term 'lipid rafts' was coined to describe a collection of small plasma membrane microdomains rich in cholesterol, sphingomyelin, and glycolipids such as GM1 ganglioside (Simons and van Meer, 1988; Simons and Ikonen, 1997; Smart *et al.*, 1999). In lipid rafts, the orientation and tight packing of the long, largely unsaturated, acyl chains of sphingolipids are thought to create liquid ordered membrane patches, surrounded by, or 'floating' in, the liquid disordered plasma membrane phospholipid 'sea' (Brown and London, 1997; Brown and London, 2000), hence the name lipid rafts.

To date, components of lipid rafts have typically been biochemically isolated via their insolubility in various concentrations of nonionic detergents, such as Triton X-100 or Brij, at 4°C (Simons and Ikonen, 1997). These coalescing detergent-resistant remnants of the lipid rafts have also been referred to as detergent-resistant membranes (DRMs), and likely represent a subset of the endogenous lipid rafts, dependent upon the extraction procedure used for their isolation. Proteins within DRMs are then typically recovered by flotation into low-density fractions from sucrose gradients (Nebl *et al.*, 2002; Pike *et al.*, 2002). It is in this form that lipid rafts and their associated proteins are typically isolated so that, in terms of their biochemical characterization, 'lipid rafts' and 'DRMs' are, in effect, interchangeable terms.

While, lipid rafts are most abundant at the plasma membrane, they are also found in the biosynthetic and endocytic pathways. Cholesterol-sphingolipid rafts are first assembled in the Golgi apparatus, move towards the plasma membrane (Simons and Ikonen, 1997), and once at the cell surface are continuously endocytosed (Mukherjee and Maxfield, 2000). Rafts either recycle back from early endosomes to the surface or return indirectly through recycling endosomes (Puri *et al.*, 1999). Spectroscopic analyses of cell surface rafts *in vivo* has revealed that these domains consist of clusters of ~25 proteins and ~3500 sphingholipid molecules and are ~10–50 nm in diameter, below the resolution limit of light microscopy (Varma and Mayor, 1998; Pralle *et al.*, 2000). Rafts can also alter in size and composition in response to intra- or extracellular stimuli. Upon aggregation of raft components induced by, for example, receptor cross-linking, the dimension of surface lipid rafts can be increased into the submicron to micron range (Thomas *et al.*, 1994; Friedrichson and Kurzchalia, 1998; Simons and Toomre, 2000). Furthermore, the induced clustering of rafts exposes their proteins to new membrane environments with different interaction partners, thus facilitating the modulation of certain signaling cascades. It is also worth noting that raft density and distribution on the cell surface can vary, depending on the cell type (Brown and Rose, 1992). For example, in lymphocytes, rafts are distributed randomly over the cell surface without polarity, whereas in epithelial cells, rafts accumulate in the apical plasma membrane.

3. EARLY STRATEGIES TO PURIFY AND ISOLATE LIPID RAFT PROTEINS

Definable subcellular entities or membrane compartments, such as ribosomes, mitochondria, peroxisomes, Golgi membranes etc., have generally accepted procedures established for their isolation (Ford *et al.*, 1994; Graham *et al.*, 1994). This is not currently the case for lipid rafts. As a result of this, the protein component of lipid rafts isolated will vary according to the cell type used, the activation status of the cells, and the strategy used for raft isolation. Since *de novo* and immuno-identification of proteins still requires a substantial amount of protein, membrane raft fractions must also be isolated from pools of cells. Thus an average picture of raft components emerges from a mixture of many structures. Resistance to solubilization by Triton X-100 at 4°C has been the most widely utilized approach for extracting non-native, aggregated lipid rafts/DRMs. However, it is important to note that usage of different detergents, or different concentrations of the same detergent will make a difference to the proteins that co-isolate with lipid rafts.

In order to separate these raft/DRM isolates from other membrane and cytosolic proteins, sucrose density centrifugation is usually used to separate away the DRMs on account of their relatively high buoyancy. Cholera toxin B subunit, a specific ligand for the ganglioside GM1 that is enriched in rafts, can be used to test the sucrose gradient fractions for the presence of GM1-containing DRMs (Harder *et al.*, 1998). This general isolation strategy has opened the door for the identification of potential raft-associated proteins via comparative immunoblotting using antibodies against previously identified proteins. Using such methodology, it was soon recognized that rafts were enriched in signaling molecules and functioned as if they were centers for receptor-mediated signaling. This suggested that rafts represented specialized membrane domains directly involved in the coupling of receptor aggregation to downstream signaling cascades (Field *et al.*, 1995; Montixi *et al.*, 1998; Xavier *et al.*, 1998). Thus a considerable amount of subsequent recent research interest has focused on the involvement of lipid rafts in cell signaling, particularly in the area of lymphocyte membranes and their associated signaling pathways.

4. HYPOTHESIS-DRIVEN IDENTIFICATION OF RAFT PROTEINS IN LYMPHOCYTES

To understand lipid raft biology, it is essential to characterize them biochemically, and in particular to identify their associated proteins. Established biochemical methods such as immunoblotting have been employed to characterize proteins co-purifying with lipid rafts. Such hypothesis-driven methodology necessarily focuses on the identification of one protein at a time, and cannot readily address the complexity and the dynamic nature of lipid rafts. However, such approaches have been highly successful, providing significant insight into the biological processes with which lipid rafts are associated, and the proteins that are involved in them in lymphocytes, as well as other cell types. For example, using an array of antibodies against known membrane proteins, or proteins involved in receptor-mediated signaling, several groups have defined many as being associated with lipid rafts, either constitutively or inducibly. In doing so, some common features have emerged that, at least in part, have bearing on whether or not a protein might localize to lipid rafts.

A feature many constitutively associated lipid raft/DRM proteins had in common was their modification with saturated chain lipid groups (Melkonian *et al.*, 1999). In addition to their role as hydrophobic anchors for otherwise soluble proteins, covalent lipid modifications were also found to have an effect on the lateral distribution of proteins within the microdomains. Polypeptides modified with myristate N-acyl or palmitate S-acyl groups (acylation) localized preferentially to the cytoplasmic leaflet on the plasma membrane, whereas the addition of a Glycosylphosphaditylinositol (GPI) anchor resulted in localization to the outer leaflet. On the other hand, farnesylated proteins were not commonly found in DRMs, since the unsaturated prenyl group does not have a high enough affinity for their liquid ordered lipid phase (Melkonian *et al.*, 1999).

Of course, lipid modification is not the exclusive mechanism that has been identified for targeting of proteins to rafts. Amino acid residues close to the exoplasmic leaflet in the transmembrane domains of certain proteins seem are able to determine either inclusion in, or exclusion from of rafts for these proteins (Wang *et al.*, 2002). However, precisely why some proteins localize to lipid rafts while others do not, remains unclear. While lipid modifications do seem to be important for the localization of some raft associated proteins, the lipid modifications alone are not necessarily sufficient. For example, the transferrin receptor, which is also multiply acylated, is excluded from rafts. (Harder *et al.*, 1998). Thus, a combination of static features, in conjugation with controllable and dynamic signals, seem to be necessary to correctly localize individual proteins to lipid rafts, either constitutively, or in a regulated, time-dependant fashion. Indeed, in support of this idea, it is worth noting that many raft-associated proteins do not exclusively localize to rafts. Instead, many exhibit a distribution within the cell membrane, with a the spatial distribution and proportion of protein partitioning into the lipid raft membrane compartment varying from one protein to the next.

As was stated above, there has been considerable interest in lipid raft associated proteins in lymphocytes, particularly in T cells. Indeed, many known T cell signaling molecules have already been identified as partitioning into rafts. These proteins, essential either for the initiation or amplification of signaling processes, include the Src family protein tyrosine kinases (PTKs), protein phosphatases, receptors, co-receptors, transmembrane adapters, as well as G-proteins. Additionally, it has also been found that cytoskeletal components are pre-associated with lipid rafts, or become increasingly associated with them following cross-linking of the T cell receptor (TCR) (Montixi *et al.*, 1998; Xavier *et al.*, 1998; Drake and Braciale, 2001; Drevot *et al.*, 2002). The following paragraphs discuss these major classes of proteins that localize to or are enriched in lipid rafts, along with their potential functional implications.

4.1. GPI-anchored proteins

Many proteins on the surface of eukaryotic cells, including differentiation antigens, adhesion molecules, surface coat proteins, and ectoenzymes, have a posttranslational GPI modification that serves as a membrane anchor (Ikezawa, 2002). These extracellularly oriented GPI-anchored proteins (GPI-APs) are also known to be characteristic markers for lipid rafts (Brown, 1994; Moran and Miceli, 1998). Although GPI-APs have been implicated in a variety of cellular processes, such as transmembrane signaling and intracellular sorting, a clear understanding of the biological role of this protein modification has yet to emerge. (Muniz and Riezman, 2000; Fivaz *et al.*, 2002; Loertscher and Lavery, 2002). GPI-APs such as CD59, CD90 (Thy-1) or CD230 (prion protein), while they lack transmembrane and cytoplasmic domains, are anchored to the extracellular leaflet of the plasma membrane through a GPI linkage. Nevertheless, upon cross-linking they have been found capable if activating intracellular signaling pathways (Loertscher and Lavery, 2002). GPI-AP-mediated activities can be modulated by the action of specific phospholipases that remove the GPI group and release the protein from the cell surface (Lehto and Sharom, 2002). It has also been suggested that the lipid moiety of the GPI-anchor may serve as second messenger in the plasma membrane under certain conditions, since its structure is similar to common second messengers such as inositol-phosphate, diacyl-glycerol, phosphatidic acid and ceramide (Chawla and Vishwakarma, 2003). Another interesting consequence of GPI-anchoring has been observed in polarized epithelial cells. In most of these cells, GPI-APs are found almost exclusively on the apical surface (Brown *et al.*, 1989; Lisanti *et al.*, 1989). This implies that in such cells, the GPI-anchor is responsible for the apical targeting.

The localization of GPI-anchored proteins to lipid rafts and their extracellular accessibility has also made rafts accessible to characterization by optical techniques (van den Berg *et al.*, 1995; Kenworthy *et al.*, 2000; Hiscox *et al.*, 2002). Fluorescence resonance energy transfer (FRET) microscopy can be utilized to study the associations between and organization of raft components on the nanometer scale, in living cells. Using FRET it has been shown, for example, that cell surface GPI-anchored folate receptors are organized in cholesterol-dependent submicron-sized domains (Wu *et al.*, 1997). In other experiments, also using cells expressing GPI-anchored fluorescent proteins, it was shown that raft-localized GPI-APs were distributed homogenously over the cell surface, supporting the notion that individual raft islets are very small and below the discrimination limit of optical microscopy (Hiscox *et al.*, 2002).

4.2. Transmembrane adapter proteins

Adapter proteins have no intrinsic catalytic or transcriptional activation domains, but instead comprise of (usually multiple) protein binding/interaction domains, hence their name. Membrane-associated adapter proteins play an important role in many receptor-mediated signaling pathways, including TCR activation, by coupling receptor ligation at the membrane to distal signaling cascades. Members of this family of proteins may have multiple protein binding domains, particularly SH2, SH3, and PTB domains, while others may contain multiple target sequences for the binding of these same domains on other proteins. In this way, adapter proteins have the capability to recruit proteins and complexes required for signaling via the receptor, when called upon.

Transmembrane adaptor proteins such as LAT (linker of activated T cells), PAG/Cbp (phosphoprotein associated with glycosphingolipid-enriched microdomains/Csk binding protein) are key functional elements of T cell lipid rafts, and interactions with and modifications of their cytoplasmic tails are critical to T-cell activation (Samelson *et al.*, 1999; Brdicka *et al.*, 2000; Kawabuchi *et al.*, 2000). LAT is a 36–38 kDa phosphoprotein that is constitutively found in rafts (Zhang *et al.*, 1998) and consists of a short extracellular domain, a transmembrane domain immediately followed by two conserved cysteine residues, and a long cytoplasmic tail. Both of these cysteine residues in LAT are palmitylated *in vivo*, which determine its localization to lipid rafts. Interestingly, it has been shown that mutagenesis of Cys-29 does not adversely affect LAT function and localization, whereas mutation of Cys-26 abolishes LAT localization to rafts, and also results in loss of TCR signaling capability (Zhang *et al.*, 1998). This indicates that palmitoylation and raft localization of LAT are not only essential for LAT function, but that the presence of and correct membrane compartmentalization of this adapter protein is also essential for TCR function.

PAG/Cbp is an 80 kDa lipid raft associated transmembrane adaptor protein, expressed in most cell types, including T cells (Brdicka *et al.*, 2000, 2001). Tyrosine phosphorylation of PAG allows for recruitment and activation of the cytoplasmic PTK Csk, a critical negative regulator of Src-family PTKs, such as Lck and Fyn. Upon TCR cross-linking, a transient dephosphorylation of PAG/Cbp results in dissociation of the Csk-PAG/Cbp complex, and release of Csk from the lipid raft compartment. Similarly, tyrosine phosphorylation of LAT allows for recruitment of various cytoplasmic proteins, including PLC-γ1 and Grb-2, facilitating their redistribution to rafts for a distinct period of time (Zhang *et al.*, 2000). This indirect and inducible recruitment of molecules not constitutively able to associate with rafts in their own right allows the cell to control and regulate signaling processes. For example, the recruitment of phospholipase PLC-γ1 to rafts allows it to gain access to its substrate, phosphatidylinositol (4,5)-bisphosphate (PIP2), which is also preferentially localized to lipid rafts (Hilgemann *et al.*, 2001). On the other hand, the interaction of Grb-2 with LAT allows for a second level of indirect recruitment of proteins to rafts, since Grb-2 itself is also an adapter protein. The SH2 domain and two SH3 domains of Grb-2 provide additional docking sites to LAT and lipid rafts, thus facilitating the assembly of task specific signaling complexes. For example, Grb-2, upon binding to raft-associated LAT, can additionally recruit the protein Sos to the complex. Sos is an effector molecule for the small G protein Ras, which in turn is a key regulator of cell cycle progression (Takuwa and Takuwa, 2001). This modular nature for transient localization to the site of receptor activation within lipid rafts may thus be a primary biological mechanism for regulating signal specificity, while maintaining spatial segregation of competing signaling complexes, in general.

4.3. Src-family PTKs

Protein phosphorylation has long been known to be a key mechanism for the regulation intracellular signaling, through its ability to determine protein-protein interactions and to regulate enzymatic activities. It has also long been known that signal transduction from many cell surface receptors, in all cell types, requires the activation of protein kinases, and in particular PTKs. Indeed, there is a whole class of protein kinases known as receptor kinases, where the receptor molecules themselves have one (or more) active cytoplasmic kinase domains. However, many receptor complexes, including the TCR, have no intrinsic catalytic activity, and must therefore associate with non-receptor kinases in order to acquire the catalytic activity required for signal initiation.

A major group of such non-receptor kinases is the family of Src-related PTKs. Src family kinases interact with and are activated following engagement of many different classes of cellular receptors, participating in signaling pathways that control a diverse spectrum biological activities (Superti-Furga, 1995; Brown and Cooper, 1996). The Src family is (currently) composed of nine members in vertebrates: Src, Yes, Fgr, Yrk, Fyn, Lyn, Hck, Lck ans Blk. Several members are expressed ubiquitously, whereas others show selective tissue expression (Thomas and Brugge, 1997; Schlessinger, 2000). Members of the Src family share a common structure, including an SH2, SH3, and a catalytic (SH1) domain (Williams *et al.*, 1998). Src family PTKs also have in common N-terminal sequences that allow them to anchor to cell membranes and enables their partitioning into lipid rafts (Kovarova *et al.*, 2001). However, the lateral movement of these kinases within the membrane, in and out of rafts, as well as their regulation of kinase activity is dependent upon additional interactions.

In T cells, the Src-family PTK Lck is known to be essential for TCR-induced cell differentiation, and the initiation of signaling via phosphorylation of the TCR ζsubunit (Kabouridis *et al.*, 1997). Localization of Lck to the inner leaflet of the lipid raft plasma membrane is dependent on the acylation of three N-terminal amino acids: Cys-3 and Cys-5, which are both posttranslationally

S-acylated (palmitoylated), and Gly-2, which is co-translationally myristoylatd (Shenoy-Scaria *et al.*, 1993). Mutation of these S-acylation sites renders Lck cytosolic, and its ability to support TCR signaling is lost (Kabouridis *et al.*, 1997). Fyn, another Src-family PTK, is similarly palmitoylated, also resulting in its localization to T cell rafts. (Liang *et al.*, 2001; Stuermer *et al.*, 2001). Indeed, the general lipid modification of all Src-family PTKs is known to result in their raft association in multiple cell types, where they perform similar signaling functions via a wide range of receptor complexes. For example, the kinase activity required by the B cell receptor in B cells is instead provided by the lipid raft-associated Src PTKs Lyn and Blk (Gauld *et al.*).

4.4. G proteins

Another well documented and key mechanism for the regulation of receptor-mediated intracellular signaling has been the coupling of a substantial number of receptor molecules to membrane-associated GTPases. These so-called G proteins regulate a variety of signal transduction pathways via the (reversible) conversion of GTP to GDP, such that the activity of the signaling pathway is dependent upon whether the G protein is in its GTP- or GDP-bound form. There are two primary classes of G proteins: the so-called 'heterotrimeric' G proteins, which comprise of a functional grouping of three proteins, and the so-called 'small' G proteins, which are primarily lower molecular weight, independent GTPases. It is also known that both heterotrimeric and small G proteins are among the more abundant lipid modified signaling components enriched in DRMs (Moffett *et al.*, 2000; Oh and Schnitzer, 2001; Prior *et al.*, 2001; Stulnig *et al.*, 2001; Niv *et al.*, 2002; Waheed and Jones, 2002).

Heterotrimeric G protein complexes are each made up of three subunits: an α, β and γ subunit, and are typically sub-classed into one of 4 families (G_s, G_i, G_q, G_{12}) as defined by their α subunit (Morris and Malbon, 1999; Neves *et al.*, 2002). Currently, there are sixteen different Gα subunits, five Gβ subunits, and ten Gγ subunits known. Many of these subunits are expressed ubiquitously, while others have temporal and/or differential tissue expression. The G protein heterotrimer is attached to the inside of the cell plasma membrane, and is responsible for transducing a signal from a G protein-coupled receptor to intracellular effectors that regulate second messengers, which in turn change the metabolic state of the cell. Gα subunits are fatty acylated with either an amide linked myristate, a thioester-linked palmitate, or both (Mumby, 1997). Gγ subunits, on the other hand, are modified with farnesyl or geranylgeranyl lipid groups (Wedegaertner *et al.*, 1995). The Gα subunits of the G_s, G_i and G_q classes are known to be enriched in lipid rafts/DRMs isolated from various cell types, though whether the Gβ or Gγ subunits are also enriched remains unclear (Pike and Miller, 1998; Galbiati *et al.*, 1999; Moffett *et al.*, 2000; Miura *et al.*, 2001). For example, the compartmentalization of G proteins such as $G\alpha_{12}$ and $G\alpha_{13}$ is of great biological importance, since they are potent inducers of cell growth, differentiation and apoptosis. It is also known that the palmitoylation of $G\alpha_{12}$ and $G\alpha_{13}$ is important for correct membrane compartmentalization. However it seems that other interactions may assist in determining their specific localization, since the interaction of $G\alpha_{12}$ with Hsp90 is required for its partitioning into lipid rafts, in contrast to $G\alpha_{13}$ (Waheed and Jones, 2002).

In contrast to the heterotrimeric G proteins, various small G protein family members are known to play a role in TCR signaling. Various members of the Ras superfamily of small G proteins have also been found in DRM preparations (Prior and Hancock, 2001; Niv *et al.*, 2002). The small G proteins consist of only a single GTPase domain, similar to the Gα domain of the heterotrimeric G proteins, with their regulation being effected via a wide range of associating proteins. For example, members of the Rho sub-family of Ras-related GTPases form a functional association with Rac, Cdc42 and Rho, and regulate actin remodeling of the cytoskeleton (Henning and Cantrell, 1998; Blanchard, 2000). Many changes in the shape of cells, such as stress fiber formation, filopodia and lamelipodia, have been attributed to regulation by these GTPases (Ridley, 2001; Wherlock and Mellor, 2002). They facilitate regulation of this actin remodeling through effector complexes, which include actin related proteins 2 and 3 (Arp2/3) and Wiskott-Aldrich syndrome protein (WASP) family members (Rozelle *et al.*, 2000; Pollard and Beltzner, 2002). Like the Src family PTKs, Ras proteins are also lipid modified (only in this case by farnesylation or geranylgeranylation) at a C-terminal cysteine, contained in a CAAX motif recognized by a cytoplasmic farnesyltransferase (Shields *et al.*, 2000). In addition, those with a farnesyl group are also palmitoylated, with the exception of K-Ras, that instead has a polybasic region, and Rho proteins that have a geranylgeranyl group without additional modification. For Ras family members, both GDP and GTP bound forms are known to reside at the membrane, with evidence that unactivated H-Ras localizes mainly to lipid rafts (Roy *et al.*, 1999; Prior *et al.*, 2001; Niv *et al.*, 2002). In contrast, K-Ras4B is found in non-raft areas of the membrane (Prior *et al.*, 2001).

4.5. Cytoskeletal proteins

It has recently become clear that TCR signaling triggers a rapid initiation of the actin polymerization (Bunnell *et al.*, 2001), followed by an apparent clustering of lipid rafts at the site of T cell and the highly ordered, long-lasting, and directional assembly of signaling complexes, as well as a major cytoskeletal reorganizations (Das *et al.*, 2002; Tseng and Dustin, 2002). It is also known that the ability of the TCR to initiate this process is dependent on both the expression and raft-localization of LAT (Samelson *et al.*, 1999; Bunnell *et al.*, 2001). Thus there is clearly a tight, functional link between lipid-raft associated mediators of cell signaling, and proteins that interact with and regulate the actin cytoskeleton. Indeed, TCR signaling has been also linked to actin cytoskeleton rearrangements via the TCR co-receptor molecules CD4 and CD8, both of which have been shown to partition into lipid rafts and associate constitutively with the cytoskeleton (Arcaro *et al.*, 2000; Fragoso *et al.*, 2003).

A number of other lipid raft associating proteins and molecules have been shown to physically link to the cytoskeleton. For example, the generation of the raft associated phospholipid PIP2 is known to recruit WASP to the raft membrane. WASP is crucial for the translation of G-protein signaling into actin polymerization via its interactions with Slp-76 (which also binds phosphorylated LAT), Nck, profilin, and the Arp2/3 complex (Rozelle *et al.*, 2000). Indeed, the Arp2/3 complex has also been shown to co-purify with lipid rafts (von Haller, P.D., Yi, E., Donohoe, S., Vaughn, K., Keller, A., Nesvizhskii, A.I., Eng, J., Li, X., Goodlett, D.R., Aebersold, R., Watts, J.D., 2003, submitted) and allows for the polarized growth and branching of F-actin. Also, the constitutively lipid raft associated transmembrane adapter protein PAG/Cbp is coupled to F-actin through two classes of cytoskeletal adapter proteins: the ERM-binding phosphoprotein of 50 kDA (EBP50), and the ezrin, radixin and moesin (ERM) family of proteins (Brdickova *et al.*, 2001; Itoh *et al.*, 2002). ERMs localize via their N-termini to the membrane upon threonine phosphorylation or PIP2 binding, while their C-termini contain an F-actin binding domain, and EBP50 couples PAG/Cbp to ERMs via its PDZ domain (Brdickova *et al.*, 2001). Thus while it is clear that the link between lipid-raft associated receptors and signaling proteins to the actin-cytoskeletal network is critical to many cellular responses, many of the proteins involved in these processes, and the precise mechanisms for their regulation, have yet to be determined.

4.6. The TCR

As has been extensively discussed above, there is good evidence that many proteins vital to TCR signaling localize to lipid rafts. These include the PTKs Lck and Fyn, the adapter protein LAT, and several TCR co-receptor molecules, including CD4 and CD8, to name but a few. However, uncertainties still remain as to whether the TCR itself also localizes to lipid rafts. While imaging of the TCR at the cell surface has suggested TCR presence in rafts (Janes *et al.*, 1999), other biochemical data has contradicted this view, instead suggesting that only a small fraction of TCR is raft-associated, or that it's association is dependent on cell activation state (Montixi *et al.*, 1998; Kosugi *et al.*, 1999; Janes *et al.*, 2000; Balamuth *et al.*, 2001; Drevot *et al.*, 2002). A possible explanation for this conflicting evidence is that these studies used different detergents, concentrations of detergents and cell types, which would likely affect which proteins co-isolate with the rafts, as was discussed above.

One explanation for these conflicting observations is that the segregation of the signaling process occurs at the raft boundary with the plasma membrane. The TCR, upon activation, may localize to the edge of the coalescing rafts, where it would be able to interact with the various proteins it requires for signaling. A weak interaction with rafts such as this might then explain why varying amounts of TCR have been detected in rafts, and why its co-isolation is particularly detergent sensitive. This model also fits with another suggested role for lipid rafts: the transient and selective displacement of negative regulators of signaling, such as the PTK Csk (Davidson *et al.*, 2003), as discussed above. Indeed, the transmembrane phosphatase CD45, a known negative regulator for Src-family PTKs including Lck and Fyn, also seems to be largely excluded from T cell rafts, though a small amount of peripheral association may occur (Janes *et al.*, 1999; Thomas, 1999; Edmonds and Ostergaard, 2002).

5. THE NEED FOR SYSTEMATIC APPROACHES FOR IDENTIFICATION OF LIPID RAFT PROTEINS

From the data described above, it seems clear that the modular architecture of signaling complexes allows cells to assemble specific signaling modules for specific tasks, using a common set of signaling mediators. The data also suggest that partitioning of environment-sensing membrane proteins into small microdomains, such as lipid rafts, might also be a mechanism for the segregation of pre- or post-assembled signaling modules. Thus the identification of the those proteins that comprise these signaling complexes, be they constitutively or inducibly associated with rafts, is paramount to the understanding of lipid raft biology.

To date, lipid raft-associated proteins have generally been identified through conventional, hypothesis-driven, biochemical means. Such methodologies are often based on immunoblotting, to test suspicions on what observed gel bands might be, or the speculative probing of lipid raft isolates to see if the given protein of interest might, at least in part, partition into this fraction. While many raft-associated proteins have been identified by such hypothesis-driven approaches, to which the many papers cited in section 4 attest, the methodology is necessarily limited by the ability of the researcher to predict which proteins to test for, and/or the availability of suitable reagents for use in testing. This also means *de facto* that only previously known proteins can be determined in this fashion. Some studies have succeeded in identifying new and important raft-associate proteins by conventional biochemical means (purification, sequencing, cloning and verification). Examples of new raft-associated proteins identified *de novo* in this way include the T cell adapter protein PAG/Cbp (Brdicka *et al.*, 2000) and the B cell adapter protein NTAL (Brdicka *et al.*, 2002), also known as LAB (Janssen *et al.*, 2003).

This approach however, while highly effective, is slow to execute and typically works one protein at a time. A systematic approach, on the other hand, such as the emerging field of proteomics, is ideally suited to the rapid identification of many proteins contained in a complex mixture, such as those that associate with lipid rafts. Another advantage of the proteomic over the hypothesis-driven approach is that no assumptions need be made about the content of the protein samples analyzed. Instead, proteomics identifies proteins, both known and unknown, provided that sequence information on them (e.g. genomic, cDNA, mRNA, hypothetical proteins etc.) exists in a sequence database.

6. THE APPLICATION OF PROTEOMICS TO LIPID RAFT PROTEINS

Proteomics is the large-scale identification of multiple proteins in complex mixtures, and has been revolutionized over the past few years by significant advances in mass spectrometry (MS) instrumentation, analytical methodologies, and software applications for database searching and analysis of results (Goodlett and Yi, 2002; Mo and Karger, 2002; Aebersold and Mann, 2003; Patterson and Aebersold, 2003). These advances have enabled attempts at describing many proteins that co-isolated with various membrane compartments, traditionally difficult to analyze biochemically due to protein solubility issues. Such studies have included identification of proteins that co-isolated with Golgi membranes (Bell *et al.*, 2001), phagosomes (Dermine *et al.*, 2001) and with lipid rafts isolated from neutrophils (Nebl *et al.*, 2002) and T cells (von Haller *et al.*, 2001). The recent emergence of quantitative proteomic technology via stable isotope labeling of proteins now also allows for both the identification of proteins and the determination of their relative abundances in related samples, in the same experiment. This in turn has opened up new opportunities for the systematic analysis of the dynamic protein changes induced by biological processes in complex protein mixtures (Goodlett and Yi, 2002; Mo and Karger, 2002; Aebersold and Mann, 2003).

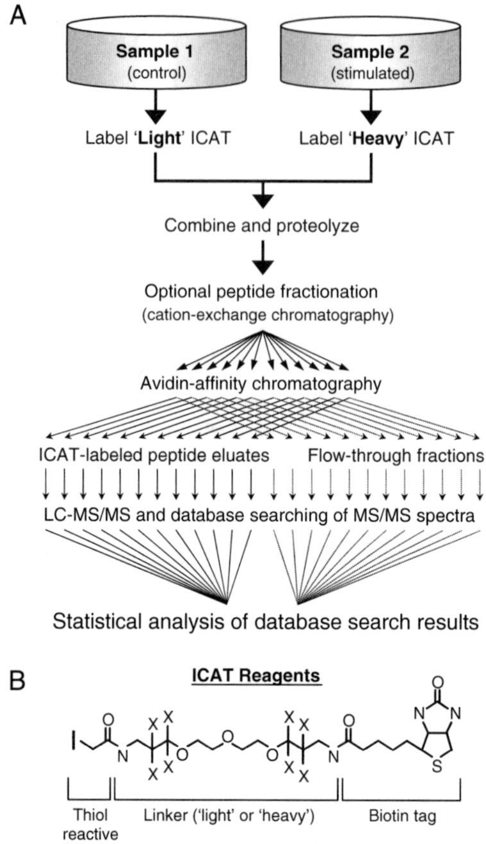

Figure 1. Schematic of a generic ICAT quantitative proteomic experiment. (A) Basic experimental strategy for an ICAT experiment. Two related protein samples (e.g. isolates from control and stimulated cell cultures) are labeled with isotopically normal ('light') and 'heavy' ICAT reagents (see panel B). Samples are combined, proteolyzed, and resultant peptides typically fractionated via one of more chromatographic steps, often including ion-exchange. Individual peptide fractions are separately subjected to avidin affinity chromatography, to enrich for the ICAT-labeled peptides, which are subjected to automated LC-MS/MS for peptide (hence protein) identification and quantification. Optional additional LC-MS/MS analyses of avidin affinity flow-through fractions will yield additional protein identifications, but no quantitative information. Peptide identification is achieved by database searching of recorded MS/MS data. Search results are then analyzed using statistical programs for determination of likely 'correct' and 'incorrect' peptide and protein identifications. (B) Structure of ICAT reagents used. A thiol reactive group allows for covalent coupling to (reduced) cysteinyl residues. A biotin tag allows for subsequent avidin affinity enrichment of labeled peptides. These groups are joined by a linker, which allows for better binding to avidin. Additionally, the linker exists in 2 forms one with eight hydrogens and the other eight deuterons where indicated (X). These two chemically identical reagents, referred to as the d0 and d8 versions, can be distinguished by MS, and thus allow for relative peptide/protein quantification.

One such approach to quantitative proteomics involves the use of isotope coded affinity tag (ICAT) reagents (Gygi *et al.*, 1999, 2002; Flory *et al.*, 2002), and is summarized in Figure 1A. In short, two related protein extracts are prepared (for example, from control and stimulated cell cultures). These protein preparations are labeled respectively with two versions of the same ICAT reagents (Figure 1B). ICAT reagents have a reactive group that will covalently couple to (reduced) cysteinyl groups, and a biotin moiety, which facilitates subsequent affinity enrichment of labeled (i.e. Cys-containing) peptides. These functional groups are joined by a linker that incorporates stable isotopes into one version of the reagent, yielding an isotopically 'light' (normal) and 'heavy' version of the same reagent. Following separate labeling of the two related protein samples with the two ICAT reagents, the samples are combined for subsequent proteolysis, peptide fractionation and purification steps, including avidin-affinity enrichment of the ICAT-labeled Cys-containing peptides. Finally, automated liquid chromatography tandem mass spectrometry (LC-MS/MS) and subsequent protein sequence database searching allows for peptide (and hence protein) sequence identification. The relative abundance of each protein thus identified in the original samples is then determined from the observed signal intensity ratios for the 'light' and 'heavy' versions of the observed ICAT-labeled peptides, readily distinguishable by MS, which were derived from the protein in question.

The key feature of this process, the isotopic tagging itself, which allows for the ultimate relative quantification of proteins in related samples is, in fact, quite adaptable. Indeed, several variations on such an approach have been successfully applied to a number of experimental systems. For example, ICAT reagents themselves have been successfully applied to a range of biological questions. These include the determination of changes in membrane protein components occurring as a result of induced cellular differentiation (Han *et al.*, 2001), the study of protein expression regulated by the Myc oncogene (Shiio *et al.*, 2002), the identification of new protein components of the yeast RNA polymerase II preinitiation complex (Ranish *et al.*, 2003) and the quantitave proteomic analysis of *Pseudomonas aeruginosa* isolates from cystic fibrosis airways (Guina *et al.*, 2003). ICAT reagents have also been produced in soluble or solid phase bound forms, either ^{2}H- or ^{13}C-labeled versions, and as acid-cleavable forms that either allow for removal of or dispense with completely the need for the biotin group, resulting in superior MS/MS data quality and increased protein identifications (Zhou *et al.*, 2002). Alternatively, *in vivo* metabolic labeling using ^{15}N-labeled amino acids, also referred to as stable isotope labeling by amino acids in cell culture (SILAC) (Ong *et al.*, 2002), has been used for the identification of epidermal growth factor signaling pathway components (Blagoev *et al.*, 2003). Other labeling approaches include methyl esterification of aspartic and glutamic acid side chains with deuterated methanol (Goodlett *et al.*, 2001), the use of

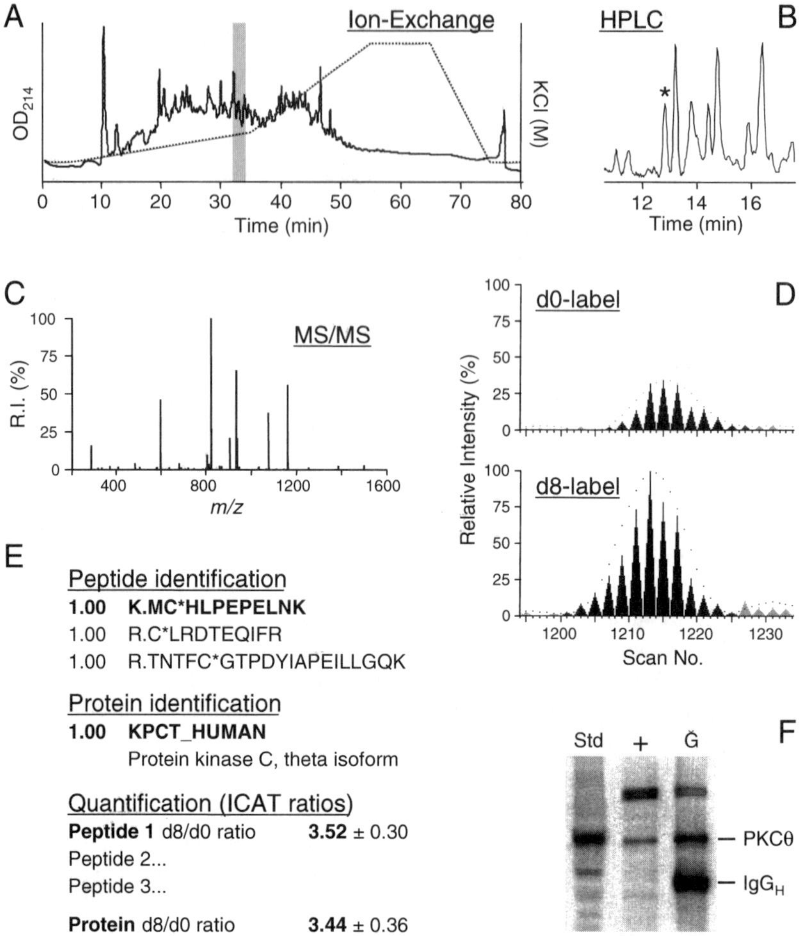

Figure 2. Identification and quantification of a single example protein (PKCθ) as part of a large proteomic study of Jurkat T cell lipid raft preparations. Following lipid raft sample isolation from control and stimulated Jurkat T cells and ICAT labeling of proteins (as described in Figure 1 and (von Haller *et al.*, 2003, submitted), samples were combined and digested with trypsin. (A) UV absorbance trace (solid line) of example cation exchange fractionation of pooled ICAT-labeled peptides, measured at 214 nm. Salt gradient used to elute peptides was 0 to 0.6 M KCl as indicated (dotted line). Individual cation exchange fractions (shaded) were separately processed over avidin affinity cartridges, collecting both eluate (ICAT-labeled) and flow-through (unlabeled) peptide fractions for subsequent LC-MS/MS. (B) Segment of an example LC total ion current trace, showing several peptide ion peaks, detected on-line by MS. Individual peptide ions thus detected (*) are automatically selected for fragmentation, and (C) the resultant MS/MS spectra are recorded. (D) Relative quantification of peptides is done by reconstructing ion chromatograms separately for the known masses of the d0 and d8 versions of each peptide selected for fragmentation, from the raw

^{18}O-labeled water during tryptic cleavage to label peptide C-termini (Shevchenko *et al.*, 1997; Stewart *et al.*, 2001) or labeling of primary amines with deuterated succinic anhydride (Wang and Regnier, 2001). All of these have been shown to be capable of yielding quantitative information.

Figure 2 illustrates how the ICAT approach, outlined in Figure 1A, has also been applied to Jurkat T cell lipid raft proteins (von Haller *et al.*, 2003, submitted). Following preparation of cell lysates (in 0.1% Triton-X 100) from control and stimulated cells (via TCR/CD28 cross-linking) the insoluble, low-density lipid raft membrane fractions were isolated via sucrose density ultracentrifugation. The control and stimulated preparations were separately labeled with isotopically normal (d0) and heavy (d8: containing 8 deuterons) ICAT reagents respectively, and then combined prior to tryptic proteolysis. Resultant peptides were next fractionated via ion-exchange chromatography (Figure 2A), and individual fractions (shaded) separately processed over avidin-affinity cartridges to enrich for the ICAT-labeled peptides. Since the d0 and d8 versions of the same peptide are chemically identical, they co-purify. Individual fractions are then analyzed via automated LC-MS/MS to both identify and quantify peptides. Figure 2B shows part of one such LC trace. Individually detected peptide peaks ions (*) were automatically selected for fragmentation (and hence sequence identification) via MS/MS (Figure 2C). The relative abundances of the same peptide in the two original samples were determined by reconstruction of the ion chromatograms for the unfragmented peptides (Figure 2D) and automated calculation of peak area ratios (the d8/d0 ratios). Peptide sequence information was obtained by searching the recorded MS/MS spectra (Figure 2C) against sequence databases (in this case human). Once the data for every fraction has

Figure 2. Continued.
data collected (panel B). Since these two versions of the peptide are chemically identical, the ratio between their signals is indicative of the relative abundances of the protein from which the peptide was derived in the original samples. (E) Following database searching of all MS/MS data, peptide and protein identifications are achieved by automated statistical modeling of the database search results (Keller *et al.*, 2002; Nesvizhskii *et al.*, 2003). Peptides and the proteins from which they are derived are assigned a probability of being correct (on a scale of 0 to 1, where 0 is 'incorrect' and 1 'correct'). Finally, automated software for determining ICAT d8/d0 ratios (from panel D) assigns quantitative information to protein identifications, where determinable. (F) Immunoblotting of the same lipid raft preparations used for proteomic analyses, retained at the start of the experiment, confirmed the MS determined identification of PKCθ and its ~3.5-fold enrichment in stimulated raft isolates (+) versus control isolates (−). Also shown is a PKCθ containing standard (Std) provided by the provider of the antibody. IgG_H: Immunoglobulin heavy chain detected by the secondary antibody, resulting from antibodies used to stimulate the TCR, present only in the stimulated raft isolates. More details on protein identification and quantification by MS can be found elsewhere (Flory *et al.*, 2002; Patton, 2002; Aebersold and Mann, 2003; Tao and Aebersold, 2003).

been collected, the entire set of search results was automatically analyzed using a suite of statistical data analysis tools, recently developed for interpretation of MS/MS search results. These software tools, by a process of statistical data modeling, assist in distinguishing between 'correct' and 'incorrect' peptide and protein identifications, their output being a computed probability that any given peptide or protein identification is 'correct', on a scale of 0 (for 'incorrect') to 1 (for 'correct') (Figure 2E). These tools are described extensively elsewhere (Keller *et al.*, 2002; Nesvizhskii *et al.*, 2003; von Haller *et al.*, 2003, submitted).

Figure 2 shows example data for the identification of the protein kinase C (PKC) theta isoform, which was identified via multiple high probability peptide matches. Furthermore, the ICAT ratio data indicated this protein was increased in amount in the stimulated lipid rafts versus control. ICAT peptide and protein ratios were also automatically determined, using another recently developed software tool for this process (Li, X., Zhang, H., Ranish, J. A., Aebersold, R., 2003, submitted). Finally, the identification of PKCθ and its ~3.5-fold increase in relative abundance upon cell stimulation, as determined by MS, were confirmed via immunoblotting (Figure 2F). These observations were consistent with previous reports of the rapid recruitment of PKCθ to lipid rafts in response to TCR/CD28 co-stimulation (Bi *et al.*, 2001; Dienz *et al.*, 2003).

6.1. Proteomic identification of raft proteins in T lymphocytes

Using the general approach summarized in Figures 1 and 2, we identified almost 700 proteins that co-purified with lipid rafts from Jurkat T lymphocytes. Furthermore, the adoption of the ICAT labeling approach allowed relative quantification of over half of these proteins, 50 of which were observed to be regulated following stimulation of the cells for 2 mins via TCR/CD28 cross-linking. These data are summarized in Table 1. Table 2 lists some examples of the protein complexes and individual proteins identified, some of which were also successfully quantified. When considering these data, it should be stressed that by the very nature of lipid raft isolation methods, it is not possible to isolate lipid rafts homogeneously. Thus the precise localization of the proteins prior to isolation cannot be known with certainty. However, despite this caveat, a considerable amount of valuable information can be obtained from large data sets of this nature.

As might be expected, a fairly large proportion of the proteins identified (81, 11.8%) were known antigens, receptors, or other known membrane proteins. Unsurprisingly, several TCR subunits were identified, and those

Table 1. Profile of proteins classes identified and quantified from Jurkat T cell lipid raft preparations

Protein class	Quantified proteins (with d8/d0 ratios)				
	Total[1]	Total[2]	Most up	Most down	Unch.
Total number of identified proteins	**685**	**355**	**43**	**7**	**305**
Antigen/receptor	36	18	3	–	15
ATPase	9	9	1	–	8
Contamination	6	2	–	–	2
Cytoskeleton	57	38	19	1	18
Cytosolic	23	13	2	–	11
ER/Golgi	44	20	–	–	20
G-protein, heterotrimeric	13	11	–	–	11
G-protein, other	26	16	–	–	16
G-protein, regulator	5	4	2	–	2
Heat shock/chaperone	14	7	–	–	7
Immunoglobulin	2	2	1	1	–
Membrane	45	24	2	–	22
Mitochondrial	135	67	2	1	64
Nuclear	38	15	–	–	15
Peroxisomal	7	3	1	–	2
Proteasome	4	1	–	–	1
Protein kinase, phosphatase & regulators	20	15	5	–	10
Ribosomal	52	31	–	–	31
Transporter	9	–	–	–	–
Unknown	112	45	3	4	38
Vesicle & vesicle sorting	14	9	1	–	8
VATP synthase	11	5	1	–	4
Viral	3	–	–	–	–

[1]Total protein identifications were made via LC-MS/MS analyses of both avidin affinity eluate (ICAT-labeled) and flow-through (unlabeled) peptides (see Figure 1).

[2]Of the 685 proteins identified, 355 yielded ICAT-labeled peptide data of sufficient quality to determine d8/d0 ratios. A ratio of >1 indicates raft *association*, and <1 indicates raft *dissociation* upon TCR/CD28 co-stimulation. Those quantified proteins deemed most strongly regulated (up or down versus control) were those that yielded d8/d0 ratios greater or less than 1 by 2.5x the relative deviation of ratios observed for a subset of 62 unchanging proteins, as described elsewhere (von Haller *et al.*, 2003, submitted). All remaining quantified proteins were considered as unchanged (Unch.). Protein classes were chosen arbitarily, with classification based upon database entry annotation, combined with additional literature searches when necessary. Database entries (proteins) annotated solely as 'unknown', 'hypothetical' or 'predicted' protein, cDNA, mRNA or partial coding sequence, with no indication of function, were collectively classified as 'unknown'. For all such gene sequences, BLAST searches were performed (http://www.ncbi.nlm.nih.gov/blast) and proteins reclassified to the appropriate group if a match to a known protein was made. Those sequences that failed to yield a BLAST match remained classified as 'unknown', as indicated in the table.

Table 2. Selected proteins and protein complexes identified and quantified from Jurkat T cell lipid raft preparations

Complex name	Components Quant/Ident	Regulated	Ratio/range
60s Ribosome	14/21	No	0.6 to 1.3
ARP2/3 complex	**5/7**	**Yes**	**1.6 to 2.0**
Mitochondrial inner membrane translocase	2/7	No	0.6 to 0.8
Myosin motor	**2/3**	**Yes**	**1.4 to 2.2**
NADH ubiquinone oxidoreductase	12/35	No	0.6 to 1.1
Nucleosome (histones)	0/5	ND	ND
TCR	**2/3**	**Yes**	**2.9 to 13.7**
VATP synthase	5/11	No	1.0 to 1.5
Protein name	**Identified**	**Regulated**	**Ratio/range**
Actins	**Multiple**	**Yes**	**1.5 to 2.1**
Calmodulin		ND	ND
Casein kinase I		No	0.8
CD47		ND	ND
CD81	Multiple	No	1.1
Coronins	**Multiple**	**Yes**	**1.5 to 2.2**
GTP-binding protein, H-Ras	Multiple	No	0.9 to 1.0
GTP-binding protein, Rac1	Multiple	No	0.7 to 1.0
GTP-binding protein, RhoA	Multiple	No	1.3
ICAM3		**Yes**	**1.9**
nGAP-like protein		**Yes**	**0.2 to 0.3**
Phosphoprotein associated with GEMs (PAG)		ND	ND
PKCθ	**Multiple**	**Yes**	**3.4 to 5.3**
Protein phosphatase, PP1-α1	**Multiple**	**Yes**	**1.3 to 1.6**
Protein tyrosine kinase, Fyn	Multiple	No	0.9 to 1.1
Protein tyrosine kinase, Itk		**Yes**	**2.3**
Protein tyrosine kinase, Lck	Multiple	No	1.0 to 1.4
RAS-GAP2		**Yes**	**2.1**
SNAP-23		No	1.0
α-SNAP		ND	ND
Spectrins	Multiple	No	0.5 to 0.8
Thy-1		ND	ND
Vimentin	**Multiple**	**Yes**	**0.5 to 0.6**
Unknown	**112**	**4 down, 3 up**	**0.2 to 5.7**

A selection of specific proteins and protein complexes identified and quantified, chosen from those summarized in Table 1. For protein complexes, the number of individual complex subunits identified (Ident), and the number of these additionally quantified (Quant), are given. Individual proteins identified in multiple experiments, or via multiple genes (e.g. Actin A, Actin B etc.) are indicated as such (Multiple). Those proteins/complexes that were regulated upon cell stimulation, determined via ICAT quantification, are indicated (**Yes**) and are also highlighted in **bold**. ND = not determinable (i.e. protein/complex identified only, did not yield quantitative ICAT data). The range of ICAT d8/d0 ratios observed (where A ratio of >1 indicates raft *association*, and <1 indicates raft *dissociation* upon TCR/CD28 co-stimulation) for each protein/complex quantified are also given where appropriate (i.e. lowest and highest value observed for multiple protein subunits of the same complex and/or quantifications of the same protein). For the 'unknown' protein class, 112 different database entries (proteins) were identified, of which 3 were up-regulated and 4 down-regulated, as determined by ICAT quantification.

quantified were also among the most strongly up-regulated in rafts post-stimulation (i.e. were rapidly recruited to rafts following TCR/CD28 co-stimulation). These observations were consistent with those of other groups who have also reported TCR redistribution to lipid rafts following various modes of TCR activation (Janes *et al.*, 2000; Drevot *et al.*, 2002). Other known T cell markers identified included CD2, CD59, Thy-1 and CD81. CD81 was notable since it is thought to be the target for both hepatitis C virus (Flint *et al.*, 2001) and malaria parasite (Silvie *et al.*, 2003) entry into liver cells. Since the biogenesis of lipid rafts is also known to originate from Golgi membranes in a process related to the endocytic and exocytic pathways (Ikonen, 2001; Nichols *et al.*, 2001), it was also not surprising to observe an additional 44 proteins (6.4%) known to localize to the ER or Golgi membranes. Also, thought it is known that various G-proteins (heterotrimeric and small) localize to membrane lipid rafts, as discussed above, the large number of such G-proteins (and their regulators) that were identified in the lipid raft isolates was still notable (43, 6.3%).

Another significant class of proteins found to co-purify with the lipid rafts was an array of cytoskeletal proteins (57, 8.3%), including actin and various actin-binding proteins. This was also unsurprising since lipid rafts have been shown to be localized to (though not limited to) the points on the cell surface where actin filaments attach (Babiychuk and Draeger, 2000; Rodgers and Zavzavadjian, 2001) and that signaling through the TCR involves many raft-associated proteins and initiates a rapid reorganization of actin fibers around the site of stimulation (Bunnell *et al.*, 2001; Foger *et al.*, 2001). Indeed, 20 of the 50 proteins whose raft association was regulated by TCR/CD28 stimulation were cytoskeletal proteins, thus confirming the significant functional importance of the lipid raft/TCR/cytoskeleton association. In particular, we observed all seven components of the Arp2/3 complex, known to link TCR signaling to actin remodeling (McGavin *et al.*, 2001). Furthermore, this complex was enriched in activated cell rafts, thus suggesting a mechanism for TCR and lipid raft modulated actin polymerization. We also found that another major protein complex associated with lipid rafts was the myosin motor, which was also further recruited to the rafts upon stimulation. One possible explanation for this is that the myosin motor may provide the energetic mechanism for the bringing of rafts together to form larger rafts, as has been observed by fluorescent imaging of T cell surface rafts upon TCR stimulation (Janes *et al.*, 1999). In this context, the detection of calmodulin in the rafts was also of interest, since it is also known that the activation of the myosin motor is mediated by the calmodulin-dependent myosin light chain kinase (Kamm and Stull, 2001). Other cytoskeletal proteins identified in the lipid raft isolates also included known raft proteins, including flotillins 1 and 2, coronins, and filamentous proteins such as tubulin, spectrins and vimentin. Indeed, a recent non-quantitative proteomic study of neutrophil lipid rafts identified a number of cytoskeletal proteins in common with those observed in T cells, including actin, myosin, α-actinin, and vimentin (Nebl *et al.*, 2002). Such observations support the notion that linkage of lipid rafts to the cytoskeleton is a feature common to most, if not all rafts, rather than being cell or tissue specific feature, or one restricted to only a small fraction of total cellular rafts.

As was discussed above, it is reasonable to assume that lipid rafts present on the cell surface differ significantly from one raft to the next, with each raft being specialized for participation in one or more signaling processes, based upon its protein components. The large number of potential raft proteins suggested in Table 1 would support this assumption. In the same way, the abundance of receptor and G-protein type molecules in lipid rafts also suggests that the signaling processes these molecules are involved with most likely represent a major functional class of lipid rafts on the cell surface. Similarly, the high abundance of actin and actin binding proteins like the ARP2/3 complex, as well as the predominance of myosin motor complexes, suggest that linkage of rafts to actin remodeling pathways, possibly in conjunction with myosin-mediated raft redistribution, may represent a significant functional element common to many, if not all rafts (at least in T cells).

Another class of proteins identified in T cell rafts worthy of note was proteins involved in vesicle sorting pathways (14, 2%). This may be related to the fact that lipid raft biogenesis originates in the Golgi apparatus and is related to the exocytosis pathway, as mentioned above. Fluorescent imaging of lipid rafts on T cells has indicated that TCR stimulation induces the formation of larger rafts than present on unstimulated cells (Janes *et al.*, 1999), possibly by a fusion process. An intriguing possibility from these observations is that vesicle fusion proteins may be functionally required for this process. Of the family of SNARE and SNAP protein families known to regulate vesicle fusion (Blumenthal *et al.*, 2003) we have consistently identified SNAP-23 in lipid raft isolates, a molecule previously shown to localize to the plasma membrane (Guo *et al.*, 1998). We also observed α-SNAP in the same isolates. This is notable since the NSF protein, an ATPase that regulates the vesicle fusion process, interacts with the SNARE complex via α-SNAP (May *et al.*, 1999), and that previous work had also identified NSF as an unexpected binding partner for the TCR co-receptor CD28 (Heller *et al.*, 2000).

A somewhat unexpected, but significant class of proteins also identified in T cell raft isolates was those proteins of no known function (112, 16.4%). These are represented by sequences in the database annotated us 'unknown' or 'hypothetical' protein, or as cDNA, mRNA or partial coding sequences only. This large number might be explained by the fact that lipid raft-associated proteins are, by definition, the most insoluble fraction of the already larger group of relatively insoluble membrane associated proteins. Historically, the most accessible proteins (based on high abundance and/or solubility) have been easiest and thus the first to be extracted, studied and characterized. Thus the lipid raft protein fraction could prove to be a rich source of new and interesting receptor, signaling and structural molecules, which many of these 112 'unknowns' might represent, especially those seen to change in relative abundance upon cell stimulation.

Finally, it is worth noting that we also observed many already known lipid raft associating proteins, including various members of the acylated Src-family of PTKs (including Lck and Fyn: both involved in TCR signaling), the recently described PAG/Cbp protein (Brdicka *et al.*, 2000), and the T cell protein LAT (B. Wollscheid, unpublished observations). However we also observed contamination coming from highly abundant protein complexes derived from other organelles, including 40s and 60s ribosomal proteins, various mitochondrial enzyme complexes, such as NADH ubiquinone oxidoreductase, and even nucleosome core histones from the nucleus. Such protein complexes are likely very much more abundant in the cell, in comparison with many true raft-associated proteins. Thus, as discussed above, since we know that it is not possible to isolate lipid rafts homogeneously and that lipid raft isolates vary significantly according to isolation procedure, some small contamination from such highly abundant complexes seems inevitable. Indeed, it is worth noting that ICAT data showed that nearly every such protein/complex quantified was at a ratio at or close to 1:1 (Table 1 and (von Haller *et al.*, 2003, submitted)). This would be expected, since stimulation of the TCR would not be expected to affect proteins in such intracellular compartments.

7. CONCLUSIONS AND FUTURE DIRECTIONS

The last decade has seen a considerable and growing interest in the microdomains, residing in the plasma membranes of eukaryotic cells, commonly referred to as lipid rafts. It has become clear that these structures are involved in the regulation of a wide and diverse range of biological functions, involving both sensing of the extracellular environment as well as the initiation of intracellular signaling pathways. Imaging of cell surface rafts has also shown them to be quite small (<50 nm) in resting cells. Together with the range of processes with which rafts are thought to be associated, this suggests that individual rafts likely differ in functional capability from each other, primarily determined at the level of their associated protein component. Thus a continued expansion of our knowledge of the protein components of lipid rafts, particularly those proteins that transiently associate with them in response to cellular stimulation, will provide valuable insight into both raft and protein function. Indeed, as we have described here, the emerging field of proteomics has considerable potential for the identification of proteins in complex mixtures, such as those that co-isolate with lipid rafts. Furthermore, quantitative proteomics has the capability to identify that class of more biologically interesting proteins that can transiently associate with rafts under certain biological conditions. However, as has also become clear over the last decade, the current inability to homogeneously isolate lipid rafts will continue to make the determination of proteins that genuinely associate with rafts problematic.

In order to advance the field of lipid raft research, it will be important to not only identify potential raft-associating proteins, but to verify their subcellular localization/distribution and determine with which raft-mediated processes they are linked to. Clearly, quantitative proteomics has the capability of identifying new potential raft-associate proteins, as well as to determine if this association is regulated by biological or chemical stimulus. Determination of which proteins are truly raft associated, however these proteins are identified, currently presents more of a challenge, and can certainly not be readily accomplished of the scale that proteomics can identify candidates. Presently, a few approaches lend themselves to answering these questions. Since lipid rafts have a different lipid composition from the bulk plasma membrane, various methods for membrane cholesterol depletion from cultured cell membranes can be used. Lipid rafts are rich in cholesterol, and upon cholesterol depletion, the lipid raft domains seem to dissipate. This process has been and can be used to distinguish raft proteins from other membrane proteins. Fluorescent imaging techniques have also proved most powerful for the determination of membrane compartmentalization of proteins, sometimes in conjunction with cholesterol depletion. Such studies can be used to determine whether candidate raft proteins co-localize with known raft proteins, or raft-specific lipids such as GM1. Perhaps, with the influx of information coming from approaches like proteomics, faster and more automated screens for such verification will be developed. Nevertheless, when used in conjunction with the identification of potentially new lipid raft proteins, such techniques will undoubtedly lead to a far greater understanding of these important membrane domains, the functions of the proteins that associate with them, and the biologically processes that they control.

REFERENCES

Aebersold, R., and Mann, M. [2003]. Mass spectrometry-based proteomics. *Nature*. **422**:198-207.

Anderson, R.G., and Jacobson, K. [2002]. A role for lipid shells in targeting proteins to caveolae, rafts, and other lipid domains. *Science*. **296**:1821-1825.

Arcaro, A., Gregoire, C., Boucheron, N., Stotz, S., Palmer, E., Malissen, B., and Luescher, I.F. [2000]. Essential role of CD8 palmitoylation in CD8 coreceptor function. *J Immunol*. **165**:2068-2076.

Babiychuk, E.B., and Draeger, A. [2000]. Annexins in cell membrane dynamics. Ca(2+)-regulated association of lipid microdomains. *J Cell Biol.* **150**:1113-1124.

Balamuth, F., Leitenberg, D., Unternaehrer, J., Mellman, I., and Bottomly, K. [2001]. Distinct patterns of membrane microdomain partitioning in Th1 and th2 cells. *Immunity.* **15**:729-738.

Bell, A.W., Ward, M.A., Blackstock, W.P., Freeman, H.N., Choudhary, J.S., Lewis, A.P., Chotai, D., Fazel, A., Gushue, J.N., Paiement, J., Palcy, S., Chevet, E., Lafreniere-Roula, M., Solari, R., Thomas, D.Y., Rowley, A., and Bergeron, J.J. [2001]. Proteomics characterization of abundant Golgi membrane proteins. *J Biol Chem.* **276**:5152-5165.

Bi, K., Tanaka, Y., Coudronniere, N., Sugie, K., Hong, S., van Stipdonk, M.J., and Altman, A. [2001]. Antigen-induced translocation of PKC-theta to membrane rafts is required for T cell activation. *Nat Immunol.* **2**:556-563.

Blagoev, B., Kratchmarova, I., Ong, S.E., Nielsen, M., Foster, L.J., and Mann, M. [2003]. A proteomics strategy to elucidate functional protein-protein interactions applied to EGF signaling. *Nat Biotechnol.* **21**:315-318.

Blanchard, J. [2000]. Small GTPases, adhesion, cell cycle control and proliferation. *Pathologie et Biologie.* **48**(3):318-327.

Blumenthal, R., Clague, M.J., Durell, S.R., and Epand, R.M. [2003]. Membrane fusion. *Chem Rev.* **103**:53-69.

Brdicka, T., Imrich, M., Angelisova, P., Brdickova, N., Horvath, O., Spicka, J., Hilgert, I., Luskova, P., Draber, P., Novak, P., Engels, N., Wienands, J., Simeoni, L., Osterreicher, J., Aguado, E., Malissen, M., Schraven, B., and Horejsi, V. [2002]. Non-T cell activation linker (NTAL): a transmembrane adaptor protein involved in immunoreceptor signaling. *J Exp Med.* **196**:1617-1626.

Brdicka, T., Pavlistova, D., Leo, A., Bruyns, E., Korinek, V., Angelisova, P., Scherer, J., Shevchenko, A., Hilgert, I., Cerny, J., Drbal, K., Kuramitsu, Y., Kornacker, B., Horejsi, V., and Schraven, B. [2000]. Phosphoprotein associated with glycosphingolipid-enriched microdomains (PAG), a novel ubiquitously expressed transmembrane adaptor protein, binds the protein tyrosine kinase csk and is involved in regulation of T cell activation. *J Exp Med.* **191**:1591-1604.

Brdickova, N., Brdicka, T., Andera, L., Spicka, J., Angelisova, P., Milgram, S.L., and Horejsi, V. [2001]. Interaction between two adapter proteins, PAG and EBP50: a possible link between membrane rafts and actin cytoskeleton. *FEBS Lett.* **507**:133-136.

Brodsky, F., Chen, C., Knuehl, C., Towler, M., and Wakeham, D. [2001]. Biological basket weaving: Formation and function of clathrin-coated vesicles. *Annual Review of Cell & Developmental Biology.* **12**:517-568.

Brown, D. [1994]. GPI-anchored proteins and detergent-resistant membrane domains. *Braz J Med Biol Res.* **27**:309-315.

Brown, D.A., Crise, B., and Rose, J.K. [1989]. Mechanism of membrane anchoring affects polarized expression of two proteins in MDCK cells. *Science.* **245**:1499-1501.

Brown, D.A., and London, E. [1997]. Structure of detergent-resistant membrane domains: does phase separation occur in biological membranes? *Biochem Biophys Res Commun.* **240**:1-7.

Brown, D.A., and London, E. [2000]. Structure and function of sphingolipid- and cholesterol-rich membrane rafts. *J Biol Chem.* **275**:17221-17224.

Brown, D.A., and Rose, J.K. [1992]. Sorting of GPI-anchored proteins to glycolipid-enriched membrane subdomains during transport to the apical cell surface. *Cell.* **68**: 533-544.

Brown, M.T., and Cooper, J.A. [1996]. Regulation, substrates and functions of src. *Biochim Biophys Acta.* **1287**:121-149.

Bunnell, S.C., Kapoor, V., Trible, R.P., Zhang, W., and Samelson, L.E. [2001]. Dynamic actin polymerization drives T cell receptor-induced spreading: a role for the signal transduction adaptor LAT. *Immunity*. **14**:315-329.

Carter, W.G., and Hakomori, S. [1981]. A new cell surface, detergent-insoluble glycoprotein matrix of human and hamster fibroblasts. The role of disulfide bonds in stabilization of the matrix. *J Biol Chem.* **256**:6953-6960.

Chawla, M., and Vishwakarma, R.A. [2003]. Alkylacylglycerolipid domain of GPI molecules of Leishmania is responsible for inhibition of PKC-mediated c-fos expression. *J Lipid Res.* **44**:594-600.

Das, V., Nal, B., Roumier, A., Meas-Yedid, V., Zimmer, C., Olivo-Marin, J.C., Roux, P., Ferrier, P., Dautry-Varsat, A., and Alcover, A. [2002]. Membrane-cytoskeleton interactions during the formation of the immunological synapse and subsequent T-cell activation. *Immunol Rev.* **189**:123-135.

Davidson, D., Bakinowski, M., Thomas, M.L., Horejsi, V., and Veillette, A. [2003]. Phosphorylation-dependent regulation of T-cell activation by PAG/Cbp, a lipid raft-associated transmembrane adaptor. *Mol Cell Biol.* **23**:2017-2028.

Dermine, J.F., Duclos, S., Garin, J., St-Louis, F., Rea, S., Parton, R.G., and Desjardins, M. [2001]. Flotillin-1-enriched lipid raft domains accumulate on maturing phagosomes. *J Biol Chem.* **276**:18507-18512.

Drake, D.R., 3rd, and Braciale, T.J. [2001]. Cutting edge: lipid raft integrity affects the efficiency of MHC class I tetramer binding and cell surface TCR arrangement on CD8+ T cells. *J Immunol.* **166**:7009-7013.

Drevot, P., Langlet, C., Guo, X.J., Bernard, A.M., Colard, O., Chauvin, J.P., Lasserre, R., and He, H.T. [2002]. TCR signal initiation machinery is pre-assembled and activated in a subset of membrane rafts. *Embo J.* **21**:1899-1908.

Edidin, M. [2001]. Shrinking patches and slippery rafts: scales of domains in the plasma membrane. *Trends Cell Biol.* **11**:492-496.

Edmonds, S.D., and Ostergaard, H.L. [2002]. Dynamic association of CD45 with detergent-insoluble microdomains in T lymphocytes. *J Immunol.* **169**:5036-5042.

Field, K.A., Holowka, D., and Baird, B. [1995]. Fc epsilon RI-mediated recruitment of p53/56lyn to detergent-resistant membrane domains accompanies cellular signaling. *Proc Natl Acad Sci USA*. **92**:9201-9205.

Fivaz, M., Vilbois, F., Thurnheer, S., Pasquali, C., Abrami, L., Bickel, P.E., Parton, R.G., and van der Goot, F.G. [2002]. Differential sorting and fate of endocytosed GPI-anchored proteins. *Embo J.* **21**:3989-4000.

Flint, M., Quinn, E.R., and Levy, S. [2001]. In search of hepatitis C virus receptor(s). *Clin Liver Dis.* **5**:873-893.

Flory, M.R., Griffin, T.J., Martin, D., and Aebersold, R. [2002]. Advances in quantitative proteomics using stable isotope tags. *Trends Biotechnol.* **20**:S23-29.

Foger, N., Marhaba, R., and Zoller, M. [2001]. Involvement of CD44 in cytoskeleton rearrangement and raft reorganization in T cells. *J Cell Sci.* **114**:1169-1178.

Ford, T., Graham, J., and Rickwood, D. [1994]. Iodixanol: a nonionic iso-osmotic centrifugation medium for the formation of self-generated gradients. *Anal Biochem.* **220**:360-366.

Fragoso, R., Ren, D., Zhang, X., Su, M.W., Burakoff, S.J., and Jin, Y.J. [2003]. Lipid raft distribution of CD4 depends on its palmitoylation and association with Lck, and evidence for CD4-induced lipid raft aggregation as an additional mechanism to enhance CD3 signaling. *J Immunol.* **170**:913-921.

Friedrichson, T., and Kurzchalia, T.V. [1998]. Microdomains of GPI-anchored proteins in living cells revealed by crosslinking. *Nature*. **394**:802-805.

Galbiati, F., Volonte, D., Meani, D., Milligan, G., Lublin, D.M., Lisanti, M.P., and Parenti, M. [1999]. The dually acylated NH2-terminal domain of gi$_1\alpha$ is sufficient to target a green fluorescent protein reporter to caveolin-enriched plasma membrane domains. Palmitoylation of caveolin-1 is required for the recognition of dually acylated g-protein alpha subunits in vivo. *J Biol Chem.* **274**:5843-5850.

Gauld, S., Dal, P.J., and Cambier, J. [2002]. B cell antigen receptor signaling: Roles in cell development and disease. *Science.* **296**:1641-1642.

Goodlett, D.R., Keller, A., Watts, J.D., Newitt, R., Yi, E.C., Purvine, S., Eng, J.K., von Haller, P., Aebersold, R., and Kolker, E. [2001]. Differential stable isotope labeling of peptides for quantitation and de novo sequence derivation. *Rapid Commun Mass Spectrom.* **15**:1214-1221.

Goodlett, D.R., and Yi, E.C. [2002]. Proteomics without polyacrylamide: qualitative and quantitative uses of tandem mass spectrometry in proteome analysis. *Funct Integr Genomics.* **2**:138-153.

Graham, J., Ford, T., and Rickwood, D. [1994]. The preparation of subcellular organelles from mouse liver in self-generated gradients of iodixanol. *Anal Biochem.* **220**:367-373.

Guina, T., Purvine, S.O., Yi, E.C., Eng, J., Goodlett, D.R., Aebersold, R., and Miller, S.I. [2003]. Quantitative proteomic analysis indicates increased synthesis of a quinolone by *Pseudomonas aeruginosa* isolates from cystic fibrosis airways. *Proc Natl Acad Sci USA.* **100**:2771-2776.

Guo, Z., Turner, C., and Castle, D. [1998]. Relocation of the t-SNARE SNAP-23 from lamellipodia-like cell surface projections regulates compound exocytosis in mast cells. *Cell.* **94**:537-548.

Gygi, S.P., Rist, B., Gerber, S.A., Turecek, F., Gelb, M.H., and Aebersold, R. [1999]. Quantitative analysis of complex protein mixtures using isotope-coded affinity tags. *Nat Biotechnol.* **17**:994-999.

Gygi, S.P., Rist, B., Griffin, T.J., Eng, J., and Aebersold, R. [2002]. Proteome analysis of low-abundance proteins using multidimensional chromatography and isotope-coded affinity tags. *J Proteome Res.* **1**:47-54.

Han, D.K., Eng, J., Zhou, H., and Aebersold, R. [2001]. Quantitative profiling of differentiation-induced microsomal proteins using isotope-coded affinity tags and mass spectrometry. *Nat Biotechnol.* **19**:946-951.

Harder, T., Scheiffele, P., Verkade, P., and Simons, K. [1998]. Lipid domain structure of the plasma membrane revealed by patching of membrane components. *J Cell Biol.* **141**: 929-942.

Harris, J., Werling, D., Hope, J., Taylor, G., and Howard, C. [2002]. Caveolae and caveolin in immune cells: distribution and functions. *Trends in Immunology.* **23**:158-164.

Heller, M., Goodlett, D.R., Watts, J.D., and Aebersold, R. [2000]. A comprehensive characterization of the T-cell antigen receptor complex composition by microcapillary liquid chromatography-tandem mass spectrometry. *Electrophoresis.* **21**:2180-2195.

Henning, S., and Cantrell, D. [1998]. GTPases in Antigen Receptor Signalling. *Current Opinion in Immunology.* **10**:322-329.

Hilgemann, D.W., Feng, S., and Nasuhoglu, C. [2001]. The complex and intriguing lives of PIP2 with ion channels and transporters. *Sci STKE.* **2001**:RE19.

Hiscox, S., Hallett, M.B., Morgan, B.P., and van den Berg, C.W. [2002]. GPI-anchored GFP signals Ca^{2+} but is homogeneously distributed on the cell surface. *Biochem Biophys Res Commun.* **293**:714-721.

Ikezawa, H. [2002]. Glycosylphosphatidylinositol (GPI)-anchored proteins. *Biol Pharm Bull.* **25**:409-417.

Ikonen, E. [2001]. Roles of lipid rafts in membrane transport. *Curr Opin Cell Biol.* **13**: 470-477.

Itoh, K., Sakakibara, M., Yamasaki, S., Takeuchi, A., Arase, H., Miyazaki, M., Nakajima, N., Okada, M., and Saito, T. [2002]. Cutting edge: negative regulation of immune synapse formation by anchoring lipid raft to cytoskeleton through Cbp-EBP50-ERM assembly. *J Immunol.* **168**:541-544.

Janes, P.W., Ley, S.C., and Magee, A.I. [1999]. Aggregation of lipid rafts accompanies signaling via the T cell antigen receptor. *J Cell Biol.* **147**:447-461.

Janes, P.W., Ley, S.C., Magee, A.I., and Kabouridis, P.S. [2000]. The role of lipid rafts in T cell antigen receptor (TCR) signalling. *Semin Immunol.* **12**:23-34.

Janssen, E., Zhu, M., Zhang, W., and Koonpaew, S. [2003]. LAB: a new membrane-associated adaptor molecule in B cell activation. *Nat Immunol.* **4**:117-123.

Kabouridis, P.S., Magee, A.I., and Ley, S.C. [1997]. S-acylation of LCK protein tyrosine kinase is essential for its signalling function in T lymphocytes. *Embo J.* **16**:4983-4998.

Kamm, K., and Stull, J. [2001]. Dedicated myosin light chain kinases with diverse cellular functions. *Journal of Biological Chemistry.* **276**:4527-4530.

Kawabuchi, M., Satomi, Y., Takao, T., Shimonishi, Y., Nada, S., Nagai, K., Tarakhovsky, A., and Okada, M. [2000]. Transmembrane phosphoprotein Cbp regulates the activities of Src-family tyrosine kinases. *Nature.* **404**:999-1003.

Keller, A., Nesvizhskii, A.I., Kolker, E., and Aebersold, R. [2002]. Empirical statistical model to estimate the accuracy of peptide identifications made by MS/MS and database search. *Anal Chem.* **74**:5383-5392.

Kenworthy, A.K., Petranova, N., and Edidin, M. [2000]. High-resolution FRET microscopy of cholera toxin B-subunit and GPI-anchored proteins in cell plasma membranes. *Mol Biol Cell.* **11**:1645-1655.

Kosugi, A., Saitoh, S., Noda, S., Yasuda, K., Hayashi, F., Ogata, M., and Hamaoka, T. [1999]. Translocation of tyrosine-phosphorylated TCRzeta chain to glycolipid-enriched membrane domains upon T cell activation. *Int Immunol.* **11**:1395-1401.

Kovarova, M., Tolar, P., Arudchandran, R., Draberova, L., Rivera, J., and Draber, P. [2001]. Structure-function analysis of Lyn kinase association with lipid rafts and initiation of early signaling events after Fcepsilon receptor I aggregation. *Mol Cell Biol.* **21**: 8318-8328.

Lehto, M.T., and Sharom, F.J. [2002]. PI-specific phospholipase C cleavage of a reconstituted GPI-anchored protein: modulation by the lipid bilayer. *Biochemistry.* **41**: 1398-1408.

Liang, X., Nazarian, A., Erdjument-Bromage, H., Bornmann, W., Tempst, P., and Resh, M.D. [2001]. Heterogeneous fatty acylation of Src family kinases with polyunsaturated fatty acids regulates raft localization and signal transduction. *J Biol Chem.* **276**:30987-30994.

Lisanti, M.P., Caras, I.W., Davitz, M.A., and Rodriguez-Boulan, E. [1989]. A glycophospholipid membrane anchor acts as an apical targeting signal in polarized epithelial cells. *J Cell Biol.* **109**:2145-2156.

Loertscher, R., and Lavery, P. [2002]. The role of glycosyl phosphatidyl inositol (GPI)-anchored cell surface proteins in T-cell activation. *Transpl Immunol.* **9**:93-96.

May, A.P., Misura, K.M., Whiteheart, S.W., and Weis, W.I. [1999]. Crystal structure of the amino-terminal domain of N-ethylmaleimide-sensitive fusion protein. *Nat Cell Biol.* **1**:175-182.

McGavin, M.K., Badour, K., Hardy, L.A., Kubiseski, T.J., Zhang, J., and Siminovitch, K.A. [2001]. The intersectin 2 adaptor links Wiskott Aldrich Syndrome protein (WASp)-mediated actin polymerization to T cell antigen receptor endocytosis. *J Exp Med.* **194**: 1777-1787.

Melkonian, K.A., Ostermeyer, A.G., Chen, J.Z., Roth, M.G., and Brown, D.A. [1999]. Role of lipid modifications in targeting proteins to detergent-resistant membrane rafts. Many raft proteins are acylated, while few are prenylated. *J Biol Chem.* **274**:3910-3917.

Miura, Y., Hanada, K., and Jones, T.L. [2001]. G(s) signaling is intact after disruption of lipid rafts. *Biochemistry.* **40**:15418-15423.

Mo, W., and Karger, B.L. [2002]. Analytical aspects of mass spectrometry and proteomics. *Curr Opin Chem Biol.* **6**:666-675.

Moffett, S., Brown, D.A., and Linder, M.E. [2000]. Lipid-dependent targeting of G proteins into rafts. *J Biol Chem.* **275**:2191-2198.

Montixi, C., Langlet, C., Bernard, A.M., Thimonier, J., Dubois, C., Wurbel, M.A., Chauvin, J.P., Pierres, M., and He, H.T. [1998]. Engagement of T cell receptor triggers its recruitment to low-density detergent-insoluble membrane domains. *Embo J.* **17**:5334-5348.

Moran, M., and Miceli, M.C. [1998]. Engagement of GPI-linked CD48 contributes to TCR signals and cytoskeletal reorganization: a role for lipid rafts in T cell activation. *Immunity.* **9**:787-796.

Morris, A., and Malbon, C. [1999]. Physiological regulation of G protein-linked signaling. *Physiological Reviews.* **79**:1373-1430.

Mukherjee, S., and Maxfield, F.R. [2000]. Role of membrane organization and membrane domains in endocytic lipid trafficking. *Traffic.* **1**:203-211.

Mumby, S.M. [1997]. Reversible palmitoylation of signaling proteins. *Curr Opin Cell Biol.* **9**:148-154.

Muniz, M., and Riezman, H. [2000]. Intracellular transport of GPI-anchored proteins. *Embo J.* **19**:10-15.

Nebl, T., Pestonjamasp, K.N., Leszyk, J.D., Crowley, J.L., Oh, S.W., and Luna, E.J. [2002]. Proteomic analysis of a detergent-resistant membrane skeleton from neutrophil plasma membranes. *J Biol Chem.* **277**:43399-43409.

Nesvizhskii, A.I., Keller, A., Kolker, E., and Aebersold, R. [2003]. A statistical model for identifying proteins by tandem mass spectrometry. *Anal.Chem., in press.*

Neves, S., Ram, P., and Iyengar, R. [2002]. G protein pathways. *Science.* **296**:1636-1639.

Nichols, B.J., Kenworthy, A.K., Polishchuk, R.S., Lodge, R., Roberts, T.H., Hirschberg, K., Phair, R.D., and Lippincott-Schwartz, J. [2001]. Rapid cycling of lipid raft markers between the cell surface and Golgi complex. *J Cell Biol.* **153**:529-541.

Niv, H., Gutman, O., Kloog, Y., and Henis, Y.I. [2002]. Activated K-Ras and H-Ras display different interactions with saturable nonraft sites at the surface of live cells. *J Cell Biol.* **157**:865-872.

Ong, S.E., Blagoev, B., Kratchmarova, I., Kristensen, D.B., Steen, H., Pandey, A., and Mann, M. [2002]. Stable isotope labeling by amino acids in cell culture, SILAC, as a simple and accurate approach to expression proteomics. *Mol Cell Proteomics.* **1**:376-386.

Patterson, S.D., and Aebersold, R.H. [2003]. Proteomics: the first decade and beyond. *Nat Genet.* **33 Suppl**:311-323.

Patton, W.F. [2002]. Detection technologies in proteome analysis. *J Chromatogr B Analyt Technol Biomed Life Sci.* **771**:3-31.

Pike, L.J., Han, X., Chung, K.N., and Gross, R.W. [2002]. Lipid rafts are enriched in arachidonic acid and plasmenylethanolamine and their composition is independent of caveolin-1 expression: a quantitative electrospray ionization/mass spectrometric analysis. *Biochemistry.* **41**:2075-2088.

Pike, L.J., and Miller, J.M. [1998]. Cholesterol depletion delocalizes phosphatidylinositol bisphosphate and inhibits hormone-stimulated phosphatidylinositol turnover. *J Biol Chem.* **273**:22298-22304.

Pollard, T., and Beltzner, C. [2002]. Structure and function of the Arp2/3 complex. *Current Opinion in Structural Biology, 2002 Dec.* **12**:768-774.

Pralle, A., Keller, P., Florin, E.L., Simons, K., and Horber, J.K. [2000]. Sphingolipid-cholesterol rafts diffuse as small entities in the plasma membrane of mammalian cells. *J Cell Biol.* **148**:997-1008.

Prior, I.A., and Hancock, J.F. [2001]. Compartmentalization of Ras proteins. *J Cell Sci.* **114**:1603-1608.

Prior, I.A., Harding, A., Yan, J., Sluimer, J., Parton, R.G., and Hancock, J.F. [2001]. GTP-dependent segregation of H-ras from lipid rafts is required for biological activity. *Nat Cell Biol.* **3**:368-375.

Puri, V., Watanabe, R., Dominguez, M., Sun, X., Wheatley, C.L., Marks, D.L., and Pagano, R.E. [1999]. Cholesterol modulates membrane traffic along the endocytic pathway in sphingolipid-storage diseases. *Nat Cell Biol.* **1**:386-388.

Ranish, J.A., Yi, E.C., Leslie, D.M., Purvine, S.O., Goodlett, D.R., Eng, J., and Aebersold, R. [2003]. The study of macromolecular complexes by quantitative proteomics. *Nat Genet.* **33**:349-355.

Razani, B., Woodman, S., and Lisanti, M. [2002]. Caveolae: From cell biology to animal physiology. *Pharmacological Reviews.* **54**:431-467.

Ridley, A.J. [2001]. Rho family proteins: coordinating cell responses. *Trends Cell Biol.* **11**:471-477.

Rodgers, W., and Zavzavadjian, J. [2001]. Glycolipid-enriched membrane domains are assembled into membrane patches by associating with the actin cytoskeleton. *Exp Cell Res.* **267**:173-183.

Roy, S., Luetterforst, R., Harding, A., Apolloni, A., Etheridge, M., Stang, E., Rolls, B., Hancock, J.F., and Parton, R.G. [1999]. Dominant-negative caveolin inhibits H-Ras function by disrupting cholesterol-rich plasma membrane domains. *Nat Cell Biol.* **1**:98-105.

Rozelle, A.L., Machesky, L.M., Yamamoto, M., Driessens, M.H., Insall, R.H., Roth, M.G., Luby-Phelps, K., Marriott, G., Hall, A., and Yin, H.L. [2000]. Phosphatidylinositol 4,5-bisphosphate induces actin-based movement of raft-enriched vesicles through WASP-Arp2/3. *Curr Biol.* **10**:311-320.

Samelson, L.E., Bunnell, S.C., Trible, R.P., Yamazaki, T., and Zhang, W. [1999]. Studies on the adapter molecule LAT. *Cold Spring Harb Symp Quant Biol.* **64**:259-263.

Schlessinger, J. [2000]. New roles for Src kinases in control of cell survival and angiogenesis. *Cell.* **100**:293-296.

Shenoy-Scaria, A.M., Gauen, L.K., Kwong, J., Shaw, A.S., and Lublin, D.M. [1993]. Palmitylation of an amino-terminal cysteine motif of protein tyrosine kinases p56lck and p59fyn mediates interaction with glycosyl-phosphatidylinositol-anchored proteins. *Mol Cell Biol.* **13**:6385-6392.

Shevchenko, A., Chernushevich, I., Ens, W., Standing, K.G., Thomson, B., Wilm, M., and Mann, M. [1997]. Rapid 'de novo' peptide sequencing by a combination of nanoelectrospray, isotopic labeling and a quadrupole/time-of-flight mass spectrometer. *Rapid Commun Mass Spectrom.* **11**:1015-1024.

Shields, J.M., Pruitt, K., McFall, A., Shaub, A., and Der, C.J. [2000]. Understanding Ras: 'it ain't over 'til it's over'. *Trends Cell Biol.* **10**:147-154.

Shiio, Y., Donohoe, S., Yi, E.C., Goodlett, D.R., Aebersold, R., and Eisenman, R.N. [2002]. Quantitative proteomic analysis of Myc oncoprotein function. *Embo J.* **21**:5088-5096.

Silvie, O., Rubinstein, E., Franetich, J.F., Prenant, M., Belnoue, E., Renia, L., Hannoun, L., Eling, W., Levy, S., Boucheix, C., and Mazier, D. [2003]. Hepatocyte CD81 is required for *Plasmodium falciparum* and *Plasmodium yoelii* sporozoite infectivity. *Nat Med.* **9**:93-96.

Simons, K., and Ikonen, E. [1997]. Functional rafts in cell membranes. *Nature.* **387**: 569-572.

Simons, K., and Toomre, D. [2000]. Lipid rafts and signal transduction. *Nat Rev Mol Cell Biol.* **1**:31-39.
Simons, K., and van Meer, G. [1988]. Lipid sorting in epithelial cells. *Biochemistry.* **27**:6197-6202.
Smart, E.J., Graf, G.A., McNiven, M.A., Sessa, W.C., Engelman, J.A., Scherer, P.E., Okamoto, T., and Lisanti, M.P. [1999]. Caveolins, liquid-ordered domains, and signal transduction. *Mol Cell Biol.* **19**:7289-7304.
Stewart, II, Thomson, T., and Figeys, D. [2001]. ^{18}O labeling: a tool for proteomics. *Rapid Commun Mass Spectrom.* **15**:2456-2465.
Stuermer, C.A., Lang, D.M., Kirsch, F., Wiechers, M., Deininger, S.O., and Plattner, H. [2001]. Glycosylphosphatidyl inositol-anchored proteins and Fyn kinase assemble in noncaveolar plasma membrane microdomains defined by reggie-1 and -2. *Mol Biol Cell.* **12**:3031-3045.
Stulnig, T.M., Huber, J., Leitinger, N., Imre, E.M., Angelisova, P., Nowotny, P., and Waldhausl, W. [2001]. Polyunsaturated eicosapentaenoic acid displaces proteins from membrane rafts by altering raft lipid composition. *J Biol Chem.* **276**:37335-37340.
Superti-Furga, G. [1995]. Regulation of the Src protein tyrosine kinase. *FEBS Lett.* **369**: 62-66.
Takuwa, N., and Takuwa, Y. [2001]. Regulation of cell cycle molecules by the Ras effector system. [2001] *Molecular & Cellular Endocrinology.* **177**(1-2):25-33.
Tao, W.A., and Aebersold, R. [2003]. Advances in quantitative proteomics via stable isotope tagging and mass spectrometry. *Curr Opin Biotechnol.* **14**:110-118.
Thomas, J.L., Holowka, D., Baird, B., and Webb, W.W. [1994]. Large-scale co-aggregation of fluorescent lipid probes with cell surface proteins. *J Cell Biol.* **125**:795-802.
Thomas, M.L. [1999]. The regulation of antigen-receptor signaling by protein tyrosine phosphatases: a hole in the story. *Curr Opin Immunol.* **11**:270-276.
Thomas, S.M., and Brugge, J.S. [1997]. Cellular functions regulated by Src family kinases. *Annu Rev Cell Dev Biol.* **13**:513-609.
Tseng, S.Y., and Dustin, M.L. [2002]. T-cell activation: a multidimensional signaling network. *Curr Opin Cell Biol.* **14**:575-580.
van den Berg, C.W., Cinek, T., Hallett, M.B., Horejsi, V., and Morgan, B.P. [1995]. Exogenous glycosyl phosphatidylinositol-anchored CD59 associates with kinases in membrane clusters on U937 cells and becomes Ca(2+)-signaling competent. *J Cell Biol.* **131**:669-677.
Varma, R., and Mayor, S. [1998]. GPI-anchored proteins are organized in submicron domains at the cell surface. *Nature.* **394**:798-801.
von Haller, P.D., Donohoe, S., Goodlett, D.R., Aebersold, R., and Watts, J.D. [2001]. Mass spectrometric characterization of proteins extracted from Jurkat T cell detergent-resistant membrane domains. *Proteomics.* **1**:1010-1021.
von Haller, P.D., Yi, E., Donohoe, S., Vaughn, K., Keller, A., Nesvizhskii, A.I., Eng, J., Li, X., Goodlett, D.R., Aebersold, R., and Watts, J.D. [2003]. Evaluation of ICAT and tandem mass spectrometry methodologies for large scale protein analysis and the application of statistical modeling approaches for data interpretation. Submitted.
Waheed, A.A., and Jones, T.L. [2002]. Hsp90 interactions and acylation target the G protein Galpha 12 but not Galpha 13 to lipid rafts. *J Biol Chem.* **277**:32409-32412.
Wang, S., and Regnier, F.E. [2001]. Proteomics based on selecting and quantifying cysteine containing peptides by covalent chromatography. *J Chromatogr A.* **924**:345-357.
Wang, X.M., Djordjevic, J.T., Bender, V., and Manolios, N. [2002]. T cell antigen receptor (TCR) transmembrane peptides colocalize with TCR, not lipid rafts, in surface membranes. *Cell Immunol.* **215**:12-19.

Wedegaertner, P.B., Wilson, P.T., and Bourne, H.R. [1995]. Lipid modifications of trimeric G proteins. *J Biol Chem.* **270**:503-506.

Wherlock, M., and Mellor, H. [2002]. The Rho GTPase family: a Racs to Wrchs story. *J Cell Sci.* **115**:239-240.

Williams, J.C., Wierenga, R.K., and Saraste, M. [1998]. Insights into Src kinase functions: structural comparisons. *Trends Biochem Sci.* **23**:179-184.

Wu, M., Fan, J., Gunning, W., and Ratnam, M. [1997]. Clustering of GPI-anchored folate receptor independent of both cross-linking and association with caveolin. *J Membr Biol.* **159**:137-147.

Xavier, R., Brennan, T., Li, Q., McCormack, C., and Seed, B. [1998]. Membrane compartmentation is required for efficient T cell activation. *Immunity.* **8**:723-732.

Zhang, W., Trible, R.P., and Samelson, L.E. [1998]. LAT palmitoylation: its essential role in membrane microdomain targeting and tyrosine phosphorylation during T cell activation. *Immunity.* **9**:239-246.

Zhang, W., Trible, R.P., Zhu, M., Liu, S.K., McGlade, C.J., and Samelson, L.E. [2000]. Association of Grb2, Gads, and phospholipase C-gamma 1 with phosphorylated LAT tyrosine residues. Effect of LAT tyrosine mutations on T cell angigen receptor-mediated signaling. *J Biol Chem.* **275**:23355-23361.

Zhou, H., Ranish, J.A., Watts, J.D., and Aebersold, R. [2002]. Quantitative proteome analysis by solid-phase isotope tagging and mass spectrometry. *Nat Biotechnol.* **20**:512-515.

Chapter 4

Lipid Composition of Membrane Domains

Kamen S. Koumanov[*,❖], Claude Wolf[❖], and Peter J. Quinn[♦]
*Institute of Biophysics, Bulgarian Academy of Sciences, 1113 Sofia, Bulgaria;
❖Faculte de Medecine Saint Antoine, Universite Pierre et Marie Curie, 75571 Paris Cedex 12, France;
♦Department of Life Sciences, King's College London, 150 Stamford Street, London SW1 9NN, UK

Abstract
The isolation of subfractions of cell membranes on the basis of their solubility in non-ionic detergents has led to the discovery of lipid domain structure in membranes. Detergents used for this purpose include Triton, Brij, Lubrol and CHAPS. Different lipid constituents are known to resist solubilization by different detergents and the resulting fractions may associate with different membrane proteins. In general, the detergent-resistant membrane fractions tend to be dominated by saturated molecular species of sphingomyelin and phosphatidylcholine and invariably include significant proportions of cholesterol. The lipid composition is consistent with formation of liquid-ordered phases. The present evidence favours a model in which the lateral segregation of membrane proteins takes place on the basis of their affinity for liquid-ordered lipid domains within the membrane.

1. INTRODUCTION

The existence of domains in the surface membrane of cells has been postulated by observing the motion of constituent molecules. Much of the evidence demonstrating constraints on motion include single particle tracking of colloidal gold labelled molecules on the surface of living cells (Daumas *et al.*, 2003), the fluorescence observed on partitioning of fluorescent probes into restricted domains (Gousset *et al.*, 2002), single dye tracing (Schutz *et al.*, 2000; Dietrich *et al.*, 2002), fluorescence microscopy (Radhakrishnan *et al.*, 2000; Dietrich *et al.*, 2001a,b; Dietrich et al., 2001; Anderson and McConnell 2002) and fluorescence quenching of membrane probes (Wang and Silvius 2000, 2001).

The creation of domains of restricted motion where the properties are distinct from domains where the motion of constituents is relatively fluid implies that the composition in lateral regions are different. While this is obvious in the case of specialized membrane domains like intercellular junctions and the like, where particular membrane proteins dominate the organisation of the constituents, the detection of membrane rafts is not so clear cut. In raft domains, it is suggested that the interaction between membrane lipids is responsible for segregation of the constituents so it might be anticipated that the lipid composition would be characteristic of such fractions.

The major challenge to research in this area has been to isolate membrane domains and to compare the respective composition of the motionally restricted regions with those of the fluid domains. The use of non-ionic detergents like Triton, Brij, Lubrol and 3-[(3-cholamidopropyl)dimethylamonio]-1-propyl sulphonate (CHAPS) to separate soluble and insoluble fractions of membranes has provided an operational basis for the isolation of rafts which resist solubilization (detergent-resistant membranes; DRM). The influence of the detergent in promoting domain formation by segregating components present in the parent membrane remains somewhat conjectural. This chapter will examine the lipid composition of detergent-resistant membrane preparations examine the evidence as to whether detergents segregate lipids into rafts and the extent to which these detergent-resistant membrane fractions resemble domains present in the original membrane.

2. FRACTIONATION OF MEMBRANE DOMAINS

Kirkpatrick and colleagues (Kirkpatrick *et al.*, 1974) were amongst the first to report selective solubilization of membrane lipids by treatment with Triton X-100. They showed that membrane proteins and lipids were solubilized in human erythrocyte membranes with increasing concentrations of detergent up to 5 mM. The insoluble fractions under these conditions were found to preferentially retain sphingomyelin and cholesterol.

Since then many studies have shown that treatment of a range of cell membranes with mild detergents at low temperatures (0–4°C) results in selective solubilization of the structure (Ahmed *et al.*, 1997; Schroeder *et al.*, 1998; Janes *et al.*, 2000; Xu and London 2000; Li *et al.*, 2001; Wang *et al.*, 2001; Xu *et al.*, 2001).

The insoluble fraction floats at low density upon gradient centrifugation and the fraction can be easily harvested (Brown and Rose, 1992; Parton *et al.*, 2002). Resudual detergent is then removed the resulting membrane can be further purified or subfractionated.

Analysis of the proteins in such fractions shows that they are predominantly attached to the membrane by lipid anchors. Such proteins include glycosylphosphatidylinositol (GPI)-anchored proteins originating from the outer surface of the plasma membrane and diacylated cytoplasmic proteins such as certain Src family kinases originating from the inner surface of the membrane (Bagnat *et al.*, 2000; Moffett *et al.*, 2000). The lipid composition of rafts, like the protein constituents, differs from the composition of the parent membrane providing evidence that fractionation depends on the solubility of the lipid in detergent solution.

3. EFFECT OF DETERGENT STRUCTURE

The lipid composition of detergent-resistant membrane fractions isolated from a particular membrane, especially the proportion of cholesterol, depends to a certain extent on the type of detergent used. Membrane fractions, for example, that are insoluble in 1% (w/v) Lubrol WX are found to be soluble in Triton X-100 at the same concentration (Roper *et al.*, 2000). Analysis of the lipid composition of detergent-resistant membrane fractions from human macrophages and fibroblasts showed that Lubrol WX insoluble membranes had more unsaturated phosphatidylcholine and less cholesterol per phosphatidylcholine than fractions insoluble in Triton X-100 (Drobnik *et al.*, 2002). Differential solubility in nonionic detergents such as Triton X-100, CHAPS and Lubrol has been exploited to define trafficking pathways of membrane proteins in cells (Mairhofer *et al.*, 2002; Sakyo and Kitagawa, 2002; Slimane *et al.*, 2003). The implication of such studies is that membrane lipid domains are primarily responsible for directing the traffic.

Preparations of DRM fractions insoluble in 1% Nonidet P-40 have been used to examine signal transduction pathways in lymphocytes (Zubiaur *et al.*, 2002). CD38 signal transduction in T-lymphocytes was said to be due to formation of supramolecular signalling complexes of CD3 molecules bearing immunoreceptor tyrosine-based activation motifs together with protein tyrosine kinases. In other studies insoluble fractions in 1% Brij 58 at 4°C were used to demonstrate that T-cell receptors are associated into cell-surface domains upon receptor stimulation providing a mechanism for coupling receptor location with downstream signalling cascades (Montixi *et al.*, 1998). Some concern has been expressed as to whether the low-temperature conditions (0–4°C) used to prepare DRMs by solubility in non-ionic detergents converts fluid lipid domains into liquid-ordered phases that resist solubilization by detergent. This question has been addressed by preparing DRM fractions at different temperatures and using different types of detergent. A recent study has been reported on preparations of detergent-resistant membrane fractions from microvilli membranes from pig small intestine (Braccia *et al.*, 2003). The lipids present in DRMs prepared using Triton X-100 at low temperatures were compared with DRMs isolated from the same membrane using Brij 98 at physiological temperatures. The lipid profiles of the DRMs were closely alike suggesting that the overall solubilizing properties of the two detergents were similar but the insoluble fraction obtained with Brij 98 was more enriched with phospholipids. These results indicate that rafts are not an artefact arising from thermotropic phase behaviour of saturated molecular species of lipid in the temperature range 0°–37°C and consistent with other studies of DRM fractions isolated with Brij 98 (Drevot *et al.*, 2002). This infers that insolubility in non-ionic detergent depends on the phase structure formed by association of particular lipid constituents that is stable over a temperature range 0–37°C rather than thermotropic phases that are generated within this temperature range.

4. LIPIDS OF DETERGENT-RESISTANT MEMBRANES

The dominant lipid constituents of detergent resistant membrane preparations from a variety of cells are sphingomyelin, cholesterol and saturated molecular species of phospholipids and glycolipids. While sphingomyelin and cholesterol have tended to achieve prominence in raft lipid composition more recent studies have indicated that glycerophospholipids and diacylglycerols are also constituents of raft fractions. One method that has been used to identify such components is the detection of lipids from radioactive precursors that are isolated in raft fractions (Rouquette-Jazdanian *et al.*, 2002). The method involves feeding cells with radiolabelled glycerol, fatty acids or water-soluble polar groups and identifying the complex lipid into which they are biosynthetically incorporated. Using this approach it was shown that DRM fractions derived from human leukaemic T-cell line Jurkat had a considerably higher cholesterol content than the parent membrane and that polar lipids incorporating [^{3}H]-glycerol were present. These glycerophospholipids included choline, ethanolamine, serine and inositol phosphoglycerides with a preponderance of phosphatidylcholine and phosphatidylserine. Incorporation of radiolabelled fatty acid precursors into the phospholipids showed preferential labelling of DRM lipids with saturated fatty acids such as palmitic and stearic acids rather than oleic, linoleic and arachidonic acids. These results are consistent with other studies indicating that the lipids isolated in the DRM fractions contain

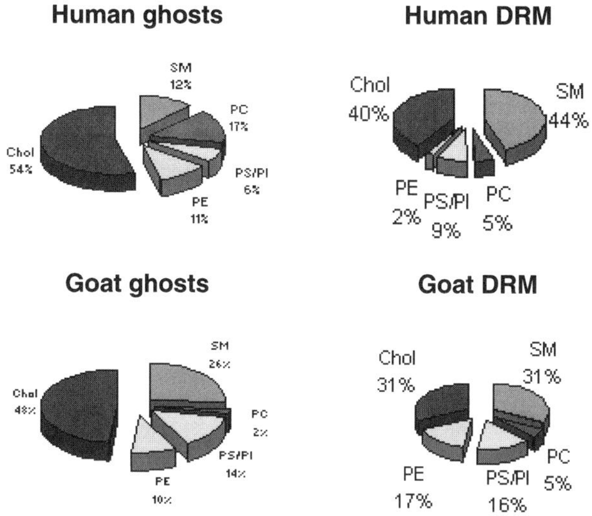

Figure 1. Major lipid composition of human and goat erythrocyte membranes and composition of corresponding detergent-resistant membrane fractions.

predominantly saturated molecular species of membrane lipids (Bunnell *et al.*, 2000; Dietrich *et al.*, 2001).

Analysis of the lipid composition of detergent-resistant membrane fractions prepared by treatment of human and goat (ruminant) erythrocyte membranes with Triton X-100 at low temperature has been used to determine the type of lipids that are solubilized and those that remain in the insoluble membrane fraction (Koumanov *et al.*, unpublished). The major lipid classes in each of the DRM fractions and the parent membranes from which they were derived are presented in Figure 1.

The relative proportion of major lipids did not differ greatly from one preparation to another suggesting that detergent treatment leads to reproducible DRM fractions, at least in respect of their lipid composition. The most notable features of the lipid composition are firstly, the relatively high proportion of sphingomyelin in goat erythrocyte ghosts which, in common with other ruminants, results from a replacement of phosphatidylcholine by sphingomyelin. Secondly, sphingomyelin appeared to be the most resistant of the polar lipids to solubilization by the detergent and the most abundant lipid found in the DRM fractions. The fact that the sphingomyelin content of human erythrocyte ghosts was low relative to that of goat was reflected in the proportion of the ghost membrane that is insoluble in Triton X-100. The human DRM fraction represented only 5% of the ghost membrane whereas 27% of the lipids of the goat erythrocyte membrane was insoluble in Triton X-100. The ratio of sphingomyelin : cholesterol is of particular interest. Cholesterol represents approximately half the total lipids of erythrocytes of both species but the proportion of chosterol recovered in the DRM fraction is significantly less than this. Indeed cholesterol is present in about the same proportion as sphingomyelin in each case. This suggests that the amount of cholesterol that resists solubilization by detergent is determined by the extent to which sphingomyelin remains insoluble in Triton X-100. The stoichiometry of sphingomyelin:cholesterol in the goat erythrocyte membrane compared to that in the DRM fraction suggests that in this membrane cholesterol is excluded from the domain that remains insoluble in Triton X-100. The result implies that it is the presence of saturated molecular species of sphingomyelin that determines the detergent-resistant domain and not the cholesterol.

To investigate the factors that govern the partition of sphingomyelin into the DRM fraction of human and goat erythrocyte membranes the molecular species of sphingomyelin recovered in the DRM fractions were compared with those in the membrane ghosts from which they were derived. The fatty acids in amide ester linkage to the sphingosine isolated from human and goat erythrocyte ghost membranes and corresponding DRM fractions were analysed and the results are shown in Table 1. In both human and goat the saturated molecular species of sphingomyelin dominate the DRM fractions at the expense of monoenoic fatty acids. This is achieved in the human membranes by an approximately three-fold enrichment of C22:0 and C24:0 molecular species of sphingomyelin and

Table 1. The fatty acids amide ester linked to the sphingosine of sphingomyelin isolated from human and goat erythrocyte ghosts and corresponding DRM fractions

	Human				Goat			
Fatty acid	Ghosts		DRM		Ghosts		DRM	
	M	SD	M	SD	M	SD	M	SD
C16:0	41.45	5.68	8.63	0.74	27.94	1.35	44.26	5.21
C18:0	6.25	0.74	6.36	0.66	3.27	0.42	3.44	0.29
C18:1	1.15	0.21	1.49	0.21	7.80	0.62	0.86	0.09
C18:2	0.10	0.01	0.18	0.03	0.79	0.09	0.15	0.01
C20:0	0.70	0.00	2.22	0.03	0.07	0.01	0.06	0.00
C22:0	5.78	0.48	14.72	2.02	0.76	0.09	0.65	0.05
C24:0	20.02	2.35	62.10	5.69	1.59	0.10	1.92	0.18
C24:1	24.02	1.87	1.47	0.19	55.58	4.85	47.07	6.25
C26:0	0.33	0.04	2.78	0.21	0.11	0.02	0.09	0.01
C26:1	0.19	0.01	0.05	0.00	2.10	0.18	1.52	0.28
SAT	74.54		96.81		33.74		50.41	
MONO	25.36		3.01		65.48		49.44	
PUFA	0.10		0.18		0.79		0.15	

the almost complete exclusion of molecular species of sphingomyelin associated with C24:1 fatty acid. Similarly, in the goat membranes the DRM fraction is enriched in the C16:0 molecular species and the exclusion of C18:1 and C24:1 sphingomyelin. The overall effect is that the more saturated molecular species of sphingomyelin dominate the DRM fractions and these molecular species appear to preferentially interact with cholesterol in a stoichiometry of 1:1.

5. PROTEIN TARGETTING TO DRM DOMAINS

The lipid domain concept of protein sorting involves segregation of membrane proteins according to their inclusion or exclusion from domains created, in turn, by the phase separation of liquid-ordered from fluid regions of the lipid bilayer. The model implies that boundaries between such domains are relatively distinct but this may not necessarily be the case.

Recent evidence suggests that subtle differences in lipid phase structure can bring about the localization of different lipid-anchored protein in separated cell surface domains. This is evidenced from analysis of the lipid composition of detergent-resistant membranes to which different GPI-anchored proteins are associated (Brugger *et al.*, submitted). The conclusions were based on segregation of detergent-resistant membrane fractions of rat brain prepared with Brij 96 using immunoaffinity purification of membranes containing lipid-anchored Thy-1 and prion protein (PrP), respectively. Unlike Triton X-100, preparation of detergent-resistant membrane fractions with Brij 96 does not result in mixing of domains from different membranes (Madore *et al.*, 1999). A summary of the analyses of fatty acids associated with sphingomyelin and phosphatidylcholine are presented in Table 2. This shows that both detergent-resistant membrane fractions contained significantly more saturated than mono- and polyunsaturated molecular species of the phospholipids compared with

Table 2. Analysis of the fatty acids associated with the major phospholipids of detergent resistant membranes prepared with Brij 98 from brain neuronal microsomes subfractionated by immunoaffinity for Thy-1 and PrP primary antibodies

Lipid	Membrane	Saturated	Monounsaturated	Polyunsaturated
SM	Whole membrane	65	33	1.8
	Thy-1	61	38	0.7
	PrP	47	51	2.1
PC	Whole membrane	13	49	34
	Thy-1	53	41	3
	PrP	38	55	6

the parent membrane from which they were derived. Equally conspicuous is the fact that the detergent-resistant membrane with which Thy-1 is associated contains molecular species of phospholipid that are significantly more saturated than their counterparts associated with PrP. This provides evidence that different GPI-anchored proteins from the same cell associate with different membrane microenvironments presumably representing domains differing in lipid phase structure.

6. CHANGES IN DRMs RELATED TO FUNCTION

There are likely to be a number of ways in which signal transduction may be modulated by rafts. Localization of receptors and associated signalling components into rafts, for example, may be a necessary process in production or transduction of a signal. Sequestration of signalling molecules, on the other hand, could suppress the generation of signals or prevent non-specific interactions required to complete a signalling pathway. Underlying the localization of signalling molecules is the formation of the lipid domains themselves and there is evidence that the lipid composition of detergent-resistant membranes are different when isolated from cells in different functional states.

The effect of light on the retinal rod outer segment disk membrane on DRM composition has been reported by (Boesze-Battaglia *et al.*, 2002). Detergent resistant membranes were prepared from retinal rod outer disc membranes either in the dark or under illumination and an analysis of the lipid composition undertaken. Lipids present in detergent-resistant membranes prepared in the dark and light using Triton X-100 were compared with fractions prepared with the non-ionic detergent, octyl glucopyranoside. A summary of the results is presented in Table 3. This shows that the lipid composition of

Table 3. Lipid composition of membrane fractions insoluble in Triton X-100 and octyl glucopyranosyl (OG) prepared from retinal rod outer segment disk membranes in the dark or under illumination

Treatment	Phospholipid*	PL/SM	SM/cholesterol
Light			
Triton X-100	17	9	2.1
OG	30	126	16.4
Dark			
Triton X-100	19	10	2.4
OG	26	111	13.4

*% of total phospholipid in disk membrane; PL, phospholipid; SM, sphingomyelin; data derived from Boesze-Battaglia *et al.*, 2002.

the detergent-resistant membranes differs markedly depending on the detergent used. While no significant differences in lipid composition could be detected in fractions from light and dark membranes prepared with Triton X-100 differences are observed in the octyl glucopyranosyl-insoluble fractions. Membranes from illuminated preparations represented a greater proportion of the total membrane and they tended to be more enriched in sphingomyelin compared with cholesterol.

7. CONCLUSIONS

The existance of lipid domains on the cell surface is considered to be a dynamic process. The creation and dissipation of liquid-ordered phases is coupled to the response of cells to specific agonists and there is evidence that the process is a component of the mechanism of transducing the signals generated. As is generally the case, signals which are generated at one point in time need to be attenuated once the response is no longer appropriate. The dynamics of lipid domains must therefore be considered in the context of membrane lipid homeostasis and in particular the activation of enzymes responsible for tailoring the lipid molecular species present at any time. One model of this process is the activation of sphingomyelinases which depend on phase structure (substrate presentation) of the lipid bilayer upon which these enzymes act. Further studies will doubtless reveal the role of these enzymes in homeostatic regulation of membrane domain dynamics.

REFERENCES

Ahmed, S. N., Brown, D. A. and London, E., 1997, On the origin of sphingolipid/cholesterol-rich detergent-insoluble cell membranes: physiological concentrations of cholesterol and sphingolipid induce formation of a detergent-insoluble, liquid-ordered lipid phase in model membranes, *Biochemistry*, **36:** 10944-10953.

Anderson, T. G. and McConnell, H. M., 2002, A thermodynamic model for extended complexes of cholesterol and phospholipid, *Biophys J*, **83:** 2039-2052.

Bagnat, M., Keranen, S., Shevchenko, A. and Simons, K., 2000, Lipid rafts function in biosynthetic delivery of proteins to the cell surface in yeast, *Proc Natl Acad Sci USA*, **97:** 3254-3259.

Boesze-Battaglia, K., Dispoto, J. and Kahoe, M. A., 2002, Association of a photoreceptor-specific tetraspanin protein, ROM-1, with triton X-100-resistant membrane rafts from rod outer segment disk membranes, *J Biol Chem*, **277:** 41843-41849.

Braccia, A., Villani, M., Immerdal, L., Niels-Christiansen, L. L., Nystrom, B. T., Hansen, G. H. and Danielsen, E. M., 2003, Microvillar membrane microdomains exist at physiological temperature. Role of galectin-4 as lipid raft stabilizer revealed by "superrafts", *J Biol Chem*, **278:** 15679-15684.

Brown, D. A. and Rose, J. K., 1992, Sorting of GPI-anchored proteins to glycolipid-enriched membrane subdomains during transport to the apical cell surface, *Cell*, **68:** 533-544.

Bunnell, S. C., Diehn, M., Yaffe, M. B., Findell, P. R., Cantley, L. C. and Berg, L. J., 2000, Biochemical interactions integrating Itk with the T cell receptor-initiated signaling cascade, *J Biol Chem*, **275:** 2219-2230.

Daumas, F., Destainville, N., Millot, C., Lopez, A., Dean, D. and Salome, L., 2003, Confined diffusion without fences of a g-protein-coupled receptor as revealed by single particle tracking, *Biophys J*, **84:** 356-366.

Dietrich, C., Bagatolli, L. A., Volovyk, Z. N., Thompson, N. L., Levi, M., Jacobson, K. and Gratton, E., 2001a, Lipid rafts reconstituted in model membranes, *Biophys J*, **80:** 1417-1428.

Dietrich, C., Volovyk, Z. N., Levi, M., Thompson, N. L. and Jacobson, K., 2001b, Partitioning of Thy-1, GM1, and cross-linked phospholipid analogs into lipid rafts reconstituted in supported model membrane monolayers, *Proc Natl Acad Sci USA*, **98:** 10642-10647.

Dietrich, C., Yang, B., Fujiwara, T., Kusumi, A. and Jacobson, K., 2002, Relationship of lipid rafts to transient confinement zones detected by single particle tracking, *Biophys J*, **82:** 274-284.

Drevot, P., Langlet, C., Guo, X. J., Bernard, A. M., Colard, O., Chauvin, J. P., Lasserre, R. and He, H. T., 2002, TCR signal initiation machinery is pre-assembled and activated in a subset of membrane rafts, *Embo J*, **21:** 1899-1908.

Drobnik, W., Borsukova, H., Bottcher, A., Pfeiffer, A., Liebisch, G., Schutz, G. J., Schindler, H. and Schmitz, G., 2002, Apo AI/ABCA1-dependent and HDL3-mediated lipid efflux from compositionally distinct cholesterol-based microdomains, *Traffic*, **3:** 268-278.

Gousset, K., Wolkers, W. F., Tsvetkova, N. M., Oliver, A. E., Field, C. L., Walker, N. J., Crowe, J. H. and Tablin, F., 2002, Evidence for a physiological role for membrane rafts in human platelets, *J Cell Physiol*, **190:** 117-128.

Janes, P. W., Ley, S. C., Magee, A. I. and Kabouridis, P. S., 2000, The role of lipid rafts in T cell antigen receptor (TCR) signalling, *Semin Immunol*, **12:** 23-34.

Kirkpatrick, F. H., Gordesky, S. E. and Marinetti, G. V., 1974, Differential solubilization of proteins, phospholipids, and cholesterol of erythrocyte membranes by detergents, *Biochim Biophys Acta*, **345:** 154-161.

Li, X. M., Momsen, M. M., Smaby, J. M., Brockman, H. L. and Brown, R. E., 2001, Cholesterol decreases the interfacial elasticity and detergent solubility of sphingomyelins, *Biochemistry*, **40:** 5954-5963.

Madore, N., Smith, K. L., Graham, C. H., Jen, A., Brady, K., Hall, S. and Morris, R., 1999, Functionally different GPI proteins are organized in different domains on the neuronal surface, *Embo J*, **18:** 6917-6926.

Mairhofer, M., Steiner, M., Mosgoeller, W., Prohaska, R. and Salzer, U., 2002, Stomatin is a major lipid-raft component of platelet alpha granules, *Blood*, **100:** 897-904.

Moffett, S., Brown, D. A. and Linder, M. E., 2000, Lipid-dependent targeting of G proteins into rafts, *J Biol Chem*, **275:** 2191-2198.

Montixi, C., Langlet, C., Bernard, A. M., Thimonier, J., Dubois, C., Wurbel, M. A., Chauvin, J. P., Pierres, M. and He, H. T., 1998, Engagement of T cell receptor triggers its recruitment to low-density detergent-insoluble membrane domains, *Embo J*, **17:** 5334-5348.

Parton, R. G., Molero, J. C., Floetenmeyer, M., Green, K. M. and James, D. E., 2002, Characterization of a distinct plasma membrane macrodomain in differentiated adipocytes, *J Biol Chem*, **277:** 46769-46778.

Radhakrishnan, A., Anderson, T. G. and McConnell, H. M., 2000, Condensed complexes, rafts, and the chemical activity of cholesterol in membranes, *Proc Natl Acad Sci USA*, **97:** 12422-12427.

Roper, K., Corbeil, D. and Huttner, W. B., 2000, Retention of prominin in microvilli reveals distinct cholesterol-based lipid micro-domains in the apical plasma membrane, *Nat Cell Biol*, **2:** 582-592.

Rouquette-Jazdanian, A. K., Pelassy, C., Breittmayer, J. P., Cousin, J. L. and Aussel, C., 2002, Metabolic labelling of membrane microdomains/rafts in Jurkat cells indicates the presence of glycerophospholipids implicated in signal transduction by the CD3 T-cell receptor, *Biochem J*, **363:** 645-655.

Sakyo, T. and Kitagawa, T., 2002, Differential localization of glucose transporter isoforms in non-polarized mammalian cells: distribution of GLUT1 but not GLUT3 to detergent-resistant membrane domains, *Biochim Biophys Acta*, **1567:** 165-175.

Schroeder, R. J., Ahmed, S. N., Zhu, Y., London, E. and Brown, D. A., 1998, Cholesterol and sphingolipid enhance the Triton X-100 insolubility of glycosylphosphatidylinositol-anchored proteins by promoting the formation of detergent-insoluble ordered membrane domains, *J Biol Chem*, **273:** 1150-1157.

Schutz, G. J., Kada, G., Pastushenko, V. P. and Schindler, H., 2000, Properties of lipid microdomains in a muscle cell membrane visualized by single molecule microscopy, *Embo J*, **19:** 892-901.

Slimane, T. A., Trugnan, G., Van, I. S. C. and Hoekstra, D., 2003, Raft-mediated Trafficking of Apical Resident Proteins Occurs in Both Direct and Transcytotic Pathways in Polarized Hepatic Cells: Role of Distinct Lipid Microdomains, *Mol Biol Cell*, **14:** 611-624.

Wang, T. Y., Leventis, R. and Silvius, J. R., 2001, Partitioning of lipidated peptide sequences into liquid-ordered lipid domains in model and biological membranes, *Biochemistry*, **40:** 13031-13040.

Wang, T. Y. and Silvius, J. R., 2000, Different sphingolipids show differential partitioning into sphingolipid/cholesterol-rich domains in lipid bilayers, *Biophys J*, **79:** 1478-1489.

Wang, T. Y. and Silvius, J. R., 2001, Cholesterol does not induce segregation of liquid-ordered domains in bilayers modeling the inner leaflet of the plasma membrane, *Biophys J*, **81:** 2762-2773.

Xu, X., Bittman, R., Duportail, G., Heissler, D., Vilcheze, C. and London, E., 2001, Effect of the structure of natural sterols and sphingolipids on the formation of ordered sphingolipid/sterol domains (rafts). Comparison of cholesterol to plant, fungal, and disease-associated sterols and comparison of sphingomyelin, cerebrosides, and ceramide, *J Biol Chem*, **276:** 33540-33546.

Xu, X. and London, E., 2000, The effect of sterol structure on membrane lipid domains reveals how cholesterol can induce lipid domain formation, *Biochemistry*, **39:** 843-849.

Zubiaur, M., Fernandez, O., Ferrero, E., Salmeron, J., Malissen, B., Malavasi, F. and Sancho, J., 2002, CD38 is associated with lipid rafts and upon receptor stimulation leads to Akt/protein kinase B and Erk activation in the absence of the CD3-zeta immune receptor tyrosine-based activation motifs, *J Biol Chem*, **277:** 13-22.

Part III

Creation of Membrane Microdomains

Chapter 5

Sphingomyelin and Cholesterol: From Membrane Biophysics and Rafts to Potential Medical Applications

Yechezkel Barenholz
Laboratory of Membrane and Liposome Research, The Hebrew University – Hadassah Medical School, Jerusalem, Israel

Abstract

The preferential sphingomyelin–cholesterol interaction which results from the structure and the molecular properties of these two lipids seems to be the physicochemical basis for the formation and maintenance of cholesterol/sphingolipid-enriched nano- and micro-domains (referred to as membrane "rafts") in the plane of plasma and other organelle (i.e., Golgi) membranes. This claim is supported by much experimental evidence and also by theoretical considerations. However, although there is a large volume of information about these rafts regarding their lipid and protein composition, their size, and their dynamics, there is still much to be clarified on these issues, as well as on how rafts are formed and maintained. It is well accepted now that the lipid phase of the rafts is the liquid ordered (LO) phase. However, other (non-raft) parts of the membrane may also be in the LO phase.

There are indications that the raft LO phase domains are more tightly packed than the non-raft LO phase, possibly due to intermolecular hydrogen bonding involving sphingolipid and cholesterol. This also explains why the former are detergent-resistant membranes (DRM), while the non-raft LO phase domains are detergent-soluble (sensitive) membranes (DSM).

Subcellular Biochemistry, Volume 37: Membrane Dynamics and Domains
Edited by Peter J. Quinn. Kluwer Academic / Plenum Publishers, New York, 2004

Recent findings suggest that protein–protein interactions such as cross-linking can be controlled by protein distribution between raft and non-raft domains, and, as well, these interactions affect raft size distribution.

The cholesterol/sphingomyelin-enriched rafts seem to be involved in many biological processes, mediated by various receptors, as exemplified by various lipidated glycosylphosphatidylinositol (GPI)- and acyl chain-anchored proteins that reside in the rafts. The rafts serve as signaling platforms in the cell. Various pathogens (viruses and toxins) utilize the raft domains on the host cell membrane as a port of entry, site of assembly (viruses), and port of exit (viral budding). Existence and maintenance of cholesterol–sphingomyelin rafts are dependent on the level of membrane cholesterol and sphingomyelin. This explains why reduction of cholesterol level – either through reverse cholesterol transport, using cholesterol acceptors such as β-cyclodextrin, or through cholesterol biosynthesis inhibition using statins – interferes with many processes which involve rafts and can be applied to treating raft-related infections and diseases.

Detailed elucidation of raft structure and function will improve understanding of biological membrane composition–structure–function relationships and also may serve as a new avenue for the development of novel treatments for major diseases, including viral infections, neurodegenerative diseases (Alzheimer's), atherosclerosis, and tumors.

1. GENERAL INTRODUCTION

1.1. Cholesterol, sphingomyelin, and membrane evolution

Cholesterol (cholest-5-*en*-3-β-ol) and sphingomyelin (*N*-acyl-sphingosyl-1-phosphocholine) are well-characterized lipids derived from natural sources. Both are major components of mammalian plasma and organelle membranes and are absent from most archaebacterial and eubacterial membranes (Barenholz and Cevc, 2000).

Chemically and physically these two lipids have little in common. Cholesterol is a neutral, uncharged, unsaponifiable sterol, based structurally on planar fused steroid rings and a tail, which together constitute a large hydrophobic part, and a small, polar 3-β-hydroxyl group. Sphingomyelin (SM) is a sphingophospholipid having two long hydrocarbon chains that are hydrophobic and a large headgroup of a phosphocholine moiety that is highly charged but, being zwitterionic, is neutral.

These two lipids also differ in their aqueous phase critical aggregation concentration (CAC), the concentration at which various types of self-assembled aggregates are formed in addition to monomers. Cholesterol, based on various measures such as hydrophilic/lipophilic balance (HLB) or on the "packing parameter" (see below), is much less polar than sphingomyelin; therefore it is surprising that it has a much higher (>10-fold) CAC (25 nM) (Gilbert and Reynolds, 1976). Similar differences exist for maximum solubility in water, which for cholesterol approaches 1 μM. These anomalies are explained by the "hydrophobic effect", according to which Yeagle (1993b) suggested that cholesterol in aqueous phase below its CAC exists as dimers and not as monomers, and therefore exposes only about half the hydrophobic surface of its molecules to the aqueous phase (Feher, 1974; Tanford, 1980). Above the solubility limit, cholesterol forms typical crystals in the aqueous phase (Small, 1986).

Cholesterol's high CAC (25 nM) and high maximal solubility (1 μM) in aqueous phase may also explain its relatively high rate of inter-membrane exchange or transfer, determined by desorption from liposomes and biological membranes ($t_{1/2}$ of less than 1 hour for membranes above the gel to liquid crystalline (solid ordered to liquid disordered or SO to LD) phase transition temperature range of the membrane-forming lipids). This rate is much faster than the desorption rate of almost all membrane-forming lipids (rev. in Yeagle, 1988; Shmeeda *et al.*, 1994a). Sphingomyelin is a "liposome-forming" lipid, which means that in aqueous phase it can form liposomes by itself, while cholesterol, having a much larger packing parameter (see 3.2.1 below), is not a liposome-forming lipid. However, biological membranes and liposomes (which serve as the main membrane model system) can accommodate a high fraction of cholesterol, in many cases up to 50 mole%, which will distribute in the membrane bilayer's two leaflets monomolecularly (Kumar, 1991). In liposomes (Simberg *et al.*, 2003) and biological membranes, such as of macrophage-derived foam cells (Tangirala *et al.*, 1994) and aged (in culture) newborn rat heart cardiomyocytes (Shmeeda *et al.*, 1995), when the level of cholesterol exceeds 50 mole%, cholesterol-rich or pure cholesterol phases will be formed, followed by separation and appearance of typical nano- and microcrystals of cholesterol that are excluded from the membranes.

In spite of the above large differences, cholesterol and sphingomyelin share a similar fate in the evolution of biological membranes. Through 3–3.5 billion years of evolution, biological membranes maintained the feature of being a lipid bilayer in which proteins are embedded, electrostatically associated, and/or covalently (via lipid) attached (lipidated proteins) (rev. in Barenholz and Cevc, 2000). From the functional perspective it is well-established that the lipid bilayer, which serves as a basis for biological membranes, is indeed an optimal solution to create the separation between the "external world" and the internal environment ("milieu interne"), which is a prerequisite for life of any living cell. This internal environment is dynamic and controlled through selectivity and signal transduction. The obligatory requirement for the internal environment is the basis of development of one of the theories concerning the origin of life (the "membrane came first" hypothesis, rev. in Barenholz and Cevc, 2000).

The lipids responsible for the formation and maintenance of a bilayer structure all share the common denominators of being amphipathic and of having the cross-sectional area of their hydrophobic region similar to that of their hydrophilic region (see packing parameter below). These lipids also form liposomes by themselves and therefore are classified as "liposome-forming lipids". However, lipid composition of biological membranes has changed dramatically through the evolutionary ladder, starting with archeabacteria, in which the bilayer is actually composed of a monolayer of polar lipid molecules, each having two different polar headgroups and two very long hydrocarbon chains connecting the two headgroups. These unique lipids are referred to as bola lipids. In eubacteria the membrane is already a real lipid bilayer composed of two phospholipid monolayers in which the hydrocarbon tails of these two leaflets oppose each other, and the polar headgroups face the external and internal aqueous phases. In eubacteria the number of lipid species is still small, the membranes lack sterols and sphingolipids, and are based mainly on glycero-polar lipids (phospho- and glycolipids). Mammalian organisms, which evolved much later, are very rich in lipid species, having more than 2000 in their membranes (Barenholz, 1984; Barenholz and Cevc, 2000) – sphingolipids and cholesterol being major and recent additions to the mammalian cell membranes.

The appearance of sterols in biological membranes is, without doubt, one of the most important steps in membrane evolution. It is clear today that large changes in environment (or medium) in which cells live, and the formation of multicellular organisms, in which cells face more than a single environment at one time, as well as the changes in environment during the lifetime of cells, led to their ability to modify the lipid composition of membranes by one or more mechanisms in order to optimize their interaction with the environment and survive. This led to the development of a huge repertoire of membrane lipids that differ in their headgroups and hydrocarbon chains, as well as of "membrane-active" sterols (Barenholz and Cevc, 2000; Barenholz, 2002).

The evolution of biological membranes has a lot to tell us as biologists, biochemists, and physical chemists. However, only recently, after the "discovery" of archeabacteria, has this field started to evolve. This fascinating field of research deserves much more attention than it gets in this review.

Not only in the evolution of biological membranes do cholesterol and sphingomyelin share a similar fate and have cross-talk, as we and others have demonstrated over the last 30 years. These lipids have an interaction with each other that gives biological membranes unique properties, including "raft" formation (rev. in Barenholz and Thompson, 1980, 1999; Barenholz and Gatt, 1982; Barenholz, 1984, 2002; Slotte, 1999; Ridgeway, 2000; Ohvo-Rekila *et al.*, 2002; Edidin, 2003). Other examples of similar fate and of cross-talk include:

(a) The high positive correlation in the level of these two lipids in biological membranes, which was noticed already in 1970 (Patton, 1970), and confirmed later by many others (rev. by Barenholz and Thompson, 1980; Barenholz, 1984; Shmeeda *et al.*, 1994a).

Of special relevance to this correlation are the "gradients" of these two lipids between plasma and subcellular organelles' membranes, reaching their highest concentration in the plasma membrane, while their lowest concentration is in the internal membranes of mitochondria and the nucleus (rev. in Barenholz and Thompson, 1980; Shmeeda *et al.*, 1994a). How can one explain such co-localization, especially since in higher eukaryotes syntheses of glycerophospholipids and cholesterol occur mainly in the endoplasmic reticulum, whereas sphingomyelin and higher glycosphingolipids are synthesized in the Golgi apparatus? It seems that cholesterol and sphingomyelin are gradually enriched along the secretory pathway, with their highest concentrations in the plasma membrane (Wieland *et al.*, 2000). How does a cell succeed in maintaining such an organelle-specific lipid composition, despite a steady flow of incoming and outgoing transport carriers along the secretory pathway? This is not yet clear. Transport and sorting along the secretory pathway of proteins and most lipids are thought to be mediated by vesicular transport, with coat protein I (COPI) vesicles operating in the early secretory pathway. Although the protein constituents of these transport intermediates are characterized in great detail, much less is known about their lipid content. Using nano-electrospray ionization tandem mass spectrometry for quantitative lipid analysis of COPI-coated vesicles and their parental Golgi membranes, Wieland *et al.* (2000) found only low amounts of sphingomyelin and cholesterol in COPI-coated vesicles compared with their donor Golgi membranes, providing evidence for a significant segregation of these lipids from COPI vesicles. In addition, these data of Wieland *et al.* (2000) indicate a sorting of individual sphingomyelin molecular species.

Another important observation is the parallel increase in the level of these two lipids in nondividing cells having a slow rate of polar lipid metabolism during aging (rev. in Shmeeda *et al.*, 1994a), during the development of atherosclerosis (rev. in Barenholz and Thompson, 1980; Barenholz, 1984), and in various types of Niemann–Pick disease (rev. in Garver *et al.*, 2002; Brady, 2003; Vincent *et al.*, 2003).

(b) The metabolism of both lipids has a high degree of cross-talk. There is evidence from human metabolic disorders such as the various types of Niemann–Pick disease (Garver *et al.*, 2002; Brady, 2003; Vincent *et al.*, 2003) and from various experimental models of Niemann–Pick disease, including a mouse model of type A/B disease (Sarna *et al.*, 2001) that cholesterol and sphingomyelin levels are coordinately regulated.

It seems that both biosynthesis and trafficking are involved in the coordination of sphingomyelin and cholesterol metabolism (Shmeeda *et al.*, 1994b; Slotte, 1999; Ridgeway, 2000; Wieland *et al.*, 2000).

(c) Both lipids are present in membrane rafts and caveolae and are involved in the creation and maintenance of these assemblies (see 4 below and, e.g., reviews in Simons and Toomre, 2000; Edidin, 2003).

(d) Each of these two lipids when present in membranes reduces membrane permeability to sugars, other nonelectrolytes, and electrolytes (Papahadjopoulos *et al.*, 1973; Grunze and Deuticke, 1974; Hertz and Barenholz, 1975; Zeidel and Hill, 2000) (see 2.5 below).

(e) Each of these two lipids when present in a membrane reduces oxidative damage to other membrane components (Tirosh *et al.*, 1997; Samuni *et al.*, 2000; Oborina and Yappert, 2003). This protective effect against oxidative damage to membranes may be an important factor responsible for the introduction of membrane-active sterols and sphingomyelin in the evolution of biological membranes (Yechiel *et al.*, 1986b).

1.2. Review objectives

This review is dedicated (a) to describe and summarize sphingomyelin and cholesterol membrane-relevant physicochemical properties, as well as sphingomyelin–cholesterol membrane-relevant interactions in physicochemical terms according to our current understanding of biological membrane structure and dynamics, and (b) to relate the implications of these physicochemical interactions to cell biology and to potential medical applications.

2. CHOLESTEROL AND BIOLOGICAL MEMBRANES

2.1. Introducing cholesterol

There is no better way to introduce cholesterol than to cite part of the 1985 Nobel Laureate lecture of Michael Brown and Joseph Goldstein:

"*Cholesterol is the most highly decorated small molecule in biology. Thirteen Nobel Prizes have been awarded to scientists who devoted major parts of their careers to cholesterol. Ever since it was isolated from gallstones in 1784, cholesterol has exerted an almost hypnotic fascination for scientists from the most diverse areas of science and medicine*". Our review will focus on biophysical aspects of membrane-active sterols (of which cholesterol is the most studied representative) that are highly relevant to membrane and plasma lipoprotein structure and function. However, it is clear that in addition to all the above straightforward physical aspects, cholesterol may have other roles. This is exemplified by:

(i) Mammalian cells need cholesterol for growth, and there is a coupling between level of cholesterol, its synthesis, and cell growth. It seems such coupling involves a mechanism at the molecular biology level in addition to the biophysical aspects discussed below (rev. in Barenholz and Cevc, 2000).
(ii) Cholesterol level in membranes seems to have an important effect on the function of various membrane proteins such as Na^+/K^+-ATPase and rhodopsin (rev. in Yeagle, 1985, 1988, 1993a). Although these effects can be attributed to cholesterol's biophysical contribution to membrane properties, there are also indications that direct effects (not yet clear) on the proteins may also be involved (Shmeeda *et al.*, 1994a; Yeagle, 1988, 1993a; Barenholz and Cevc, 2000; Mitchell and Litman, 2001).

2.2. Defining a "membrane-active sterol"

Certain sterols (defined as "membrane-active sterols" serve as a major means by which eukaryotic cells modulate and refine structural, physicochemical, and biological membrane properties, including barrier properties and lateral membrane organization (rev. in Yeagle, 1988, 1993a; Barenholz and Cevc, 2000; Barenholz, 2002; Jedlovsky and Mezei, 2003, and references therein). Only sterols that affect membrane properties in specific ways (described below) are referred to as membrane-active sterols. The minimal structural requirements for a sterol to be a membrane-active sterol include a flat (planar) steroid fused-ring system, a β-hydroxyl (or other very small polar group in β configuration) at position 3, a cholesterol-like flexible, iso-octyl hydrocarbon tail, and a relatively small minimal area (<40 Å^2/molecule) at the air/water interface (Yeagle, 1988, 1993a; Ben-Yashar and Barenholz, 1989; Bittman, 1997, and references therein).

These minimal requirements agree well with the membrane-active sterols found along the evolution ladder (Nes, 1974; Ben-Yashar and Barenholz, 1989). Being classified as a membrane-active sterol requires that the sterol will: (i) have special effects on short- and long-range membrane order and organization and on membrane dynamics (rev. in Yeagle, 1988, 1993a; Barenholz and Cevc, 2000); (ii) interact with phospholipids in the bilayer in such a way that it efficiently abolishes their solid ordered to liquid disordered (SO–LD) phase transition (previously referred to as gel to liquid crystalline) in a mole-fraction-dependent manner in the range 0–0.33 mole%; (iii) induce the formation of a new, liquid ordered (LO) phase; (iv) reduce efficiently permeability across phospholipid bilayers; and (v) support growth of microorganisms which are sterol auxotrophs (Ben Yashar and Barenholz, 1989; Yeagle, 1993a). Cholesterol is the main mammalian membrane-active sterol, and also the most studied one (Yeagle, 1985, 1988, 1993a; Barenholz and Cevc, 2000; Barenholz, 2002).

2.3. Physicochemical effects on membrane structure, organization, and dynamics

2.3.1. Molecular mechanism

It is accepted that cholesterol is oriented in lipid bilayers with its hydrophobic steroid ring "buried" parallel to the upper ~9 carbons of the membrane-forming lipid hydrocarbon chains, its hydrophobic tail parallel to the lower part of the hydrocarbon chains, and its small polar region (3-β-hydroxyl moiety) in the lipid/water interface at the same plane as the glycerol–acyl ester bond (Huang, 1977; Yeagle, 1993a; see also Figure 1). These and many other

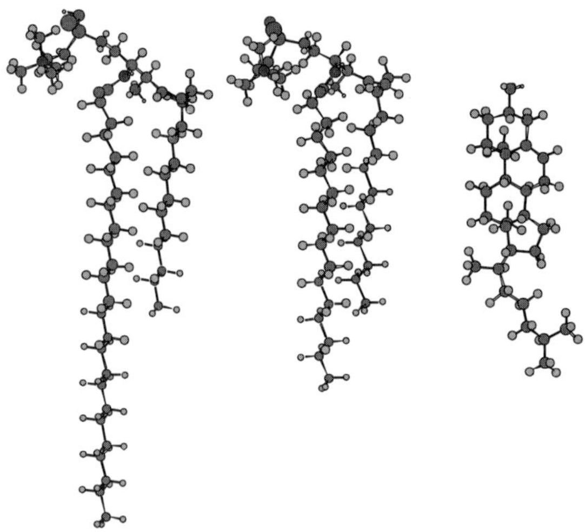

1 2 3

1: C24:0 sphingosylphosphorylcholine
2: C16:0 sphingosylphosphorylcholine
3: cholesterol

Figure 1. Molecular models of D-*erythro*-sphingomyelins (1) C24:0 and (2) C16:0 of sphingosylphosphorylcholines, and (3) cholesterol.

studies indicate that membrane-active sterols affect both the "translational" degree of freedom and the "internal" degree of freedom that correspond to chain conformational states. This results in affecting intermembrane short-range order and long-range lateral organization (Nielsen *et al.*, 1999).

2.3.2. Effect on lipid phase

The physical effects of membrane-active sterols, and especially of cholesterol, on lipid bilayers were studied using almost all physical techniques available for membrane research and are reviewed elsewhere (Barenholz, 1984; Yeagle, 1985, 1988, 1993a; Mouritsen and Jorgensen, 1994; Nielsen *et al.*, 1999; Barenholz and Cevc, 2000, and many others). These studies show that cholesterol when present in a bilayer of phospholipid which undergoes SO to LD phase transition reduces the number of phospholipid molecules participating in the cooperative process of this phase transition in a manner dependent on mole% cholesterol (Rubenstein *et al.*, 1979; Snyder and Freire, 1980; de Kruijff *et al.*, 1985; Small, 1986). When reaching a mole fraction of approx. 33 mole%, cholesterol completely abolishes the SO to LD phase transition. Following the degree of order of the phospholipid chains using ^{2}H NMR, ESR, or fluorescence depolarization techniques indicates that cholesterol acts as an "averager" by reducing the order of the phospholipid molecules in the lipid bilayer in their gel (SO) state by perturbing the hexagonal chain packing. In the liquid crystalline (LD) state cholesterol increases the bilayer order by limiting the probability of *trans–gauche* isomerization in the phospholipid acyl chains. The latter effect is related to the interaction of the rigid cholesterol ring with the acyl chains (Barenholz, 1984; Dufourc *et al.*, 1984; Shinitzky, 1984; de Kruijff *et al.*, 1985; Yeagle, 1985; Small, 1986). In other words, cholesterol has dual effects on other membrane lipids: on one hand, acting like an "ice-breaker", interfering with interlipid interactions responsible for crystallization, and, on the other hand, like a "chain liquidifier". Combining the two effects results in the role of cholesterol as "averager".

Two parameters well-established as factors controlled by membrane lipid composition in general, and by the level of membrane-active sterol in particular, were identified, both related to structural discontinuities and defects in the lipid bilayer and both related to the "free volume" (Straume and Litman, 1987) in which small permeating molecules can reside and move (see below). At the molecular level (short range) there are "kinks", formed due to *trans–gauche* isomerization, which run along the hydrocarbon chain (Ipsen *et al.*, 1987, 1990; Yeagle, 1993a; Mouritsen and Jorgensen, 1994; rev. in Barenholz and Cevc, 2000). At the lateral organization level there are membrane defects formed at the phase boundary under conditions that lateral phase separation occurs (Ipsen *et al.*, 1990). The presence of membrane-active sterols (such as cholesterol) in the membrane decreases free volume, thereby reducing membrane permeability (Mouritsen and Jorgensen, 1994, 1998). As described above, membranes composed of phospholipids alone (without membrane-active sterols) can be either in an LD phase, or in an SO phase, depending on the phospholipid composition, which determines the temperature range at which the phase transition between these two states occurs (Ipsen *et al.*, 1990; Mouritsen and Jorgensen, 1994, 1998; Barenholz and Cevc, 2000). When the level of membrane cholesterol is high (approaching 33 mole%), most membrane lipids are in a new phase referred to as liquid-ordered (LO), having a low free volume (Mouritsen and Jorgensen, 1994, 1998; Mitchell and Litman, 2001). The existence of the LO phase is highly dependent on the presence of membrane-active sterol.

The simplest physicochemical approach to follow the appearance of, and to quantify, the level of LO is to determine the reduction in number of molecules which undergo phase transition from the SO to LD phase (rev. in Barenholz and Cevc, 2000; Yeagle, 1993a; Mouritsen and Jorgensen, 1994, 1998). There are two other important effects of membrane-active sterols on membrane properties:

(i) It is now evident that cholesterol "dries" the lipid/water interface of membranes (Parasassi *et al.*, 1994; Samuni *et al.*, 2000). This drying effect enhances close contact and van der Waals interaction between adjacent lipid molecules, thereby contributing to the reduced membrane permeability and to the increased chemical stability of membrane lipids to oxidative damage and hydrolysis (Samuni *et al.*, 2000).
(ii) In addition, cholesterol induces straightening of lipid hydrocarbon tails, causing increase of bilayer thickness (Israelachvili, 1992; Yeagle, 1993a; Deamer and Chakrabarti, 1996; Jedlovsky and Mezei, 2003), and therefore improving the barrier properties of the lipid bilayer (reducing membrane permeability).

It is also clear that the degree of acyl chain order as well as the lateral organization in lipid bilayers and biological membranes containing cholesterol are strongly influenced by the exact membrane phospholipid composition (Barenholz *et al.*, 1981; Barenholz, 1984; Yechiel *et al.*, 1985, 1986b; Yeagle, 1988; Barenholz and Thompson, 1999; Nielsen *et al.*, 1999; Barenholz and Cevc, 2000; Mitchell and Litman, 2001; Barenholz, 2002).

2.3.3. Theoretical approaches

Various theoretical approaches were used in order to model or to explain the results obtained experimentally.

Huang (1977) proposed a model that, based on the structure of cholesterol and phosphatidylcholine, tried to explain the molecular interactions between the two leading to a complex formation between them in the lipid bilayer. The Huang (1977) model is based to a large extent on a linear hydrogen bonding between the β-hydroxyl group of the cholesterol (as hydrogen acceptor) and the PC acyl ester carbonyl (as hydrogen donor). This model is still unproven and controversial.

Various models were proposed to explain how cholesterol affects lateral organization of membranes.

Snyder and Freire (1980) suggested a model for predicting the amount of distinct phases of pure phospholipid, of cholesterol–phospholipid mixed domains (cholesterol-rich domains), and of the interphases between them. They also dealt specifically with two SM/cholesterol systems based on fully synthetic D,L-*erythro* SMs (N-palmitoyl (C16) SM and N-lignoceryl (C24) SM) and compared them with DPPC/Chol. Based on differential scanning calorimetry (DSC) of Barenholz and coworkers (Figures 3 and 4 in Barenholz, 1984), Snyder and Freire's calculations suggest that cholesterol-rich domain size is the smallest for C16 SM and the largest for C24 SM, with DPPC/Cholesterol having intermediate size of cholesterol-rich domains. Rubenstein *et al.* (1979) suggested that a sudden lateral connectivity of the mixed domains at about 20 mole% cholesterol causes an abrupt decrease in the total interphase area and therefore has a major impact on the lateral diffusion of membrane lipids. Presti *et al.* (1982) tried to relate the detailed molecular mechanism for cholesterol–phospholipid interaction with the lateral organization of the bilayer. Later models were proposed by Mouritsen and coworkers (Ipsen *et al.*, 1990; Mouritsen and Jorgensen, 1994; Nielsen *et al.*, 1999).

All these and other models are part of the major efforts that have been directed to characterize the unique effect of cholesterol on membrane physicochemical properties and especially on the lateral distribution of cholesterol in the bilayer plane and how it affects the bilayer lateral organization. However, in spite of the many years of intensive experimental and theoretical studies, still many questions related to the exact physicochemical effects of membrane-active sterols remain open.

2.4. Preferential interaction of membrane-active sterols with other membrane lipids

Cholesterol (like other membrane-active sterol) interaction with the membrane-forming polar lipids is selective and preferential (Demel *et al.*, 1997; Barenholz and Thompson, 1999; Barenholz and Cevc, 2000; Veiga *et al.*, 2001). All regions of the polar lipid: the headgroup, the interface, and the hydrocarbon chains (see Figure 2) are involved and affect the preference, and therefore the effect of cholesterol on membranes is not as simple as was thought in the past, but rather highly dependent on membrane lipid composition. In addition, and no less important, the presence of cholesterol (or other membrane-active sterol) and its level in a multilipid component bilayer is a major factor in determining the lateral organization in the plane of this bilayer, and seems to be highly relevant to formation and maintenance of membrane rafts (as discussed below).

There are many ways to follow this preferential interaction using both physicochemical and biological approaches. For example, one of the biological approaches is related to the observation that phospholipids reduce the exposure of cholesterol to oxidation by the enzyme cholesterol oxidase. However, this protection is dependent on the type of phospholipid, with sphingomyelin being the most effective; the protection effect is in the order: sphingomyelin>phosphatidylcholine>phosphatidylserine (Patzer *et al.*, 1978a). Physicochemical approaches study how biological membranes and liposome lipid compositions containing membrane-active sterols affect (a) rate and extent of their membrane-active sterol desorption (kinetic approach, 2.4.1. below), and (b) membrane-active sterol partition coefficient (thermodynamic approach, 2.4.2. below).

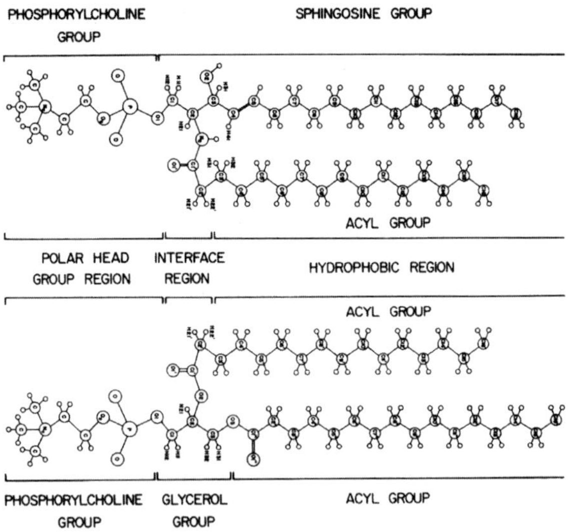

Figure 2. Molecular structure of D-*erythro*-*N*-palmitoyl-sphingosyl phosphocholine (C16 SM) (*top*) and *sn*-L-dipalmitoylglycerophosphocholine (DPPC) (*bottom*). (Based on Barenholz and Gatt, 1982.)

Through many years of research it became evident that the effect of membrane-active sterols on membrane properties is dependent on other membrane lipids (Cevc and Barenholz, 2000) and that the interaction of cholesterol with other membrane lipids is selective. Cholesterol prefers to interact with certain membrane lipid species more than with others. This preferential interaction of cholesterol with membrane lipids is rather complicated since it is multifactorial and involves, in addition to preference for saturated hydrocarbon chains (Mitchell and Litman, 2001), also preferential interaction based on membrane lipid headgroup (Barenholz, 2002; Niu and Litman, 2002, and references therein), and probably polar lipid interface region (Barenholz and Thompson, 1980, 1999; Barenholz and Cevc, 2000). In addition, phospholipid–cholesterol interaction also seems to be affected by the level, type, and mode of membrane protein–lipid association (rev. in Edidin, 2003). In this review we will focus on aspects related mostly to membrane lipids, and not to membrane proteins, for which the reader is referred to Edidin (2003).

Therefore, the exact molecular details of the lipids that form the membrane will determine cholesterol lateral distribution. At low cholesterol mole%, cholesterol will preferentially interact with certain lipids, as discussed above and below, while at saturating mole% ($\sim$33 mole%) it may interact with most membrane lipids, but even under such conditions the influence on various membrane lipids will not be identical, and certain lipids (those that are preferred by membrane-active sterols) will be affected more than others. This may be one of the main reasons for lateral phase separation in membranes of eukaryotes.

Wolf *et al.* (2001) demonstrated that in a membrane composed of a mixture of lipids, cholesterol favors separation of sphingomyelin/cholesterol phase. These authors studied, using X-ray diffraction, the phase behavior of mixed lipid dispersions representing the inner leaflet of the cell membrane. It was found that aqueous dispersions of phosphatidylethanolamine : phosphatidylserine (4 : 1 mole/mole) have a heterogeneous structure comprising an inverted hexagonal phase H_{II} and a lamellar phase. Both phases coexist in the temperature range 20–45°C. The LD to SO mid-transition temperature of the lamellar phase assigned to phosphatidylserine is decreased from 27 to 24°C in the presence of calcium ions. Addition of sphingomyelin to phosphatidylethanolamine/phosphatidylserine prevents phase separation of the hexagonal H_{II} phase of phosphatidylethanolamine, while the ternary mixture phase separates into two lamellar phases of periodicity 6.2 and 5.6 nm, respectively. The 6.2 nm periodicity is assigned to the SO phase enriched in sphingomyelin of molecular species comprising predominantly long saturated hydrocarbon chains because it undergoes SO to LD phase transition above 40°C. The coexisting fluid LD phase is assigned to phosphatidylethanolamine and phosphatidylserine and low-melting-point molecular species of sphingomyelin, which suppress the tendency of phosphatidylethanolamine to phase-separate into a hexagonal H_{II} structure. There is evidence for considerable hysteresis in the separation of lamellar fluid and gel phases during cooling. The addition of cholesterol prevents phase separation of the gel phase of high-melting-point sphingomyelin in mixtures with phosphatidylserine and phosphatidylethanolamine. In the quaternary mixture, the lamellar fluid phase, however, is separated into two lamellar phases of periodicities of 6.3 and 5.6 nm (20°C). The lamellar phase of periodicity 5.6 nm is assigned to a phase enriched in aminoglycerophospholipids and the periodicity 6.3 nm to an LO phase formed from cholesterol and high-melting-point molecular species of sphingomyelin characterized previously by ESR. Substituting 7-dehydrocholesterol for cholesterol did not result in evidence for lamellar phase separation in the mixture within the temperature range 20–40°C. The specificity of cholesterol in creation of an LO lamellar phase is inferred.

2.4.1. Effect of membrane lipid composition on rate and extent of desorption of membrane-active sterol

The existence of two kinetically distinct cholesterol pools seems to be one of the main reasons for differences in rate and extent of membrane-active sterol desorption of liposomes of different phospholipid composition (Bar *et al.*, 1986, 1987, 1989). One of the two pools transfers spontaneously with a relatively fast first-order rate and a relatively short half-life on a time scale of minutes; the second is referred to as a nonexchangeable pool, as it transfers spontaneously on a time scale of many (≥12) hours. The size of these two pools and the rate of transfer from the first pool depend on membrane lipid composition and temperature (see Table 1A and B). At 37°C, 90% of the cholesterol present in vesicles composed of bovine brain sphingomyelin resides in the nonexchangeable pool, compared with only ~20% in the nonexchangeable pool in 1-palmitoyl-2-oleyl phosphatidylcholine vesicles. Such a difference may explain the positive correlation between the levels of sphingomyelin and cholesterol in biological membranes (Patton, 1970; rev. in Barenholz and Thompson, 1980; Shmeeda *et al.*, 1994a; Barenholz and Thompson, 1999). There is a good possibility that at least part of this preferential interaction of cholesterol with sphingomyelin is related to the fact that all natural sphingomyelins are much more saturated than most naturally occurring phosphatidylcholines (Barenholz and Thompson, 1980; Shmeeda *et al.*, 1994a). However, this is not the complete explanation, since the relative size of the nonexchangeable cholesterol pool in sphingomyelin bilayers is significantly higher than in dipalmitoyl phosphatidylcholine (Table 1B and Bar *et al.*, 1987), although their LD to SO phase transition occurs at a similar temperature range (Barenholz and Thompson, 1980).

Table 1A. Effect of the liposome-forming phospholipid of SUV donor on the [^{3}H]cholesterol-transfer parameters (based on Bar *et al.*, 1987)

			Initial cholesterol concentration			
			1 mole%		10 mole%	
Liposome-forming choline phospholipid	Temperature (°C)	Physical state	$t_{1/2}$ (min)[a]	% Nonexchangeable cholesterol	$t_{1/2}$ (min)	% Nonexchangeable cholesterol
POPC	37	LD	46 ± 7	18.4 ± 2.3	71 ± 9	23.0 ± 3.0
DMPC	37	LD	121 ± 30	18.0 ± 5.0	153 ± 10	20.1 ± 3.0
SM (*bovine brain*)	37	LD/SO	202 ± 70	88.0 ± 3.0	190 ± 100	90.0 ± 2.0
SM (*bovine brain*)	50	LD	–	–	135 ± 20	60.0 ± 7
SM (*egg yolk*)	50	LD	–	–	373 ± 100	64.0 ± 3
SM (*bovine erythrocyte*)	50	LD	–	–	424 ± 100	38.0 ± 4

Table 1B. Effect of temperature on kinetic parameters of [^{3}H]cholesterol exchange from SUV containing 10 mole% cholesterol and 90 mole% choline phospholipid (based on Bar *et al.*, 1987)

Liposome-forming choline phospholipid	Temperature (°C)	Physical state	$t_{1/2}$ (min)[a]	% Nonexchangeable cholesterol
POPC	37	LD	71 ± 9	23.0 ± 3.0
	50	LD	13 ± 4	12.7 ± 1.9
DMPC	37	LD	153 ± 10	20.1 ± 3.0
	50	LD	38 ± 2	20.0 ± 4.0
DPPC	50	LD	62 ± 6	31.0 ± 1.5
DHPC	50	LD	62 ± 6	31.0 ± 1.5
SM (*bovine brain*)	37	LD/SO	190 ± 100	90.0 ± 2.0
	50	LD	135 ± 20	60.0 ± 7.0

[a]The $t_{1/2}$ and percent of nonexchangeable cholesterol values given in Tables 1A and 1B are the mean ± SE of different vesicle preparations.
Abbreviations: SUV, small unilamellar vesicles; POPC, 1-palmitoyl-2-oleoylphosphatidylglycerol; DMPC, dimyristoylphosphatidylcholine; DPPC, dipalmitoylphosphatidylcholine; DHPC, dihexadecylphosphatidylcholine (an ether lipid analogue of DPPC).

2.4.2. Effect of acceptor liposome lipid composition on cholesterol partition between the acceptor liposome and the reference donor

In addition to the kinetic approach described above, one can assess the differences in affinity of the various membrane-forming lipids to cholesterol by determining the cholesterol partition coefficient between two vesicle (LUV) populations. One LUV population serves as cholesterol donor and the other as an acceptor, both having defined liposome-forming lipids. This is a thermodynamic approach, which, in order to be valid, requires overcoming two technical difficulties: the physical separation between the donor and analyte acceptor LUV has to be complete, and the equilibrium between the donor and analyte acceptor has to be reached reasonably fast. Recently Niu and Litman (2002) described such a system in which these obstacles were overcome. Their approach is based on the use of methyl-β-cyclodextrin (CD) cholesterol complexes of 2 nm size as the identical cholesterol donor and LUV of different defined liposome composition as analyte acceptor. As cholesterol exchange betwen CD and membranes follows an equilibrium model, by determining the CD/LUV partition coefficients using LUV of various defined compositions (LUV A) and LUV of identical composition POPC as LUV B, the partition coefficient between LUV populations of defined composition ($K_p^{A/B}$) can be determined. Preferential interactions of cholesterol with various defined membrane phospholipid types, and for the same phospholipid type, varying in acyl chain unsaturation, were studied by measuring the putative partition coefficient for cholesterol in the above LUV–cyclodextrin system. Variation in phospholipid type resulted in cholesterol partitioning in the order of sphingomyelin (SM)>>phosphatidylserine (PS)>phosphatidylcholine (PC)>>phosphatidylethanol-amine (PE), spanning a range of partition free energy change (ΔG) of −1181 cal/mol for SM to +683 cal/mol for PE. In LUV composed of mixtures of phospholipids of different types (PC/SM, PS/PC, and PE/PC), cholesterol partition coefficients have intermediate values slightly in favor of the PC (Niu and Litman, 2002). Among the PC acyl chains examined using di18:1 ng PC (DOPC) as identical LUV B, the order of cholesterol partitioning between the various LUV A to the same DOPC LUVB for the different LUV B was 18:0.18:1n-9PC > di-18:1n-12PC > di-18:2n-6PC > 16:0.22:6n-3PC > di-18:3n-3PC > di-22:6n-3PC, with a range in partition ΔG of 913 (−207 to 706) cal/mol. These results agree well with the kinetic approach described in 2.4.1. above (Table 2).

These preferential interactions of cholesterol with different membrane lipids as expressed through both the kinetic and thermodynamic approaches, described above, support the suggestion that both kinetics and thermodynamics may play an important role in formation and maintenance of lateral domain

Table 2. Summary of the effect of LUV lipid composition on the cholesterol partition coefficient $K_p^{A/B}$ and partition free energy (ΔG)* (Based on Niu and Litman, 2002)

LUV composition			
LUV A	LUV B	$K_p^{A/B}$	ΔG, cal/mol
		Effect of phospholipid type	
16:0, 18:1n-9PC	16:0, 18.1n-9PC	1	0
16:0-SM	16:0, 18:1n-9PC	6.78 (±0.88)	−1181 (±156)
16:0,18:1n-9PS	16:0, 18:1n-9PC	1.91 (±0.58)	−400 (±123)
16:0,18:1n-9PE	16:0, 18:1n-9PC	0.33 (±0.04)	683 (±88)
		Effect of acyl chain	
18:0,18:1n-9PC	di18:1n-9PC	1.40 (±0.11)	−207 (±19)
di18:1n-9PC	di18:1n-9PC	1	0
di18:1n-12PC	di18:1n-9PC	0.67 (±0.13)	243 (±47)
di18:2n-6PC	di18:1n-9PC	0.55 (±0.04)	369 (±28)
di18:3n-3PC	di18:1n-9PC	0.51 (±0.04)	416 (±30)
16:0,22:6n-3PC	di18:1n-9PC	0.54 (±0.10)	377 (±30)
di22:6n-3PC	di18:1n-9PC	0.32 (±0.039)	706 (±87)

*All values were measured at 37°C. $K_p^{A/B}$ and ΔG associated with headgroup compositions were referenced to 16:0,18:1n-9 PC = POPC (LUV B), while those with acyl chain compositions were referenced to di18:1n-9 PC (LUV B).

formation in the membrane plane. These differences may also explain in part the heterogeneous cholesterol distribution between the various cellular membranes, described as the sphingomyelin and cholesterol gradient. Sphingomyelin and cholesterol are highly enriched in plasma membranes and relatively depleted in intracellular membranes, and especially in the inner mitochondrial and nuclear membranes (1.1 above).

2.4.3. Relevance to membrane rafts

This preferential interaction of membrane-active sterols with other membrane lipids serves as one of the unique roles of membrane-active sterols in biological membranes, the modulation of membrane lateral organization, especially lateral heterogeneity in the membrane plane. Polyunsaturated chains, such as the docosahexanoic acyl chain (C22:6), have almost no affinity for cholesterol at all (Mitchell and Litman, 1998, 2001), and regions enriched in such lipids will be less affected by this sterol. As an example, in membranes composed of various lipid species, cholesterol preferentially interacts with sphingomyelins (Barenholz and Thompson, 1999; Barenholz and Cevc, 2000; Veiga *et al.*, 2001; Niu and Litman, 2002), and among the PCs, those species with saturated acyl chains (Mitchell and Litman, 2001; Niu and Litman, 2002).

Therefore, regions enriched with these lipids will be affected more by the membrane-active sterols. In this respect, it is worth noting that cholesterol and probably other membrane-active sterols induce formation of LO domains in the LD continuum (or LD domains in the LO continuum) in all cases that these sterols interact with the bilayer-forming lipids. However, the LO phase based on 1-saturated, 2-unsaturated phosphatidylcholines (as in most membrane phospholipids) is less ordered and less "dry" than the LO phase in which sphingolipids are the matrix lipids (2.3.2 above). This is explained by the much higher temperature of SO to LD phase transition (T_m) of the sphingolipids and possibly the involvement of sphingolipid–sterol hydrogen bonding (Barenholz and Thompson, 1980, 1999; Shmeeda *et al.*, 1994a. Such differences may explain the immiscibility in the LO phase between domains rich and poor in sphingolipids. Such LO/LO immiscibility may have important implications for biological membrane function (discussed in **4** below, and exemplified in London and Brown, 2000; Brown and London, 2000; Yechiel *et al.*, 1985; Yechiel and Barenholz, 1985). This immiscibility may be one of the main factors leading to the formation and maintenance of (controversial) rafts in the plane of biological membranes (see 4.2 and London and Brown, 2000; Brown and London, 2000; Edidin, 2003).

In addition, these preferential interactions of cholesterol with sphingomyelin monitored through kinetics of slower desorption and higher partition coefficient may be relevant to designing strategies for therapies to slow down cholesterol absorption in the gastrointestinal tract or in treatment of a situation which involves hypercholesterolemia and/or cholesterol accumulation such as in atheroma in the blood vessels (see 5, Medical Applications, and references listed therein).

2.5. Effect on membrane permeability: mechanism

The effect of membrane-active sterols on membrane permeability is well-established (Yeagle, 1988, 1993a, and references therein). Since the introduction of liposomes as a preferred membrane model, this property has been used routinely by liposomologists to control the rate of liposome permeability (de Gier *et al.*, 1969; Demel *et al.*, 1972; Papahadjopoulos *et al.*, 1973; Lichtenberg and Barenholz, 1988; Haran *et al.*, 1993; Yeagle, 1993a; Barenholz and Cevc, 2000; Barenholz, 2001; Grant *et al.*, 2001; Haines, 2001).

Membrane-active sterols reduce membrane permeability to polar nonelectrolytes such as sugars, as well as to ions (rev. Yeagle *et al.*, 1988, 1993a; Barenholz and Cevc, 2000; Haines, 2001). Cholesterol (~30 mole%) reduces the partition coefficient of ethanol into lipid membranes by a factor of 3 (Mouritsen *et al.*, 1999), in agreement with reducing free volume as well as the effect of cholesterol of the interphase region of phospholipid molecules. The presence of proteins (such as cytochrome *c*) enhances Na^+ permeability across membranes of liposomes lacking membrane-active sterols (Papahadjopoulos *et al.*, 1973). Addition of cholesterol protects against this protein-induced enhancement of membrane leakage. In general, membrane permeability to ions is lower than to sugars, due to the lower "solubility" (a concept related to membrane defects and free volume (see below)) of the ions in the hydrophobic part of the membrane. Assuming permeability through the membrane is affected by solute partition into the membrane (membrane "solubility") and by diffusion across the membrane (Yeagle, 1988; Barenholz and Cevc, 2000), cholesterol should affect both since it also increases membrane thickness (as determined with small angle X-ray scattering (SAXS) (Jedlovsky and Mezei, 2003)), thereby reducing diffusion across the membrane, as well as reducing "solubility" of the solute in the membrane by reducing free volume and defects (see 2.3.2 and Straume and Litman, 1987). The "solubility" effect is more pronounced for ions, which explains the unique effect of membrane-active sterols on reduction of sodium and potassium leaks. This very important effect of cholesterol in preventing sodium and potassium ion leakage (influx and efflux, respectively) gives the membrane-active sterol a special physiological role in conserving metabolic ATP energy that would be lost as a result of such ion leakage (Haines, 2001).

A recent study (Zeidel and Hill, 2000), aiming to define aspects of lipid composition and bilayer asymmetry critical to barrier function, examines the permeabilities of liposomes that model individual leaflets of the apical membrane of a barrier epithelium, Madin–Darby canine kidney type 1 cells. Using published lipid compositions, Zeidel and Hill (2000) prepared exofacial liposomes containing phosphatidylcholine, sphingomyelin, glycosphingolipids, and cholesterol; and cytoplasmic liposomes containing phosphatidylethanolamine, phosphatidylserine, and cholesterol. The osmotic permeability of cytoplasmic (C(cy)) liposomes to water (Pf (cy)), solutes, and NH_3 was 18–90-fold higher than that of the exofacial liposomes (Pf(ex)): (Pf(ex)) = $2.4 \pm 0.4 \times 10^{-4}$ cm/s, Pf(cy) = $4.4 \pm 0.3 \times 10^{-3}$ cm/s; Pglycerol(ex) = $2.5 \pm 0.3 \times 10^{-8}$ cm/s, Pglycerol(cy) = $2.2 \pm 0.02 \times 10^{-6}$ cm/s; PNH_3(ex) = $0.13 \pm 0.4 \times 10^{-4}$ cm/s, PNH_3(cy) = $7.9 \pm 1.0 \times 10^{-3}$ cm/s). By contrast, the apparent proton permeability (PH) of exofacial liposomes was 4-fold higher than cytoplasmic liposomes (PH^+(ex) = $1.1 \pm 0.1 \times 10^{-2}$ cm/s, PH^+(cy) = $2.7 \pm 0.6 \times 10^{-3}$ cm/s). By adding single leaflet permeabilities the authors calculated a theoretical Pf for a Madin–Darby canine kidney apical membrane of 4.6×10^{-4} cm/s, which compares favorably with experimentally determined values. In exofacial liposomes lacking glycosphingolipids or sphingomyelin, permeabilities were 2–7-fold higher, indicating that both species play a role in barrier function. Removal of cholesterol resulted in 40–280-fold increases in permeability. The conclusions of this study are: (1) that using liposomes, the biophysical properties of a barrier membrane were reconstituted; (2) that the barrier resides in the exofacial leaflet; (3) that both sphingomyelin and glycosphingolipids play a role in reducing membrane permeability, but that cholesterol is much more crucial to the barrier properties as there is an absolute requirement for cholesterol to achieve a good barrier; (4) that these results further validate the hypothesis that each leaflet offers an independent resistance to permeation; and (5) that proton permeation is a special case, as it was enhanced by sphingolipid/cholesterol interactions, possibly due to intermolecular hydrogen bonds (described below, 4.1). This last aspect of proton leakage requires special attention as it may explain why the mitochondrial inner membrane, which is responsible for ATP production by means of a proton gradient, is lacking sphingolipids and cholesterol (see 1.1).

3. SPHINGOMYELIN AND BIOLOGICAL MEMBRANES

3.1. General introduction – sphingomyelin

Sphingomyelin (*N*-acyl-sphingosine-1-phosphocholine) seems to be the antithesis of cholesterol.

It took more than 100 years after its discovery by Thudicum (1874, 1962) for sphingomyelin to reach headlines in the life sciences. Until then, interest in sphingomyelin was limited to a small group of lipid biochemists, biophysicists, and geneticists. The latter were mainly interested in the various types of Niemann–Pick disease (for review see Barenholz and Thompson, 1980; Barenholz and Gatt, 1982; Barenholz, 1984). Two findings in the last 15 years made sphingomyelin a "hot topic" in life sciences and medicine. The first one is the finding that sphingomyelin, being a direct and indirect source of lipid second messengers (ceramide, ceramide-1-phosphate, sphingosine-1-phosphate, sphingosine), is involved in signal transduction (Kolesnick and Golde, 1994; Obeid and Hannun, 1995; Merrill *et al.*, 1997).

The sphingomyelin-derived second messengers, which are initiated by the "sphingomyelin pathway" (also referred to as sphingomyelin cycle), affect major events related to cell growth, differentiation, and cell death. Being generated in most cell types, the sphingomyelin cycle may have general biological implications, which are now under intensive investigations. Many academic and pharmaceutical industry laboratories are studying the basic biochemistry and molecular and cell biology of these phenomena.

The second finding is the major role of sphingomyelins, together with cholesterol and glycosphingolipids, in the formation and maintenance of "rafts" and "caveolae" in the plane of plasma membranes of many cell types; these are laterally-separated domains having distinct roles in many cell functions (see 5 below and rev. in Edidin, 2003).

It was not clear if and to what extent membrane biophysics, and especially the role of sphingomyelin in membrane "raft" formation, has relevance to sphingomyelin involvement in signal transduction. Recent studies suggest that such connection does exist, as "rafts", which are sphingomyelin-enriched domains, may serve as a preferred substrate for sphingomyelinase, and the ceramide microdomains formed are involved in signal transduction (Dobrowsky, 2000; Simons and Toomre, 2000).

Other important indirect evidence for the connection between sphingomyelin involvement in signal transduction and in membrane biophysics is the change that occurs in lipid membrane composition as a result of initiating the sphingomyelin pathway by TNFα, which leads to as much as 50% hydrolysis of membrane sphingomyelin, replacing it with ceramide. It is expected that such a massive change in the level of SM, which occurs over a relatively short time, should affect many membrane-related properties. Recently we found that such massive hydrolysis of sphingomyelin leads to transient increased leakage of lactic dehydrogenase from HL-60 cells in culture (Samet and Barenholz, unpublished). It is expected that changes in lateral membrane organization will also occur. All this remains to be studied.

A good demonstration of the high interest in sphingomyelin is the fully dedicated special issue of *Chemistry and Physics of Lipids*, "Sphingomyelin: Chemistry, Biophysics, Metabolism, Genetics and Signaling", edited by Gatt and Barenholz, 1999.

3.2. Sphingomyelin: biophysical aspects

3.2.1. Molecular structure and its implications

Sphingomyelins (SMs) form one of the major classes of membrane phospholipids in eukaryotic organisms, constituting, together with the class of phosphatidylcholines (PCs), more than 50% of membrane phospholipids. Both SMs and PCs are located mainly in the outer leaflet of plasma membranes (rev. in Barenholz and Thompson, 1980; Barenholz, 1984). In order to understand the contribution of SMs to membrane structure and function, they have to be looked upon in the framework of the updated "fluid mosaic" model of the membrane (rev. in Barenholz and Cevc, 2000). According to this, biological membranes are multicomponent lipid bilayer assemblies in which the two leaflets (monolayers) forming the bilayer are compositionally distinct, and laterally heterogeneous due to existence of microdomains and nanodomains such as rafts and caveolae in the membrane plane.

The structures of SMs and PCs present in biological membranes (Figure 2) reveal gross similarities between the two species. Both have the identical highly-charged and zwitterionic phosphocholine hydrophilic headgroup, an interface region, and two long ($\geq$16 carbon atoms) hydrocarbon chains that form the hydrophobic bulk domain of the amphiphiles. Both are classified as membrane (liposome)-forming lipids. According to one approach of amphiphile classification both PCs and SMs are classified as nonsoluble swelling amphiphiles (Small, 1970, 1986) that form a stable monomolecular layer at the air/water interface. They have a similar "packing parameter" (PP). A concept introduced by Israelachvili and coworkers (1980, 1992), PP is a "simple" geometrical parameter which relates the cross-sectional areas of the hydrophobic and hydrophilic domains of the amphiphile: $PP = v/al_c$, where a is the average surface area occupied by the polar region of the amphiphile at the air/water interface, v is the volume of the hydrophobic part of the amphiphile, and l_c is the fully extended length of the hydrophobic region of the amphiphile. v/l_c actually describes the cross-sectional area of the hydrophobic part of the amphiphile, so PP is the ratio between the latter and the cross-sectional area of the hydrophilic region of the amphiphile.

When a liposome or a biological membrane is a multicomponent system, its stability can be predicted from its "additive PP" value, which is calculated as the sum of the PP of each of the components multiplied by the component mole% (Kumar, 1991). Based on the following: (i) stable bilayers exist at the additive PP range of 0.74–1.0, (ii) PCs have a PP of ~0.80, and (iii) cholesterol has a PP value of ~1.2, then the PC bilayer can accommodate cholesterol up to a mole fraction of 0.5 (Kumar, 1991). It is important to note that the application of the "additive PP" concept to bilayers in which there is lateral phase separation has not yet been validated.

Most membrane-forming SMs and PCs that originate from natural sources are amphiphiles with PP value in the range of 0.75–0.85, that is, molecules having a nearly cylinder-like shape. While such amphiphiles, when present in an aqueous phase tend to form stable bilayers (Kumar, 1991), amphiphiles with a much lower or higher PP value will not form stable bilayers and will self-associate into micelles ($PP < 0.74$) or inverted hexagonal structures (H_{II} phase) ($PP > 1.0$). Also, for most SMs and PCs occurring in biological membranes, changing the temperature, pH, or ionic strength of the suspension medium will not change the packing parameter to an extent that affects their state of aggregation, and they will remain as lipid bilayers. In this respect, both SMs and PCs represent stable systems when compared with the rich thermotropic and lyotropic polymorphism of other biological amphiphiles such as phosphatidylethanolamine (PE) and phosphatidylserine (PS) (Israelachvili *et al.*, 1980; Lichtenberg and Barenholz, 1988; Kumar, 1991; Israelachvili, 1992).

The similarity in PP and the fact that both lipids are zwitterionic may explain why PC and SM are quite interchangeable in biological membranes and liposomes. However, the structural resemblance of PC and SM is limited to the phosphocholine zwitterionic headgroups only which are responsible for the relatively large level of hydration of PC and SM (Jendrasiak and Smith, 2001). The other regions of these two phospholipid types differ substantially, and so, also many of the physicochemical properties of these two choline phospholipids. For the hydrophobic region (the hydrocarbon chains): the PC acyl or alkyl chains are bound by acyl ester or alkyl ether bonds, respectively, to the glycerol (Figure 2). For most naturally-occurring PCs, these hydrocarbon chains are almost identical in length; the acyl chain attached to the C1 carbon of the glycerol (*sn* nomenclature) is normally saturated, while the other aliphatic residue, in the *sn*-2-position, is in most cases unsaturated. This latter residue normally contains between 1 and 6 double bonds, all in the *cis*-configuration. The hydrophobic part of SM is different, its only one acyl chain is bound via an amide linkage to the primary amino group at C2 of the sphingosine (see Figure 2), while the second hydrocarbon chain (~13 carbon atoms) is contributed by the sphingoid base D-*erythro* (2*S*,3*R*) sphingosine (4-sphingenine) of 18 carbons. The other 5 carbons of the sphingosine form the SM interface region (see below and Figure 2). Naturally-occurring SMs are much more saturated than naturally-occurring PCs. The dominant acyl chain of most SMs is C_{18} (stearoyl), C_{24} (lignoceryl), or *cis*-15-tetracosenyl ($C_{24:1}$, nervonyl); however, in red blood cells C_{16} (palmitoyl) is also a major species (Barenholz and Thompson, 1980; Barenholz and Gatt, 1982; Barenholz, 1984; Barenholz and Cevc, 2000). More than 50% of the SM species present in most biological membranes have very long acyl chains (>20 carbons). Generally, in biological membranes of eukariotic organisms there are fewer SM species ($\leq$20) than PC species (~60) (Shmeeda *et al.*, 1994a). These dissimilarities in the hydrocarbon chains lead to two main implications:

(a) Naturally-occurring SMs have their SO to LD phase transition at a much higher temperature range (25–40°C) than naturally-occurring PCs (<15°C) (Barenholz and Thompson, 1980; Frank *et al.*, 1983; Barenholz, 1984; Marsh, 1990). Even the unsaturated *N*-nervonyl (24:1) SM has a broad SO to LD phase transition with a T_m of 27.5°C (Cohen *et al.*, 1984); this high T_m is related in part to the position of its *cis* double bond, which is not in the middle of the chain but between C15 and C16 in the same plane as the omega methyl of the sphingosyl moiety (Barenholz and Thompson, 1980, 1999; Barenholz, 1984). Their counterpart PCs have T_m much below the physiological temperature range, as most PCs have an unsaturated acyl chain at *sn* 2 and at least one of the *cis* double bonds is located in the middle of the glycerol C2 hydrocarbon chain.

(b) In the SM molecule there is a large chain disparity (also referred to as asymmetry or chain mismatch) (Schmidt *et al.*, 1978; Barenholz, 1984; Levin *et al.*, 1985; Huang and Mason, 1986; Bar *et al.*, 1997; Barenholz and Cevc, 2000). Such chain mismatch may lead to interdigitation between the hydrocarbons of the two opposing monolayers that form the lipid bilayer (Schmidt *et al.*, 1978). Two possibilities for such interdigitation are presented in Figure 3 (Levin *et al.*, 1985). In cases in which chain mismatch is borderline, small changes may lead to formation of metastable phases in addition to the well-characterized SO and LD phases (Estep *et al.*, 1980).

The interface region of the two molecules also shows major differences. In the SM molecules it is based on carbons 1–3 of the sphingoid base. It contains the components of the amide linkage (between the acyl chain and the primary amino group at C2 of the sphingosine), as well as the free hydroxyl group attached to C3; the *trans* double bond between C4 and C5 probably also is part of the interface (Figure 2). These amide and hydroxyl groups allow hydrogen bonding in the interfacial region of SM, with both moieties serving as donors as well as acceptors of hydrogens, while the SM carbonyl serves only as a hydrogen acceptor (rev. in Barenholz and Thompson, 1980, 1999). The phosphate of the phosphocholine headgroup may also be involved in the H bonding with the amide and/or hydroxyl groups at the interface region, and such interaction may restrict the headgroup motion through inter- or intramolecular interactions (Talbott *et al.*, 2000; Ferguson-Yankey *et al.*, 2000). In comparison, PC, through its two acyl ester carbonyls can serve only as hydrogen acceptor (with respect to hydrogen bonds). The *trans* double bond between the C4 and C5 atoms of the sphingosine moiety (4-sphingenine) has the ability to induce dipoles in the interfacial region (Barenholz and Thompson, 1980). Better chain stacking and closer

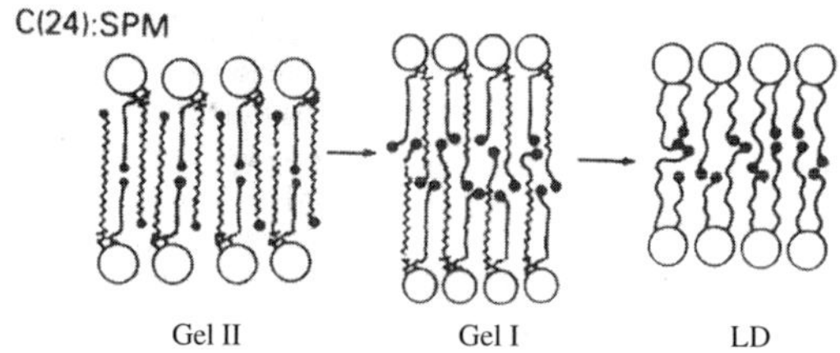

Figure 3. Proposed models describing molecular packing of fully hydrated *N*-lignoceryl (C24:0) D-*erythro* SM. The models from left to right represent gel II (mixed fully interdigitated), gel I (unmixed interdigitated), and a liquid disordered (LD) phase. It is expected that the gel II phase will be the most ordered phase. Bilayer width is in the order: gel I > gel II > LD. (For more details see Levin *et al.*, 1985.)

lipid packing may result from this, and this may be an additional explanation for the higher LD to SO phase transition temperature range of SMs than PCs (Barenholz *et al.*, 1976; Frank *et al.*, 1983). A larger degree of intermolecular hydrogen bonding (Ferguson-Yankey *et al.*, 2000; Talbott *et al.*, 2000), as well as tighter membrane packing and lower membrane permeability are further consequences (Abrahamsson *et al.*, 1977) of an increased mole fraction of SMs in membranes at the expense of PCs. The presence of 10 mole% enhances the partition coefficient of ethanol into lipid membranes and this effect is related, to a large extent, to the interface region of SMs and is much less affected by the headgroup (Mouritsen *et al.*, 1999).

In most biological membranes *N*-acyl sphingosyl-1-phosphocholine is the dominant sphingolipid, while *N*-acyl dihydrosphingosyl-1-phosphocholine (which lacks the *trans* double bond between the C4 and C5 of the sphingoid base) is a minor component. The exception is the eye lens membrane, where the dihydrosphingomyelin is the major sphingolipid. Comparison between *N*-palmitoyl sphingomyelin (SM) and *N*-palmitoyl dihydrosphingomyelin (DHSM) reveals that, as expected, the latter has its SO $\rightarrow$ LD phase transition at a slightly (6–7°C) higher temperature (Barenholz *et al.*, 1976). Proton and ^{31}P NMR studies suggest the existence of both intra- and intermolecular H bonds, involving the C3 OH moiety and the amide NH on C2 of the sphingoid base as hydrogen donors. In hydrated lipids, the amide proton is involved mostly in an intermolecular H-bond that links adjacent DHSM molecules, while in the absence of water the NH moiety forms an intramolecular H-bond with the phosphate moiety, which restricts the mobility of the latter. In excess water the latter intramolecular H-bond is disrupted, thereby changing the balance in favor of water-mediated intermolecular H-bonding (Ferguson-Yankey *et al.*, 2000).

The difference between SM and DHSM is that in DHSM the intermolecular H-bonding is dominant, while in SM the C4–C5 *trans* bond interferes with the ability of the molecules to come closer to each other, and therefore the intramolecular H-bonding is more pronounced in SM than in DHSP (Talbott *et al.*, 2000).

3.2.2. Physicochemical effect on membrane structure, organization, and dynamics

Physical studies done with SM are much fewer than with PC (Barenholz and Thompson, 1980; Barenholz, 1984; Slotte, 1999; Barenholz and Cevc, 2000; Ohvo-Rekila *et al.*, 2002). While single-crystal X-ray diffraction data are available for dimyristoyl-PC (DMPC) (Pearson and Pascher, 1979; Hauser *et al.*, 1981), the corresponding information pertaining to SM can only be deduced from the data available for ceramides, cerebrosides, and SM model compounds (Sundaralingam, 1972; Pascher, 1976; Lofgren and Pascher, 1977).

Most SMs present in biological membranes have their T_m close to 37°C (Frank *et al.*, 1983; Barenholz, 1984; Cohen *et al.*, 1984; Bar *et al.*, 1997), and therefore these SMs are good candidates as molecules which have the potential to introduce lateral heterogeneity and immiscibility into the membrane plane. The role of SMs in generating domains in biological membranes is supported by studies in rat cardiomyocytes in culture (Yechiel and Barenholz, 1985; Yechiel *et al.*, 1986a,b), by our results on the relationships between level of membrane sphingomyelin and scrapie prion formation in neuroblastoma cells infected with prions (Naslavsky *et al.*, 1999), and by prion protein hydrophobic interaction with raft-like membranes enriched in SM and cholesterol in Syrian hamsters (Pinheiro and Sanghera, 2002).

Phase behavior, lateral diffusion, and percolation in the range of coexisting gel (SO) and fluid (LD) phases in binary mixtures of bovine brain SM and egg PC were studied, and lateral phase separation of SM-enriched domains was demonstrated (rev. in Barenholz, 1984) such as the racemic DL-*erythro-N*-palmitoyl (C_{16}) SM with DMPC (Lentz *et al.*, 1981). Mixtures of synthetic SMs or D-*erythro-N*-palmitoyl (C_{16}) SM or D-*erythro-N*-lignoceryl (C_{24}) SM both with DMPC were studied (Bar *et al.*, 1997). The phase diagrams obtained for DL-*erythro* and DMPC systems and for the D-*erythro* SM/DMPC system were similar although they were studied using different methodologies: diphenyl hexatriene fluorescence polarization (Shinitzky and Barenholz, 1974, 1978) and freeze fraction electron microscopy for DL-*erythro* C16/DMPC (Lentz *et al.*, 1981), and DSC together with FRAP for D-*erythro* C16 SM/DMPC (Bar *et al.*, 1997), suggesting the SM chirality has minimal effect on the phase diagram. In both systems the mixing of the PC and SM is close to ideal (Lentz *et al.*, 1981; Bar *et al.*, 1997). The study of Bar *et al.* (1997) is especially interesting since both D-*erythro* SMs investigated have a similar T_m (41°C and 45°C, respectively). However the level of chain mismatch in these two SMs is considerably different, being much larger for the D-*erythro-N*-lignoceryl (C_{24}) SM (highly asymmetric) than for the D-*erythro-N*-palmitoyl (C_{16}) SM (symmetric SM). The C_{24}-SM exhibits more complex thermotropic behavior than the C_{16} SM (Bar *et al.*, 1997).

DMPC (di-C_{14}) and D-*erythro* C_{16} mix nearly ideally; their percolation threshold locus is close to the liquidus on the phase diagram. In contrast, DMPC and D-*erythro* C_{24}-SM mix non-ideally, with the percolation threshold locus close to the solidus. The latter mixture also forms particles, at least some of which are not multilamellar vesicles (Bar *et al.*, 1997).

This study shows clearly that the mixing of DMPC with C_{24}-SM in bilayers either of the various assemblies present in aqueous dispersion or the multi-bilayers used in the fluorescence recovery after photobleaching (FRAP) experiments is very non-ideal. This may be related to the two forms of gel phase of C_{24}-SM (gel I and gel II), which differ in the extent of interdigitation of opposing monolayers (see Figure 3 and Levin *et al.*, 1985). It is also apparent that the non-ideality must be primarily due to the intramolecular methylene chain mismatch in this sphingomyelin and, to a much lesser extent, to the SM T_m. Since this sort of chain mismatch exists in plasma membranes of mammalian cells, the mixing of such a sphingomyelin with other, mostly unsaturated and nearly symmetric (lacking chain mismatch), lipid components of these membranes must also be highly non-ideal, as was observed for multilamellar vesicles (MLV) composed of mixtures of egg PC and bovine brain SM (which is enriched with long acyl chain SM (Barenholz, 1984)). Therefore the presence of SMs having chain mismatch in biological membranes may be a necessary condition for phase domain formation, as shown in plasma membranes of rat cardiomyocytes (Yechiel and Barenholz, 1985). As lipid molecules having chain mismatch affect membrane curvature, such a composition is also a good example of the ability to couple phase separation and curvature, which may explain, in part, cell shapes and shape transitions including budding (Sackman, 1994; Lipowsky, 2002).

Although it is possible that in many aspects asymmetric SMs and PCs behave similarly (Bar *et al.*, 1997), the fact that only the asymmetric SMs are abundant in biological membranes may have a functional significance, for example in involvement in signal transduction via the sphingomyelin cycle (Kan and Kolesnick, 1993; Riboni *et al.*, 1997), as well as in the formation and/or stabilization of microdomains (rafts) and related structures (Futerman, 1995; Naslavsky *et al.*, 1999; Edidin, 2003).

4. MEMBRANE RAFTS

4.1. Definitions

The Random House Dictionary of the English Language (2nd ed., unabridged, 1987) defines the noun "raft" in a number of ways, of which the 3 definitions most relevant to the concept of membrane rafts are (a) "*a more-or-less rigid floating platform made of buoyant materials*; (b) *a collection of logs, planks, casks, etc., fastened together for floating on water*; (c) *a slab of reinforced concrete providing a footing on yielding soil*". None of these definitions explains in full the way scientists of the different disciplines involved in membrane research visualize membrane rafts. However, the one thing in common to all the dictionary definitions of raft which fits the concept of lipid membrane raft is a small platform floating in a more fluid continuum. Although this still does not explain thc dctails of what rafts actually are, it does fit the gross picture of rafts we have today. The details and the discrepancies are related to some extent to the fact that scientists involved in membrane raft research belong to very different disciplines: some of them, like cell biologists (rev. in Edidin, 2003), are mainly descriptive, and others, like biophysicists and physical chemists (rev. in Barenholz, 2002; Edidin, 2003), are highly quantitative. Even theoreticians (rev. in Lipowsky, 2002) are trying to go into dimensions, forces, and mechanisms involving raft formation. The theoreticians say that it is intuitively clear that when a membrane is composed of many components [lipid species and proteins – most membranes of eukaryotic cells contain >2000 lipid species (Shmeeda *et al.*, 1994a)], there are many ways that these components can interact with each other. As their properties are not identical, there is a high probability that interactions between certain components will be preferred and lead to phase separation and/or intramembrane domain formation in the membrane plane. The physicochemical (and therefore also the biological) properties of the various domains will also differ based on differences in their composition. An important question is what will determine the size distribution of such domains. Various theoretical considerations suggest that they can reach a certain maximal size determined by the exact membrane composition (Kumar *et al.*, 2001). Rafts are a result of such lateral phase separation. Based on current knowledge, in cell biology and membrane biophysics rafts are defined as lateral nano- and/or micro-domains in plasma membrane that are enriched with cholesterol, sphingolipids, and specific proteins (in particular, glycosylphosphatidylinositol (GPI)-anchored proteins and acyl chain-lipidated proteins (Simons and Ikonen, 1997; Kurzchalia and Parton, 1999; Brown and London, 2000; Fielding and Fielding, 2000; Migata, 2001; Edidin, 2003). The limited raft size and high concentration of specific proteins in such rafts results in protein crowding and therefore protein–protein interactions, such as the cross-linkage required for inducing various biological activities including signal transduction (discussed in Thompson *et al.*, 1995 and Naslavsky *et al.*, 1999).

The membrane rafts and caveolae serve as one of the cell's ports of entry for various agents and a platform for various signal transduction processes, in addition to being involved in protein sorting and cell movement (rev. in Edidin, 2003). It seems that these supermolecular assemblies are also responsible for in-plane lateral heterogeneity in protein distribution in plasma membrane, as some proteins are targeted while others are excluded from these domains (rev. in Edidin, 2003).

4.2. Open questions about rafts

Still, in spite of the extensive intradisciplinary research on membrane rafts in the last 15 years, they remain a controversial topic, as demonstrated by the dispute on cholesterol involvement. While most studies indicate the major and cardinal role of cholesterol, a few studies, like the one by Milhiet *et al.* (2002), suggest that this is not the case. Many membrane rafts in eukaryotic cells are characterized by a mosaic of ordered microdomains (rafts) enriched in sphingolipids and cholesterol and of fluid domains (or continuum). Milhiet *et al.* (2002), using differential scanning calorimetry (DSC) and atomic force microscopy, studied the thermotropic behavior and the topology of bilayers made of a phosphatidylcholine/sphingomyelin mixture that mimics the lipid composition of the external leaflet of renal brush-border membranes. In the absence of cholesterol, a broad phase separation process occurred where SO (gel) phase domains of size varying from the mesoscopic to the microscopic scale, enriched in sphingomyelin, occupied half of the bilayer surface at room temperature. Increasing amounts of cholesterol progressively decreased the enthalpy of the transition and modified the topology of membrane domains, up to a concentration of 33 mole%, for which no membrane domains were detected (similar to the cholesterol effect on SO $\rightarrow$ LD phase transitions (see 2.3.2 above). The results of Milhiet *et al.* (2002) strongly suggest that in membranes highly enriched in sphingolipids, like renal and intestinal brush borders, there is a threshold close to the physiological concentration, above which cholesterol acts as a suppressor rather than as a promoter of membrane domains. They also suggest that cholesterol depletion does not abolish the lateral heterogeneity in brush-border membranes. However, the DSC approach used by these authors can neither detect directly the LO phase nor the immiscibility in the LO phase, and therefore it may be misleading.

Many basic issues related to membrane rafts are not clear and remain unresolved. These include:

(a) What are the actual forces leading to the formation and maintenance of the rafts? Especially, what is the contribution of biological processes such as membrane trafficking versus physical forces such as thermodynamically and/or kinetically preferred interactions, as both types of factors have the potential to be the cause of raft formation?
(b) Among the physicochemical factors that may be involved in raft formation and maintenance, what is the contribution of thermodynamics versus kinetics? These two are best exemplified as ways to explain why rafts are enriched in cholesterol. Is this determined by cholesterol partition coefficient (thermodynamics) versus its rate of desorption (kinetics)? (For more details see 2.4.1 and 2.4.2 above.)
(c) Are these physicochemical factors by themselves sufficient to explain the lipid composition of the rafts, or is there a need for help by some specific proteins like sterol-binding proteins, other lipid transfer proteins, or other biological processes?
(d) What is the relevancy of detergent-insoluble (resistant) membranes (DRM) to the rafts? Or, in other words, do they represent the actual plasma

membrane rafts? An important related question is if the isolated DRM represent native domains in the membranes or are a result of partial solubilization and reassembly, namely an artifact of the method of isolation (Sot *et al.*, 2002).

(e) What is raft size distribution, and how stable is this distribution with time and with changing membrane composition and external conditions? If there are changes, on what time scale do they occur?

(f) Another important question related to the physical state of the rafts relative to the whole membrane is whether membrane rafts represent a situation of immiscibility in the LO phase? It is accepted that membrane rafts are in an LO phase (rev. in Edidin, 2003). However, as plasma membrane as a whole is enriched in cholesterol, the rest of the plasma membrane (the "non-raft" continuum) must also be in the LO phase. Therefore, we are dealing with immiscibility within the LO phase. Does this mean that having an immiscibility in the LO phase is a requirement for having rafts?

Although researchers are trying to answer these questions, and there is suggestive information for some of them (rev. in Edidin, 2003), most of these issues still remain unresolved.

4.3. Brief historical perspective

The term "membrane rafts" was first coined by cell biologists about 15 years ago to explain the selective delivery of lipids and proteins to the apical and basolateral surfaces of polarized epithelial cells (Simons and van Meer, 1988, rev. in Edidin, 2003). This phenomenon is related to the selective co-sorting of specific lipidated proteins and sphingolipids and may also involve the intermolecular hydrogen bonding typical of the sphingolipids (Pascher, 1976; Boggs, 1987; Patra *et al.*, 1999; Ferguson-Yankey *et al.*, 2000; Talbott *et al.*, 2000). However, the phenomenon itself of in-plane lipid microdomains was described before. Research done in the 1970s suggested, in general, the existence of lipid composition-dependent lateral phase separation and in-plane domain formation (Jain and White, 1977; Barenholz and Thompson, 1980; Klausner *et al.*, 1980; Barenholz, 1984).

Early studies on the lipid composition of viral membranes revealed that for many virions the lipid composition of viral membrane is enriched with sphingomyelin and cholesterol when compared to the plasma membrane of their host cells. This led to the suggestion that virion budding occurs from specific regions of the host cell plasma membranes (Patzer *et al.*, 1978b). This was recently confirmed for many viruses. For a comprehensive review on the historical perspective see Edidin (2003).

Recently it was shown that "rafts" are not limited to plasma membranes and are present also in the Golgi membranes. Helms *et al.* (2001) described Golgi-derived detergent resistant membranes (DRM) (see 4.4 below); these have a low buoyant density and high content of phospholipids. Twenty-five percent of Golgi phospholipids is present in the DRM, including 67% of Golgi sphingomyelin. In addition, 43% of Golgi cholesterol is in the DRM. In contrast to total cell DRM, which is rich in different proteins, the Golgi DRM contain only 10 major proteins, present in nearly stoichiometric amounts, including the alpha- and beta-subunits of heterotrimeric G proteins, flotillin-1, caveolin, and subunits of the vacuolar ATPase. Morphological data show a brefeldin A-sensitive and temperature-sensitive localization to the Golgi complex. Strikingly, their stability does not depend on their membrane environment, because after addition of brefeldin A to cells, DRM can be isolated from a fused Golgi-endoplasmic reticulum organelle. This indicates that these microdomains are not in a dynamic equilibrium with neighboring membrane proteins and lipids. After disruption of the microdomains by cholesterol extraction with β-cyclodextrin, a subcomplex of several Golgi DRM proteins including the B-subunit of the vacuolar ATPase, flotillin-1, caveolin, and p17 could still be isolated by immunoprecipitation. This indicates that several of the identified Golgi DRM proteins localize to the same microdomains and that the microdomain scaffold is not required for protein interactions between these Golgi DRM proteins, but instead might modulate their affinity (Helms *et al.*, 2001).

The major contribution to the understanding and characterization of the membrane raft concept came from the differential membrane solubilization by detergents resulting in detergent-soluble membrane (DSM) and detergent-resistant membrane (DRM) (Brown and Rose, 1992) fractions, which can be separated after the solubilization by various physical approaches, as discussed in 4.4 below.

4.4. Detergent solubilization as the approach to isolate and characterize membrane rafts

The raft concept of today is based on studies performed in biological membranes and in defined model membranes such as liposomes and lipid monolayers. One of the major tools used to identify and characterize membrane raft lipid and protein composition is detergent solubilization under various well-defined conditions. These studies involved interaction of a specific detergent with liposomes of defined composition or with biological membranes under defined conditions of: detergent/membrane-lipid mole ratio, temperature, and time of exposure. Following the interaction, a separation of DSM and DRM fractions according to density is performed, and their lipid and protein compositions determined. The results are highly dependent on the exact conditions of the solubilization procedure used. The use of Triton X-100 at 4°C in a standardized way was introduced by Brown and coworkers (Brown and Rose, 1992) and this is now considered the gold standard of the studies used to separate detergent insoluble and soluble membrane fractions.

The studies of Brown and coworkers demonstrated that an insoluble lipid fraction (DRM) can be isolated after exposure of eukaryotic cells to Triton X-100 in the cold (4°C), which is enriched in sphingolipids and cholesterol, as well as in certain proteins such as GPI-anchored enzymes (alkaline phosphatase). These insoluble membranes (DRM) appear to originate from sphingolipid- and cholesterol-rich membrane domains, which the authors described as membrane rafts of tightly packed LO phase (London and Brown, 2000). In these DRM the mole ratio of glycerophospholipids : sphingolipids : cholesterol was almost 1 : 1 : 1. This "magic" ratio was used by many to imitate lipid composition of membrane rafts (rev. in Edidin, 2003).

Lipidic membrane systems that have been reported to be composed of sphingomyelin (SM)-cholesterol (Chol) microdomains or "rafts" by Dietrich *et al.* (2001), palmitoyloleoyl-phosphatidylcholine (POPC)/SM/Chol, 1/1/1, and of SCRL: liver-PC/liver-phosphatidylethanolamine/SM/Cerebrosides/Chol, 1/1/1/1/2 (Schroeder *et al.*, 1994) were investigated by Aussenac *et al.* (2003) under the form of fully hydrated liposomes by the noninvasive solid-state ^{31}P and ^{2}H NMR. Liposomes of binary lipid composition POPC/Chol and SM/Chol were also studied as boundary/control systems. All systems are found to be in the LO phase at physiological temperatures. Use of deuterium-labeled cholesterol shows the axis of anisotropic rotation of cholesterol such that the molecule is, on average, quasiperpendicular to the membrane plane, in all of the four systems investigated. Cholesterol order parameters greater than 0.8 are observed, indicating that the sterol is in a very motionally restricted environment in the temperature range 0–60°C. The binary mixtures present "boundary" situations with the lowest values for POPC/Chol and the highest for SM/Chol. The SCRL raft mixture has the same ordering as the SM/Chol, i.e., the highest order parameter values over the temperature range. It demonstrates that in the SCRL mixture cholesterol dynamics is as in the binary system SM/Chol, therefore suggesting that it might be depleted from the rest of the membrane to form complexes as if it were alone with SM. On the other hand, the mixture POPC/SM/Chol exhibits an intermediate ordering situation between those of SM/Chol and POPC/Chol. This strongly suggests that cholesterol could be in fast exchange, at the NMR time scale (milli- to microseconds), between two or more membrane regions of different dynamics and questions the statement of "rigid domains" made of SM and cholesterol in the model "raft" system POPC/SM/Chol.

Parkin *et al.* (2001) demonstrated that the insolubility of GPI-anchored proteins in certain detergents is an intrinsic property of their association with sphingolipids and cholesterol in lipid rafts. For example, the GPI-anchored membrane protein dipeptidase is localized in detergent-insoluble (DRM) lipid rafts isolated from porcine kidney microvillar membranes, and these rafts, which lack caveolin, are enhanced not only in sphingomyelin and cholesterol, but also in the glycosphingolipid lactosylceramide (LacCer). Dipeptidase purified from porcine kidney was reconstituted into artificial liposomes in order to investigate the relationship between glycosphingolipids and GPI-anchored protein detergent-insolubility. The enzyme dipeptidase was detergent-insoluble in liposomes containing extremely low concentrations of LacCer. In contrast, identical concentrations of glucosylceramide or galactosylceramide failed to promote significant detergent-insolubility. Dipeptidase was insoluble in liposomes containing extremely low concentrations of LacCer. Cholesterol was shown to enhance the detergent-insoluble effect of LacCer. GC-MS analysis revealed dramatic differences between the fatty acyl chain compositions of LacCer and those of the other glycosphingolipids. However, despite these differences, the authors show that the unusually marked effect of LacCer to promote the detergent-insolubility of dipeptidase cannot be singularly attributed to the fatty acyl chain composition of this glycosphingolipid molecule. Instead, it was suggested that the ability of LacCer to confer detergent-insolubility on this GPI-anchored protein is dependent on the structure of the lipid molecule in its entirety, and that this glycosphingolipid may have an important role to play in the stabilization of lipid rafts, particularly the caveolin-free glycosphingolipid signaling domains (Parkin *et al.*, 2001).

However, although the separation between DRM and DSM has been used operationally to define membrane domains, the analyses of the results are not straightforward. Given the multitude of detergents used for this purpose, Schuck *et al.* (2003) investigated whether extraction with different detergents might reflect the same underlying principle of domain formation. When they compared the protein and lipid content of DRM prepared with a variety of detergents from two cell lines, they found that the detergents differed considerably in their ability to selectively solubilize membrane proteins and to enrich sphingolipids and cholesterol over glycerophospholipids, as well as saturated over unsaturated phosphatidylcholines. In addition, they studied the cell type-dependent variations of the molecular characteristics of DRM and the effectiveness of particular detergents. The results make it unlikely that different detergents reflect the same aspects of membrane lateral organization.

4.5. Why DRM can be isolated

The tight packing of the lipids in the membrane raft domains prevents the Triton X-100 from penetrating into these domains. On the other hand, the continuum (or other domains), less densely packed, enables the Triton X-100 to penetrate and to accumulate; then when reaching a critical concentration at which the additive packing parameter is decreased to a value of <0.74, a solubilization of this region occurs leaving the DRM intact. As Triton X-100 by itself has a higher density than phospholipids (Yedgar *et al.*, 1974), Triton-enriched membrane fractions will have a higher buoyant density than the DRM. This is the basis for the density gradient separation used routinely to separate DRM from DSM. The resistance to solubilization by Triton X-100 was directly demonstrated using atomic force microscopy (Rinia *et al.*, 2001). This differential solubilization pattern of different membrane lipids with Triton X-100 and especially the differences between SM and PC were demonstrated more than 25 years ago for liposomes (Hertz and Barenholz, 1977), and recently confirmed in the same and many other systems by Goñi and coworkers (Patra *et al.*, 1999).

It is possible that intermolecular SM–SM and SM–cholesterol hydrogen bonding is involved in the formation of tightly packed SM-enriched LO domains. This is supported by the lack of isolation of DRM when 3 M urea is added to the Triton X-100 during the solubilization (Patra *et al.*, 1999).

Recently, using detergent solubilization studies combined with fluorescence quenching, Xu *et al.* (2001) demonstrated that the sterol has to be a "membrane-active" sterol (see 2.2 above) in order to allow the tight packing of the LO phase needed to obtain DRM. The membrane-active sterols ergosterol and 7-dehydrocholesterol were both superior to cholesterol in promoting DRM. Among the sphingolipids, SM is much superior to cerebroside in ability to form DRM, while low levels (3 mole%) of ceramide enhanced membrane raft formation (Xu *et al.*, 2001). This group also demonstrated that the membrane-forming lipid has to be saturated in order to allow sufficient tightness of the LO phase (Xu and London, 2000).

Formation of rafts and their characterization in human platelets in response to cell activation were studied by Gousset *et al.* (2002). Lipid phase separation and domain formation were detected using the fluorescent dye 1,1′-dioctadecyl-3,3,3′,3′-tetramethyl-indocarbocyanine perchlorate (dil-C18), which preferentially partitions into SO-like lipid domains. It was shown that when human platelets are activated by cold and physiological agonists, rafts coalesce into visible aggregates. These events were disrupted by depletion of membrane cholesterol. Using Fourier transform infrared spectroscopy (FTIR), a thermal phase transition at around 30°C in intact platelets was found, which was assigned as the LO to the LD phase transition of rafts. Phase separation of the phospholipid and the sphingomyelin-enriched rafts could be observed as two phase transitions at around 15 and 30°C, respectively. The higher transition, assigned to the rafts, was greatly enhanced with removal of membrane cholesterol. DRM was isolated and was enriched in cholesterol (50%) and sphingomyelin (20%). The multi-functional platelet receptor CD36 selectively partitioned into DRM, whereas the GPI-linked protein CD55 and the major platelet integrin αIIbβ3a did not, which suggests that the clustering of proteins within rafts is a selective and regulated process dependent on specific lipid–protein interactions. The authors suggest that raft aggregation is a dynamic, reversible physiological event triggered by cell activation.

Another relevant example (Mueller *et al.*, 2001) deals with rafts in brain membranes. Lipid membrane rafts were isolated from purified synaptosome and synaptic plasma membranes of mice using Triton X-100. Additionally, lipid rafts were isolated from purified synaptosomes using a detergent-free preparation. Purity of lipid raft fractions was determined using the marker enzyme alkaline aminopeptidase and Western blot analysis of flotillin. Lipid raft fractions were highly enriched in both protein markers as compared with the non-lipid raft fractions. These membrane rafts were also enriched with cholesterol and sphingomyelin. Both the detergent and detergent-free methods of fractionation showed qualitatively the same results; however, the procedure using Triton X-100 had higher yields of lipid rafts compared to the detergent-free method. Mueller *et al.* (2001) show that lipid rafts can be isolated from purified brain membranes, and this procedure can be used to study the complex function of these structures in aging and Alzheimer's disease, two conditions in which cholesterol homeostasis is disrupted (Dufour *et al.*, 2003; Koudinova *et al.*, 2003).

One of the main criticisms raised against the use of detergent solubilization is based on potential artifacts due to the need to perform the solubilization at the nonphysiological conditions of 4°C (at this temperature the membranes' phase structure may be irrelevant to the situation at 37°C). Possibly this criticism has now been overcome, as it was shown that Triton X-100 solubilization may be replaced with a similar solubilization profile at 37°C which makes use of a similar nonionic detergent, the polyoxyethylene ether Brij 98. This detergent has a bulky headgroup and a monounsaturated ether chain (Drevot *et al.*, 2002).

5. MEDICAL APPLICATIONS RELEVANT TO SPHINGOMYELIN–CHOLESTEROL INTERACTIONS AND MEMBRANE RAFTS

5.1. Liposomes for drug delivery

Liposomes composed of sphingomyelin and cholesterol are used as drug carriers. Drug loading and loading stability require a low-pH intraliposome aqueous phase, and an extended circulation time requires addition of a lipopolymer (polyethyleneglycol attached to the liposome lipid). The following drugs are examples: the antibiotic ciprofloxacin and the anticancer alkaloid drugs swainsonine, vincristine, and vinblastin (Webb *et al.*, 1998). The main advantage of these liposomes (referred to as sphingosomes) is their high chemical and physical stability. The chemical stability is related to sphingomyelins lacking the acyl ester bonds of the phosphatidylcholines and having the much more chemically stable amide bond (between the acyl moiety and the primary amino group at C2 of the sphingosine base). The superior stability of the liposome pH gradient during liposome storage is due to the excellent barrier properties of the sphingomyelin/cholesterol membrane.

5.2. Reduction of intestinal cholesterol absorption in order to reduce serum cholesterol levels

The preferential interaction between sphingomyelin and cholesterol (see 2.4 above) is applied as a means to reduce intestinal absorption of cholesterol. In one study Carey *et al.* (2002) hypothesized that cholesterol thermodynamic activity (At) would predict bioavailability of cholesterol monomers in intestinal content, and that natural dietary phospholipids exhibiting high affinity for cholesterol would reduce its intestinal absorption and therefore help to reduce serum cholesterol level. They determined cholesterol At by measuring partitioning of monomeric cholesterol from aqueous solutions of taurocholate, cholesterol, and either milk sphingomyelin (MSM), dipalmitoyl phosphatidylcholine (DPPC), or egg yolk phosphatidylcholine (EYPC) into wafers of silicone. Cholesterol absorption from the same mixtures was tested with monolayers of Caco-2 cells. For *in vivo* absorption studies (employing male C57L/J mice), the authors used the dual-isotope fecal method during dietary enrichment with MSM, DPPC, or EYPC at varying dose levels. Results show that cholesterol At values were reduced significantly in MSM- and DPPC-containing systems compared with EYPC, and correlated positively with reduced uptake and esterification of cholesterol by Caco-2 cells. Mice fed chow absorbed 31.4% ± 6.9% (mean ± SEM) cholesterol, whereas enrichment with MSM or DPPC led to dose-dependent decreases in cholesterol absorption; even at 0.1% MSM, cholesterol absorption was reduced by 20.4% ± 15.4% ($P < 0.05$, $n = 6$). Therefore the authors concluded that different phospholipids differ in their effects on micellar cholesterol At, which predicts cholesterol uptake by enterocytes *in vitro* as well as *in vivo*. Natural phospholipids having high affinity for cholesterol, as evidenced particularly by sphingomyelin, decrease At and reduce intestinal cholesterol absorption.

In another study, Nyberg *et al.* (2000) examined the effects of SM on cholesterol absorption in the rat and compared them with those of phosphatidylcholine (PC). Cholesterol absorption was studied by use of the dual-isotope plasma ratio method. The authors also studied the effect of sterols on the fecal excretion of undigested SM and its metabolites after a single meal of ^{3}H-dihydrosphingosine-labeled SM. When cholesterol was given dissolved in soybean oil, without addition of SM or other phospholipids, absorption was 68 ± 12%. Absorption was less efficient from the cholesterol/phospholipid dispersions. In dispersions with cholesterol and SM, the lowest cholesterol absorption (9 ± 2%) was seen with a cholesterol : SM molar ratio of 1 : 1. With dispersions of cholesterol and different PC substrates the absorption of cholesterol was lower with saturated PC (16 ± 8%) than with soybean-PC (22 ± 4%) or dioleoyl PC (23 ± 8%). Uptake of SM in the rat intestine was reduced by sterols. For example, percentage recovery of ^{3}H radioactivity in fecal lipids was 38 ± 8% when SM was given with cholesterol and 16 ± 3% without any sterol. One-third of the radioactivity in feces was present as ceramide. Sitostanol had the same effect on uptake of SM as cholesterol. This study shows that when rats are fed mixtures of SM and cholesterol, intestinal uptake is less than when fed only one of them.

These findings have two different implications: (1) Feeding mixtures of SM and cholesterol reduces cholesterol absorption and therefore cholesterol serum level (e.g., Carey *et al.*, 2002) and (2) Feeding mixtures of SM and sterols induces increased exposure of the colon to ceramide, which may be relevant to treatment of colon cancer (Merrill *et al.*, 1995).

5.3. Reverse cholesterol transport

This application was already used in the 1950s by Friedman and coworkers (Friedman *et al.*, 1957; rev. in Williams *et al.*, 1984). The principle of this approach is to intravenously inject or infuse into a mammal a suspension (high dose) of unilamellar liposomes composed of phospholipids and no sterols. As long as liposomes are below 150 nm in size they circulate long enough. Being cholesterol-free, these liposomes will serve as a cholesterol acceptor, and when they are infused intravenously in high dose they remove cholesterol from other tissues using plasma lipoproteins and red blood cells as mediators (rev. in Williams *et al.*, 1984; Williams and Tall, 1988). To complete the reverse cholesterol transport it is also required that these liposomes, after acquiring cholesterol, will be taken up by the liver, where the cholesterol is metabolized to bile acids, which will then be excreted in the feces (Williams *et al.*, 1984; Williams and Tall, 1988). Much research was done using such an approach with encouraging results in animals (rev. in Williams *et al.*, 1984, Williams and Tall, 1988; Barenholz and Yechiel, 1989; Zomber *et al.*, 1996; Barenholz and Greenfeld, 1997). Most of these studies made use of PC liposomes to accomplish the reverse cholesterol transport, concomitantly replacing excess sphingomyelin with PC (Yechiel and Barenholz, 1985; Barenholz and Yechiel, 1989; Barenholz and Greenfeld, 1997; Barenholz and Shmeeda, 2003). Recently similar experiments were also performed with sphingomyelin as a cholesterol acceptor, taking advantage of the superior affinity of cholesterol to SM. Some of these developments are now in clinical trials. In some of these treatments the researchers combined the use of liposomes with additional drugs which affect cholesterol metabolism (Williams, 2002).

5.4. Identifying anti-disease targets by lipid raft immunization and proteomics

This approach focuses on a method to identify anti-disease targets localized in membrane rafts (exemplified by anti-tumor and anti-viral targets) by isolating and examining lipid rafts, comparing the desired cells derived from normal and diseased individuals. In one example, Tso and Green (2002) (U.S. Pat. Appl. 269010/10) identified an anti-tumor agent by lipid rafts immunization. This invention also provided hybridomas produced by the method of identifying anti-tumor agents by lipid rafts immunization and antibodies produced by the hybridomas.

5.5. Modulation of composition, level, and/or distribution of lipid rafts in order to affect signaling to treat various diseases

5.5.1. Rationale

Recently it was demonstrated that for at least a few enveloped viruses, including measles, influenza, and HIV-1, some of their membrane proteins are targeted to the host cell membrane rafts (Ali *et al.*, 2000; Nguyen and Hildreth, 2000; Vincent *et al.*, 2000; Zhang *et al.*, 2000; Lindwasser and Resh, 2001; Freed, 2002). Disruption of the raft interferes with virus production (Ono and Freed, 2001). This suggests that this virion assembly requires host cell membrane rafts for infectivity. It was found that rafts serve as a port of entry for several classes of viruses, since the plasma membrane receptors for these viruses reside in these rafts. Therefore, disruption of host cell membrane rafts inhibits virus infectivity (Liao *et al.*, 2001), and this may provide a strategy to prevent viral infection.

Bavari *et al.* (2002) show that the highly pathogenic Ebola and Marburg viruses (both are filoviruses) also enter and exit host cells using lipid rafts. Treating host cells with nystatin and filipin – two agents which, respectively, interact with, or extract, cholesterol with high affinity – inhibit filovirus infectivity. Thus, as with HIV-1, both virion assembly and host cell entry by filoviruses take place in a raft-rich domain of the host cell plasma membrane (Bavari *et al.*, 2002).

Already 25 years ago it was demonstrated that the vesicular stomatitis virus (VSV) requires a high cholesterol level in its membrane for infectivity, as extraction of cholesterol by lipoproteins or liposomes reduced VSV infectivity 300-fold, while after replenishment of these cholesterol-depleted virions with cholesterol, infectivity was partially recovered (Pal *et al.*, 1981). If this cholesterol is related to rafts, this raises a question of whether each of the virions represent a single raft or whether the virions are highly enriched with rafts?

Rafts are also involved in the development of malaria. Haldar *et al.* (2000) show that erythrocytes (RBC), which are incapable of endocytosis or phagocytosis, can be infected by the malaria parasite *Plasmodium falciparum*. It was found in RBC that Duffy (a transmembrane protein), GPI-lipidated proteins, and cytoplasmic proteins associated with DRMs that are characteristic of microdomains in host RBC membranes, are internalized by vacuolar parasites, while the major integral RBC membrane proteins and cytoskeletal proteins are not. The internalized host proteins and a plasmodial transmembrane resident parasitophorous vacuolar membrane (PVM) protein are detected in DRMs associated with vacuolar parasites. This paper (Haldar *et al.*, 2000), is the first report of a host transmembrane protein being recruited into an apicomplexan vacuole and of the presence of vacuolar DRM; it establishes that integral association does not preclude protein internalization into the *P. falciparum* vacuole. Rather, as shown for Duffy, intracellular accumulation occurs at the same rate as that seen for a DRM-associated GPI-anchored protein. Furthermore, novel mechanisms regulated by the DRM lipids sphingomyelin and cholesterol mediate (i) the uptake of host DRM proteins and (ii) maintenance of the intracellular vacuole in the non-endocytic red cell, which may have implications for intracellular parasitism and pathogenesis.

In agreement with Haldar *et al.* (2000) it was found that depletion of cholesterol markedly inhibited the ability of the malaria parasite to invade red blood cells, while repletion of cholesterol completely restored normal parasite invasion (Samuel *et al.*, 2001).

There are indications that rafts are also involved in neurodegenerative diseases such as Alzheimer's (Burns and Duff, 2002; Koudinova *et al.*, 2003) and prion-related diseases (Borchelt *et al.*, 1990; Taraboulos *et al.*, 1995; Naslavsky *et al.*, 1999). In the latter it was demonstrated that the pathogenic scrapie prion protein drops dramatically in cholesterol-depleted cells (Taraboulos *et al.*, 1995). However, inhibition of sphingolipid synthesis has the opposite effect, accelerating scrapie prion formation. The discrepancy between the two effects is not yet clear and is discussed in more detail in Naslavsky *et al.* (1999). The main implication of this discrepancy is that we have to better understand the involvement of cholesterol and SM in raft formation and maintenance, as well as their effects on protein–protein interactions in rafts.

5.5.2. Examples of application

One good example is the invention of Lu *et al.*, 2002 (U.S. Pat. Appl. 080273/10), based on the involvement of lipid rafts in immune cell signaling using a protein or a ligand that modifies the immune cell responses through affecting composition and/or distribution of lipid rafts in the membrane of the immune cell. This is exemplified in a few inventions by using β-cyclodextrin compositions to remove cholesterol from cells, thereby disrupting membrane rafts, and as a result blocking passage of the pathogen through the lipid rafts. Hildreth, 2001 (U.S. Pat. Appl. 801393/09) uses this approach to prevent the transmission of sexually transmitted diseases.

ACKNOWLEDGMENTS

The author's work discussed in this review was supported in part by various grants from the United States–Israel Binational Science Foundation (BSF), the Israel National Science Foundation (ISF), PHS NIH grant HL17576, and the Barenholz Fund.

The help of Beryl Levene in typing and Sigmund Geller in editing this review is acknowledged with pleasure.

REFERENCES

Abrahamsson, S., Dahlen, B., Lofgren, H., Pascher, I., Sundell, S. 1977. In: Abrahamsson, S., Pascher, I., Eds. Structure of Biological Membranes. Plenum Press, New York, p. 1.

Ali, A., Avalos, R.T., Ponimaskin, E., Nayak, D.P. 2000. Influenza virus assembly: effect of influenza virus glycoproteins on the membrane association of M1 protein. J. Virol. 74:8709-8719.

Aussenac, F., Tavares, M., Dufourc, E.J. 2003. Cholesterol dynamics in membranes of raft composition: a molecular point of view from ^{2}H and ^{31}P solid-state NMR. Biochemistry 42, 1383-1390.

Bar, L.K., Barenholz, Y., Thompson, T.E. 1986. The fraction of cholesterol undergoing spontaneous exchange between small unilamellar phosphatidylcholine vesicles. Biochemistry 25:6701-6705.

Bar, L.K., Barenholz, Y., Thompson, T.E. 1987. Dependence of the fraction of cholesterol undergoing spontaneous exchange between small unilamellar vesicles and their phospholipid composition. Biochemistry 26:5460-5465.

Bar, L.K., Chong, P.C., Barenholz, Y., Thompson, T.E. 1989. Spontaneous transfer between phospholipid bilayers of dehydroergosterol, a fluorescent cholesterol analog. Biochim. Biophys. Acta 983:109-112.

Bar, L.K., Barenholz, Y., Thompson, T.E. 1997. The effect of sphingomyelin composition on the phase structure of phosphatidylcholine-sphingomyelin bilayers. Biochemistry 36:2507-2516.

Barenholz, Y. 1984. Sphingomyelin-lecithin balance in membranes: composition, structure, and function relationships. In: Shinitzky, M., Ed. Physiology of Membrane Fluidity, Vol. I. CRC Press, Boca Raton, FL, pp. 131-172.

Barenholz, Y. 2001. Liposome application: problems and prospects. Curr. Opin. Colloid Interface Sci. 6:66-77.

Barenholz, Y. 2002. Cholesterol and other membrane active sterols: from membrane evolution to "rafts". Prog. Lipid Res. 41:1-5.

Barenholz, Y., Cevc, G. 2000. Structure and properties of membranes. In: Baszkin, A., Norde, W., Eds. Physical Chemistry of Biological Surfaces. Marcel Dekker, New York, pp. 171-241.

Barenholz, Y., Gatt, S. 1982. Sphingomyelin: metabolism, chemical synthesis, chemical and physical properties. In: Hawthorne, J.N., Ansell, G.B., Eds. Phospholipids. Comprehensive Biochemistry. Elsevier, New York, pp. 129-177.

Barenholz, Y., Greenfeld, Z. 1997. Method of improving renal function. U.S. Patent 5,622,715 (April 22).

Barenholz, Y., Shmeeda, H. 2003. Involvement of membrane and lipoprotein lipid in aging processes. In: Rosin, A., Ed. Aging and Gerontology in Israel. JDC-Eshel, Jerusalem, pp. 173-200.

Barenholz, Y., Thompson, T.E. 1980. Sphingomyelins in bilayers and biological membranes. Biochim. Biophys. Acta 604:129-158.

Barenholz, Y., Thompson, T.E. 1999. Sphingomyelin: biophysical aspects. Chem. Phys. Lipids 102:29-34.

Barenholz, Y., Yechiel, E. 1989. Lipid replacement therapy. U.S. Patent 4,812,314, March 14.

Barenholz, Y., Suurkuusk, J., Mountcastle, D., Thompson, T.E., Biltonen, R.L. 1976. A calorimetric study of the thermotropic behavior of aqueous dispersions of natural and synthetic sphingomyelins. Biochemistry 15:2441-2447.

Barenholz, Y., Yechiel, E., Cohen, R., Deckelbaum, R.J. 1981. Importance of cholesterol-phospholipid interaction in determining dynamics of normal and abetalipoproteinemia red blood cell membrane. Cell Biophys. 3:115-126.

Bavari, S., Bosio, C.M., Wiegand, E., Ruthel, G., Will, A.B. *et al.* 2002. Lipid raft microdomains: a gateway for compartmentalized trafficking of Ebola and Marburg viruses. J. Exp. Med. 195:593-602.

Ben-Yashar, V., Barenholz, Y. 1989. The interaction of cholesterol and cholest-4-en-3-one with dipalmitoylphosphatidylcholine: comparison based on the use of three fluorophores. Biochim. Biophys. Acta 985:271-278.

Bittman, R. 1997. Has nature designed the cholesterol side chain for optimal interaction with phospholipids? Subcell. Biochem. 28:145-171.

Boggs, J.M. 1987. Lipid intermolecular hydrogen bonding: influence on structural organization and membrane function. Biochim. Biophys. Acta 906:353-404.

Borchelt, D.R., Scott, M., Taraboulos, A., Stahl, N., Prusiner, S.B. 1990. Scrapie and cellular prion proteins differ in their kinetics of synthesis and topology in cultured cells. J. Cell Biol. 110:743-752.

Brady, R.O. 2003. Enzyme replacement therapy: conception, chaos and culmination. Philos. Trans. R. Soc. London B. Biol. Sci. 358:915-919.

Brown, D.A., London, E. 2000. Structure and function of sphingolipid- and cholesterol-rich rafts. J. Biol. Chem. 275:17221-17224.

Brown, D.A., Rose, J.K. 1992. Sorting of GPI-anchored proteins to glycolipids-enriched membrane subdomains during transport to the apical cell surface. Cell 68:533-544.

Burns, M., Duff, K. 2002. Cholesterol in Alzheimer's disease and tauopathy. Ann. N.Y. Acad. Sci. 977:367-675.

Carey, M.C., Donovan, J.M., Eckhardt, E.R.M., Wang, D.Q-H. 2002. Dietary sphingomyelin suppresses intestinal cholesterol absorption by decreasing thermodynamic activity of cholesterol monomers. Gastroenterology 122:948-956.

Cohen, R., Barenholz, Y., Gatt, S., Dagan, A. 1984. Preparation and characterization of well defined D-*erythro* sphingomyelins. Chem. Phys. Lipids 35:371-384.

Cooper, R.A., Strauss, J.F. 1984. In: Shinitizky, M., Ed. Physiology of Membrane Fluidity, Vol. I. CRC Press, Boca Raton, FL, pp. 73-97.

de Gier, J., Manderslot, J.G., van Deenen, L.L.M. 1969. The role of cholesterol in lipid membranes. Biochim. Biophys. Acta 173:143-152.

de Kruijff, B., Cullis, P.R., Verkeleij, A.J., Hope, M.J., Van Echteld, C.J.A., Taraschi, T.F. 1985. In: Marnetosi, A.N., Ed. The Enzymes of Biological Membranes, 2nd ed., Vol. I. Plenum Press, New York, pp. 131-204.

Deamer, D.W., Chakrabarti, A. 1996. In: Barenholz, Y., Lasic, D.D., Eds. Handbook of Nonmedical Applications of Liposomes, Vol. II (Models for biological phenomena). CRC Press, Boca Raton, FL, pp. 303-313.

Demel, R.A., Bruckdorfer, K.R., van Deenen, L.L.M. 1972. The effect of sterol structure on the permeability of liposomes to glucose, glycerol and Rb^+. Biochim. Biophys. Acta 255:321-330.

Demel, R.A., Jansen, J.W.C.M., Van Dijck, P.W.M., Van Deenen, L.L.M. 1977. The preferential interaction of cholesterol with different classes of phospholipids. Biochim. Biophys. Acta 465:1-10.

Dietrich, C., Bagatolli, L.A., Volovyk, Z.N., Thompson, N.L., Levi, M., Jacobson, K., Gratton, E. 2001. Lipid rafts reconstituted in model membranes. Biophys. J. 80:1417-1428.

Dobrowsky R.T. 2000. Sphingolipid signaling domains floating on rafts or buried in caves? Cell. Signal. 12:81-90.

Drevot, P., Langlet, C., Guo, X.J., Bernard, A.M., Colard, O., Chauvin, J.P., Lasserre, R., He, H.T. 2002. TCR signal initiation machinery is pre-assembled and activated in a subset of membrane rafts. EMBO J. 21:1899-1908.

Dufourc, E.J., Parish, E.J., Chitrakorn, S., Smith, I.C.P. 1984. Structural and dynamical details of cholesterol lipid interaction as revealed by deuterium NMR. Biochemistry 23:6062-6071.

Dufour, F., Zhao, W., Ravindranath, L., Alkon, D. 2003. Abnormal cholesterol processing in Alzheimer's disease patient's fibroblasts. Neurobiol. Lipids Vol. 1, article 7. Published online March 14, 2003. Available at: http://neurobiologyoflipids.org/ content/1/7/

Edidin, M. 2003. The state of lipid rafts: from model membranes to cells. Annu. Rev. Biophys. Biomol. Struct. 32:257-283.

Estep, T.N., Calhoun, W.I., Barenholz, Y., Biltonen, R.L., Shipley, G.G., Thompson, T.E. 1980. Evidence for metastability in N-stearoyl sphingomyelin in bilayer. Biochemistry 19:20-24.

Feher, J.J. 1974. Studies of self-association and solvent-association of cholesterol and other 3β-hydroxysteroids in nonpolar media. J. Phys. Chem. 78:250-255.

Ferguson-Yankey, S.R., Borchman, D., Taylor, K.G., Du Pre, A.B., Yappert, M.C. 2000. Conformational studies of sphingolipid by NMR spectroscopy. I. Dihydrosphingomyelin. Biochim. Biophys. Acta 1467:307-325.

Fielding, C.J., Fielding, P.E. 2000. Cholesterol and caveolae: structural and functional relationships. Biochim. Biophys. Acta 1529:210-222.

Frank, A., Barenholz, Y., Lichtenberg, D., Thompson, T.E. 1983. Spontaneous transfer of sphingomyelin between phospholipid bilayers. Biochemistry 22:5647-5651.

Freed, E.O. 2002. Rafting with Ebola. Science 296:279.
Friedman, M., Byers, S.O., Rosenman, R.H. 1957. Resolution of aortic atherosclerotic infiltration in the rabbit by phosphatide infusion. Proc. Soc. Exp. Biol. Med. 95:586-588.
Futerman, A.H. 1995. Inhibition of sphingolipid synthesis - effects on glycosphingolipid-GPI-anchored protein microdomains. Trends Cell Biol. 5:377-380.
Garver, W.S., Heidenreich, R.A. 2002. The Niemann-Pick C proteins and trafficking of cholesterol through the endosomal/lysosomal system. Curr. Mol. Med. 2:485-505.
Gatt, S., Barenholz, Y. 1999. Editors of "Sphingomyelin: Chemistry, Biophysics, Metabolism, Genetics and Signaling", Chem. Phys. Lipids 102: issues 1 and 2.
Gilbert, D.B., Reynolds, J.A. 1976. Thermodynamic equilibria of cholesterol–detergent–water. Biochemistry 15:71-74.
Gousset, K., Crowe, J.H., Field, C.L., Oliver, A.E., Tablin, F., Tsvetkova, N.M., Walker, N.J., Wolkers, W.F. 2002. Evidence for a physiological role for membrane rafts in human platelets. J. Cell. Physiol. 190:117-128.
Grant, G.J., Barenholz, Y., Piskoun, B., Bansinath, M., Turndorf, H., Bolotin, E. 2001. DRV liposomal bupivacaine: preparation, characterization and *in vivo* evaluation in mice. Pharm. Res. 18:336-343.
Grunze, M., Deuticke, B. 1974. Changes of membrane permeability due to extensive cholesterol depletion in mammalian erythrocytes. Biochim. Biophys. Acta 356:125-136.
Haines, T.H. 2001. Do sterols reduce proton and sodium leaks through lipid bilayers? Prog. Lipid Res. 299-324.
Haldar, K., Harison, T., Hiller, N.L., Lauer, S., McManus, H., Mohandas, N., Samuel, B.U., VanWye, J. 2000. Vacuolar uptake of host components, and a role for cholesterol and sphingomyelin in malarial infection. EMBO 19:3556-3564.
Haran, G., Cohen, R., Bar, L.K., Barenholz, Y. 1993. Transmembrane ammonium sulfate gradients in liposomes produce efficient and stable entrapment of amphipathic weak bases. Biochim. Biophys. Acta 1151:201-215.
Hauser, H., Pascher, I., Pearson, R.H., Sundell, S. 1981. Preferred conformation and molecular packing of phosphatidylethanolamine and phosphatidylcholine. Biochim. Biophys. Acta 650:25-51.
Helms, J.B., Bruegger, B., Gkantiragas, I., Kaloyanova, D., Li, X.-Y., Loehr, K., Lottspeich, F., Stueven, E., Wieland, F.T. 2001. Sphingomyelin-enriched microdomains at the Golgi complex. Mol. Biol. Cell 12:1819-1833.
Hertz, R., Barenholz, Y. 1975. Permeability and integrity properties of lecithin-sphingomyelin liposomes. Chem. Phys. Lipids 15:138-156.
Hertz, R., Barenholz, Y. 1977. The relations between the composition of liposomes and their interaction with Triton X-100. J. Colloid Interface Sci. 60:188-200.
Hildreth, J.E. 2001. Beta-cyclodextrin compositions and use to prevent transmission of sexually transmitted diseases. U.S. Pat. Appl. 801393/09, July 3.
Huang, C. 1977. A structural model for the cholesterol–phosphatidyl choline complexes in bilayer membranes. Lipids 12:348-358.
Huang, C., Mason, J.T. 1986. Structure and properties of mixed-chain phospholipid assemblies. Biochim. Biophys. Acta 864:423-470.
Ipsen, J.H., Karlstrom, G., Mouritsen, O.G., Wennerstrom, H., Zuckerman, M.J. 1987. Phase equilibria in the phosphatidylcholine-cholesterol system. Biochim. Biophys. Acta 905:162-172.
Ipsen, J.H., Mouritsen, O.G., Bloom, M. 1990. Relationships between lipid membrane area, hydrophobic thickness, and acyl-chain order. The effects of cholesterol. Biophys. J. 57:405-412.
Israelachvili, J.N. 1992. Intermolecular and Surface Forces. 2nd ed. Academic Press, London, pp. 366-399.

Israelachvili, J.N., Marcelja, S., Horn, R.G. 1980. Physical principles of membrane organization. Q. Rev. Biophys. 13:121-200.
Jain, M.K., White, H.B. 3rd. 1977. Long-range order in biomembranes. Adv. Lipid Res. 15:1-60.
Jedlovszky, P., Mezei, M. 2003. Effect of cholesterol on properties of phospholipid membrane. 1: Structural features. J. Phys. Chem. B107:5311-5321.
Jendrasiak, G.L., Smith, R.L. 2001. The effect of choline head group on phospholipid hydration. Chem. Phys. Lipids 113:55-66.
Kan, C.C., Kolesnick, R. 1993. Signal transduction via the sphingomyelin pathway. Trends Glycosci. Glycotechnol. 5:99-106.
Klausner, R.D., Kleinfeld, A.M., Hoover, R.L., Karnovsky, M.J. 1980. Lipid domains in membranes. Evidence derived from structural perturbations induced by free fatty acids and lifetime heterogeneity analysis. J. Biol. Chem. 255:1286-1295.
Kolesnick, R., Golde, D.W. 1994. The sphingomyelin pathway in tumor necrosis factor and interleukin-1 signaling. Cell 77:325-328.
Koudinova, N.V., Kontush, A., Berezov, T.T., Koudinov, A.R. 2003. Amyloid beta, neural lipids, cholesterol and Alzheimer's disease. Neurobiol. Lipids Vol. 1, article 6. Published online March 3, 2003. Available at http://neurobiologyoflipids.org/content/1/6.
Kumar, S., Gompper, G., Lipowsky, K. 2001. Budding dynamics of multicomponent membranes. Phys. Rev. Lett. 86:3911-3914.
Kumar, V.V. 1991. Complementary molecular shapes and additivity of the packing parameter of lipids. Proc. Natl. Acad. Sci. USA 88:444-448.
Kurzchalia, T.V., Parton, R.G. 1999. Membrane microdomains and caveolae. Curr. Opin. Cell Biol. 11:424-431.
Lentz, B.R., Hoechli, M., Barenholz, Y. 1981. Acyl chain order and lateral domain formation in mixed phosphatidylcholine-sphingomyelin multilamellar and unilamellar vesicles. Biochemistry 20:6803-6809.
Levin, I.W., Thompson, T.E., Barenholz, Y., Huang, C. 1985. Two types of hydrocarbon chain. Interdigitation in sphingomyelin bilayers. Biochemistry 24:6282-6286.
Liao, Z., Graham, D.R., Hildreth, J.E.K. 2003. Lipid rafts and HIV pathogenesis: viron-associated cholesterol is required for fusion and infection of susceptible cells. Aids Res. Hum. Retroviruses 19:675-687.
Liao, Z., Cimakasky, L.M., Hampton, R., Nguyen, D.H., Hildreth, J.E.K. 2001. Lipid rafts and HIV pathogenesis: host membrane cholesterol is required for infection by HIV type I. Aids Res. Hum. Retroviruses 17:1009-1019.
Lichtenberg, D., Barenholz, Y. 1988. Liposomes: Preparation, characterization and preservation. In: Glick, D., Ed. Methods of Biochemical Analysis. Wiley, New York, Vol. 33, pp. 337-462.
Lindwasser, O.W. and Resh, M.D. 2001. Multimerization of human immunodeficiency virus type 1 Gag promotes its localization to barges, raft-like membrane microdomains. J. Virol. 75:7913-7924.
Lipowsky, R. 2002. Domains and rafts in membranes – hidden dimensions of self-organization. J. Biol. Phys. 28:195-210.
Lofgren, H., Pascher, I. 1977. Molecular arrangements of sphingolipids. The monolayer behaviour of ceramides. Chem. Phys. Lipids 20: 273-284.
London E, Brown, D.A. 2000. Insolubility of lipids in Triton X-100: physical origin and relationship in sphingolipid/cholesterol membranes domains (rafts). Biochim. Biophys. Acta 1508:182-195.
Lu, P.S., Diaz-Sarmineto, C.S., Seed, B.A., Xavier, R., Irving, B. 2002. PDZ domain interactions and lipid rafts. U.S. Pat. Appl. 080273/10, Feb. 19.

Marsh, D. 1990. CRC Handbook of Lipid Bilayers. CRC Press, Boca Raton, FL.
Merrill, A.H. Jr., Schmelz, E.M., Wang, E., Schroeder, J.J., Dillehay, D.L. 1995. Role of dietary sphingolipids and inhibitors of sphingolipid metabolism in cancer and other diseases. J. Nutr. 125:1677S-1682S.
Merrill, A.H. Jr., Schmelz, E.M., Dillehay, D.L., Spiegel, S., Shayman, J.A., Schroeder, J.J., Riley, R.T., Voss, K.A., Wang. E. 1997. Sphingolipids – the enigmatic lipid class: biochemistry, physiology, and pathophysiology. Toxicol. Appl. Pharmacol. 142:208-225.
Migata, H. 2001. Microstructure in cell membranes and their roles in cellular activities. Membrane 26:134-140.
Milhiet, P.E., Giocondi, M-C., Le Grimellec, C. 2002. Cholesterol is not crucial for the existence of microdomains in kidney brush-border membrane models. J. Biol. Chem. 277:875-878.
Mitchell, D.C., Litman, B.J. 1998. Effect of cholesterol on molecular order and dynamics in highly polyunsaturated phospholipid bilayers. Biophys. J. 75:896-908.
Mitchell, D.C., Litman, B.J. 2001. Modulation of receptor signaling by phospholipid acyl chain composition, In: Mostofsky, D.I., Yehuda, S., Salem, N., Eds. Fatty acids: From Neuronal Membrane to Physiological and Behavioral Functions, Part I, Chapter 2. Humana Press.
Mouritsen, O.G., Jørgensen, K. 1995. Micro, nano and meso-scale heterogeneity of lipid bilayers and its influence on macrospopic membrane properties. Mol. Membr. Biol. 12:15-20.
Mouritsen, O.G., Jørgensen, K. 1994. Dynamical order and disorder in lipid bilayers. Chem. Phys. Lipids 73:3-25.
Mouritsen, O.G., Jørgensen, K. 1998. A new look at lipid-membrane structure in relation to drug research. Pharm. Res. 15:1507-1519.
Mouritsen, O.G., Jørgensen, K., Trandum, C., Werth, P. 1999. Association of ethanol with lipid membranes containing cholesterol, sphingomyelin and gangliosides. A titration calorimetry study. Biochim. Biophys. Acta 1420:179-188.
Mueller, W.E., Eckert, G.P., Igbavoboa, U., Wood, W.G. 2001. Characterization of lipid rafts isolated from purified brain membranes of C57BL/6J mice. Soc. Neurosci. Abstracts 27.
Naslavsky, N., Shmeeda, H., Friedlander, G., Yanai, A., Futerman, A.H., Barenholz, Y., Taraboulos, A. 1999. Sphingolipid depletion increases formation of the scrapie prion protein in neuroblastoma cells infected with prions. J. Biol. Chem. 274:20763-20771.
Nes, W.R. 1974. Role of sterols in membranes. Lipids 9:596-612.
Nguyen, D.H., Hildreth, J.F. 2000. Evidence for budding of human immunodeficiency virus type 1 from glycolipid-enriched membrane rafts. J. Virol. 74:3264-3272.
Nielsen, M., Miao, L., Ipsen, J.H., Zuckerman, M., Mouristen, O. 1999. Off lattice model for phase behavior of lipid cholesterol bilayers. Phys. Rev. 59:5790-5803.
Niu, S., Litman, B.J. 2002. Determination of membrane cholesterol partition coefficient using a lipid vesicle-cyclodextrin binary system: effect of phospholipid acyl chain unsaturation and headgroup composition. Biophys. J. 83:3408-3415.
Nyberg, L., Duan, R., Nilsson, A. 2000. A mutual inhibitory effect on absorption of sphingomyelin and cholesterol. J. Nutr. Biochem. 11:244-249.
Obeid, L.M., Hannun, Y.A. 1995. Ceramide: a stress signal and mediator of growth suppression and apoptosis. J. Cell Biochem. 58:191-198.
Oborina, E.M., Yapert, M.C. 2003. Effect of sphingomyelin versus dipalmitoyl phosphatidyl choline on extent of lipid oxidation. Chem. Phys. Lipids 123:223-232.
Ohvo-Rekila, H., Ramstedt, B., Leppimaki, P., Slotte, J.P. 2002. Cholesterol interaction with phospholipids in membranes. Prog. Lipid Res. 41:66-97.
Ono, A., Freed, E.O. 2001. Plasma membrane rafts play a critical role in HIV-assembly and release. Proc. Natl. Acad. Sci. USA 98:13925-13930.

Pal, R., Barenholz, Y., Wagner, R.R. 1981. Depletion and exchange of cholesterol from the membrane of vesicular stomatitis virus by interaction with serum lipoproteins or poly(vinylpyrrolidone) complexed with bovine serum albumin. Biochemistry 20:530-539.

Paphadjopoulos, D., Lowden, M., Kimelberg, H. 1973. Role of cholesterol in membranes: effects on phospholipid–protein interactions, membrane permeability and enzymatic activity. Biochim. Biophys. Acta 330:8-26.

Parasassi, T., Distfano, M., Loiero, M., Ravagnan, G., Gratton, E. 1994. Biophys. J. 66:1895-1902.

Parkin, E.T., Hooper, N.M., Turner, A.J. 2001. Differential effects of glycosphingolipids on the detergent insolubility of the glycosylphosphatidyl-inositol-anchored membrane dipeptidase. Biochem. J. 358:209-216.

Pascher, I. 1976. Molecular arrangements in sphingolipids conformation on membrane stability and permeability. Biochim. Biophys. Acta 455:433-451.

Patra, S.K., Alonso, A., Arrondo, J.L.R., Goñi, F.M. 1999. Liposomes containing sphingomyelin and cholesterol: detergent solubilization and infrared spectroscopic studies. J. Liposome Res. 9:247-260.

Patton, S. 1970. Correlative relationship of cholesterol and sphingomyelin cell membrane. J. Theor. Biol. 29:489-491.

Patzer, E.J., Wagner, R.R., Barenholz, Y. 1978a. Cholesterol oxidase as a probe for studying membrane organization. Nature 274:394-395.

Patzer, E.J., Moore, N.F., Barenholz, Y., Shaw, J.M., Wagner, RR. 1978b. Lipid organization of the membrane of vesicular Stomatitis virus. J. Biol. Chem. 253:4544-4550.

Pearson, R.H., Pascher, R.H. 1979. The molecular structure of lecithin dihydrate. Nature 281:499-501.

Pinheiro, T.J., Sanghera, N. 2002. Binding of prion protein to lipid membranes and implications for prion conversion. J. Mol. Biol. 315:1241-1256.

Presti, F.T., Pace, R.J., Chan, S.I. 1982. Cholesterol–phospholipid interaction in membranes. 2. Stoichiometry and molecular packing of cholesterol-rich domains. Biochemistry 21:3831-3835.

Riboni, R.H., Riboni, L., Viani, P., Rosseria, B., Rinetti, A., Tettamenti, G. 1997. The role of sphingolipids in the process of signal transduction. Prog. Lipid Res. 36:153-195.

Ridgeway, N.D. 2000. Interaction between metabolism and intracellular distribution of cholesterol and sphingomyelin. Biochim. Biophys. Acta 1484:129-141.

Rinia, H.A., Snel, M.M., van der Eerden, J.P., de Kruijff, B. 2001. Visualizing detergent resistant domains in model membranes with atomic force microscopy. FEBS Lett. 501:92-96.

Rubenstein, J.L.R., Smith, B.A., McConnell, H.M. 1979. Lateral diffusion in binary mixtures of cholesterol and phosphatidylcholines. Proc. Natl. Acad. Sci. USA 76:15-18.

Sackman, E. 1994. Membrane bending energy concept of vesicle- and shape-transitions. FEBS Lett. 346:3-16.

Samuel, B., McManus, H., Hiller, L., Harrison, T., Halder, K., Lanes, S., Mohandas, M., Van Wye, J. 2001. Red cell microdomain lipids play a critical role in malarial parasite invasion and intracellular development. Blood 96:5919.

Samuni, A., Lipman, A., Barenholz, Y. 2000. Damage to liposomal lipids: protection by antioxidants and cholesterol-mediated dehydration. Chem. Phys. Lipids 105:121-134.

Sarna, J., Miranda, S.R.P., Schuchman, E.H., Hawkes, R. 2001. Patterned cerebellar Purkinije cell death in transgenic mouse model of Niemann-Pick type A/B disease. Eur. J. Neurosci. 13:1873-1880.

Schmidt, C.F., Barenholz, Y., Huang, C., Thompson, T.E. 1978. Monolayer coupling in sphingomyelin bilayer systems. Nature 271:775-777.

Schroeder, R., London, E., Brown, D. 1994. Interactions between saturated acyl chains confer detergent resistance on lipids and glycosylphosphatidylinositol (GPI)-anchored proteins: GPI-anchored proteins in liposomes and cells show similar behavior. Proc. Natl. Acad. Sci. USA 91:12130-12134.

Schuck, S., Honsho, M., Ekroos, K., Shevchenko, A., Simons, K. 2003. Resistance of cell membranes to different detergents. Proc. Natl. Acad. Sci. USA 100:5795-5800.

Shinitzky, M. 1984. In: Shinitzky, M., Ed. Physiology of Membrane Fluidity, Vol. I. CRC Press, Boca Raton, FL, pp. 1-51.

Shinitzky, M., Barenholz, Y. 1974. Dynamics of the hydrocarbon layer in liposomes of lecithin and sphingomyelin containing dicetylphosphate. J. Biol. Chem. 249:2652-2657.

Shinitzky, M., Barenholz, Y. 1978. Fluidity parameters of lipid regions determined by fluorescence polarization. Biochim. Biophys. Acta 515:367-394.

Shmeeda, H.R., Golden, E.B., Barenholz, Y. 1994a. In: Shinitzky, M., Ed. Handbook of Biomembranes: Structural and Functional Aspects. VCH, Weinheim, Balaban Publ., pp. 1-82.

Shmeeda, H., Petkova, D., Barenholz, Y. 1994b. Cholesterol homeostasis in cultures of rat heart myocytes: relationship to cellular hypertrophy. Am. J. Physiol. 267 (Heart Circ. Physiol. 36):H1689-H1697.

Shmeeda, H., Petkova, D., Barenholz, Y. 1995. Cholesterol distribution in rat-heart myocytes. Am. J. Physiol. 268 (Heart Circ. Physiol.):H759-H766.

Simberg, D., Weisman, S., Talmon, Y., Faerman, A., Shoshani, T., Barenholz, Y. 2003. The role of organ vascularization and lipoplex-serum initial contact in intravenous murine lipofection. J. Biol. Chem. 278:39858-39865.

Simons, K., Ikonen, E. 1997. Functional rafts in cell membranes. Nature 387:564-572.

Simons, K., van Meer, G. 1988. Lipid sorting in epithelial cells. Biochemistry 27:6197-6202.

Simons, K., Toomre, J. 2000. Lipid raft and signal transduction. Nat. Rev. Mol. Cell Biol. 1:31-39.

Slotte, J.P. 1999. Sphingomyelin–cholesterol interaction in biological and model membranes. Chem. Phys. Lipids 102:13-27.

Small, D.M. 1970. Surface and bulk interactions of lipids and water with a classification of biologically active lipids based on these interactions. Fed. Proc. 29:1320-1326.

Small, D.M. 1986. The Physical Chemistry of Lipids: Handbook of Lipid Research Vol. 4, Plenum Press, New York.

Snyder, B., Freire, E. 1980. Compositional domain structure in phosphatidylcholine–cholesterol and sphingomyelin–bilayers. Proc. Natl. Acad. Sci. USA 77:4055-4059.

Sot, J., Collado, M.I., Arrondo, J.L.R., Alonso, A., Goñi, F.M. 2002. Triton X-100-resistant bilayers: effect of lipid composition and relevance to the raft phenomenon. Langmuir 18:2828-2835.

Straume, M., Litman, B.J. 1987. Influence of cholesterol on equilibrium and dynamic bilayer structure of unsaturated acyl chain phosphatidylcholine vesicles as determined from higher order analysis of fluorescence anisotropy decay. Biochemistry 26:5121-5126.

Sundaralingam, M. 1972. Molecular structures and conformations of the phospholipids and sphingomyelins. Ann. NY Acad. Sci. 195:324-355.

Talbott, C.M., Vorobyov, I., Borchman, D., Taylor, K.G., Du Pre, D.B., Yappert, M.C. 2000. Conformational studies of sphingolipids by NMR spectroscopy. II: Sphingomyelin. Biochim. Biophys. Acta 1467:326-337.

Tanford, C. 1980. The Hydrophobic Effect. John Wiley and Sons, New York.

Tangirala, R., Jerome, W.G., Jones, N.L., Small, D.M., Johnson, W.J., Glick, J.M., Mahlberg, F.H., Rothblat, G.H. 1994. Formation of cholesterol monohydrate crystals in macrophage-derived foam cells. J. Lipid Res. 35:93-104.

Taraboulos, A., Scott, M., Semenov, A., Avrahami, D., Laszlo, L., Prusiner, S.B., Avraham, D. 1995. Cholesterol depletion and modification of COOH-terminal targeting sequences of the prion protein inhibit formation of the scrapie isoform. J. Cell Biol. 129:121-132, Erratum in J. Cell Biol. 130:501.

Thompson, T.E., Sankarin, M.B., Biltonen, R.L., Marsh, D., Vas, W.L.C. 1995. Effects of domain structure on in-plane reactions and interactions. Mol. Membr. Biol. 12:157-162.

Thudicum, J.L.W. 1874. Researches in the chemical constitution of the brain. Rept. Med. Off. Privy Council New Series NO3 Ap6. Eyre and Spottiswoode, London, p. 113.

Thudicum, J.L.W. 1962. A treatise on chemical constitution of the brain (reprinted, with new historical introduction by Drabkin, D.L.). Archon, Hamden, CT.

Tirosh, O., Kohen, R., Katzhendler, J., Alon, A., Barenholz, Y. 1997. Oxidative stress effect on the integrity of lipid bilayers is modulated by cholesterol level of bilayers. Chem. Phys. Lipids 87:17-22.

Tso, J.Y., Green, J.M. 2002. Identifying anti-tumor targets or agents by lipid raft immunization and proteomics. U.S. Pat. Appl. 269010/10, Oct. 10.

Veiga, M.P., Arondo, J.L.R., Goñi, F.M., Alonso, A., Marsh, D. 2001. Interaction of cholesterol with sphingomyelin in mixed membranes containing phosphatidylcholine, studied by spin-label ESR and IR spectroscopies. A possible stabilization of gel-phase sphingolipid domains by cholesterol. Biochemistry 40:2614-2622.

Vincent, S., Gerlier, D., Manie, S.N. 2000. Measles virus assembly within membrane rafts. J. Virol. 74:9911-9915.

Vincent, I., Bu, B., Erickson, R.P. 2003. Understanding Niemann-Pick type C disease: a fat problem. Curr. Opin. Neurol. 16:155-161.

Webb, M.S., Bally, M.B., Mayer, L.D., Miller, J.J., Gardi, P.G. 1998. Patent assignee: Inex Pharmaceuticals Corporation. Sphingosomes for enhanced drug delivery. U.S. Patent 05,814,335, Sept. 29.

Wieland, F.T., Bruegger, B., Gorgas, K., Helms, J.B., Lehmann, W.-D., Malsam, J., Nickel, W., Sandhoff, R., Wegehingel, S. 2000. Evidence for segregation of sphingomyelin and cholesterol during formation of COPI-coated vesicles. J. Cell Biol. 151:507-517.

Williams, K.J. 2002. U.S. Patent 06,367,479.

Williams, K.J., Tall, A.R. 1988. Interactions of liposomes with lipoproteins: relevance to drug delivery systems and to the treatment of atherosclerosis. In: Gregoriadis, G., Ed. Liposomes as Drug Carriers, John Wiley and Sons Ltd., pp. 93-111.

Williams, K.J., Werth, V.P., Wolff, J.A. 1984. Intravenously administered lecithin liposomes: a synthetic antiatherogenic lipid particle. Perspect. Biol. Med. 27:417-431.

Wolf, C., Koumanov, K., Quinn, P.J., Tenchov, B. 2001. Cholesterol favors phase separation of sphingomyelin. Biophys. J. 89:163-172.

Xu, X., London, E. 2000. The effect of sterol structure on membrane lipid domains reveals how cholesterol can induce lipid domain formation. Biochemistry 39:843-849.

Xu, X.L., Bittman, R., Duportail, G., Heissler, D., Vilcheze, C., London, E. 2001. Effect of the structure of natural sterols and sphingolipids on the formation of ordered sphingolipid/sterol domains (rafts). J. Biol. Chem. 276:33540-33546.

Yeagle, P.L. 1985. Cholesterol and the cell membrane. Biochim. Biophys. Acta 822:267-287.

Yeagle, P.L. 1988. Cholesterol and the cell membrane. In: Yeagle, P.L., Ed. Biology of Cholesterol. CRC Press, Boca Raton, Chapter 6, pp. 121-145.

Yeagle, P.L. 1993a. The Membrane of Cells, 2nd ed. Academic Press, San Diego, pp. 69-165.

Yeagle, P.L. 1993b. The Membrane of Cells, 2nd ed. Academic Press, San Diego, pp. 13-17.

Yechiel, E., Barenholz, Y. 1985. Relationships between membrane lipid composition and biological properties of rat myocytes: effects of aging and manipulation of lipid composition. J. Biol. Chem. 260:9132-9136.

Yechiel, E., Henis, Y.I., Barenholz, Y. 1986a. Aging of rat heart myocytes and fibroblasts: relationship between lipid composition, membrane organization and biological properties. In: Freysz, L., Dreyfus, H., Massarelli, R., Gatt, S., Eds. Enzymes of Lipid Metabolism II. Plenum Press, New York, pp. 519-533.

Yechiel, E., Henis, Y.I., Barenholz, Y. 1986b. Aging of rat heart fibroblasts: relationship between lipid composition membrane organization and biological properties. Biochim. Biophys. Acta 859:95-104.

Yedgar, S., Barenholz, Y., Cooper, V.G. 1974. Molecular weight, shape and structure of mixed micelles of Triton X-100 and sphingomyelin. Biochim. Biophys. Acta 363:98-111.

Zeidel, L., Hill, W.G. 2000. Reconstituting the barrier properties of a water-tight epithelial membrane by design of leaflet-specific liposomes. J. Biol. Chem. 275:30176-30185.

Zhang, J., Pekosz, A., Lamb, R.A. 2000. Influenza virus assembly and lipid raft microdomains: a role for the cytoplasmic tails of the spike glycoproteins. J. Virol. 74:4634-4644.

Zomber, G., Bogin, E., Barenholz, Y. 1996. Effect of i.v. injection of small unilamellar liposomes of egg phosphatidylcholine on cholesterol in plasma and erythrocytes, serum enzymes and liver function in dogs. J. Liposome Res. 6:455-477.

Chapter 6

Membrane Targeting of Lipid Modified Signal Transduction Proteins

Marilyn D. Resh
Member and Professor, Cell Biology Program, Memorial Sloan-Kettering Cancer Center, 1275 York Avenue, Box 143, New York, NY 10021, USA

Abstract
Covalent attachment of lipophilic moieties to proteins influences interaction with membranes and membrane microdomains, as well as signal transduction. The most common forms of fatty acylation include modification of the N-terminal glycine of proteins by N-myristoylation and/or attachment of palmitate to internal cysteine residues. Protein prenylation involves attachment of farnesyl or geranylgeranyl moieties via thio-ether linkage to cysteine residues at or near the C-terminus. Attachment of each of these lipophilic groups is catalyzed by a distinct enzyme or set of enzymes: N-myristoyl transferase for N-myristoylation, palmitoyl acyl transferases for palmitoylation, and farnesyl or geranylgeranyl transferases for prenylation. The distinct nature of the lipid modification determines the strength of membrane interaction of the modified protein as well as the specificity of membrane targeting. Clusters of basic residues can also synergize with the lipophilic group to promote membrane binding and targeting. The final destination of the modified protein is influenced by multiple factors, including the localization of the modifying enzymes, protein/protein interactions, and the lipid composition of the acceptor membrane. In particular, much interest has been focused on the ability of fatty acylated proteins to preferentially partition into membrane rafts, subdomains of the plasma membrane that are enriched in cholesterol and glycosphingolipids. Lipid raft localization is necessary for efficient

signal transduction in a wide variety of systems, including signaling by T and B cell receptors, Ras, and growth factor receptors. However, certain membrane subdomains, such as caveolae, can serve as reservoirs for inactive signaling proteins. Heterogeneity in the types of membrane subdomains, as well as in the types of lipophilic groups that are attached to proteins, provide an additional level of complexity in the regulation of signaling by membrane bound proteins.

1. INTRODUCTION

Intracellular protein distribution is regulated by sequences that serve as molecular "zipcodes". These zipcodes function in several different contexts: (1) as direct signals that target a protein to a specific subcellular organelle (e.g. nuclear localization signals), (2) as signals for interaction with other proteins that contain targeting information (e.g. SH2 domains), or (3) as signals for co- or post-translational protein modifications that promote protein targeting. This chapter will examine signals that direct protein localization to membranes, and in particular, will focus on covalent modification of proteins by attachment of lipophilic moieties. The abilities of these signals to modulate protein interaction with membranes and membrane microdomains, as well as signal transduction, will be explored.

2. MODIFICATION OF PROTEINS WITH FATTY ACIDS AND PRENYL GROUPS

2.1. Fatty acylation of proteins

Covalent attachment of fatty acids occurs on a wide variety of proteins. In general, fatty acylated proteins can be divided into three classes: those modified by the 14-carbon fatty acid myristate, those modified by the 16-carbon fatty acid palmitate, and those modified by both fatty acids (Resh, 1999; Resh, 1994). N-myristoylated proteins can be distinguished by the presence of the sequence Met-Gly at their N-terminus. The initiating Met is removed by methionine aminopeptidase and myristate is then linked to the N-terminal glycine via an amide bond. This reaction occurs co-translationally and is catalyzed by N-myristoyl transferase (NMT), a 50–60 kDa cytosolic protein. N-myristoylated proteins comprise approximately 0.5% of all proteins in the human genome (Maurer-Stroh *et al.*, 2002a). However, since 15% of all proteins contain Met-Gly at their N-terminii, additional features must be involved in substrate recognition by NMT. An algorithm for predicting N-myristoylation of proteins was recently developed (Maurer-Stroh *et al.*, 2002b) and is available on the Web: http://mendel.imp.univie.ac.at/myristate/.

Protein palmitoylation generally involves covalent attachment of palmitate to a cysteine residue via thio-ester linkage. In contrast to N-myristoylation, palmitoylation of proteins is not well understood and therefore prediction of palmitoylated sequences is more difficult. The reasons for this state of affairs are several. First, a wide variety of proteins are palmitoylated, including transmembrane proteins, N-myristoylated proteins and prenylated proteins (Resh, 1999). Second, the enzymology of the palmitoylation process is not well understood. Non-enzymatic incorporation of palmitate into peptides and proteins can occur *in vitro* (Duncan and Gilman, 1996; Bano *et al.*, 1998) in a time and sequence dependent manner. Difficulties in purifying palmitoyl acyl transferase (PAT) activities to homogeneity combined with the propensity of the cysteine sulfhydryl group to react with palmitate in a variety of sequence contexts has led many investigators to propose that non-enzymatic palmitoylation of proteins occurs *in vivo* (Navarro-Lerida *et al.*, 2002). However, the rate of non-enymatic palmitoylation *in vivo* is predicted to be very slow, due to the presence of Acyl-CoA binding proteins (Leventis *et al.*, 1997). In contrast, PAT activity is retained in the presence of physiological concentrations of Acyl-CoA binding protein (Dunphy *et al.*, 2000). Moreover, two recent reports have identified and purified PATs that palmitoylate Ras (Lobo *et al.*, 2002) and casein kinase (Roth *et al.*, 2002). It is therefore likely that identification of additional PATs will follow.

The sonic hedgehog protein (Shh) has been shown to contain palmitate that is covalently linked via amide bond to the N-terminus (Pepinsky *et al.*, 1998). The acyltransferase that palmitoylates Shh has recently been identified (ski or skinny hedgehog) and is a segment polarity gene in Drosophila. It has been proposed that the acylation reaction occurs in 2 steps: initial formation of a thioester with the N-terminal cysteine, followed by rearrangement via a cyclic intermediate to form the N-terminal amide (Pepinsky *et al.*, 1998). Shh also contains an additional lipophilic modification; a molecule of cholesterol is covalently linked to its C-terminus (Porter, 1996).

2.2. Protein prenylation

Many proteins, especially small G proteins, contain farnesyl or geranylgeranyl moieties that are attached via thio-ether linkage to cysteine residues at or near the C-terminus. A C-terminal "CAAX" box is a common feature of proteins that will be farnesylated or geranylgeranylated (Casey *et al.*, 1991). Depending on the C-terminal amino acid, the CAAX box is recognized by farnesyl transferase (FTase) (X = met, ser) or geranylgeranyl transferase I

(GGTase I) (X = leu). Both enzymes are αβ heterodimers with a common α subunit. Once the cysteine residue within the CAAX box is prenylated, additional modifications occur. These include proteolysis of the 3 terminal amino acids by the endoprotease hRce1, and carboxymethylation of the prenylated C-terminal cysteine by the carboxymethyltransferase pcCMT/Icmt. A third prenyl transferase, Rab GGTase or GGTase II, recognizes proteins that terminate with –CC, CXC, CCX or CCXX motifs; these substrates are primarily Rab proteins. Rab GGTase differs from the other prenyl transferases in that it can only prenylate Rab that is complexed with Rab escort protein. In addition, the crystal structure of Rab GGTase reveals a deeper prenyl binding pocket than FTase (Zhang *et al.*, 2000).

2.3. Detection of fatty acylated and prenylated proteins

There are several methods available to determine whether a particular protein is fatty acylated. As indicated above, an extremely accurate algorithm has been developed for predicting N-myristoylation (Maurer-Stroh *et al.*, 2002a). Alternatively, or in the case of palmitoylation, empirical methods must be employed (Berthiaume *et al.*, 1995). Cells can be radiolabeled with ^{3}H-myristate or ^{3}H-palmitate, which are commercially available, or with synthetic ^{125}I-labeled iodo-fatty acid analogs (Peseckis *et al.*, 1993; Alland *et al.*, 1994). Cell lysates are then immunoprecipitated with antibody to the protein of interest, and radiolabel incorporation is detected by fluorography or phosphorimaging. Radiolabeling times are usually limited to 4 hours in order to prevent metabolism and incorporation of radiolabel into other fatty acids or into the polypeptide backbone. The identity of the fatty acylated species can then be determined by acid (for N-myristoylation) or base (for thio-ester palmitoylation) hydrolysis followed by TLC or HPLC. A more precise, although technically demanding method is to isolate the modified protein and perform mass spectrometric analysis. For example, MALDI-TOF MS has recently been used successfully to identify both amide and thio-ester linked fatty acids on Src family kinases and GAP43 (Liang *et al.*, 2001; Liang *et al.*, 2002).

Prenylated proteins can often be recognized by the presence of a C-terminal CAAX box or other cysteine-rich sequences near the C-terminus (see above). The prenyl groups can be radiolabeled by incubating cells with ^{3}H-mevalonate, an isoprenoid precursor, and isolating the radiolabeled protein by immunoprecipitation. Verification that the radiolabel was incorporated into a cysteine residue via thio-ether bond is typically obtained by treating the labeled protein with Raney nickel, which should release the radiolabel (Fukada, 1995). Alternatively, mass spectrometric methods can be used to identify the modifying group(s) on prenylated peptides derived from the modified protein.

3. ROLE OF LIPID MODIFICATIONS IN PROMOTING MEMBRANE BINDING

Lipophilic moieties play critical roles in promoting interaction of proteins with lipid bilayers. The specific nature of the lipid modification determines the strength of membrane interaction. For example, palmitate or geranylgeranyl alone are sufficiently hydrophobic to provide strong association of the modified proteins with membranes (Shahinian and Silvius, 1995). In contrast, neither myristate nor farnesyl attachment provides stable membrane interaction. The apparent K_d for interaction of a myristoylated or farnesylated peptide with a phospholipid bilayer is approximately 10^{-4} M (Peitzsch and McLaughlin, 1993; Silvius and L'Heureux, 1994); this binding energy is barely sufficient to retain a modified protein in the bilayer. Since many myristoylated and farnesylated proteins are membrane bound, a second signal is required to enhance membrane association. This second signal is either a cluster of basic residues or palmitate (Resh, 1999). Basic residues promote electrostatic interactions with acidic membrane phospholipids whereas palmitate provides hydrophobic interactions; both mechanisms provide the additional membrane binding energy that is needed to achieve stable membrane interaction. Examples of proteins that use membrane binding motifs consisting of "myristate + basic" (Src, HIV-1Gag), "myristate + palmitate" (Src family kinases, Gα subunits), "farnesyl + basic" (K-Ras4B), "farnesyl + palmitate" (H-Ras), geranylgeranyl + palmitate (RhoB, TC10) and geranylgeranyl + polybasic (Rac1, RhoA) have all been described in the literature.

4. ROLE OF LIPID MODIFICATIONS IN PROMOTING MEMBRANE TARGETING

The presence of a single lipophilic modification is not sufficient to promote targeting to a specific membrane. For example, N-myristoylated proteins are found in many different subcellular membrane locations, as well as the cytosol (Resh, 1999; McCabe and Berthiaume, 1999). As with membrane binding, a second signal is required to promote membrane targeting of modified proteins. The presence of a myristate + palmitate motif is both necessary and sufficient to promote plasma membrane localization of native proteins as well as peptides and chimeric GFP fusion proteins (Schroeder *et al.*, 1996; Wolven *et al.*, 1997; McCabe and Berthiaume, 1999; Zacharias *et al.*, 2002). It is not intrinsically obvious how one or two saturated fatty acids can promote targeting specificity solely by hydrophobic effects. The surface area of internal membranes, especially the endoplasmic reticulum (ER), is vastly greater than that of the plasma membrane. Thus, if simple hydrophobicity was driving membrane association, one would expect fatty acylated proteins to be primarily localized to the ER. One explanation for the ability of palmitoyation to promote plasma membrane targeting is the presence of a palmitoyl transferase at a particular membrane, e.g. the plasma membrane (Dunphy *et al.*, 1996). Once a protein is dually fatty acylated at the plasma membrane, the hydrophobic interaction with the lipid bilayer becomes so strong that the modified protein effectively remains "trapped" in the membrane where it was palmitoylated (Shahinian and Silvius, 1995). Based on this reasoning, it is likely that the targeting of palmitoylated proteins will be influenced by the localization of the palmitoyl transferase that modifies them.

Proteins with myristate + basic or farnesyl + basic motifs also localize to the plasma membrane. The mechanism that accounts for targeting specificity likely involves electrostatic interactions between the positively charged amino acid residues and negatively charged membrane phospholipids (Zhou *et al.*, 1994; Roy *et al.*, 2000). It is possible that the concentration of acidic phospholipids in the plasma membrane is higher than in other intracellular membranes, thereby allowing proteins with basic motifs to be preferentially retained at the plasma membrane (Murray *et al.*, 1997). Alternatively, preferential binding to a specific plasma membrane anionic phospholipid could promote targeting. This is the case for MARCKS, a myristoylated protein containing a cluster of basic residues within its effector domain. Strong binding of the MARCKS effector domain to phosphatidylinositol 4,5-bisphosphate, a phospholipid enriched at the plasma membrane, could contribute to plasma membrane targeting of MARCKS (Wang *et al.*, 2001).

An alternative mechanism to control specific targeting of lipohilic modified proteins is transport through the secretory or exocytic pathway. This model is best illustrated for Ras proteins (Choy *et al.*, 1999; Apolloni *et al.*, 2000; Prior and Hancock, 2001). The prenylated CAAX motif directs Ras proteins to the endoplasmic reticulum (ER), where they are further modified by ER resident enzymes Rce1p and pcCMT/Icmt. Proteins that contain a palmitoylation motif, such as H-Ras and N-Ras, then proceed through the secretory pathway via vesicular trafficking, passing through the Golgi and ending up at the plasma membrane.

Protein-protein interactions can also strongly influence the ultimate localization of a modified protein. When geranylgeranylated RhoA is bound to RhoGDI, the complex is inactive and cytosolic. Activation of RhoA results in its release from RhoGDI and localization of RhoA to the plasma membrane, as well as internal membranes (Michaelson *et al.*, 2001). Recent studies of heterotrimeric G proteins have revealed the importance of protein–protein interactions for subcellular localization. Interaction of Gα subunits with $\beta\gamma$ appears to regulate the final destination of the α subunit, and in some instances, may even supersede the need for fatty acylation. For example, overexpression of βγ can rescue the palmitoylation and plasma membrane localization defects of non-myristoylated G2A α subunits of G_i or G_z (Degtyarev *et al.*, 1994; Morales *et al.*, 1998). In addition, mutation of the βγ binding site on Gα results in mislocalization of α subunits to intracellular membranes (Fishburn *et al.*, 2000; Evanko *et al.*, 2001). Moreover, artificial targeting of Gβγ to mitochondria induces a redistribution of wild-type as well as non-palmitoylated α_z to the mitochondria (Fishburn *et al.*, 2000). A model to explain the cooperative nature of palmitoylation and Gα/βγ interactions was recently proposed (Michaelson *et al.*, 2002). Prenylation of the CAAX box within the γ subunit directs the βγ complex from the cytosol to the ER. Following proteolytic cleavage and carboxymethylation, the complex proceeds to the Golgi, where it associates with the α subunit. Palmitoylation of the α subunit occurs on the Golgi, and this acylation event is required for trafficking of the heterotrimeric complex to the plasma membrane. This model provides additional proof of principle for the "two signal" hypothesis of membrane targeting, with each signal being localized on two different proteins.

5. ROLE OF LIPID MODIFICATIONS IN PROMOTING TARGETING TO LIPID RAFTS

In the past several years, the literature has been deluged with studies examining the structure and function of membrane rafts. Rafts are membrane microdomains that are enriched in cholesterol and glycosphingolipids (reviewed in (Brown and London, 2000; Simons and Toomre, 2000; Simons and Ehehalt, 2002)). The sphinoglipids and phospholipids in rafts primarily contain saturated fatty acid chains, allowing the lipid molecules to pack tightly together and form a "liquid ordered" phase. Lipophilic modification plays an important role in protein targeting to rafts. For example, GPI-anchored proteins are highly enriched in the outer leaflet of membrane rafts. In addition, fatty acylated proteins partition into the cytoplasmic leaflet of plasma membrane rafts. Myristoylation and palmitoylation, or dual palmitoylation are necessary and sufficient to promote protein targeting to rafts (Wolven *et al.*, 1997; Galbiati *et al.*, 1999; Zacharias *et al.*, 2002). It is likely that the saturated nature of the fatty acyl modification promotes insertion into the liquid ordered membrane domain. This hypothesis predicts that introduction of unsaturated fatty acids with *cis* double bonds into proteins would result in exclusion of the modified proteins from rafts, since it is difficult to pack these bulky acyl chains into a highly ordered lipid domain (Figure 1). Indeed, several recent studies have shown that fatty acids other than myristate or palmitate can be attached to fatty acylated

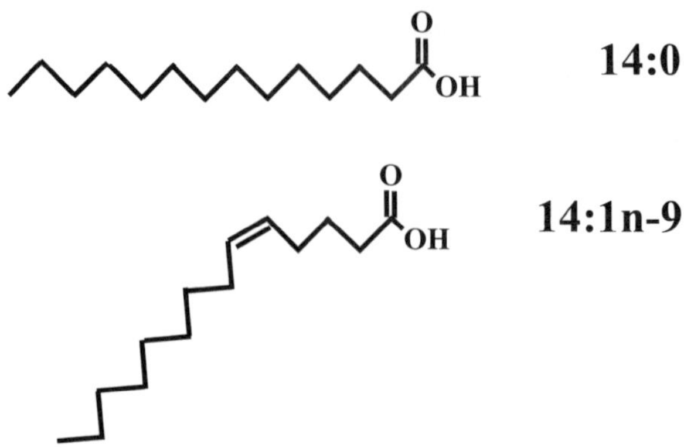

Figure 1. The presence of a *cis* double bond introduces a kink into the fatty acid and generates a more bulky chain with decreased partitioning to liquid ordered raft domains.

proteins, and that acylation with unsaturated fatty acids reduces partitioning of the modified proteins into rafts *in vitro* (Moffett *et al.*, 2000) and *in vivo* (Webb *et al.*, 2000; Liang *et al.*, 2002; Lindwasser and Resh, 2002).

Since fatty acylation promotes raft association, one would predict that removal of the fatty acid(s) would result in release of the deacylated protein from membranes and/or rafts. This prediction has been verified in a number of different systems. For example, mutation of fatty acylation site(s) or treatment of cells with fatty acylation inhibitors prevents raft association of Src family kinases (Shenoy-Scaria *et al.*, 1994; Robbins *et al.*, 1995; Wolven *et al.*, 1997; Webb *et al.*, 2000). Moreover, protein palmitoylation is a dynamic process consisting of cycles of palmitoylation and depalmitoylation. Regulated depalmitoylation has been shown to occur for Gα subunits, G protein coupled receptors, eNOS and PSD-95 (Robinson *et al.*, 1995; Mumby, 1997; El-Husseini *et al.*, 2002), resulting in release of these proteins from membranes and membrane microdomains. At least one palmitoyl acyl thioesterase that removes palmitate from Ga proteins and eNOS has been identified (Duncan and Gilman, 1998; Yeh *et al.*, 1999).

Proteins that are modified with bulky, branched prenyl groups (farnesyl, geranylgeranyl) generally tend to be excluded from rafts (Melkonian *et al.*, 1999; Zacharias *et al.*, 2002). However, the situation is more complex, as evidenced by recent studies of Ras protein localization. K-Ras4B is an isoform of Ras that is bound to the membrane via a prenyl + basic motif (Hancock *et al.*, 1991); this protein is excluded from rafts (Prior *et al.*, 2001). In contrast, H-Ras, which is modified by farnesyl and palmitate groups, is targeted to membrane rafts. However, activation of H-Ras by binding of GTP results in redistribution of the protein into the bulk plasma membrane. Release of activated H-Ras from rafts has been shown to be essential for signal transduction (Prior and Hancock, 2001; Prior *et al.*, 2001). The molecular mechanism for H-Ras redistribution has not been determined, but is likely to involve a conformational change in the molecule that regulates the accessibility of the palmitate moiety for insertion into raft domains.

Some transmembrane proteins have also been shown to be enriched in plasma membrane rafts. In several cases, the mechanism responsible for raft association appears to be palmitoylation. For example, LAT is a T cell transmembrane protein that is modified by 2 palmitates. While palmitoylation is not required for plasma membrane localization *per se*, it is necessary for localization of LAT to membrane rafts and for LAT-mediated signaling through the T cell receptor (Zhang *et al.*, 1998; 1999). Similarly, palmitoylation of the transmembrane proteins CD8, influenza virus HA, and HIV-1 gp160 is required for these proteins to associate with lipid rafts (Arcaro *et al.*, 2000; Rousso *et al.*, 2000). In contrast, there are examples of transmembrane proteins that are palmitoylated and raft associated, but do not require palmitoylation for raft association, eg LMP1 of Epstein-Barr virus (Higuchi *et al.*, 2001) and the tetraspanin protein CD151 (Yang *et al.*, 2002). Moreover, there are transmembrane proteins that are raft associated but are not palmitoylated. The mechanism that promotes raft partitioning of the latter class of proteins is not clear (van Duyl *et al.*, 2002), but appears to be related to the amino acid composition of the transmembrane domain (Scheiffele *et al.*, 1997). These proteins have been proposed to be encased by a lipid shell consisting of cholesterol and sphingolipid; the lipid shells would then promote association with liquid ordered raft domains (Anderson and Jacobson, 2002).

Clustering of proteins into larger complexes often serves to enhance raft localization. This mechanism may be particularly relevant for proteins that exhibit weak intrinsic affinity for rafts when the proteins are monomeric. For example, activation of the T cell receptor results in receptor crosslinking and recruitment and more stable association with rafts (Montixi *et al.*, 1998; Janes *et al.*, 1999). Likewise, multimerization of myristoylated HIV-1 Gag induces localization into large raft-like microdomains termed "barges" that promote viral particle assembly (Lindwasser and Resh, 2001).

Protein–protein interactions have also been shown to enhance interactions with membrane microdomains. For example, caveolae are flask shaped invaginations of the plasma membrane that share many of the same lipid characteristics with rafts, but are distinguished by the presence of a marker protein caveolin-1 (Anderson, 1998). Interaction of fatty acylated proteins such as endothelial nitric oxide synthase (eNOS) with caveolin promotes localization to caveolae (Feron *et al.*, 1998). As described below, interaction of eNOS with caveolin keeps eNOS in a repressed state.

When and where do proteins associate with rafts? Cholesterol and sphingolipid enriched rafts are first formed in the Golgi and trafficked to the plasma membrane (Brown and London, 1998; Simons and Toomre, 2000). Proteins that travel through the secretory pathway most likely encounter raft domains during their transit through trafficking vesicles. However, many fatty acylated proteins are synthesized on soluble polysomes and post-translationally target to the plasma membrane independently of the classic exocytic pathway. Kinetic studies of Fyn and HIV-1 Gag have revealed that these proteins bind to the plasma membrane first, and then associate with detergent-insoluble raft domains (van't Hof and Resh, 1997; Ono and Freed, 2001).

6. LIPID MODIFICATION, RAFT LOCALIZATION AND SIGNAL TRANSDUCTION

6.1. Signaling systems that are dependent on rafts

There is now extensive evidence in the literature that raft integrity is necessary for efficient signal transduction in a wide variety of systems, including T and B cell receptor signaling, Ras signaling, and growth factor receptor signaling. These topics are covered in a number of chapters in this volume, as well as by several recent reviews (Brown and London, 2000; Simons and Toomre, 2000; Pierce, 2002; Tsui-Pierchala *et al.*, 2002). The partitioning of proteins into rafts enhances protein/protein and protein/lipid interactions that are important for efficient signal transduction. In many instances, the signaling is dependent on raft localized fatty acylated proteins, such as Src family kinases, or by GPI-anchored proteins on the exoplasmic leaflet. Evidence for the requirement of rafts derives primarily from three types of experiments. First, inhibition of raft targeting of proteins by mutation of fatty acylation sites, by acylation with unsaturated fatty acids, or by incubation of cells with palmitoylation inhibitors blocks signaling (Zhang *et al.*, 1998; Janes *et al.*, 1999; Webb *et al.*, 2000; Liang *et al.*, 2001). Second, targeting of key proteins, such as Src family kinases, to the bulk plasma membrane rather than rafts is insufficient for propagating signal transduction, thereby emphasizing the functional significance of membrane microdomain localization (Kabouridis *et al.*, 1997; Janes *et al.*, 1999; van't Hof and Resh, 1999). Third, treatment of cells with cholesterol extraction agents such as cyclodextrin or filipin disrupts lipid rafts and inhibits signaling (Xavier *et al.*, 1998; Schade and Levine, 2002). However, caution must be observed in interpreting experiments involving cyclodextrins, as cholesterol depletion from cells has additional effects, eg EGF Receptor activation (Chen and Resh, 2002; Pike and Casey, 2002).

6.2. Rafts/caveolae may exert negative effects on signaling

Despite the abundance of reports documenting a positive role for rafts in signal transduction, raft localization also has been shown to exert a negative effect on a number of signaling proteins. For example, as noted above, activation of H-Ras requires movement of the protein out of rafts into the bulk plasma membrane where signal transduction is initiated (Prior and Hancock, 2001). Likewise, caveolae can serve as reservoirs for inactive proteins (Okamoto *et al.*, 1998). Myristoylated and palmitoylated eNOS is in an inactive state when it is bound to caveolin in caveolae. Agonists induce a Ca^{2+}/calmodulin dependent disruption of the eNOS/caveolin complex as well as eNOS depalmitoylation, resulting in release from caveolae into the cytosol and eNOS activation (Robinson *et al.*, 1995; Feron *et al.*, 1998). The inhibitory effect of caveolin on eNOS is exerted only when both proteins are present in caveolae rather than membrane rafts (Sowa *et al.*, 2001).

Finally, B cells provide examples where the same protein, the B cell receptor (BCR) can signal in or out of rafts with vastly different consequences. BCR signaling in mature B cells involves translocation of the BCR into membrane rafts, resulting in activation of tyrosine kinases and ultimately stimulation of B cell proliferation and differentiation (Cheng *et al.*, 1999). In contrast, in immature B cells, the BCR does not translocate into rafts. Instead, signaling is initiated outside of rafts, resulting in a dramatically different outcome – apoptosis (Sproul *et al.*, 2000).

7. CONCLUSIONS AND PERSPECTIVES

In conclusion, fatty acylation and/or prenylation of proteins can serve to modulate protein–protein as well as protein–lipid interactions and thereby regulate the ultimate subcellular localization of the modified protein. It is now clear that the biochemistry of lipid modification is fairly complex. Multiple acyl and prenyl transferases exist, and it is likely that a variety of thioesterases will also be found. In addition, heterogeneous fatty acylation of proteins occurs, and the type of fatty acid that is attached influences the ability of the modified protein to interact with membrane microdomains. Finally, membrane microdomains are themselves heterogeneous. Multiple types of cholesterol and sphingolipid domains exist within the plasma membrane of mammalian cells and these different domains may serve to segregate membrane bound proteins from one another on a molecular level. It is anticipated that continued study of fatty acylated and prenylated proteins will aid in understanding the basic biochemical mechanisms that regulate signal transduction and cell growth.

REFERENCES

Alland, L., Peseckis, S.M., Atherton, R.E., Berthiaume, L., and Resh, M.D. (1994). Dual Myristylation and Palmitylation of Src Family Member p59fyn Affects Subcellular Localization. J. Biol. Chem. *269*, 16701-16705.

Anderson, R.G. and Jacobson, K. (2002). A role for lipid shells in targeting proteins to caveolae, rafts, and other lipid domains. Science *296*, 1821-1825.

Anderson, R.G.W. (1998). The Caveolae Membrane System. Annu. Rev. Biochem. *67*, 199-225.

Apolloni, A., Prior, I.A., Lindsay, M., Parton, R.G., and Hancock, J.F. (2000). H-ras but not K-ras traffics to the plasma membrane through the exocytic pathway. Mol. Cell. Biol. *20*, 2475-87.

Arcaro, A., Grégoire, C., Boucheron, N., Palmer, E., Malissen, B., and Luescher, I.F. (2000). Essential Role of CD8 Palmitoylation in CD8 Coreceptor Function. J. Immunol. *165*, 2068-2076.

Bano, M.D., Jackson, D.S., and Magee, A.I. (1998). Pseudo-enzymatic S-acylation of a myristoylated Yes protein tyrosine kinase peptide in vitro may reflect non-enzymatic S-acylation *in vivo*. Biochem. J. *330*, 723-731.

Berthiaume, L., Peseckis, S.M., and Resh, M.D. (1995). Synthesis and use of Iodo-fatty acid analogs. Meth. Enzymol. *250*, 454-466.

Brown, D.A. and London, E. (1998). Functions of Lipid Rafts in Biological Membranes. Annu. rev. Cell Dev. Biol. *14*, 111-136.

Brown, D.A. and London, E. (2000). Structure and function of Sphingolipid and Cholesterol rich Membrane Rafts. J. Biol. Chem. *275*, 17221-17224.

Casey, P.J., Thissen, J.A., and Moomaw, J.F. (1991). Enzymatic modification of proteins with a geranylgeranyl isoprenoid. Proc. Natl. Acad. Sci. USA *88*, 8631-8635.

Chen, X. and Resh, M.D. (2002). Cholesterol Depletion from the Plasma Membrane Triggers Ligand-independent Activation of the Epidermal Growth Factor Receptor. J. Biol. Chem. *277*, 49631-49637.

Cheng, P.C., Dykstra, M.L., Mitchell, R.N., and Pierce, S.K. (1999). A role for lipid rafts in B cell antigen receptor signaling and antigen targeting. J. Exp. Med. *190*, 1549-1560.

Choy, E., Chiu, V.K., Silletti, J., Feoktistov, M., Morimoto, T., Michaelson, D., Ivanov, I.E., and Philips, M.R. (1999). Endomembrane trafficking of ras: the CAAX motif targets proteins to the ER and Golgi. Cell *98*, 69-80.

Degtyarev, M.Y., Spiegel, A.M., and Jones, T.L.Z. (1994). Palmitoylation of a G Protein _i Subunit Requires Membrane Localization Not Myristoylation. J. Biol. Chem. *269*, 30898-30903.

Duncan, J.A. and Gilman, A.G. (1996). Autoacylation of G Protein α Subunits. J. Biol. Chem. *271*, 23594-23600.

Duncan, J.A. and Gilman, A.G. (1998). A Cytoplasmic Acyl-Protein Thioesterase that Removes Palmitate from G Protein α Subunits and p21Ras. J. Biol. Chem. *273*, 15830-15837.

Dunphy, J.T., Greentree, W.K., Manahan, C.L., and Linder, M.E. (1996). G-protein palmitoyltransferase activity is enriched in plasma membranes. J. Biol. Chem. *271*, 7154-7159.

Dunphy, J.T., Schroeder, H., Leventis, R., Greentree, W.K., Knudsen, J.K., Silvius, J.R., and Linder, M.E. (2000). Differential effects of acyl-CoA binding protein on enzymatic and non- enzymatic thioacylation of protein and peptide substrates. Biochim. Biophys. Acta *1485*, 185-198.

El-Husseini, A.-D., Schnell, E., Dakoji, S., Sweeney, N., Zhou, Q., Prange, O., Gauthier-Campbell, C., Aguilera-Moreno, A., Nicoll, R.A., and Bredt, D.S. (2002). Synaptic strength regulated by palmitate cycling on PSD-95. Cell *108*, 849-863.

Evanko, D.S., Thiyagarajan, M.M., Siderovski, D.P., and Wedegaertner, P.B. (2001). Gβγ isoforms selectively rescue plasma membrane localization and palmitoylation of mutant Gαs and Gαq. J. Biol. Chem. *276*, 23945-23953.

Feron, O., Michel, J.B., Sase, K., and Michel, T. (1998). Dynamic regulation of endothelial nitric oxide synthase: complementary roles of dual acylation and caveolin interactions. Biochemistry *37*, 193-200.

Fishburn, C.S., Pollitt, S.K., and Bourne, H.R. (2000). Localization of a peripheral membrane protein: Gbetagamma targets Galpha(Z). Proc Natl Acad. Sci. USA *97*, 1085-1090.

Fukada, Y. (1995). Prenylation and Carboxylmethylation of G-Protein γ Subunit. Meth. Enzymol. *250*, 91-105.

Galbiati, F., Volonte, D., Meani, D., Milligan, G., Lublin, D.M., Lisanti, M.P., and Parenti, M. (1999). The Dually Acylated NH2-terminal Domain of Gi1alpha Is Sufficient to Target a Green Fluorescent Protein Reporter to Caveolin-enriched Plasma Membrane Domains. Palmitoylation of caveolin-1 is required for the recognition of dually acylated G-protein α subunits in vivo. J. Biol. Chem. *274*, 5843-5850.

Hancock, J.R., Cadwallader, K., Paterson, H., and Marshall, C.J. (1991). A CAAX or a CAAL motif and a Second Signal are Sufficient for Plasma Membrane Targeting of Ras Proteins. EMBO J. *10*, 4033-4039.

Higuchi, M., Izumi, K.M., and Kieff, E. (2001). Epstein-Barr virus latent-infection membrane proteins are palmitoylated and raft-associated: protein 1 binds to the cytoskeleton through TNF receptor cytoplasmic factors. Proc. Natl Acad. Sci. USA *98*, 4675-4680.

Janes, P.W., Ley, S.C., and Magee, A.I. (1999). Aggregation of Lipid Rafts Accompanies Signaling Via the T Cell Antigen Receptor. J. Cell Biol. *147*, 447-461.

Kabouridis, P.S., Magee, A.I., and Ley, S.C. (1997). S-acylation of Lck protein tyrosine kinase is essential for its signalling function in T lymphocytes. EMBO J. *16*, 4983-4998.

Leventis, R., Juel, G., Knudsen, J.K., and Silvius, J.R. (1997). Acyl-CoA binding proteins inhibit the nonenzymic S-acylation of cysteinyl-containing peptide sequences by long-chain acyl-CoAs. Biochemistry *36*, 5546-5553.

Liang, X., Lu, Y., Neubert, T.A., and Resh, M.D. (2002). Mass spectrometric analysis of GAP-43/neuromodulin reveals the presence of a variety of fatty acylated species. J. Biol. Chem. *277*, 33032-33040.

Liang, X., Nazarian, A., Erdjument-Bromage, H., Bornmann, W., Tempst, P., and Resh, M.D. (2001). Heterogeneous fatty acylation of Src family kinases with polyunsaturated fatty acids regulates raft localization and signal transduction. J. Biol. Chem. *276*, 30987-30994.

Lindwasser, O.W. and Resh, M.D. (2001). Multimerization of human immunodeficiency virus type 1 Gag promotes its localization to barges, raft-like membrane microdomains. J. Virol. *75*, 7913-7924.

Lindwasser, O.W. and Resh, M.D. (2002). Myristoylation as a target for inhibiting HIV assembly: unsaturated fatty acids block viral budding. Proc. Natl Acad. Sci. USA *99*, 13037-13042.

Lobo, S., Greentree, W.K., Linder, M.E., and Deschenes, R.J. (2002). Identification of a Ras palmitoyltransferase in *Saccharomyces cerevisiae*. J. Biol. Chem. *277*, 41268-41273.

Maurer-Stroh, S., Eisenhaber, B., and Eisenhaber, F. (2002a). N-terminal N-myristoylation of proteins: prediction of substrate proteins from amino acid sequence. J. Mol. Biol. *317*, 541-557.

Maurer-Stroh, S., Eisenhaber, B., and Eisenhaber, F. (2002b). N-terminal N-myristoylation of proteins: refinement of the sequence motif and its taxon-specific differences. J. Mol. Biol. *317*, 523-540.

McCabe, J.B. and Berthiaume, L.G. (1999). Functional roles for fatty acylated amino-terminal domains in subcellular localization. Mol. Biol. Cell *10*, 3771-3786.

Melkonian, K.A., Ostermeyer, A.G., Chen, J.Z., Roth, M.G., and Brown, D.A. (1999). Role of lipid modifications in targeting proteins to detergent-resistant membrane rafts. Many raft proteins are acylated, while few are prenylated. J. Biol. Chem. *274*, 3910-3917.

Michaelson, D., Ahearn, I., Bergo, M., Young, S., and Philips, M. (2002). Membrane trafficking of heterotrimeric G proteins via the endoplasmic reticulum and Golgi. Mol. Biol. Cell *13*, 3294-3302.

Michaelson, D., Silletti, J., Murphy, G., D'Eustachio, P., Rush, M., and Philips, M.R. (2001). Differential localization of Rho GTPases in live cells: regulation by hypervariable regions and RhoGDI binding. J. Cell. Biol. *152*, 111-126.

Moffett, S., Brown, D.A., and Linder, M.E. (2000). Lipid-dependent targeting of G proteins into rafts. J. Biol. Chem. *275*, 2191-2198.

Montixi, C., Langlet, C., Bernard, A.-M., Thimonier, J., Dubois, C., Wurbel, M.-A., Chauvin, J.-P., Pierres, M., and He, H.-T. (1998). Engagement of T cell receptor triggers its recruitment to low-density detergent-insoluble membrane domains. EMBO J. *17*, 5334-5348.

Morales, J., Fishburn, C.S., Wilson, P.T., and Bourne, H.R. (1998). Plasma Membrane Localization of Gαz Requires Two Signals. Mol. Biol. Cell *9*, 1-14.

Mumby, S.M. (1997). Reversible palmitoylation of signaling proteins. Curr. Opin. Cell Biol. *9*, 148-154.

Murray, D., Ben-Tal, N., Honig, B., and McLaughlin, S. (1997). Electrostatic interaction of myristoylated proteins with membranes: simple physics, complicated biology. Structure *5*, 985-989.

Navarro-Lerida, I., Alvarez-Barrientos, A., Gavilanes, F., and Rodriguez-Crespo, I. (2002). Distance-dependent cellular palmitoylation of de-novo-designed sequences and their translocation to plasma membrane subdomains. J. Cell Sci. *115*, 3119-3130.

Okamoto, T., Schlegel, A., Schereer, P.E., and Lisanti, M.P. (1998). Caveolins, a Family of Scaffolding Proteins for Organizing "Preassembled Signaling Complexes" at the Plasma Membrane. J. Biol. Chem. *273*, 5419-5422.

Ono, A. and Freed, E.O. (2001). Plasma membrane rafts play a critical role in HIV-1 assembly and release. Proc. Natl. Acad. Sci. USA *98*, 13925-13930.

Peitzsch, R.M. and McLaughlin, S. (1993). Binding of Acylated Peptides and Fatty Acids to Phospholipid Vesicles: Pertinance to Myristoylated Proteins. Biochemistry *32*, 10436-10443.

Pepinsky, R.B., Zeng, C., Wen, D., Rayhorn, P., and Baker, D.P.e.a. (1998). Identification of a Palmitic Acid modified form of Human Sonic hedgehog. J. Biol. Chem. *273*, 14037-14045.

Peseckis, S.M., Deichaite, I., and Resh, M.D. (1993). Iodinated Fatty Acids as Probes for Myristate Processing and Function. J. Biol. Chem. *268*, 5107-5114.

Pierce, S.K. (2002). Lipid rafts and B-cell activation. Nat. Rev. Immunol. *2*, 96-105.

Pike, L.J. and Casey, L. (2002). Cholesterol levels modulate EGF receptor-mediated signaling by altering receptor function and trafficking. Biochemistry *41*, 10315-10322.

Porter, J.A, Young, K.E, and Beachy, P.A. (1996). Cholesterol Modification of Hedgehog Signaling Proteins in Animal Development. Science *274*, 255-259.

Prior, I.A. and Hancock, J.F. (2001). Compartmentalization of Ras Proteins. J. Cell Sci. *114*, 1603-1608.

Prior, I.A., Harding, A., Yan, J., Sluimer, J., Parton, R.G., and Hancock, J.F. (2001). GTP-dependent segregation of H-ras from lipid rafts is required for biological activity. Nat. Cell Biol. *3*, 368-375.

Resh, M.D. (1994). Myristylation and Palmitylation of Src Family Members: The Fats of the Matter. Cell *76*, 411-413.

Resh, M.D. (1999). Fatty acylation of proteins: new insights into membrane targeting of myristoylated and palmitoylated proteins. Biochim. Biophys. Acta *1451*, 1-16.

Robbins, S.M., Quintrell, N.A., and Bishop, J.M. (1995). Myristoylation and Differential Palmitoylation of the HCK Protein Tyrosine Kinases Govern Their Attachment to Membranes and Association with Caveolae. Mol. Cell. Biol. *15*, 3507-3515.

Robinson, L.J., Busconi, L., and Michel, T. (1995). Agonist-modulated Palmitoylation of Endothelial Nitric Oxide Synthase. J. Biol. Chem. *270*, 995-998.

Roth, A.R., Feng, Y., Chen, L., and Davis, N.G. (2002). The yeast DHC cysteine-rich domain protein Akr1p is a palmitoyl transferase. J. Cell Biol. *159*, 23-28.

Rousso, I., Mixon, M.G., Chen, B.K., and Kim, P.S. (2000). Palmitoylation of the HIV-1 envelope glycoprotein is critical for viral infectivity. Proc. Natl. Acad. Sci. USA *97*, 13523-13525.

Roy, M.O., Leventis, R., and Silvius, J.R. (2000). Mutational and biochemical analysis of plasma membrane targeting mediated by the farnesylated, polybasic carboxy terminus of K-ras4B. Biochemistry *39*, 8298-8307.

Schade, A.E. and Levine, A.D. (2002). Lipid raft heterogeneity in human peripheral blood T lymphoblasts: a mechanism for regulating the initiation of TCR signal transduction. J. Immunol. *168*, 2233-9.

Scheiffele, P., Roth, M.G., and Simons, K. (1997). Interaction of influenza virus haemagglutinin with sphingolipid-cholesterol membrane domains via its transmembrane domain. EMBO J. *16*, 5501-5508.

Schroeder, H., Leventis, R., Shahinian, S., Walton, P.A., and Silvius, J.R. (1996). Lipid-modified, Cysteinyl-containing Peptides of Diverse Structure are Efficiently S-Acylated at the Plasma Membrane of Mammalian Cells. J. Cell Biol. *134*, 647-660.

Shahinian, S. and Silvius, J.R. (1995). Doubly lipid-modified protein sequence motifs exhibit long lived anchorage to lipid bilayer membranes. Biochemistry *34*, 3813-3822.

Shenoy-Scaria, A.M., Dietzen, D.J., Kwong, J., Link, D.C., and Lublin, D.M. (1994). Cysteine-3 of Src Family Protein Tyrosine Kinases Determines Palmitoylation and Localization in Caveolae. J. Cell Biol. *126*, 353-363.

Silvius, J.R. and L'Heureux, F. (1994). Fluorometric evaluation of the affinities of isoprenylated peptides for lipid bilayers. Biochemistry *33*, 3014-3022.

Simons, K. and Ehehalt, R. (2002). Cholesterol, lipid rafts, and disease. J. Clin. Invest. *110*, 597-603.

Simons, K. and Toomre, D. (2000). Lipid rafts and signal transduction. Nat. Rev. Mol. Cell Biol. *1*, 31-39.

Sowa, G., Pypaert, M., and Sessa, W.C. (2001). Distinction between signaling mechanisms in lipid rafts vs. caveolae. Proc. Natl. Acad. Sci. USA *98*, 14072-14077.

Sproul, T.W., Malapati, S., Kim, J., and Pierce, S.K. (2000). Cutting edge: B cell antigen receptor signaling occurs outside lipid rafts in immature B cells. J. Immunol. *165*, 6020-6023.

Tsui-Pierchala, B.A., Encinas, M., Milbrandt, J., and Johnson, E.M. Jr (2002). Lipid rafts in neuronal signaling and function. Trends Neurosci. *25*, 412-417.

van Duyl, B.Y., Rijkers, D.T., de Kruijff, B., and Killian, J.A. (2002). Influence of hydrophobic mismatch and palmitoylation on the association of transmembrane alpha-helical peptides with detergent-resistant membranes. FEBS Lett *523*, 79-84.

van't Hof, W. and Resh, M.D. (1997). Rapid plasma membrane anchoring of newly synthesized p59fyn: selective requirment for NH2-terminal myristoylation and palmitoylation at cysteine-3. J. Cell Biol. *136*, 1023-1035.

van't Hof, W. and Resh, M.D. (1999). Dual fatty acylation of p59fyn is required for association with the T-cell receptor zeta chain through phosphotyrosine-SH2 interactions. J. Cell Biol. *145*, 377-389.

Wang, J., Arbuzova, A., Hangyas-Mihalyne, G., and McLaughlin, S. (2001). The effector domain of myristoylated alanine-rich C kinase substrate binds strongly to phosphatidylinositol 4,5-bisphosphate. J. Biol. Chem. *276*, 5012-5019.

Webb, Y., Hermida-Matsumoto, L., and Resh, M.D. (2000). Inhibition of protein palmitoylation, raft localization and T cell signaling by 2-bromopalmitate and polyunsaturated fatty acids. J. Biol. Chem. *275*, 261-270.

Wolven, A., Okamura, H., Rosenblatt, Y., and Resh, M.D. (1997). Palmitoylation of p59fyn is Reversible and Sufficient for Plasma Membrane Association. Mol. Biol. Cell *8*, 1159-1173.

Xavier, R., Brennan, T., Li, Q., McCormack, C., and Seed, B. (1998). Membrane Compartmentation is Required for Efficient T Cell Activation. Immunity *8*, 723-732.

Yang, X., Claas, C., Kraeft, S.K., Chen, L.B., Wang, Z., Kreidberg, J.A., and Hemler, M.E. (2002). Palmitoylation of tetraspanin proteins: modulation of CD151 lateral interactions, subcellular distribution, and integrin-dependent cell morphology. Mol. Biol. Cell. *13*, 767-81.

Yeh, D. C., Duncan, J. A., Yamashita, S., and Michel, T. (1999). Depalmitoylation of endothelial nitric-oxide synthase by acyl-protein thioesterase 1 is potentiated by Ca(2+)-calmodulin. J. Biol. Chem. *274*, 33148-33154.

Zacharias, D.A., Violin, J.D., Newton, A.C., and Tsien, R.Y. (2002). Partitioning of lipid-modified monomeric GFPs into membrane microdomains of live cells. Science *296*, 913-916.

Zhang, H., Seabra, M.C., and Deisenhofer, J. (2000). Crystal structure of Rab geranylgeranyltransferase at 2.0 A resolution. Structure *8*, 241-251.

Zhang, W., Irvin, B.J., Trible, R.P., Abraham, R.T., and Samelson, L.E. (1999). Functional analysis of LAT in TCR-mediated signaling pathways using a LAT-deficient Jurkat cell line. Int. Immunol. *11*, 943-950.

Zhang, W., Trible, R.P., and Samelson, L.E. (1998). LAT palmitoylation: its essential role in membrane microdomain targeting and tyrosine phosphorylation during T cell activation. Immunity *9*, 239-246.

Zhou, W., Parent, L.J., Wills, J.W., and Resh, M.D. (1994). Identification of a Membrane Binding Domain within the Amino-Terminal Region of Human Immunodeficiency Virus Type 1 Gag Protein Which Interacts with Acidic Phospholipids. J. Virol. *68*, 2556-2569.

Chapter 7

Role of the Membrane Skeleton in Creation of Microdomains

Ken Ritchie and Akihiro Kusumi

Department of Biological Science, Graduate School of Science, Nagoya University, Nagoya 464-8602, Japan and Kusumi Membrane Organizer Project, ERATO, Japan Science and Technology Corporation, 5-11-33, Nagoya 460-0012, Japan

Abstract

The membrane skeleton is a specialized part of the cytoskeleton that is in close proximity to the cell membrane with a protein composition and structure that differs from that of the bulk cytoskeleton. The membrane skeleton and various transmembrane proteins bound to it form a mosaic of compartments in the membrane that is responsible for the temporary confinement of membrane proteins and lipids and controls the rate of their repetitive hop movements between these membrane skeleton-based compartments, known as "hop diffusion", found by observation of single-molecule diffusion. Such hop diffusion has been found to be universal with compartment sizes that range from 30 to 700 nm, depending on the cell type. The part of the membrane skeleton that is directly involved in temporal confinement of membrane molecules has been successfully imaged by raster scanning a single membrane molecule using an optical trap (single molecule scanning optical force imaging). Such compartmentalization enables dynamic spatial regulation of signal transduction in the plasma membrane, by arresting signaling complexes of activated receptor molecules and enlarged, stabilized rafts within a compartment. Furthermore, high concentrations of the membrane skeleton and its associated immobile transmembrane proteins are involved in formation of the cell membrane polarity such as is found across the initial segment between the axon and the cell body in neurons. Argument is advanced that the creation of various membrane domains in the cell membrane must be influenced by the membrane skeleton.

1. INTRODUCTION

Initial models of the plasma membrane of cells give a simple approximation to their complexity. The existence of stable and meta-stable domains and compartments within the fluid lipid bilayer implies that the membrane has a fine structure that goes beyond that predicted by the *fluid mosaic model* (Singer and Nicolson, 1972; Jacobson *et al.*, 1995). In fact when speaking of the plasma membrane, one must be mindful of the proteins that are integral to the membrane as well as the subsurface actin-based membrane skeleton that underlies the lipid bilayer. What we wish to convey in this chapter is the importance of the membrane skeleton in the dynamic organization of the plasma membrane of cells.

2. MEMBRANE SKELETON AS AN INTEGRAL COMPONENT OF THE PLASMA MEMBRANE

When referring to the membrane skeleton we specifically mean the layer of the actin based cytoskeleton that comes in close proximity to the lipid bilayer. In fact, the skeleton comes close enough that it comes in contact with and interacts both specifically and non-specifically with the proteins embedded in the lipid matrix. Direct evidence of this interaction has been shown through observing the diffusion of individual transmembrane proteins along the plasma membrane. Observation of the motion of a single/few protein(s) using single particle tracking (SPT) revealed features of the membrane which were hidden by ensemble averaging over all molecules in bulk observation techniques (de Brabander *et al.*, 1985; Gelles *et al.*, 1988; Sheetz *et al.*, 1989; Kucik *et al.*, 1989; Kusumi *et al.*, 1993). SPT is a technique where a membrane protein is tagged with a small particle (a 200 nm latex particle or, to reduce the effect of crosslinking, a 40 nm diameter colloidal gold particle, which will be discussed here) and the protein/colloidal probe complex is directly observed through contrast enhanced (Nomarski or bright-field) video microscopy. The trajectories of many transmembrane proteins appear to be non-Brownian and reveal the underlying membrane skeleton structure. Transmembrane protein undergoes random motion in finite sized compartments, with infrequent transitions to neighboring compartments. This motion has been termed “hop diffusion” (Kusumi *et al.*, 1993; Sako and Kusumi, 1994; Saxton, 1995).

2.1. Membrane skeleton fence structure

The mechanism behind this anomalous diffusion was proposed to be a steric interaction between the cytoplasmic domain of the transmembrane protein and the membrane cytoskeleton, the actin meshwork that comes into close proximity to the membrane bilayer. This mechanism was termed the "fence" model, where the actin strands in the skeleton network act as fences to corral the membrane proteins (Kusumi *et al.*, 1993; Sako and Kusumi, 1994; Kusumi and Sako, 1996). The model is illustrated in Figure 1.

This mechanism behind motion was deemed to be due to the membrane skeleton for two reasons:

1. The requirement of the cytoplasmic tail. This was shown through cytoplasmic domain deletion mutants of E-cadherin in L cells (Sako *et al.*, 1998) as

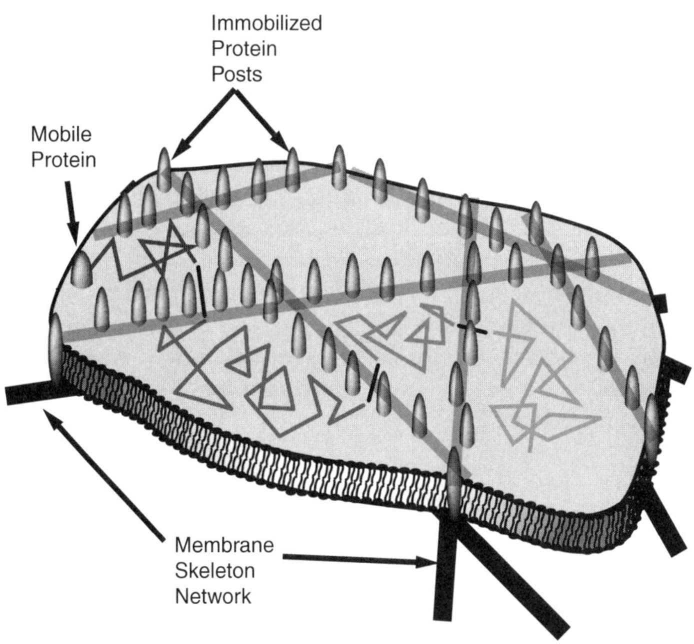

Figure 1. The fence and picket structure of the plasma membrane. Transmembrane protein is compartmentalized by specific (binding) and non-specific (steric) interactions with the actin membrane skeleton *fences*. Lipid diffusion is also compartmentalized, but this is due to the fraction of transmembrane protein immobilized through binding to the underlying membrane skeleton network. The hydrodynamic-like interaction predicted by the free area theory of lipid diffusion (Sperotto and Mouritsen, 1991; Almeida *et al.*, 1992; Dodd *et al.*, 1995) between the immobilized protein *fence pickets* acts to corral the lipids.

well as cytoplasmic-domain-cleaved band 3 in the erythrocyte membrane (Tomishige *et al.*, 1998) (where the membrane skeleton consists of mainly the soft, structural protein spectrin).

2. The requirement of the membrane skeleton meshwork. This was shown through very delicate use of actin depolymerzation drugs (such as cytocholasin D and latrunculin A) where slight depolymerzation of the skeleton tends to increase the average compartment size and through an actin-stabilizing drug (jasplakinolide), which tends to increase the average residency time in a compartment. In addition, membrane blebs where the membrane skeleton is largely lost, the majority of transmembrane proteins undergo simple Brownian diffusion with diffusion coefficients comparable to those in reconstructed membranes (Fujiwara *et al.*, 2002).

As discussed in the previous section, the diffusion of transmembrane protein looks hop-like instead of undergoing a simple Brownian random walk. Although the presence of compartments is seen by eye, analysis of their properties requires a more stringent methodology (Kusumi *et al.*, 1993). Analysis of the hop-like trajectories follows from determining the mean-squared-displacement (MSD) of the protein through its trajectory. The MSD (as a function of time) is then fit by an equation derived from the theory for diffusion through an infinite array of semi-permeable barriers (Powles *et al.*, 1992). The analysis gives the average compartment size, residency time in a compartment and diffusion coefficient inside of a compartment. Surprisingly, the diffusion coefficient inside of a compartment is on the order of 5–10 $\mu m^2/s$ in all cases. This value is similar to that expected for free diffusion in pure lipid bilayers and reconstituted membranes (Fujiwara *et al.*, 2002). This suggests that the diffusion of membrane protein is not slow because it itself is slow, but rather the slowing effect is a direct result of infrequent transitions or hops between membrane compartments.

2.2. Membrane skeleton anchored protein pickets

SPT has also recently been applied to observe the diffusion of lipid, specifically L-α-dioleoylphosphatidylethanolamine (DOPE), in the plasma membrane of NRK cells (Fujiwara *et al.*, 2002). Surprisingly, the diffusion of DOPE also reflects the compartmentalized structure of the membrane (Fujiwara *et al.*, 2002), although the lipid is restricted to the outer leaflet of the membrane and hence cannot directly interact with the membrane cytoskeleton. The compartment size found for the hop diffusion of DOPE is very similar to that found for transmembrane protein on the same cell type, but the lipid has a much increased hop frequency between compartments. As well, actin stabilizing/depolymerizing drugs have the same effects on the compartmentalization of DOPE as they do on transmembrane protein. Motion in membrane skeleton lacking blebs also shows only simple Brownian diffusion with a diffusion coefficient of 9-$\mu m^2/s$ (Fujiwara *et al.*, 2002), which is comparable to the diffusion coefficient inside the membrane compartments. Furthermore, removal of extracellular matrix proteins and extracellular domains of membrane proteins, and partial cholesterol depletion does not affect the hop parameters. It has been proposed that the membrane skeleton interacts indirectly with even the lipid on the outer leaflet through both steric and hydrodynamic-like interaction (Dodd *et al.*, 1995; Bussell *et al.*, 1995) with the transmembrane protein that is immobilized to the skeleton. These immobilized protein "posts" or "fence pickets" hold the membrane skeleton "fences" in close proximity to the membrane and hence reflect the structure of the underlying actin-based membrane skeleton meshwork (see Figure 1). The hydrodynamic-like interaction predicted by the free area theory of lipid diffusion (Sperotto and Mouritsen, 1991; Almeida *et al.*, 1992) has been shown through Monte Carlo simulation to be able to reproduce the experimental results. Only 20–30% coverage of the compartment boundaries with anchored transmembrane proteins was sufficient to reproduce the experimentally observed hop frequency (Fujiwara *et al.*, 2002), i.e., when transmembrane proteins are anchored to the membrane skeleton, the mobility of the lipid next to the immobilized protein is decreased, and this effect propagates over distances of several protein diameters (Sperotto and Mouritsen, 1991; Almeida *et al.*, 1992). It should be noted that if the transmembrane proteins are mobile, such immobilization effects on the surrounding lipids are substantially diminished (East *et al.*, 1985; Ryba *et al.*, 1987).

2.3. Cell-to-cell variations of the compartment size and hop rate

From observations of the diffusion of transmembrane protein and lipid in a variety of cell types, compartmentalization seems to be a general feature of the plasma membrane of cells. There is, though, a variance in the compartment size and residency time between cell types. The compartment size ranges from 30 nm at the smallest for CHO and FRSK cells to 200 and 700 nm for the unique, doubly compartmentalized NRK cell (see Table 1). Also shown in Table 1 are the residency times for the lipid DOPE in the various cells for comparison. They range from 1.5 ms to 410 ms for lipid diffusion and can be as slow as every 2 s for the transmembrane protein transferrin in the larger NRK cell compartments (Murase *et al.*, 2001).

Table 1. Variation in compartment size between cell types

Cell type	Compartment size[1] (nm)	Residency time for DOPE[2] (ms)
FRSK	41	2.3
CHO	32	1.5
PtK2	39	2.4
ECV304	110	8.1
NRK[3] (outer)	750	410
NRK[3] (inner)	230	12

Notes:
[1]The compartment length is determined along two orthogonal directions (L_x, L_y). The Compartment size is determined as the geometric mean of these lengths, $(L_x, L_y)^{1/2}$.
[2]L-α-dioleoylphosphatidylethanolamine.
[3]The NRK cell is unique in that it has two distinct, nested compartments (Fujiwara *et al.*, 2002).

2.4. Direct visualization of pickets and fences by raster-scanning single receptor molecules using an optical trap

Direct observation of the membrane cytoskeleton based fences and pickets present its own problems. The scale of the compartment size is from 40–200 nm, which precludes the use of direct optical imaging due to the Raleigh resolution limit. Another key problem is separating the actin skeleton that is in close proximity to the membrane from the dense actin cytoskeleton in the interior of the cell.

Direct imaging of this cortical actin layer has been achieved by trapping a colloidal gold-labeled transmembrane protein (used as a scanning "tip") in a highly focused laser beam (optical tweezers, OT) and scanning the cell beneath the trapped probe. Deviations of the probe from the center of the OT measures the resistance of the probe protein to dragging in the membrane and maps out the compartment barriers (Ritchie and Kusumi, 2002). Figure 2 shows the result of repeatedly scanning a single line along the plasma membrane of a live normal rat kidney fibroblast using the integral membrane protein CD44 as a probe. Note that light areas reveal resistances to dragging of 0.2 pN and show the locations and dynamics of barriers along the scanning path.

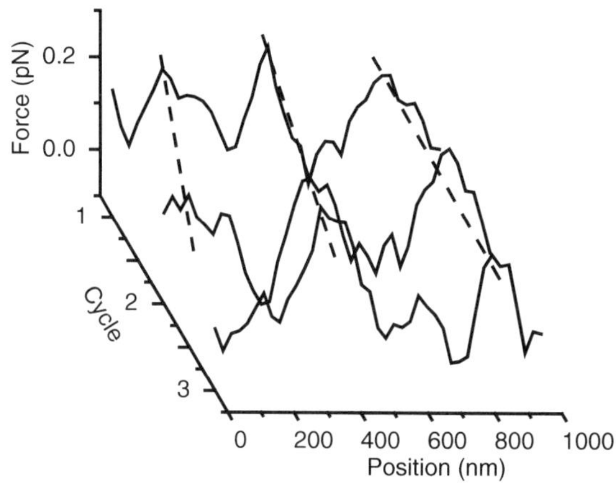

Figure 2. Forces felt by the transmembrane protein CD44 labeled by a 40 nm colloidal gold particle on repeated scanning of the same line. The array of high force regions (indicated by dashed lines, maximum force level of 0.2 pN) displays the positions of barriers to free diffusion due to the cytoskeletal meshwork. Scan speed 1.8 pN/s; 3 cycles; forward scans only.

3. PLASMA MEMBRANE COMPARTMENTALIZATION BY THE MEMBRANE SKELETON

As illustrated in Figure 3, the existence of the membrane skeleton fence and picket structure forms a compartmentalized mosaic in the membrane. As such, assemblies of proteins have to contend with the compartments when diffusing through the membrane. This gives the cell a template over which to design specialized domains. It is important to realize that the radius of a diffusing aggregate in two-dimensions only logarithmically affects the diffusion rate. The existence of the fence and picket structure allows for the slowing and arresting of large aggregates, as their size approaches the inter-picket spacing.

3.1. Diffusion barriers in the membrane for cell polarization

On the largest scale, many cell types have specialized domains that encompass a significant portion of the membrane, as is specifically found polarized cells such as neurons (between the distal end of the axon and the cell body) and endothelial cells (between the apical and basal surfaces). In these types of cells a different set of protein may reside on the apical surface than on the basal

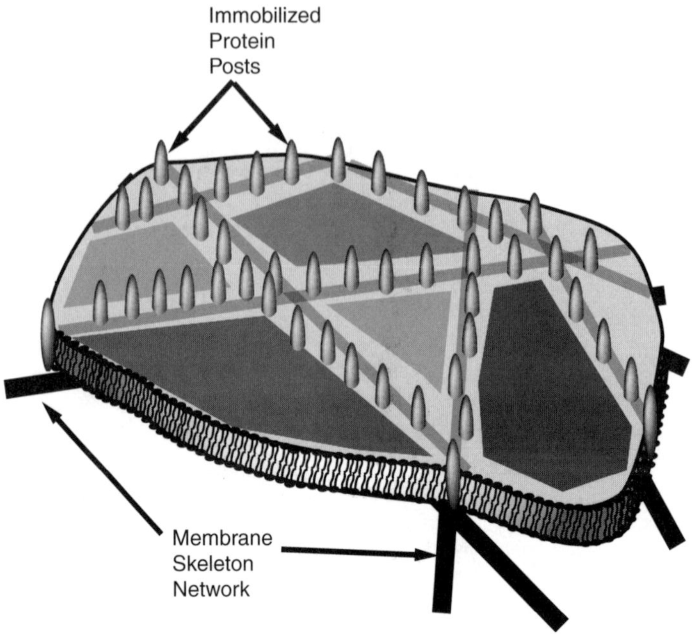

Figure 3. Compartmental mosaic of the membrane. The membrane skeleton fences and pickets form a template for domain formation in the plasma membrane of cells. The structure and dynamics of specialized domains such as lipid rafts are limited by the compartmentalized structure of the membrane.

surface. A mechanism for maintenance of this large-scale inhomogeneity in a fluid membrane is not obvious. The existence of a high density of protein pickets in the barrier region between the two dissimilar regions coupled with the reduction in diffusion rate expected due to the fence and picket model, implies that in the barrier region an effective mechanism for blocking diffusional mixing is present (Figure 4). In fact, it has been shown that a reduction in the diffusion of even a phospholipid occurs developmentally at the initial segment of the axon of hippocampal neural cells and is well correlated to the time course for accumulation of sodium channels in the same region (Nakada *et al.*, 2001).

3.2. Signal localization by the membrane compartments

The implications of an actin skeleton-based (along with its associated immobilized proteins) fine structure in the fluid membrane, which can slow long-range diffusion even for single lipids, are that an aggregate of lipid and

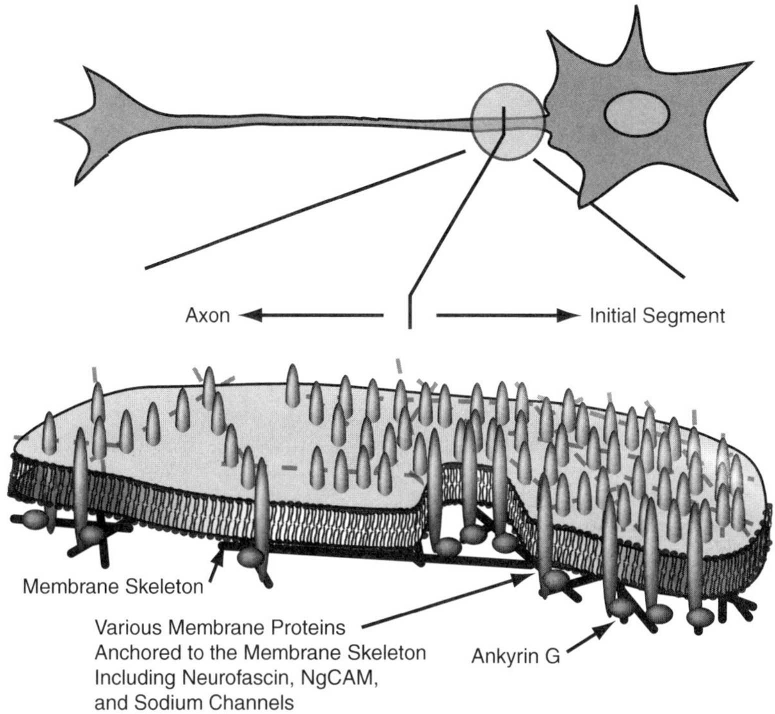

Figure 4. Fence and picket involvement in cell polarization. Polarized distributions of membrane bound molecules, such as glutamate receptors, $GABA_A$ receptors, transferrin receptors and EGF receptors in the somatodendritic region and L1, K+ channel (Kv1.4), TAG-1 and GAP-43 in the axonal region, can be maintained by the existence of a densely packed region of pickets. Such a region has been shown to develop at the initial segment of neurons, consequent with the onset of a diffusion barrier between the distal axon and the cell body (Nakada *et al.*, 2001).

protein (even GPI-anchored protein), as expected in the formation of lipid domains or rafts, would have severely restricted the long-range motion. In fact, after growing to a certain finite size the complex would be permanently trapped in its local compartment. Such effects have been seen in complex formation of E-cadherin (labeled with green fluorescent protein, GFP) in L cells through single fluorophore video imaging (SFVI) (Iino *et al.*, 2001), see Figure 5. The technique of SFVI utilizes a fluorophore tagged protein or lipid which is viewed through a total internal reflection microscope equipped with an intensified video imaging system. The advantage of this technique over SPT is the ability to tag, with certainty, a single molecule at the cost of the higher signal-to-noise ratio of fluorescence imaging that limits the observation

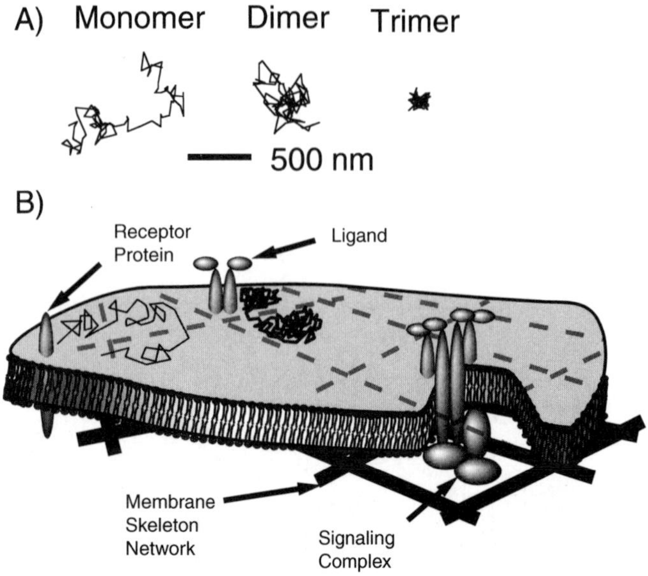

Figure 5. Trajectories of E-cadherin-GFP in L cells reveals that oligomerized molecules undergo slower long-range diffusion. E-cadherin-GFP is seen through single molecule fluorescence video imaging. The trajectories are 3 s (90 video frames) in length at 33 ms/frame. Trajectories of fluorescent spots deemed to represent a single, double and triple fluorophore molecule undergo simple diffusion, compact (i.e. their long range diffusion is slowed) and confined (i.e. no long-range diffusion) motion. (B) The oligomerization induced trapping model. A single receptor protein may freely diffuse between compartments in the membrane. After binding ligand, the receptor with ligand diffuses until it meets similar molecules to form a dimer or greater oligomer, slowing its transition rate between compartments. As the signaling complex forms, signaling molecules, both membrane-bound and cytoplasmic, begin to collect around the receptor molecule. The signaling complex is now arrested in a compartment due to its size (i.e. it can no longer escape through the fence and/or pickets) or due to direct interaction with the membrane skeleton.

to standard video rates. This slowing of molecular complexes may be an important mechanism for the localization of external signals.

Without this localization mechanism, after receiving a signal through ligand binding, the spatial information of the signal would be quickly lost due to random diffusion of the receptor molecules. A freely diffusing molecule in a pure model bilayer, devoid of a compartmentalized fine structure should have a diffusion coefficient in the range 5–10 $\mu m^2/s$ (Fujiwara *et al.*, 2002) and thus will cover an area of diameter 2–3 μm in a second. In the plasma membrane of a CHO or FRSK cell, after aggregation, the signaling complex can be restricted to a compartment of only 30 nm in extent until dissolution of the signaling complex.

3.3. Interaction of the membrane skeleton with raft molecules and other membrane domains

The slowing and eventual confinement of the motion of protein aggregates has further implications on the movement of lipid domains (also referred to as "rafts"). If the size of rafts are large enough, that is, larger than the inter picket spacing, they will be trapped inside a single domain. Surprisingly, the motion of individual raft molecules, specifically CD59, shows no qualitative difference in its motion from that of the non-raft-associated lipid DOPE in the resting cells (Suzuki *et al.*, 2001). After stimulation by liganding or antibody-induced clustering of CD59 molecules, two important changes occur to the motion. First the overall rate of diffusion is decreased as expected due to the increase in diffusive radius due to clustering of the molecules combined with the picket model. This decrease implies that the raft has grown in size, but since it may still pass through the pickets, it is thus still small in size. The second feature is a random intermittent temporary cessation of motion during the motion of the clusters. Actin destabilizing drugs, cholesterol removal from the membrane and blockage of src-family kinases reduced the frequency of stoppage, implying that the mobile raft, after collecting with the aid of cholesterol the raft-partitioning proteins required for signaling (including src-family kinases), becomes a large enough signaling complex to no longer be able to pass through the pickets or becomes temporarily tethered to the skeleton through one of the recruited proteins. Motion resumes after the signaling complex disassociates.

4. MEMBRANE SKELETON FOR SPECIALIZED MEMBRANE DOMAINS

The membrane skeleton provides through its fences and pickets a means to efficiently arrest the motion of protein/lipid complexes for the duration of their association. As such, they become a template over which dynamic domain formation and dissolution occurs. They may be required for maintenance of cellular polarity and for signal localization. The existence of the compartmentalized structure of the plasma membrane is ubiquitous among different cell types and thus must be taken into account when the dynamics of membrane components and their assemblies, such as lipid rafts are considered.

REFERENCES

Almeida, P. F., Vaz, W. L. and Thompson, T. E., 1992, Lateral diffusion and percolation in two-phase, two-component lipid bilayers. Topology of the solid-phase domains in-plane and across the lipid bilayer. *Biochemistry*, **31**, 7198-7210.

Bussell, S. J., Koch, D. L. and Hammer, D. A., 1995, Effect of Hydrodynamic Interactions on the Diffusion of Integral Membrane-Proteins – Diffusion in Plasma-Membranes. *Biophys J*, **68**, 1836-1849.

de Brabander, M., Geuens, G., Nuydens, R., Moeremans, M. and Demey, J., 1985, Probing Microtubule-Dependent Intracellular Motility with Nanometer Particle Video Ultramicroscopy (Nanovid Ultramicroscopy). *Cytobios*, **43**, 273-283.

Dodd, T. L., Hammer, D. A., Sangani, A. S. and Koch, D. L., 1995, Numerical Simulations of the Effect of Hydrodynamic Interactions on Diffusivities of Integral Membrane-Proteins. *J Fluid Mech*, **293**, 147-180.

East, J. M., Melville, D. and Lee, A. G., 1985, Exchange rates and numbers of annular lipids for the calcium and magnesium ion dependent adenosinetriphosphatase. *Biochemistry*, **24**, 2615-2623.

Fujiwara, T., Ritchie, K., Murakoshi, H., Jacobson, K. and Kusumi, A., 2002, Phospholipids undergo hop diffusion in compartmentalized cell membrane. *J Cell Biol*, **157**, 1071-1081.

Gelles, J., Schnapp, B. J. and Sheetz, M. P., 1988, Tracking Kinesin-Driven Movements with Nanometre-Scale Precision. *Nature*, **331**, 450-453.

Iino, R., Koyama, I. and Kusumi, A., 2001, Single molecule imaging of green fluorescent proteins in living cells: E-cadherin forms oligomers on the free cell surface. *Biophys J*, **80**, 2667-2677.

Jacobson, K., Sheets, E. D. and Simson, R., 1995, Revisiting the fluid mosaic model of membranes. *Science*, **268**, 1441-1442.

Kucik, D. F., Elson, E. L. and Sheetz, M. P., 1989, Forward Transport of Glycoproteins on Leading Lamellipodia in Locomoting Cells. *Nature*, **340**, 315-317.

Kusumi, A. and Sako, Y., 1996, Cell surface organization by the membrane skeleton. *Curr Opin Cell Biol*, **8**, 566-574.

Kusumi, A., Sako, Y. and Yamamoto, M., 1993, Confined Lateral Diffusion of Membrane-Receptors as Studied by Single-Particle Tracking (Nanovid Microscopy) – Effects of Calcium-Induced Differentiation in Cultured Epithelial-Cells. *Biophys J*, **65**, 2021-2040.

Murase, K., Fujiwara, T., Iino, R., Murakoshi, H., Ritchie, K. P. and Kusumi, A., 2001, Compartmentalization of the plasma membrane into 40 nm compartments which induce hop diffusion of phospholipids as visualized by single molecule techniques. *Mol Biol Cell*, **12**, 470a.

Nakada, C., Ritchie, K., Fujiwara, T., Yamaguchi, K. and Kusumi, A., 2001, Formation of a diffusion barrier in the plasma membrane of the neuronal initial segment; A single molecule study. *Biophys J*, **80**, 179a.

Powles, J. G., Mallett, M. J. D., Rickayzen, G. and Evans, W. A. B., 1992, Exact Analytic Solutions for Diffusion Impeded by an Infinite Array of Partially Permeable Barriers. *Proc. R. Soc. Lond. (A. Maths)*, **436**, 391-403.

Ritchie, K. and Kusumi, A., 2002, Single molecule probe scanning optical force imaging microscope for viewing live cells. *J Biol Phys*, **28**, 619-626.

Ryba, N. J., Horvath, L. I., Watts, A. and Marsh, D., 1987, Molecular exchange at the lipid-rhodopsin interface: spin-label electron spin resonance studies of rhodopsin-dimyristoylphosphatidylcholine recombinants. *Biochemistry*, **26**, 3234-3240.

Sako, Y. and Kusumi, A., 1994, Compartmentalized Structure of the Plasma-Membrane for Receptor Movements as Revealed by a Nanometer-Level Motion Analysis. *J Cell Biol*, **125**, 1251-1264.

Sako, Y., Nagafuchi, A., Tsukita, S., Takeichi, M. and Kusumi, A., 1998, Cytoplasmic regulation of the movement of E-cadherin on the free cell surface as studied by optical tweezers and single particle tracking: Corralling and tethering by the membrane skeleton. *J Cell Biol*, **140**, 1227-1240.

Saxton, M. J., 1995, Single-Particle Tracking – Effects of Corrals. *Biophys J*, **69**, 389-398.

Sheetz, M. P., Turney, S., Qian, H. and Elson, E. L., 1989, Nanometer-Level Analysis Demonstrates That Lipid Flow Does Not Drive Membrane Glycoprotein Movements. *Nature*, **340**, 284-288.

Singer, S. J. and Nicolson, G. L., 1972, Fluid Mosaic Model of Structure of Cell-Membranes. *Science*, **175**, 720-731.

Sperotto, M. M. and Mouritsen, O. G., 1991, Monte Carlo simulation studies of lipid order parameter profiles near integral membrane proteins. *Biophys J*, **59**, 261-270.

Suzuki, K., Sanematsu, F., Fujiwara, T., Edidin, M. and Kusumi, A., 2001, A GPI-anchored membrane protein, CD59, frequently visits/forms rafts in signaling: A single molecule study. *Biophys J*, **80**, 252a.

Tomishige, M., Sako, Y. and Kusumi, A., 1998, Regulation mechanism of the lateral diffusion of band 3 in erythrocyte membranes by the membrane skeleton. *J Cell Biol*, **142**, 989-1000.

Chapter 8

Membrane/Cytoskeleton Communication

Karina F. Meiri
Department of Anatomy and Cellular Biology, Tufts University School of Medicine, 136 Harrison Ave, Boston MA 02111, USA

Abstract
Accumulations of particular lipids in ordered arrays in the membrane (termed microdomains or lipid rafts) can attract proteins with specific targeting domains. Both the lipid and protein components of rafts communicate with the cytoskeleton directly thereby regulating cellular responses. Recent evidence implicating phosphoinositide 1,5 bisphosphate (PIP2) in cytoskeletal regulation shows that agonist sensitive regulation of PIP2 homoeostasis occurs specifically rafts, which appear to provide a major structural substrate for its function. The crucial role of PIP2 in generating cytoskeletal responses is chiefly achieved by regulating proteins that control actin dynamics directly. Many of these regulatory proteins are also specifically enriched in rafts either directly (by insertion into the lipid bilayer via acetylation motifs), or indirectly via interactions with other raft components. The notion that rafts form membrane platforms or modules that mediate signaling responses has been most extensively demonstrated in the immune synapse (IS) of T cells, a complex assemblage of rafts that integrates signaling cascades originating from the simultaneous activation of a wide variety of receptors. The IS is essential for both the amplification and maintenance of T-cell activation, and its assembly at the antigen presenting site depends on the interactions between rafts and the actin cytoskeleton that regulates coalescence of smaller raft components into the larger IS complex. Likewise the neuron, which represents the most highly polarized cell in the body, utilizes the regulation of actin dynamics in response to a plethora of extracellular signals to control

Subcellular Biochemistry, Volume 37: Membrane Dynamics and Domains
Edited by Peter J. Quinn. Kluwer Academic / Plenum Publishers, New York, 2004

axon pathfinding thereby sculpting nervous system cytoarchitecture with utmost precision. It is now becoming clear, that as in the T-cell, lipid rafts in the growing axon can assemble into highly specific, yet malleable and dynamic, signaling modules that regulate actin dynamics in a fashion that is also PIP2-dependent and that utilizes both familiar and novel regulatory mechanisms. It seems clear that raft mediated cytoskeletal regulation represents a highly conserved mechanism to integrate cellular responses to diverse signals.

1. INTRODUCTION

During the last several years it has become evident that most cellular responses to extracellular signals requires extensive communication between the membrane and the cytoskeleton. Significant progress in uncovering the signaling pathways activated by specific extracellular cues, has demonstrated the structural substrates in the membrane that act in concert with the cytoskeleton to evoke a response. Pioneering work on erythrocytes demonstrated that the cytoskeleton is specialized for communication: the juxtamembrane 'membrane skeleton' provides a structural substrate for signals transduced at the membrane to impinge on responsive cytoskeletal proteins. More recently it has become clear that the membrane too is structurally heterogeneous: by assembling components of transduction pathways into signaling modules, membrane specializations called microdomains form an important intracellular junction with the cytoskeleton. In this chapter I use the T cell and the growing neuron to illustrate how very different functional outcomes can be coordinated by membrane microdomain communication with the cytoskeleton.

It is now well accepted that the restricted distribution of many signaling pathways in modules at the cell membrane depends in large part on the underlying compartmentalization of membrane lipids. A surprisingly large proportion – up to 70% – of external membrane lipids (chiefly cholesterol and glycosphingolipids) can be localized in these liquid ordered microdomains (Mayor and Maxfield, 1995; Fridriksson *et al.*, 1999). The lipid components of complementary inner leaflet microdomains are less well characterized. The significance of these lipid heterogeneities to membrane/cytoskeletal communication lies in their ability to attract proteins with characteristic targeting motifs. Extracellularly either cholesterol binding or the presence of a glycosylphosphoinositide (GPI) linkage confers the molecular address, while intracellularly dual acylation, particularly by palmitate but also myristate performs the same function (Brown and London, 1998). The few transmembrane proteins that are also microdomain components are important because many are the receptors for extracellular signals. They may also contain a targeting motif – for example many members of the cell adhesion molecule superfamily of proteins are dually palmitoylated at the region where the cytoplasmic tail emerges from the inner leaflet of the membrane (Little *et al.*, 1998). Alternatively their tight association with microdomains may be indirectly mediated. For example, the high and low affinity neurotrophin receptors (Trks A, B, C and p75 respectively) apparently associate via extracellular interactions between glycosylated residues (Mutoh *et al.*, 2000). Our appreciation of the role microdomains play in cell signaling arose from studies directly linking integrity of the structure and the signaling process itself: That removing membrane cholesterol or interfering with its biosynthesis causes downregulation of signal transduction has now been documented for diverse pathways in a wide variety of both eukaryotic and prokaryotic cells, emphasizing that microdomains are an ancient strategy to generate signal transduction modules in both polarized and non-polarized cells (Simons and Toomre, 2000; Smart *et al.*, 1999; Draber and Draberova, 2002; Yin and Janmey, 2002). Moreover motile cells in particular are giving us insights into how microdomain mediated signaling at the membrane is integrated with cytoskeletal responses: In motile cells extracellular signals change cell behavior by affecting cell morphology. The growing neuron provides an extreme example – the highly motile growth cone at the tip of a growing axon acts as a 'response organelle' encountering significant numbers of highly diverse signals on its journey to identify its synaptic target. It responds to each either by attraction – the elaboration of ruffling lamellae and extension of filopodia that coordinate to promote axon growth, or repulsion – the collapse of lamellae and filopodia that prevents axon growth. Both responses are generated by membrane-associated complexes that differentially regulate the actin cytoskeleton. However, even cells that are not polarized to the same extent like T cells, rely on communication with the cytoskeleton to respond to extracellular signals: Antigen presentation requires that the actin cytoskeleton orchestrates redistribution of receptor complexes as a prelude to the stimulation of T cell proliferation (Magee *et al.*, 2002). In both cases we are now beginning to understand how this communication is achieved.

2. MICRODOMAINS (RAFTS) AND THEIR ASSOCIATION WITH THE CYTOSKELETON

The first cholesterol and glycosphingolipid (ganglioside)-rich microdomain structures to be studied were identified ultrastructurally as flask shaped invaginations of the plasma membrane and hence called caveolae

(Anderson and Jacobson, 2002). Their resistance to cold non-ionic detergent extraction, particularly with Triton-X100, suggested that the major lipids, cholesterol and sphingolipids, must be highly structured (liquid ordered). Antibodies to the major protein component, the cholesterol binding protein caveolin (Rothberg *et al.*, 1992), immunoprecipitated caveolae in the absence of detergent thereby identifying other molecular constituents without confounding detergent-induced aggregation (Schnitzer *et al.*, 1995). It rapidly became clear that morphologically identifiable caveolae are absent from many cell types that nonetheless express caveolin and conversely that caveolin is absent from many detergent resistant membranes that are nonetheless highly enriched in cholesterol and gangliosides. In the absence of defining ultrastructure biochemical approaches, such as operationally defined extraction with non-ionic detergents, were enlisted to investigate microdomain structure and function, but the innate deficiencies of this approach have caused many hiatuses in progress. As a result, many of the physico-chemical and biological characteristics of non-caveolar microdomains remain controversial even after 20 years investigation. An eventual consensus has emerged: the term caveola should be restricted to microdomains with specific morphological characteristics and defined lipid and protein constituents (caveolin). Biochemical isolation which produces fractions without these morphological characteristics and/or without caveolin have variously called detergent-resistant membranes (DRMs), glycosphingolipid enriched membranes (GEMs), detergent-insoluble glycolipids (DIGs) or lipid rafts (Simons and Toomre, 2000). I will use the term raft here.

Conventionally, and in the absence of morphological characteristics, rafts have been operationally defined on the basis of insolubility in cold non-ionic detergents (TX-100 or TX-114) which solubilize non-raft membranes (Brown and Rose, 1992; Fra *et al.*, 1994; Friedrichson and Kurzchalia, 1998). Separation of rafts after TX-100 extraction is usually achieved by density gradient centrifugation utilizing either Optiprep or sucrose after which the cholesterol-rich rafts float with a buoyant density of about 1.09–1.13 g/ml. Other membrane proteins normally have a buoyant density of around 1.3 g/ml in the absence of lipid. Any detergent resistant cytoskeleton that is also not associated with lipid will pellet (Timasheff, 1995). Separation of rafts after TX-114 extraction is usually achieved by raising the temperature to 37°C to induce phase separation, after which solubilized transmembrane proteins partition into the detergent phase whereas raft components are collected from the pellet. Both methods are highly operational and thus are beset with potential problems, the chief one being that the detergent to protein ratio must be determined and held constant at 5–10:1. Below this, solubilization of non-raft transmembrane proteins may be incomplete and they may remain artifactually associated with the lipid structures by virtue of mutual hydrophobicity. The propensity of detergent extraction to induce artifactual aggregation cannot be overemphasized: For instance, *in vivo* spectroscopy of individual rafts shows clusters of ~3500 sphingolipid molecules containing ~20 protein components that are between 10–300 nm (average 50 nm) in diameter, well below the limit of resolution of light microscopy (Anderson and Jacobson, 2002) – and yet fluorescence light microscopy is commonly employed to infer raft structure and function. As we shall see, some biological processes legitimately induce raft clustering but distinguishing between this and artifactual aggregation of raft components is challenging in experiments that rely solely on a detergent extraction of a cell population, especially when highly polarized cells are used. Extraction of live cells on cover slips limits the extent of aggregation that can occur.

2.1. Structurally distinct classes of rafts

We are now beginning to appreciate that rafts are structurally heterogeneous, and the experimental challenge becomes how to separate different raft components from each other. This can be approached in two ways: The first, paying careful attention to biochemical differences especially in buoyant density and solubility in conventional non-ionic detergents, is beginning to be exploited. In this regard many published methods leave the impression that raft material floats as a single band after centrifugation. This is not true, especially when a highly polarized cell is used as a source of detergent resistant material, unless a high-density cushion is included in the gradient. In fact several raft-containing bands are usually generated with a high degree of reproducibility. For instance, we have found 5 such bands, each differing with respect to the activation status of cell-adhesion molecule-mediated signal transduction mechanisms and also the association of cytoskeletal proteins (He and Meiri, 2002), are generated when a subcellular fraction of growth cones isolated from neonatal rat brain is extracted with TX-114 and fractionated on a sucrose gradient. All the fractions contain bonafide raft components such as the GPI and transmembrane associated neural cell adhesion molecules (NCAM120 and NCAM 140 respectively) however only the lightest fraction contains caveolin. In this fraction NCAM signal transduction is activated via Fyn kinase, another raft component. In contrast, in heavier fractions which do not contain caveolin and in which Fyn kinase is inactive, NCAM signal transduction mediated via PKC is upregulated. The heaviest raft fractions are also seen to be associated with actin. Similarly, rafts with different fractionation characteristics are also generated when activated bovine neutrophils are extracted with TX-100 and fractionated with Optiprep. In this case Nebl *et al.* collected detergent-resistant rafts on a sucrose cushion and then subjected the mixture to progressively shallower gradients that emphasized the small differences in buoyant density that correspond to structural differences in composition (Nebl *et al.*, 2002). The lighter fraction (DRM-L) had a buoyant density between 1.09–1.13 g/ml similar to rafts containing GPI anchored proteins, heterotrimeric G proteins and Src family kinases, whereas the slightly heavier faction (DRM-H) fraction had a buoyant density of 1.15–1.18 g/ml. It too contained cytoskeleton-associated material. Importantly both we and Nebl *et al.* took pains to establish and maintain the detergent: protein ratio, and we also both verified the ultrastructure of the DRM/cytoskeleton fraction: In both cases, electron microscopic analysis showed fragments of membrane that had resisted detergent solubilization to which fibrils of cytoskeleton (probably the juxtamembrane actin-based membrane skeleton) were associated. Both approaches therefore show a direct structural association between rafts and the cytoskeleton. Whether the increased complexity of the growth cone rafts reflects an increased functional complexity of the growth cone compared with the neutrophil is an intriguing possibility that remains to be explored.

A second approach to identify different raft subfractions is to substitute other detergents for TX-100, particularly those that are less hydrophobic. For example, extracting T-cells with the non-ionic detergent Brij 58 (no detergent:protein ratio supplied alas) produced cholesterol-enriched fractions that differed with respect to activation of the Src family kinase Lck (Giurisato *et al.*, 2003). Significantly the active Lck fraction also contained considerably more of the TCR coreceptor CD3ζ whose association with rafts has been controversial. The results suggest that different subpopulations of rafts exist. A second possibility is that single proteins may be surrounded by a shell of a few lipid molecules, and these may cofractionate with larger rafts under some circumstances (Anderson and Jacobson, 2002).

3. HOW LIPIDS OF INNER LEAFLET RAFTS REGULATE MEMBRANE/CYTOSKELETAL INTERACTIONS

3.1. PIP2 in membrane rafts

The membrane lipids of inner leaflet rafts have been less studied than their extracellular counterparts, however, recent evidence implicates membrane associated phosphoinositides (PPIs), of which PIP2 is one, in raft/cytoskeletal interactions (reviewed excellently by Yin and Janmey, 2002). The PPIs are already known to play important roles in regulating cytoskeletal dynamics that impacts on a number of fundamental cellular functions such as membrane trafficking, growth and apoptosis (Sechi and Wehland, 2000; Martin, 2001; Takenawa and Itoh, 2001). In particular agonist sensitive regulation of PIP2 homeostasis, a major means of cytoskeletal regulation, occurs specifically rafts. This is due in large part to the enrichment in rafts of the enzymes PIP5KI and PLC which synthesize and hydrolyze PIP2 respectively (Rozelle *et al.*, 2000). In general, stimulation of PIP5KI to increase synthesis of PIP2 initiates actin assembly. For instance, overexpression of PIPKI induces cholesterol dependent formation of actin comet tails that extend preferentially from Golgi-derived vesicles enriched in rafts. Conversely when rafts are disrupted the number of PIP5KI-induced actin comets is dramatically reduced (Rozelle *et al.*, 2000). In a similar vein, hydrolysis of PIP2 by PLC, causes immediate downregulation of the PIP2 signal and, in general, depletion of PIP2 triggers actin depolymerization (Takenawa and Itoh, 2001). After agonist stimulation cytoskeletal regulation is crucially dependent on the cyclical regulation of PIP2 homeostasis either directly because of the complement of PIP2-regulated actin regulatory proteins in individual microdomains, or indirectly via the metabolites generated upon hydrolysis (Ins(1,4,5)P3 and diacylglycerol) which are themselves second messengers for PKC-mediated signaling pathways and which can affect the cytoskeleton independently. Regulation of homeostasis is not the only role that rafts play in PIP2 function however since PIP2 can accumulate in rafts by electrostatic interactions that also regulate cytoskeletal behaviors (Laux *et al.*, 2000; McLaughlin *et al.*, 2002). In sum rafts have now emerged as platforms for the integration of PIP2 signaling and actin polymerization.

3.2. Stimulating PIP2 turnover in rafts

Raft involvement in the activation of PIP2 synthetic and metabolic enzymes is a key way in which cytoskeletal regulation occurs. PIP5KI is recruited to rafts by two families of GTPases, themselves raft components. First, PIP5KI can bind to both rho and rac and be activated by Rho-kinase (Chong *et al.*, 1994; Oude Weernink *et al.*, 2000; Chatah and Abrams, 2001). In platelets this Rac1 stimulation of PIP2 synthesis induces actin polymerization directly (Tolias and Carpenter, 2000). Alternatively PIP5KI can also bind to and be activated by the GTPase Arf-6, which can also stimulate comet formation from vesicles and appears to be primarily involved in vesicle recycling, (Brown *et al.*, 2001). It is therefore thought that the transient increases of PIP2 seen at sites where dynamic membrane movement and cytoskeletal remodeling converge, such as the ruffling edges of lamellae, may be dependent on rac activation of PIP2 turnover, whereas the increases seen in phagocytosis may be due to Arf-6. Likewise PLC can also drive cytoskeletal rearrangements through downregulation of PIP2. For example, in activated T cells, when PLCγ is recruited to rafts and activated by the Src family kinase Lck, PIP2 function is downregulated and a there is a localized release of PIP2-binding proteins. (Veri *et al.*, 2001).

3.3. PIP2/protein interactions

Although PIP2 influences the behavior of many proteins through direct interactions, it is becoming increasingly obvious that so called 'PIP2 binding motifs' on proteins such as plextrin homology (PH) domains do not necessarily impart PIP2 binding *in vivo*. Rather its seems that a more common feature of protein/PIP2 interactions is a strong electrostatic interaction between them (McLaughlin *et al.*, 2002). The unique role of PIPs in such interactions derives from the fact that with 3 to 5 negative charges per 702 Å area, rafts enriched in PIP2 may transiently form local areas of negative surface potential far greater than other parts of the cell membrane. This localized electrostatic docking would tend to attract PIP2-binding proteins with appropriate characteristics (Yin and Janmey, 2002).

3.4. Where does PIP2 regulation of actin interactions with membrane occur?

From pioneering studies in erythrocytes it was determined that attachments between actin networks and the plasma membrane that reinforced substrate adhesion and enabled membrane deformations were consolidated by proteins contained within the cortical membrane skeleton (Bennett and Gilligan, 1993). It now turns out that many of the membrane skeleton proteins that crosslink actin with transmembrane receptors, are activated by PIP2. Recent studies clarifying that rafts can interact with the cytoskeleton suggests that membrane skeleton/actin linkages may not provide a parallel raft-independent context for PIP2 in the inner leaflet to cause a general drive to polymerize and crosslink actin: When ultrastructural characterization of so called 'membrane skeleton' fractions (often like rafts defined purely operationally by detergent resistance) was carried out, they were often associated with detergent resistant membrane patches and conversely, subfractions from raft preparations yield material that structurally resembles the membrane skeleton (Meiri and Gordon-Weeks, 1990). Thus a significant proportion of the regulatable cytoskeletal associations with the membrane are now expected to occur within the context of rafts, as we will demonstrate below.

3.5. PIP2 regulation of actin dynamics

PIP2 affects actin polymerization by regulating the behavior of proteins that control actin treadmilling (net polymerization at the (+) end and net depolymerization at the (−) end), seen as ruffling of the membrane, particularly gelsolin. After stimulation (for instance by growth factors) Ca^{2+} dependent activation of gelsolin increases treadmilling by severing actin filaments to generate new (+) end nuclei, which stimulates polymerization. In response, depolymerization from the (−) ends and hence treadmilling and as a consequence cell membranes ruffle (Barkalow *et al.*, 1996). In gelsolin-null cells growth factor-stimulated ruffling ceases because actin treadmilling is reduced. A result of the reduction in actin depolymerization is generation of excess actin stress fibers that makes cells migrate more slowly (Witke *et al.*, 1995). PIP2 is thought to inhibit treadmilling – cells overexpressing PIP5KI (and therefore PIP2) also cannot ruffle their membranes in a Gelsolin-dependent response to growth factor stimulation, and also have more stress fibers (Yamamoto *et al.*, 2001).

A second PIP2-mediated mechanism that regulates lamellal growth involves so-called dendritic nucleation at preexisting filaments that is regulated by the Arp2/3 complex. In unstimulated fibroblasts components of the Arp2/3 complex are cytoplasmic, but on stimulation they become enriched in the lamellae (Machesky, 1997) where they can intercalate into an existing filament nucleating branching growth of filaments at 70° angles. The complex also caps the (−) end inhibiting treadmilling (Mullins *et al.*, 1998). Arp2/3 promotes (+) end polymerization in concert with profilin and capping protein. Conversely depolymerization at the branch point is signaled after (−) end hydrolysis of bound ATP by release of phosphate (Blanchoin *et al.*, 2000). By itself the activity of the Arp2/3 complex is quite low, but it can be stimulated by the Wiskott-Aldrich syndrome (WASP) family proteins, WASP itself, neuronal WASP (N-WASP) and Scar/WAVE. Under basal conditions WASP family proteins are autoinhibited and this is relieved by two major pathways involving either Cdc42 signaling which in turn activates PIP2, or tyrosine phosphorylation which creates SH3 domains. These activators act synergistically to stimulate WASP – for example, PIP2 and Cdc42 together increase N-WASP activation by more than 300-fold *in vitro* (Prehoda *et al.*, 2000). The role of PIP2 in WASP/Arp2/3 mediation of actin nucleation has also been deduced from the formation of actin comets on vesicles. PIP2 activates NWASP and the comets seen in PIP5KI overexpressing cells are N-WASP dependent. PIP5KI itself is enriched at the (+) end head of the comets (Rozelle *et al.*, 2000). The responsiveness of WASP proteins makes them very effective in generating an extended range of responses to specific signals, however there is some degree of redundancy between them, so N-WASP null mice respond to growth factor stimulation with (+) end actin nucleation and extend filopodia and lamellae (Snapper *et al.*, 2001).

3.6. PIP2 regulation of other cytoskeletal proteins

Filamin and alpha-actinin both crosslink actin with membranes and have been shown to indirectly associate with rafts (Dyson *et al.*, 2001). The behavior of alpha-actinin is straightforward – its binding to PIP2 unmasks and activates the actin binding motif in a mechanism reminiscent of N-WASP (Fukami *et al.*, 1992). Filamin is more complicated. Its ability to cross-link actin is inhibited by its interactions with PIP2, but the *in vivo* consequences of the inhibition is also dependent on its strong binding to rafts in a PIP2 independent fashion. Hence the destabilization of actin binding to membranes requires both PPI turnover and PIP2-independent regulation of filamin to occur in concert (Goldmann, 2001). Like alpha-actinin, Ezrin/radixin/moesin (ERM) proteins bifunctionally bind actin filaments to transmembrane proteins (Hirao *et al.*, 1996). In the basal state self-association inactivates the actin- and membrane binding sites, but stimulation of PIP2 production mediated by Rho signaling causes PIP2 binding to ERM which reveals the membrane binding sites. Many of the interactions are in rafts. For instance ezrin binds to ICAM (Heiska *et al.*, 1998), while moesin binds to CD44, both of which are raft components (Neame *et al.*, 1995; Tilghman and Hoover, 2002). In both cases these interactions potentially provide a platform for F-actin association with rafts (Yonemura *et al.*, 1998). Like ezrin, talin increases actin association with membranes and is activated by PIP2 (Martel *et al.*, 2001). Conversely, the binding of actin to the vinculin tail is inhibited by PIP2. The extent to which these interactions rely on raft integrity remains to be seen. Finally, gelsolin null cells have provided insights into interrelationships between cytoskeletal regulatory mechanisms. Gelsolin null cells have increased Rac expression, which in turn stimulates PIP2. This is turn stimulates ERM protein binding to actin facilitating assembly of the abnormally long actin filaments into thick stress fibers. (Witke *et al.*, 1995; Azuma *et al.*, 1998).

4. LESSONS FROM T CELLS ABOUT THE ROLE OF RAFTS IN MEMBRANE CYTOSKELETAL COMMUNICATION

The notion that rafts form membrane platforms or modules that mediate cytoskeletal responses to signaling has been perhaps most exhaustively demonstrated in T cells (there are many excellent reviews, for example (Alonso and Millan, 2001). It will therefore be of value to examine the underlying principles of cytoskeletal regulation that have emerged. When a T cell encounters a antigen presenting cell (APC), movement of the T cell receptor (TCR) to the T-cell/APC interface can only be induced if the major T cell costimulatory receptor CD28 is also engaged and activated, and if GPI linked CD4 is cross linked. This crosslinking causes formation of a juxtamembrane 'actin cap' that is coincident intracellularly with the extracellular activation site (Wulfing *et al.*, 1999). Thus the relationship between concentration of

T cell receptors at the extracellular membrane and reorganization of the cytoskeleton intracellularly occurs from the very beginning. The next event is formation of a macromolecular activation complex opposite the APC that has variously been called the supramolecular activation cluster (SMAC) or immunological synapse (IS) and that couples the T-cell receptor with signal transduction machinery. The IS integrates signaling cascades originating from the simultaneous activation of a wide variety of receptors and is essential for both the amplification and maintenance of T cell activation. It has become clear that the IS represents an extremely complex assemblage of rafts, each playing a specific role in the goal of initiating proliferation of the T cells. We now understand quite well that this concentration of rafts at the antigen presenting site and their coalescence into the much larger synaptic complex is dependent on interactions with the cytoskeleton (Janes *et al.*, 1999; Alonso and Millan, 2001).

In steady state T cell membranes, prior to antigen activation, some components of the IS such as the CD3ζ coreceptor, GPI linked CD4 and certain proteins that will be involved in early signaling events such as the inner leaflet Src family kinases Lck and Fyn, are all associated with small yet distinct rafts that are dispersed on the cell surface (Montixi *et al.*, 1998; Xavier *et al.*, 1998). Likewise the transmembrane raft protein LAT that is crucial to assembly of the IS is targeted to inner leaflet rafts by dual palmitoylation and acts as a linker with outer leaflet rafts (Zhang *et al.*, 1998a; Zhang *et al.*, 1998b). The distribution of T cell receptor complex itself (TCR) is less straightforward: On one hand it has been solubilized by the ionic detergents conventionally used to prepare rafts, on the other, either colocalization immunocytochemistry in the absence of detergent or substituting less hydrophobic detergents (such as the non-ionic polyoxyethylene ether Brij series) to extract the membranes clearly demonstrates that the TCR can codistribute with other raft components. The consensus is that the T cell receptor only 'weakly' associated with rafts (Montixi *et al.*, 1998; Janes *et al.*, 1999; Kosugi *et al.*, 1999). But what does 'weakly' mean in these terms? Two possibilities suggest themselves: first that bonafide liquid ordered rafts are in fact heterogeneous with respect to their lipid content as well as their protein constituents, and the T cell receptor may associate with rafts with different lipid compositions than those resistant to Triton extraction. A second notion is that individual transmembrane proteins, or presumably complexes like the TCR, may exist cocooned within a shell consisting of fewer than 100 molecules of raft-like lipids. This organization would confer both low buoyant density (for instance the buoyant density of a typical membrane protein would fall from about 1.3 g/ml to around 1.03 g/ml by the association of around 80 molecules of cholesterol/sphingolipid) and a propensity to associate with more substantial rafts if clustering is induced extracellularly. In sum then, both quantitative and qualitative differences in rafts may regulate their constituents and their behavior (see also van der Goot and Harder, 2001).

4.1. Dual role of the actin cytoskeleton during T-cell activation

Reorganization of the actin cytoskeleton is a central feature of T-cell activation. After the TCR has been engaged and clusters opposite the APC, rafts concentrate at the antigen presenting site and then coalesce into a much larger synaptic complex (Janes *et al.*, 1999). This coalescence of rafts which potentiates the assembly of the macromolecular IS complex crucially requires engagement between the IS and the cytoskeleton (Miceli *et al.*, 2001). Thus in general terms, treatment of T cells with drugs that interfere with different aspects of actin treadmilling or prevent activation of Rho family GTPases result in defective cap formation and inhibition of both early and late events in the TCR signaling cascade. The basic relationship between clustering of the TCR, rafts and actin was demonstrated by (Rodgers and Zavzavadjian, 2001) who used a Lck kinase sequence (Lck-NT) to provide a dual palmitoylation, raft-targeting motif for a GFP construct. When Jurkat T-cells were stimulated with anti-CD3 antibodies to induce clustering of the TCR the Lck-NT-GFP construct colocalized with actin in the membrane caps, and this colocalization was inhibited when the cytoskeleton was depolymerized, and preventing cap formation.

The first encounter with the APC ligates and activates both the TCR and the CD28 coreceptor, and this parallel activation (Viola *et al.*, 1999) causes phosphorylation of the Src family kinase Lck and the tyrosine kinase ZAP 70 which induces actin polymerization (Salojin *et al.*, 1999). This stimulates the first reorganization of the actin cytoskeleton to form a cap and it does not require raft-mediated regulation of actin dynamics. Activation of CD28 by the APC also induces phosphorylation of the Rho family GTPase exchanger Vav-1 which is constitutively associated with the TCR component CD3ζ, which affects actin dynamics through its guanine nucleotide exchange activity on Rac1 (Michel *et al.*, 2000). This in turn further boosts TCR signals through a synergistic cooperation between Vav-1 and the linker protein SLP-76 that is also recruited to the T cell–APC interface upon activation of the TCR (Wu *et al.*, 1996). Vav is also located either outside rafts or is only weakly associated with them. It seems likely that the first wave of Vav phosphorylation that initiates F-actin reorganization in turn induces aggregation of small rafts circumferentially around the TCR, which helps to stabilize them and promote their further coalescence. Hence at early stages in TCR activation regulation of actin is raft independent. However during amplification of the T cell response lipid raft reorganization does drive actin regulation (Gomez-Mouton *et al.*, 2001; Valensin *et al.*, 2002). The second stage of T-cell activation is achieved by coalescence of smaller individual rafts that recruit raft-associated tyrosine kinases to the TCR complex and thereby sustain signaling. For instance Src family kinases, particularly Lck and Fyn, are recruited into the TCR complex via cooperation between activated CD28 and CD4. CD28 then induces Lck autophosphorylation sustaining its activation (Holdorf *et al.*, 2002). This initiates a second wave of Vav phosphorylation, most likely by Fyn, that stimulates actin reorganization (Valensin *et al.*, 2002). A positive feedback loop is set up because Vav-independent actin reorganization can also stimulate Vav phosphorylation in parallel: here coengagement of GPI linked CD48 with the TCR promotes phosphorylation of TCRζ enhancing F actin binding directly to CD3ζ thereby inducing cytoskeletal reorganization that is also independent of rafts (Moran and Miceli, 1998). Even though they are not directly involved in these early stages, raft-mediated regulation does play an indirect role – cyclodextrin treatment inhibits the inducible tyrosine phosphorylation of TCRζ thereby preventing its association with actin (Moran and Miceli, 1998). Another route by which rafts indirectly initiate reorganization of the actin cytoskeleton occurs after cross-linking of GPI linked CD59 or GM1 by cholera toxin. In this instance F-actin becomes enriched in the patches formed by rafts (Miceli *et al.*, 2001).

A major question about T cell activation is how the proximal and distal portions of the TCR signaling pathway are linked so that the initial stimulation mediates a sustained response. A likely candidate is the transmembrane linker protein LAT which, by virtue of dual palmitoylation in its cytoplasmic domain, acts to bridge external and internal leaflet rafts so that they can function in concert (Zhang *et al.*, 1998). After TCR engagement LAT becomes rapidly tyrosine phosphorylated so that it directly binds the Src homology 2 (SH2) domains of other proteins, including PLCγ1, Grb2, Gads, Grap, 3BP2, and Shb, and indirectly binds SOS, c-Cbl, Vav, SLP-76, and Itk. LAT thereby plays a major role in amplifying the signaling associated with the rafts (Tanimura *et al.*, 2003). Specific exclusion of tyrosine phosphatases such as CD45 from rafts potentiates the amplification effect. Notably, in the absence of LAT, TCR engagement does not lead to activation of distal signaling events, including actin reorganization, thus supporting the notion that during the sustained portion of the response it is raft dynamics that influences actin reorganization (Wange and Samelson, 1996). This relationship between raft clustering and actin polymerization is also crucial for the eventual downstream functional consequences of T cell activation. Thus PKCθ, which is required to activate AP-1 and NFKB leading to IL2 production that precedes proliferation is located in the core of the IS in antigen stimulated cells and undergoes Vav/Rac dependent translocation to the membrane and cytoskeleton when T cells are activated (Gomez-Mouton *et al.*, 2001; Villalba *et al.*, 2001).

4.2. Control of actin polymerization at the immunological synapse

It has been proposed that the physical and functional link between lipid rafts and the actin cytoskeleton is provided by the WASP proteins that control *de novo* nucleation of actin branches on preexisting filaments through regulated activation of the seven-subunit Arp2/3 complex (Higgs and Pollard, 2000). The crucial role of WASP in T cells was demonstrated from Wiskott-Aldrich syndrome (WAS) patients and WASP-deficient mice whose T cells proliferate poorly after TCR/CD28 activation and whose capacity to cluster the lipid raft marker GM1 and to upregulate GM1 cell surface expression is impaired. T cell proliferation and lipid raft clustering can be restored by retroviral transfer of the WASP gene (Klein *et al.*, 2003). These results identify a potential mechanism underlying the T cell defect affecting WAS patients (Molina *et al.*, 1993; Gallego *et al.*, 1997; Zhang *et al.*, 1999; Snapper *et al.*, 2001). WASP integrates multiple signals to promote the formation of actin networks in contact with membrane rafts, thus sustaining assembly of the IS. However WASP is neither a transmembrane protein nor lipid-modified so its association with rafts presumably occurs via non-covalent interactions with lipids and/or proteins. In resting T cells, WASP is not detected in isolated rafts, but a few seconds after TCR/CD28 activation, WASP (and its activators Vav-1 and PIP2) can be detected in rafts at the site of contact between the T cell and the APC, (Krause *et al.*, 2000; Cannon *et al.*, 2001) indicating that it must act very early during activation. WASP recruitment to the T cell–APC interface involves the ZAP-70 kinase, which recruits a complex of CrkL WASP and its inhibitory interacting protein, WIP (Martinez-Quiles *et al.*, 2001). Autophosphorylation of ZAP-70 generates the phosphotyrosine(s) that are essential for CrkL docking. Then PKCθ-mediated phosphorylation of WIP allows WASP to dissociate, concurrently releasing it from inhibition and allowing it to be activated by membrane-bound Cdc42 and thereby initiate Arp2/3 complex-dependent actin polymerization. ZAP-70-deficient cells that also expressed dominant-negative CrkL, WIP−/− and PKC (−/−) T cells all failed to accumulate F-actin at the T cell–APC interface, or to increase total cellular F-actin content following TCR engagement (Anton *et al.*, 2002). These data suggest that WASP plays a central role in promoting the organization of actin at the IS. WIP also binds and stabilizes actin filaments directly and therefore plays an additional independent role which helps stabilize the IS. Thus the failure of WIP (−/−) T cells to increase F-actin following TCR ligation is due both to their inability to recruit WASP to ZAP-70 and their inability to bind F-actin directly (Ramesh *et al.*, 1997; Anton *et al.*, 2002).

5. LESSONS FROM GROWTH CONES ABOUT THE ROLE OF RAFTS IN MODIFYING MEMBRANE CYTOSKELETAL COMMUNICATION TO FACILITATE SPECIFIC RESPONSES TO EXTRACELLULAR SIGNALS

Although antigen presentation can induce a specific and coincident reorganization of membrane proteins, signal transduction machinery and the underlying actin cytoskeleton opposite the APC the T cell is not stably polarized. In contrast neurons are the most highly polarized cells in the body, and this polarization enables a number of highly complex functions. For instance during development a highly motile structure, the growth cone, is specifically generated at the tip of the growing axon. Each growth cone is subject to a myriad of extracellular signals that direct its trajectory through the embryo to find an appropriate target. The precise regulation of growth cone actin dynamics in response to these signals is used to sculpt nervous system cytoarchitecture with the utmost precision. Understanding the membrane events that initiate coordination between signal transduction and the cytoskeletal dynamics that are the necessary prelude to the growth cone response represents a major cell biological challenge.

5.1. The actin cytoskeleton in growth cone function

As in other migrating cells, the growth cone actin cytoskeleton takes on two major forms: The leading edge is characterized by a broad, flat lamella that is produced by the polymerization of branched actin filaments underneath the cell membrane. In contrast to lamellal actin of keratocytes and fibroblasts, which is organized, by virtue of WASP/Arp2/3 activity, as a network of short branches that form with barbed ends oriented towards the edge of the migrating cell, the growth cone lamella is less well organized (Gomez-Mouton *et al.*, 2001). Actin filaments are longer, less branched, and have barbed ends with less uniform orientation (Bear *et al.*, 2002). Radiating from the edge of the lamella are long, finger-like projections comprised of long bundles of 6–12 actin filaments called filopodia (Lewis and Bridgman, 1992) It is the membrane of the filopodia that is thought to contain the receptors for the extracellular signals that direct axon pathfinding (Gomez-Mouton *et al.*, 2001). Filopodia therefore are likely to be an important site for regulation of membrane cytoskeletal communication, and identifying actin regulatory proteins that are enriched in filopodia is a high priority in growth cone research.

The growth cone also differs from non-neuronal cells by the absence of certain actin-based structures that are abundant in non-neuronal cells, especially in culture. Thus both stress fibers and the focal adhesions to which they are attached are both absent from growth cones.

The dynamic motility of the growth cone is regulated by *de novo* generation of lamellae and filopodia as growth cones advance (Welnhofer *et al.*, 1997). Disruption of either, but especially filopodia may result in loss of growth cone function reflecting their significant roles in signal transduction (Zhou and Cohan, 2001). Our understanding of the molecular basis for growth cone regulation lags behind our understanding of T cell biology, and it is difficult to draw precise parallels with molecular precision. In part this reflects the sheer technical complexity of functionally dissecting such a highly polarized source in a biochemical fashion, and in part it reflects the fact that the growth cone has apparently superimposed several levels of regulation to accomplish the spatial and temporal specificities of responses required during target identification. Understanding how the complexities of growth cone signaling are integrated in a spatio-temporal fashion represents one of the true final frontiers in cell biology.

5.2. An important role for rafts in signaling at the growth membrane

After many years during which neurobiologists resisted the notion that rafts had functional relevance to neurons, an increasing body of evidence confirms their significance in generating signaling platforms, just as they do in non-neuronal cells. Here we illustrate that importance by discussing the varying behaviors of three different types of receptor tyrosine kinase – a family that are constitutive raft components; a family that are recruited to rafts following *cis* stimulation and a family recruited by *trans* stimulation.

5.3. Receptor tyrosine kinases that are constitutive raft components – neurotrophins and ephrins

Most growth factors signal by binding to, and activating, receptor tyrosine kinases (RTKs). Many receptors for non-neuronal growth factors – for instance EGFR (epidermal growth factor receptor) (Mineo *et al.*, 1996) PDGFR (platelet-derived growth factor receptor) (Liu *et al.*, 1996), and the insulin receptor (Yamamoto *et al.*, 1998) – are located in rafts. Upon activation RTKs autophosphorylate on tyrosine residues, thereby creating binding sites for SH2- and PTB (phosphotyrosine binding domain)-containing proteins. The activated RTKs form signaling complexes with docking proteins that are concentrated in rafts thereby stimulating multiple signaling cascades (Schlessinger, 2000). It is no surprise, therefore, that receptors for the neuronal growth factors – the neurotrophins – behave similarly. The neurotrophins, are represented by nerve growth factor (NGF), brain-derived neurotrophic factor (BDNF), neurotrophin 3 (NT3) and neurotrophin 4 (NT4) regulate neurite outgrowth and neuronal survival during development by interacting with two types of receptor. The Trk A, B and C tyrosine kinases and the low-affinity receptor p75NTR (a member of the tumor necrosis factor receptor superfamily which has no intrinsic kinase activity of its own but which, when coexpressed with appropriate Trk receptors), increases neurotrophin binding affinity (Friedman and Greene, 1999). There is good evidence that the majority of Trk signaling occurs in rafts: TrkA and B and p75NTR are all highly enriched in low buoyant density rafts prepared from synaptic plasma membranes (Bilderback *et al.*, 1999) and tyrosine phosphorylation of TrkA in rafts increases when ligands bind. NGF binding to p75NTR and TrkA occurs mainly in rafts and NGF can also be chemically crosslinked to p75NTR in rafts (Huang *et al.*, 1999). Virtually all activated TrkA is found in rafts, and in addition many of the intermediates in the signaling cascade activated by Trk receptors are also present, including PI3K, phospholipase C (PLC), Shc, Grb2, Ras, Raf-1 and mitogen-associated protein kinase (MAPK), and they can be coimmunoprecipitated by TrkA from rafts (Simons and Toomre, 2000).

In contrast with neurotrophins, which are secreted molecules, the extensive ephrin (ligand) and Eph (receptor) system mediates signaling by cell–cell contact, playing critical roles in regulating neuronal migration, axon guidance and boundary formation in the embryo (Dodelet and Pasquale, 2000; Kullander and Klein, 2002). Ephrin ligands are either anchored to the plasma membrane via a GPI link (class A) or have a transmembrane domain followed by a short cytoplasmic domain (class B). They cannot stimulate Eph receptor signaling unless they are membrane anchored or artificially clustered. Membrane-bound ephrins that are engaged by their eph receptors can also transduce signals, thus ephrin/Eph signaling is bidirectional (Cowan and Henkemeyer, 2002). Both GPI-anchored and transmembrane ephrins, as well as the Eph receptors themselves are localized to rafts (Bruckner *et al.*, 1999) and several signaling events are critically dependent on this localization. For example, when GPI-anchored ephrin-A5 binds to and activates its cognate Eph receptor in rafts it is able to generate a signaling cascade that results in the recruitment and activation of the Src family kinase Fyn (Davy *et al.*, 1999). Likewise activation of ephrin-B1 promotes the recruitment of glutamate–receptor interacting protein (GRIP)1 and (GRIP)2 to rafts which results in raft coalescence and the recruitment of a GRIP-associated serine/threonine kinase (Bruckner *et al.*, 1999). The intracellular consequences of ephrin activation with respect to cytoskeletal communication are so far largely unknown. Cross-linking of ephrin A5 induces the redistribution of the cytoskeletal protein, vinculin to membrane/substrate contact sites and increases cellular adhesion to substrate (Davy *et al.*, 1999) but how that impacts on regulation of actin dynamics has so far not been investigated. EphrinB, which is also constitutively located in rafts also recruits GRIP adaptor proteins and a serine/threonine kinase to rafts (Bruckner *et al.*, 1999; Lin *et al.*, 1999) In this case it is known that the kinase phosphorylates GRIP, thereby initiating reverse signaling. Importantly, this process is regulated by binding of ephrinB to its Eph receptors: cross-linking with EphB2-Fc also results in the co-localization of ephrinB and GRIP to large raft patches (Lin *et al.*, 1999).

5.4. GDNF family receptors c-Ret and GFR: Receptor tyrosine kinases that are recruited to rafts

The glial cell line-derived neurotrophic factor (GDNF) ligand family, which includes GDNF, neurturin, persephin and artemin, promotes the survival and differentiation of a number of neuronal subpopulations (Airaksinen and Saarma, 2002). The GDNF receptor system has two components: a ligand binding molecule (GFR) which is localized in rafts by virtue of its GPI linkage and a RTK, c-Ret, which is the product of the c-Ret proto-oncogene (Treanor *et al.*, 1996). Only low levels of c-Ret are present in rafts in the resting state, but when a GDNF family ligand binds its cognate GFR, c-Ret is recruited to the complex causing receptor dimerization, autophosphorylation and intracellular signaling (Paratcha *et al.*, 2001). GDNF activation of Ret therefore parallels the CD28 activated translocation of the TCR to rafts that is required to initiate a cellular response. There appear to be two ways in which this recruitment occurs: Rapid translocation of c-Ret to the raft compartment does not depend upon activation of the c-Ret kinase and is likely to be mediated by the interaction of its extracellular domain with the GDNF/GFR1 complex (Paratcha *et al.*, 2001). However the rapid initiation of GDNF signaling is dependent on raft integrity, because cholesterol depletion reduces GDNF-dependent activation of MAPK and Akt kinases. Likewise, co-stimulation of Ret with GDNF in the absence of endogenous GFR and in the presence of a soluble form of GFR that lacks the GPI anchor diminishes downstream Ret signaling (Tansey *et al.*, 2000). This inability of soluble GFR1 to translocate Ret into rafts without a GPI anchor suggests 'cis' mediated activation. Ret mediated signaling is not completely abolished however because there is also a second means by which c-Ret can be induced to interact with rafts: Even in the absence of the GFR the c-Ret kinase domain is slowly activated and this can also result in translocation to rafts either via the adaptor protein FRS2 (FGFR substrate 2) (Kouhara *et al.*, 1997) or p60Src (Tansey *et al.*, 2000) both of which are raft components. In direct contrast with T cells however, translocation to rafts is required to initiate signal transduction downstream of Ret activation, but not for the maintenance of Ret activity. Moreover activated c-Ret can also be detected outside rafts, suggesting a dynamic equilibrium between the two membrane compartments. However, GDNF signal transduction through c-Ret located outside rafts is different from the signaling occurring inside. Interestingly FRS2 seems to associate with Ret specifically in rafts, whereas Shc interactions appear to predominate outside (Paratcha *et al.*, 2001). This demonstrates a strategy that can be used to diversify downstream signaling, a modification apparently adapted specifically by neurons.

5.5. Cell adhesion molecules, synergy between two pathways regulates axon outgrowth

Stable contacts between axons that are growing and between axons and their targets are crucial for both axon guidance and mature nervous system function (Walsh *et al.*, 1997) and are classically mediated in either a homophilic or a heterophilic fashion by molecules of the immunoglobulin superfamily (IgCAMs). Growing evidence supports an important function of rafts in the signaling mediated by cell adhesion molecules, particularly IgCAMs. Many adhesion molecules have GPI anchors. For example TAG-1 (transiently expressed axonal glycoprotein-1), NCAM-120 (neural cell adhesion molecule) F3/contactin and Thy-1 are all GPI-anchored and located in rafts. Both TAG-1 and NCAM are IgCAMs. Homophilic adhesion between IgCAMs can transmit intracellular signals. For example, antibody cross-linking of adhesion molecules such as TAG-1 activates the Lyn kinase within lipid rafts, promoting tyrosine phosphorylation events (Henke *et al.*, 1997; Kasahara *et al.*, 2000). Conversely, depletion of TAG-1 from rafts, by blocking the synthesis of glycosphingolipids, inhibits the ability of TAG-1 to promote phosphotyrosine signaling events (Kasahara *et al.*, 2000). However signal initiation is not limited to the GPI linked proteins, many cell adhesion molecules also exist in transmembrane forms that activate different transduction mechanisms. For instance the Neural Cell Adhesion molecule NCAM has two transmembrane isoforms (NCAM140 and NCAM180) in addition to GPI linked NCAM120. Both NCAM120 and NCAM140 can localize to rafts and rafts are required to initiate Fyn mediated signaling that occurs when either NCAM120 or NCAM140 are clustered. Fyn mediated activation of the Erk pathway is not sufficient to initiate neurite outgrowth however, because NCAM also signals via the fibroblast growth factor (FGF) receptors (FGFR1-4), a family of RTKs that are normally activated by FGF. Receptor dimerization causes tyrosine autophosphorylation, which then recruits other cellular targets coupling FGFR signaling to downstream effectors such as PLCγ which produces IP3 and DAG (Doherty and Walsh, 1996). It is thought that NCAM dimerizes and activates FGFR2 by interactions between complementary Ig domains. Like T cell receptors, FGFRs do not appear to be present at high levels in rafts (Davy *et al.*, 2000), however the adaptor protein FRS2 which also recruits Eph receptors to rafts has been shown to be crucially involved in the effects of FGF on neuronal differentiation suggesting that it may also recruit the FGFR to rafts (Kouhara *et al.*, 1997). One study showing that dual stimulation of the Fyn and FGFR2 mediated pathways was required to stimulate neurite outgrowth argued that FGFR signaling occurred outside rafts (Niethammer *et al.*, 2002). Nonetheless, stimulation of neuroblastoma cells with FGF2 resulted in tyrosine phosphorylation of several proteins, including the two Src family kinases Fyn and Lyn, within rafts (Davy *et al.*, 2000). That study also showed that FGF treatment induces the recruitment of the Src kinase substrate annexin II to rafts. One reason for these apparent discrepancies may be that functionally different populations of rafts are generated in different areas of the neuron, such as growth cone. Indeed when we isolated raft components from growth cone subcellular fractions, we found 5 fractions that were differentially enriched in different elements of the NCAM signaling pathway (He and Meiri, 2002). Thus activated Fyn appeared in the lightest fraction that was also enriched in NCAM120, whereas phosphorylated GAP-43, a substrate of the FGFR signal transduction pathway, appeared in a heavier fraction that was enriched in NCAM140 as well as the FGFR itself. The Fyn in this fraction was inactive. Whether this result reflects a complex organization of rafts such as is seem to cluster around the TCR during antigen presentation remains to be seen, but seems likely.

5.6. Rho GTPases and their role in regulating actin dynamics in the growth cone

A primary goal of extracellular guidance and motility cues is to initiate signaling cascades that regulate generation of lamellae and filopodia. The rho GTPases therefore represent a nodal point in growth cone responses since they coordinate the organization of actin dynamics that is directly regulated by Arp2/3 and ADF/cofilin. The rho GTPases Rac and Cdc42 cause the formation of lamellae and filopodia respectively (Ridley and Hall, 1992; Nobes and Hall, 1995), and are therefore key regulators of growth cone actin dynamics. The Rac activity that leads to generation of lamellae also appears to provide the tensile force for neurite elongation (Lamoureux *et al.*, 1997). In fibroblasts rho causes the formation of actin stress fibers, which cause cellular contraction and focal adhesion enlargement (Ridley and Hall, 1992). However neither of these structures are present in growth cones and the generally inhibitory role of rho in neurite outgrowth has been attributed to a retraction that is achieved via activation of myosins that generate force against the actin cytoskeleton (Narumiya *et al.*, 1997). In principle then, extracellular guidance cues may steer growth cones by activating rac and cdc42 to attract the growth cone by stimulating lamella and filopodia formation, or may stop neurite growth by activating rho. In reality the complexity of the process of neuronal guidance requires modifications to allow additional spatial and temporal specificity of responses. We will first discuss how the core components of the actin regulatory response have been adapted by neurons, and follow this by examining the modifications that have been superimposed on this basal regulation.

Activation of Rho GTPases by GTP binding is facilitated by specific guanine nucleotide exchange factors (GEFs) that facilitate GDP/GTP exchange (Bateman and Van Vactor, 2001; Vetter and Wittinghofer, 2001). Conversely, inactivation of rhoGTPases can be regulated by GTPase-activating proteins (GAPs) that stimulate GTP hydrolysis (Scheffzek *et al.*, 1998) or by GDP dissociation inhibitors (GDIs), which inhibit GDP release (Sasaki and Takai, 1998; Olofsson, 1999). Regulation of rac by specific GEFs can stimulate growth cone motility and promote neurite growth. Thus Tiam-1 causes neurite elaboration in neuroblastoma cells (Leeuwen *et al.*, 1997) and neurons of the weaver mutant mouse which lack Tiam-1 fail to migrate (Ehler *et al.*, 1997). Another rac-GEF, STEF, is expressed in the developing CNS and causes rac-dependent neurite growth in N1E-115 cells. Likewise Trio is also a rac-GEF, and Drosophila lacking Trio show defects in both central and peripheral axon guidance (Bateman *et al.*, 2000; Newsome *et al.*, 2000). Because of the opposite activities of rac and rho, GEFs may be bifunctional. Thus Ephrin-stimulated growth cone collapse (Wahl *et al.*, 2000) requires a rhoGEF, ephexin (Shamah *et al.*, 2001). GAP activity represents another mechanism for controlling neurite growth. RhoGAP increases neurite length in cortical neurons by inactivating rho (Threadgill *et al.*, 1997). GAPs can also be activated by extracellular signals. The Slit receptor Robo activates a novel GAP, srGAP1, which in turn inactivates cdc42 leading to neurite retraction (Wong *et al.*, 2001).

Identifying proteins that interact with rho GTPases is the key to understanding how rho, rac, and cdc42 lead to changes in actin polymerization. Two kinases in particular are important: PAK kinase localizes to membrane ruffles in response to rac activation (Dharmawardhane *et al.*, 1999) and becomes autophosphorylated upon binding rac or cdc42. The phosphorylated PAK then released activates downstream signaling cascades in many cell types (Knaus and Bokoch, 1998). Neurons contain a novel PAK isoform, PAK5, which supports neurite growth (Dan *et al.*, 2002). Another key kinase is Rho-associated kinase, or ROCK that interacts specifically with rho/GTP (Matsui *et al.*, 1999). ROCK activity is required for agonist-induced neurite retraction in neuroblastomas (Hirose *et al.*, 1998). Both PAK and ROCK directly link to the cytoskeleton via LIM kinase (LIMK) a target of both kinases. While LIMK is clearly a key regulatory link between rhoGTPase signaling and cofilin activity its function downstream of the rho GTPases is still somewhat murky. Thus inactive LIMK blocks rac-induced lamella formation (Yang *et al.*, 1998), as well as cdc42-induced filopodia formation while activation of LIMK by Rac and cdc42 signaling via PAK leads to neurite growth via LIMK phosphorylation (Edwards *et al.*, 1999). So far so good, but LIMK can also be activated by Rho signaling via ROCK, which activates LIMK by phosphorylating the same site (Maekawa *et al.*, 1999; Ohashi *et al.*, 2000). However the end result here is neurite retraction. The mechanism by which these disparate responses are obtained is not at all clear.

5.7. Actin polymerization kinetics

5.7.1. ADF/cofilin

It appears that only a few key components are required to simulate *in vivo* actin polymerization kinetics *in vitro*. These are actin-depolymerizing factor (ADF) and cofilin which depolymerize older parts of the actin network and allow monomers to recycle to the leading edge of the network for *de novo* generation of lamellae and filopodia, and Arp2/3 which nucleates new actin formation, and acts as a capping protein to restrict the length of growing filaments (Carlier *et al.*, 1997; Rosenblatt *et al.*, 1997). ADF is largely responsible for the *in vivo* kinetics of actin treadmilling in lamellae and is a key effector of signaling mechanisms that utilize rho GTPase activity. In neurons its activation is regulated by second messenger systems and growth factors (Meberg *et al.*, 1998), which stimulate its phosphorylation and translocation to ruffles where it can promote actin reorganization associated with lamella formation. Growth factors that activate ADF phosphorylation act via the rho GTPases and LIMK. Cofilin is also a LIMK substrate and LIMK is also the regulatory link between rhoGTPase signaling and cofilin activity, inactivating cofilin and leading to inhibition of lamellae extension (Zebda *et al.*, 2000). A model for regulation of actin dynamics that includes both LIMK and cofilin suggests that ADF/cofilin activity is regulated via phosphate turnover, and that ADF/cofilin needs to be in both activated and de-activated states at specific spatial and/or temporally distinct points in the growth cone.

5.7.2. Arp 2/3 WASP

The WASP proteins couple the rho GTPases to actin polymerization by forming a direct link between cdc42 and Arp2/3. Unsurprisingly then, the WASP/Arp2/3 complex also appears to be involved in neurite growth. Neuronal WASP (N-WASP) is concentrated at nerve terminals (Fukuoka *et al.*, 1997) and mutations in N-WASP or Scar block neurite growth in certain neurons (Banzai *et al.*, 2000). Arp2/3 complex proteins are concentrated with F-actin in NGF stimulated growth cones (Goldberg *et al.*, 2000) and when mutated result in abnormal axon development (Zallen *et al.*, 2002). Another potential activator of Arp2/3 in growth cones is cortactin which in non-neuronal cells binds and activates Arp2/3, both forming and stabilizing actin branches (Uruno *et al.*, 2001; Weaver *et al.*, 2001). Cortactin is also found in growth cone lamellae (Du *et al.*, 1998).

6. REGULATION OF ACTIN INTERACTIONS WITH MEMBRANES

The major role of the growth cone is to identify a directionally suitable trajectory that results in target identification and synapse formation. The growth cone therefore requires a much higher degree of ongoing spatial control of actin dynamics than other motile or polarized cells (Meyer and Feldman, 2002). Key to this behavior is the ability of extracellular signals to spatially restrict growth cone responses by generating signaling modules in specific areas of the growth cone, and then inactivating these modules when necessary so that they will not interfere with the responses to subsequent, potentially different, extracellular signals. These control mechanisms are superimposed on the regulation of actin. Rafts play an important role in this spatial segregation. In this section we compare the localization and regulation of three actin-binding proteins that are involved with controlling how actin interacts with the growth cone membrane in a spatially restricted fashion. Each of them, when genetically deleted, is characterized by a severe *in vivo* phenotype, which leads to defects in neuronal cytoarchitecture, attesting to their importance in nervous system function.

6.1.1. Ena/VASP

The Ena/VASP family of actin-binding proteins comprise three vertebrate members: vasodilator stimulated phosphoprotein (VASP), mammalian enabled (Mena) and Ena-VASP-like (EVL) which are distributed at the tips of filopodia and at the edge of lamellae (Krause *et al.*, 2002). They are structurally and functionally similar, each containing similar membrane targeting motifs. Thus the N-terminal EVH1 domain can adopt a structure related to PH domains that is thought to confer PIP2 binding (the NWASP EVHI domain does not bind PIP2, but does cause association with rafts). Likewise the central region has an SH2 domain (Bear *et al.*, 2001). Mena predominates in mice, and mice lacking in Mena show axon guidance deficits (Lanier *et al.*, 1999). In fibroblasts with sequestered (and therefore inactive) Mena, actin filaments are highly branched and short with their barbed ends oriented towards the membrane. These cells are also less motile presumably because sequestration of Mena causes more stable lamellae. Conversely, when active Mena is targeted to the cell membrane the filaments are less branched, much longer, and their barbed ends are less uniformly oriented, like the actin meshwork observed in growth cones (Bear *et al.*, 2002). Although membrane-targeted Mena causes more dynamic lamellae with increased ruffling, forward movement of the cell is not increased, suggesting that lamellae that are more frequently extended and retracted may be less likely to translocate (Bear *et al.*, 2000; Bear *et al.*, 2002). Ena/VASP proteins can act as a profilin ligand stimulating binding to the barbed ends of filaments and competing with capping protein, allowing for longer filament extension (Bear *et al.*, 2002). However deletion of the profilin binding motif has no effect on Mena function, whereas a second F-actin binding site was required. Preliminary evidence suggests that ena/VASP may be in the rhoGTPase pathway (Krause *et al.*, 2002). In fibroblasts, an adaptor between cdc42 and Mena has been identified and activation of the complex can stimulate filopodia elaboration (Krugmann *et al.*, 2001). It thus appears that ena/Vasp function has the potential to impact growth cones in areas where dynamic extension of filopodia and lamellae need to be decoupled from forward translocation, such as in an environment where complex extracellular signals needs to be sampled by filopodia without moving the neurite forward until the correct trajectory has been identified.

6.1.2. ED domain proteins – PIP2 modulators at the growth cone membrane

One important clue about how specific regulatory mechanisms at the cell membrane overlie direct regulation of actin dynamics to provide spatial and temporal control may be provided by a group of so called 'ED' domain proteins of which Myristoylated Alanine-rich C kinase substrate (MARCKS), its Marcks related protein analog (MRP or MacMARCKS) and the nervous system specific Growth Associated Protein-43 (GAP-43) are the most prominent members (Arbuzova *et al.*, 2002). Their key feature is a highly basic ED domain, which allows electrostatic interactions with the acidic lipids of cell membranes especially, but not exclusively, PIP2 (McLaughlin and Aderem, 1995). It is now considered that electrostatic interactions rather than binding domains such as PH motifs are the best predictor of lipid interactions (see, Yin and Janmey, 2002). Localization at the inner leaflet is not random, GAP-43 is dually palmitoylated and a key component of rafts (Arni *et al.*, 1998) whereas MARCKS has one myristate and so is also likely to be only weakly associated, or associated with a different subpopulation, in much the same way as Src (Casey, 1995). Indeed these 'ED' domain proteins can concentrate PIP2 at the membrane, presumably in rafts (GAP-43) or a similar fraction (MARCKS) (Laux *et al.*, 2000). The ED domain has an independent regulatory role; it is both a binding site for calmodulin and a phosphorylation site for PKC, which are mutually exclusive, and it is therefore able to modulate cross talk between calmodulin (CaM) and PKC activated pathways (Blackshear, 1993). Thus CaM mediated processes are likely to be upregulated when the ED site is phosphorylated (Porumb *et al.*, 1997). CaM/PKC regulation also affects PIP2 distribution. The ED motif and myrisotylation are both required for MARCKS association with membranes, and the ED site in particular is required for PIP2 localization (McLaughlin and Aderem, 1995). Hence CaM binding to MARCKS, which releases it from the membrane, makes PIP2 accessible to other binding proteins. In contrast, activation of PKC in rafts which phosphorylates GAP-43, and which is stimulated by neurotrophin or IgCAM signaling would be expected to downregulate PIP2 signaling because PKC is activated by PLCγ. This mimics what is seen in T cells (Viola *et al.*, 1999). The metabolites produced (IP3 and DAG) are also expected to sustain PKC activation and PIP2 downregulation in these rafts. Besides any indirect effect on actin mediated via PIP2 such as we discussed in detail above, ED proteins affect actin directly: They bind actin and affect polymerization kinetics *in vitro*. Most significantly however, F-actin in growth cone lamellae is significantly reduced *in vivo* when they are absent, in the case of GAP-43 by about 50% (Shen *et al.*, 2002). It seems likely that the sensitivity of the ED site to both CaM and PKC represents a level of control that is superimposed on PIP2 regulation of actin dynamics, which mainly affects treadmilling. The situation is somewhat clearer with GAP-43 where activation of PKC and hence GAP-43 phosphorylation at the ED domain is likely to lead to downregulation of PIP2 signaling. This phosphorylation also causes GAP-43 to interact with F-actin and downregulate treadmilling (He *et al.*, 1997), which tends to favor forward movement of the growth cone, as we saw previously. Indeed phosphorylated GAP-43 is enriched in translocating lamellae, whereas the unphosphorylated form is found in lamellae that are motile, but not translocating (Dent and Meiri, 1998). The significance is that PKC phosphorylation of GAP-43 is stimulated by two extracellular signaling systems that both use rafts as signaling platforms – neurotrophins and the IgCAM superfamily, as described above (Meiri and Burdick, 1991). Hence GAP-43 may act as a clutch mechanism to allow these extracellular signals to override actin treadmilling and stimulate forward movement. MARCKS appears to behave differently: CaM binding or PKC activation releases it from the membrane thereby potentially uncovering PIP2 to interact with other actin binding proteins (discussed in Caroni (2001). It will be crucial to find out whether MARCKS and GAP-43 exist in the same membrane population of rafts. It will also be important to determine the relationship between GAP-43 and raft-mediated signaling in development. During early stages of axon outgrowth *in vivo* GAP 43 phosphorylation is highly dynamic, consistent with a role in raft-mediated signaling outlined above. However, as growing axons approach their targets motility of the growth cone is reduced as it prepares to dock and form a stable synaptic connection (Meiri *et al.*, 1991). At this time too GAP-43 phosphorylation becomes more sustained, and palmitoylation is also down regulated (Patterson and Skene, 1999). It will be important to determine the nature of the membrane compartment with which it is now associated and whether it involves more direct associations between the membrane and the cytoskeleton via membrane skeleton components, as was proposed previously (Meiri and Gordon-Weeks, 1990), or alternatively whether different acylation characteristics take over to target it to membrane rafts and continue communication with the cytoskeleton in this fashion, such as has been recently suggested (Liang *et al.*, 2002).

7. CONCLUSION

Communication between the membrane and the cytoskeleton is a key element in cellular responses to extracellular signals that is not restricted to polarized cells that respond by dramatic changes in morphology. In both polarized and non-polarized cells this communication critically depends on the generation of membrane signaling modules that serve to concentrate and amplify signal transduction mechanisms and that also contain actin regulatory mechanisms. Thus initiation of signal transduction and regulation of actin dynamics occur in parallel in these signaling modules. Current evidence shows that in many diverse circumstances the structural substrates for these signaling modules are lipid raft–membrane heterogeneities where specific lipids exist in a liquid ordered state and which can attract proteins with specific targeting domains. Components of lipid rafts therefore include membrane receptors for extracellular signals, downstream kinase effectors and actin regulatory proteins themselves. The coordinate activation of extracellular and intracellular raft components plays a key role when the spatial and temporal regulation of membrane/cytoskeletal communication is necessary in order for a cellular response to be generated to an extracellular signal. Current research is focused on the identification of functionally distinct rafts within cells, and a more complete understanding of how they regulate both signaling and actin cytoskeleton dynamics in a spatially and temporally appropriate manner.

REFERENCES

Airaksinen, M.S. and Saarma, M., 2002. The GDNF family: signalling, biological functions and therapeutic value, *Nat Rev Neurosci.* **3**: 383-394.

Alonso, M.A. and Millan, J., 2001. The role of lipid rafts in signalling and membrane trafficking in T lymphocytes, *J Cell Sci.* **114**: 3957-3965.

Anderson, R.G. and Jacobson, K., 2002. A role for lipid shells in targeting proteins to caveolae, rafts, and other lipid domains, *Science.* **296**: 1821-1825.

Anton, I.M., de la Fuente, M.A., Sims, T.N., Freeman, S., Ramesh, N., Hartwig, J.H., Dustin, M.L. and Geha, R.S., 2002. WIP deficiency reveals a differential role for WIP and the actin cytoskeleton in T and B cell activation, *Immunity.* **16**: 193-204.

Arbuzova, A., Schmitz, A.A. and Vergeres, G., 2002. Cross-talk unfolded: MARCKS proteins, *Biochem J.* **362**: 1-12.

Arni, S., Keilbaugh, S.A., Ostermeyer, A.G. and Brown, D.A., 1998. Association of GAP-43 with detergent-resistant membranes requires two palmitoylated cysteine residues, *J Biol Chem.* **273**: 28478-28485.

Azuma, T., Witke, W., Stossel, T.P., Hartwig, J.H. and Kwiatkowski, D.J., 1998. Gelsolin is a downstream effector of rac for fibroblast motility, *Embo J.* **17**: 1362-1370.

Banzai, Y., Miki, H., Yamaguchi, H. and Takenawa, T., 2000. Essential role of neural Wiskott-Aldrich syndrome protein in neurite extension in PC12 cells and rat hippocampal primary culture cells, *J Biol Chem.* **275**: 11987-11992.

Barkalow, K., Witke, W., Kwiatkowski, D.J. and Hartwig, J.H., 1996. Coordinated regulation of platelet actin filament barbed ends by gelsolin and capping protein, *J Cell Biol.* **134**: 389-399.

Bateman, J., Shu, H. and Van Vactor, D., 2000. The guanine nucleotide exchange factor trio mediates axonal development in the Drosophila embryo, *Neuron.* **26**: 93-106.

Bateman, J. and Van Vactor, D., 2001. The Trio family of guanine-nucleotide-exchange factors: regulators of axon guidance, *J Cell Sci.* **114**: 1973-1980.

Bear, J.E., Krause, M. and Gertler, F.B., 2001. Regulating cellular actin assembly, *Curr Opin Cell Biol.* **13**: 158-166.

Bear, J.E., Loureiro, J.J., Libova, I., Fassler, R., Wehland, J. and Gertler, F.B., 2000. Negative regulation of fibroblast motility by Ena/VASP proteins, *Cell.* **101**: 717-728.

Bear, J.E., Svitkina, T.M., Krause, M., Schafer, D.A., Loureiro, J.J., Strasser, G.A., Maly, I.V., Chaga, O.Y., Cooper, J.A., Borisy, G.G. and Gertler, F.B., 2002. Antagonism between Ena/VASP proteins and actin filament capping regulates fibroblast motility, *Cell.* **109**: 509-521.

Bennett, V. and Gilligan, D.M., 1993. The spectrin-based membrane skeleton and micronscale organization of the plasma membrane, *Annu Rev Cell Biol.* **9**: 27-66.

Bilderback, T.R., Gazula, V.R., Lisanti, M.P. and Dobrowsky, R.T., 1999. Caveolin interacts with Trk A and p75(NTR) and regulates neurotrophin signaling pathways, *J Biol Chem.* **274**: 257-263.

Blackshear, P.J., 1993. The MARCKS family of cellular protein kinase C substrates, *J Biol Chem.* **268**: 1501-1504.

Blanchoin, L., Pollard, T.D. and Mullins, R.D., 2000. Interactions of ADF/cofilin, Arp2/3 complex, capping protein and profilin in remodeling of branched actin filament networks, *Curr Biol.* **10**: 1273-1282.

Brown, D.A. and London, E., 1998. Structure and origin of ordered lipid domains in biological membranes, *J Membr Biol.* **164**: 103-114.

Brown, D.A. and Rose, J.K., 1992. Sorting of GPI-anchored proteins to glycolipid-enriched membrane subdomains during transport to the apical cell surface, *Cell.* **68**: 533-544.

Brown, F.D., Rozelle, A.L., Yin, H.L., Balla, T. and Donaldson, J.G., 2001. Phosphatidylinositol 4,5-bisphosphate and Arf6-regulated membrane traffic, *J Cell Biol.* **154**: 1007-1017.

Bruckner, K., Pablo Labrador, J., Scheiffele, P., Herb, A., Seeburg, P.H. and Klein, R., 1999. EphrinB ligands recruit GRIP family PDZ adaptor proteins into raft membrane microdomains, *Neuron.* **22**: 511-524.

Cannon, J.L., Labno, C.M., Bosco, G., Seth, A., McGavin, M.H., Siminovitch, K.A., Rosen, M.K. and Burkhardt, J.K., 2001. Wasp recruitment to the T cell:APC contact site occurs independently of Cdc42 activation, *Immunity.* **15**: 249-259.

Carlier, M.F., Laurent, V., Santolini, J., Melki, R., Didry, D., Xia, G.X., Hong, Y., Chua, N.H. and Pantaloni, D., 1997. Actin depolymerizing factor (ADF/cofilin) enhances the rate of filament turnover: implication in actin-based motility, *J Cell Biol.* **136**: 1307-1322.

Caroni, P., 2001. New EMBO members' review: actin cytoskeleton regulation through modulation of PI(4,5)P(2) rafts, *Embo J.* **20**: 4332-4336.

Casey, P.J., 1995. Protein lipidation in cell signaling, *Science.* **268**: 221-225.

Chatah, N.E. and Abrams, C.S., 2001. G-protein-coupled receptor activation induces the membrane translocation and activation of phosphatidylinositol-4-phosphate 5-kinase I alpha by a Rac- and Rho-dependent pathway, *J Biol Chem.* **276**: 34059-34065.

Chong, L.D., Traynor-Kaplan, A., Bokoch, G.M. and Schwartz, M.A., 1994. The small GTP-binding protein Rho regulates a phosphatidylinositol 4-phosphate 5-kinase in mammalian cells, *Cell.* **79**: 507-513.

Cowan, C.A. and Henkemeyer, M., 2002. Ephrins in reverse, park and drive, *Trends Cell Biol.* **12**: 339-346.

Dan, C., Nath, N., Liberto, M. and Minden, A., 2002. PAK5, a new brain-specific kinase, promotes neurite outgrowth in N1E-115 cells, *Mol Cell Biol.* **22**: 567-577.

Davy, A., Feuerstein, C. and Robbins, S.M., 2000. Signaling within a caveolae-like membrane microdomain in human neuroblastoma cells in response to fibroblast growth factor, *J Neurochem.* **74**: 676-683.

Davy, A., Gale, N.W., Murray, E.W., Klinghoffer, R.A., Soriano, P., Feuerstein, C. and Robbins, S.M., 1999. Compartmentalized signaling by GPI-anchored ephrin-A5 requires the Fyn tyrosine kinase to regulate cellular adhesion, *Genes Dev.* **13**: 3125-3135.

Dent, E.W. and Meiri, K.F., 1998. Distribution of phosphorylated GAP-43 (neuromodulin) in growth cones directly reflects growth cone behavior, *J Neurobiol.* **35**: 287-299.

Dharmawardhane, S., Brownson, D., Lennartz, M. and Bokoch, G.M., 1999. Localization of p21-activated kinase 1 (PAK1) to pseudopodia, membrane ruffles, and phagocytic cups in activated human neutrophils, *J Leukoc Biol.* **66**: 521-527.

Dodelet, V.C. and Pasquale, E.B., 2000. Eph receptors and ephrin ligands: embryogenesis to tumorigenesis, *Oncogene.* **19**: 5614-5619.

Doherty, P. and Walsh, F.S., 1996. CAM-FGF Receptor Interactions: A Model for Axonal Growth, *Mol Cell Neurosci.* **8**: 99-111.

Draber, P. and Draberova, L., 2002. Lipid rafts in mast cell signaling, *Mol Immunol.* **38**: 1247.

Du, Y., Weed, S.A., Xiong, W.C., Marshall, T.D. and Parsons, J.T., 1998. Identification of a novel cortactin SH3 domain-binding protein and its localization to growth cones of cultured neurons, *Mol Cell Biol.* **18**: 5838-5851.

Dyson, J.M., O'Malley, C.J., Becanovic, J., Munday, A.D., Berndt, M.C., Coghill, I.D., Nandurkar, H.H., Ooms, L.M. and Mitchell, C.A., 2001. The SH2-containing inositol polyphosphate 5-phosphatase, SHIP-2, binds filamin and regulates submembraneous actin, *J Cell Biol.* **155**: 1065-1079.

Edwards, D.C., Sanders, L.C., Bokoch, G.M. and Gill, G.N., 1999. Activation of LIM-kinase by Pak1 couples Rac/Cdc42 GTPase signalling to actin cytoskeletal dynamics, *Nat Cell Biol.* **1**: 253-259.

Ehler, E., van Leeuwen, F., Collard, J.G. and Salinas, P.C., 1997. Expression of Tiam-1 in the developing brain suggests a role for the Tiam-1-Rac signaling pathway in cell migration and neurite outgrowth, *Mol Cell Neurosci.* **9**: 1-12.

Fra, A.M., Williamson, E., Simons, K. and Parton, R.G., 1994. Detergent-insoluble glycolipid microdomains in lymphocytes in the absence of caveolae, *J Biol Chem.* **269**: 30745-30748.

Fridriksson, E.K., Shipkova, P.A., Sheets, E.D., Holowka, D., Baird, B. and McLafferty, F.W., 1999. Quantitative analysis of phospholipids in functionally important membrane domains from RBL-2H3 mast cells using tandem high-resolution mass spectrometry, *Biochemistry.* **38**: 8056-8063.

Friedman, W.J. and Greene, L.A., 1999. Neurotrophin signaling via Trks and p75, *Exp Cell Res.* **253**: 131-142.

Friedrichson, T. and Kurzchalia, T.V., 1998. Microdomains of GPI-anchored proteins in living cells revealed by crosslinking, *Nature.* **394**: 802-805.

Fukami, K., Furuhashi, K., Inagaki, M., Endo, T., Hatano, S. and Takenawa, T., 1992. Requirement of phosphatidylinositol 4,5-bisphosphate for alpha-actinin function, *Nature.* **359**: 150-152.

Fukuoka, M., Miki, H. and Takenawa, T., 1997. Identification of N-WASP homologs in human and rat brain, *Gene.* **196**: 43-48.

Gallego, M.D., Santamaria, M., Pena, J. and Molina, I.J., 1997. Defective actin reorganization and polymerization of Wiskott-Aldrich T cells in response to CD3-mediated stimulation, *Blood.* **90**: 3089-3097.

Giurisato, E., McIntosh, D.P., Tassi, M., Gamberucci, A. and Benedetti, A., 2003. T cell receptor can be recruited to a subset of plasma membrane rafts, independently of cell signalling and attendantly to raft clustering, *J Biol Chem.* **278**: 6771-6778.

Goldberg, D. J., Foley, M. S.,Tang, D. and Grabham, P. W., 2000. Recruitment of the Arp2/3 complex and mena for the stimulation of actin polymerization in growth cones by nerve growth factor, *J Neurosci Res.* **4**: 458-467.

Goldmann, W.H., 2001. Phosphorylation of filamin (ABP-280) regulates the binding to the lipid membrane, integrin, and actin, *Cell Biol Int.* **25**: 805-808.

Gomez-Mouton, C., Abad, J.L., Mira, E., Lacalle, R.A., Gallardo, E., Jimenez-Baranda, S., Illa, I., Bernad, A., Manes, S. and Martinez, A.C., 2001. Segregation of leading-edge and uropod components into specific lipid rafts during T cell polarization, *Proc Natl Acad Sci USA.* **98**: 9642-9647.

He, Q., Dent, E.W. and Meiri, K.F., 1997. Modulation of actin filament behavior by GAP-43 (neuromodulin) is dependent on the phosphorylation status of serine 41, the protein kinase C site, *J Neurosci.* **17**: 3515-3524.

He, Q. and Meiri, K.F., 2002. Isolation and characterization of detergent-resistant microdomains responsive to NCAM-mediated signaling from growth cones, *Mol Cell Neurosci.* **19**: 18-31.

Heiska, L., Alfthan, K., Gronholm, M., Vilja, P., Vaheri, A. and Carpen, O., 1998. Association of ezrin with intercellular adhesion molecule-1 and -2 (ICAM-1 and ICAM-2). Regulation by phosphatidylinositol 4, 5-bisphosphate, *J Biol Chem.* **273**: 21893-21900.

Henke, R.C., Seeto, G.S. and Jeffrey, P.L., 1997. Thy-1 and AvGp50 signal transduction complex in the avian nervous system: c-Fyn and G alpha i protein association and activation of signalling pathways, *J Neurosci Res.* **49**: 655-670.

Higgs, H.N. and Pollard, T.D., 2000. Activation by Cdc42 and PIP(2) of Wiskott-Aldrich syndrome protein (WASp) stimulates actin nucleation by Arp2/3 complex, *J Cell Biol.* **150**: 1311-1320.

Hirao, M., Sato, N., Kondo, T., Yonemura, S., Monden, M., Sasaki, T., Takai, Y. and Tsukita, S., 1996. Regulation mechanism of ERM (ezrin/radixin/moesin) protein/plasma membrane association: possible involvement of phosphatidylinositol turnover and Rho-dependent signaling pathway, *J Cell Biol.* **135**: 37-51.

Hirose, M., Ishizaki, T., Watanabe, N., Uehata, M., Kranenburg, O., Moolenaar, W.H., Matsumura, F., Maekawa, M., Bito, H. and Narumiya, S., 1998. Molecular dissection of the Rho-associated protein kinase (p160ROCK)-regulated neurite remodeling in neuroblastoma N1E-115 cells, *J Cell Biol.* **141**: 1625-1636.

Holdorf, A.D., Lee, K.H., Burack, W.R., Allen, P.M. and Shaw, A.S., 2002. Regulation of Lck activity by CD4 and CD28 in the immunological synapse, *Nat Immunol.* **3**: 259-264.

Huang, C.S., Zhou, J., Feng, A.K., Lynch, C.C., Klumperman, J., DeArmond, S.J. and Mobley, W.C., 1999. Nerve growth factor signaling in caveolae-like domains at the plasma membrane, *J Biol Chem.* **274**: 36707-36714.

Janes, P.W., Ley, S.C. and Magee, A.I., 1999. Aggregation of lipid rafts accompanies signaling via the T cell antigen receptor, *J Cell Biol.* **147**: 447-461.

Kasahara, K., Watanabe, K., Takeuchi, K., Kaneko, H., Oohira, A., Yamamoto, T. and Sanai, Y., 2000. Involvement of gangliosides in glycosylphosphatidylinositol-anchored neuronal cell adhesion molecule TAG-1 signaling in lipid rafts, *J Biol Chem.* **275**: 34701-34709.

Klein, C., Nguyen, D., Liu, C.H., Mizoguchi, A., Bhan, A.K., Miki, H., Takenawa, T., Rosen, F.S., Alt, F.W., Mulligan, R.C. and Snapper, S.B., 2003. Gene therapy for Wiskott Aldrich Syndrome: rescue of T-cell signaling and amelioration of colitis upon transplantation of retrovirally transduced hematopoietic stem cells in mice, *Blood.* **101**: 2159-2166.

Knaus, U.G. and Bokoch, G.M., 1998. The p21Rac/Cdc42-activated kinases (PAKs), *Int J Biochem Cell Biol.* **30**: 857-862.

Kosugi, A., Saitoh, S., Noda, S., Yasuda, K., Hayashi, F., Ogata, M. and Hamaoka, T., 1999. Translocation of tyrosine-phosphorylated TCRzeta chain to glycolipid-enriched membrane domains upon T cell activation, *Int Immunol.* **11**: 1395-1401.

Kouhara, H., Hadari, Y.R., Spivak-Kroizman, T., Schilling, J., Bar-Sagi, D., Lax, I. and Schlessinger, J., 1997. A lipid-anchored Grb2-binding protein that links FGF-receptor activation to the Ras/MAPK signaling pathway, *Cell.* **89**: 693-702.

Krause, M., Bear, J.E., Loureiro, J.J. and Gertler, F.B., 2002. The Ena/VASP enigma, *J Cell Sci.* **115**: 4721-4726.

Krause, M., Sechi, A.S., Konradt, M., Monner, D., Gertler, F.B. and Wehland, J., 2000. Fyn-binding protein (Fyb)/SLP-76-associated protein (SLAP), Ena/vasodilator-stimulated phosphoprotein (VASP) proteins and the Arp2/3 complex link T cell receptor (TCR) signaling to the actin cytoskeleton, *J Cell Biol.* **149**: 181-194.

Krugmann, S., Jordens, I., Gevaert, K., Driessens, M., Vandekerckhove, J. and Hall, A., 2001. Cdc42 induces filopodia by promoting the formation of an IRSp53: Mena complex, *Curr Biol.* **11**: 1645-1655.

Kullander, K. and Klein, R., 2002. Mechanisms and functions of Eph and ephrin signalling, *Nat Rev Mol Cell Biol.* **3**: 475-486.

Lamoureux, P., Altun-Gultekin, Z.F., Lin, C., Wagner, J.A. and Heidemann, S.R., 1997. Rac is required for growth cone function but not neurite assembly, *J Cell Sci.* **110** (Pt 5): 635-641.

Lanier, L.M., Gates, M.A., Witke, W., Menzies, A.S., Wehman, A.M., Macklis, J.D., Kwiatkowski, D., Soriano, P. and Gertler, F.B., 1999. Mena is required for neurulation and commissure formation, *Neuron.* **22**: 313-325.

Laux, T., Fukami, K., Thelen, M., Golub, T., Frey, D. and Caroni, P., 2000. GAP43, MARCKS, and CAP23 modulate PI(4,5)P(2) at plasmalemmal rafts, and regulate cell cortex actin dynamics through a common mechanism, *J Cell Biol.* **149**: 1455-1472.

Leeuwen, F.N., Kain, H.E., Kammen, R.A., Michiels, F., Kranenburg, O.W. and Collard, J.G., 1997. The guanine nucleotide exchange factor Tiam1 affects neuronal morphology; opposing roles for the small GTPases Rac and Rho, *J Cell Biol.* **139**: 797-807.

Lewis, A.K. and Bridgman, P.C., 1992. Nerve growth cone lamellipodia contain two populations of actin filaments that differ in organization and polarity, *J Cell Biol.* **119**: 1219-1243.
Liang, X., Lu, Y., Neubert, T.A. and Resh, M.D., 2002. Mass spectrometric analysis of GAP-43/neuromodulin reveals the presence of a variety of fatty acylated species, *J Biol Chem.* **277**: 33032-33040.
Lin, D., Gish, G.D., Songyang, Z. and Pawson, T., 1999. The carboxyl terminus of B class ephrins constitutes a PDZ domain binding motif, *J Biol Chem.* **274**: 3726-3733.
Little, E.B., Edelman, G.M. and Cunningham, B.A., 1998. Palmitoylation of the cytoplasmic domain of the neural cell adhesion molecule N-CAM serves as an anchor to cellular membranes, *Cell Adhes Commun.* **6**: 415-430.
Liu, P., Ying, Y., Ko, Y.G. and Anderson, R.G., 1996. Localization of platelet-derived growth factor-stimulated phosphorylation cascade to caveolae, *J Biol Chem.* **271**: 10299-10303.
Machesky, L.M., 1997. Cell motility: complex dynamics at the leading edge, *Curr Biol.* **7**: R164-167.
Maekawa, M., Ishizaki, T., Boku, S., Watanabe, N., Fujita, A., Iwamatsu, A., Obinata, T., Ohashi, K., Mizuno, K. and Narumiya, S., 1999. Signaling from Rho to the actin cytoskeleton through protein kinases ROCK and LIM-kinase, *Science.* **285**: 895-898.
Magee, T., Pirinen, N., Adler, J., Pagakis, S.N. and Parmryd, I., 2002. Lipid rafts: cell surface platforms for T cell signaling, *Biol Res.* **35**: 127-131.
Martel, V., Racaud-Sultan, C., Dupe, S., Marie, C., Paulhe, F., Galmiche, A., Block, M.R. and Albiges-Rizo, C., 2001. Conformation, localization, and integrin binding of talin depend on its interaction with phosphoinositides, *J Biol Chem.* **276**: 21217-21227.
Martin, T.F., 2001. PI(4,5)P(2) regulation of surface membrane traffic, *Curr Opin Cell Biol.* **13**: 493-499.
Martinez-Quiles, N., Rohatgi, R., Anton, I.M., Medina, M., Saville, S.P., Miki, H., Yamaguchi, H., Takenawa, T., Hartwig, J.H., Geha, R.S. and Ramesh, N., 2001. WIP regulatesN-WASP-mediated actin polymerization and filopodium formation, *Nat Cell Biol.* **3**:484-491.
Matsui, T., Yonemura, S. and Tsukita, S., 1999. Activation of ERM proteins in vivo by Rho involves phosphatidyl-inositol 4-phosphate 5-kinase and not ROCK kinases, *Curr Biol.* **9**: 1259-1262.
Mayor, S. and Maxfield, F.R., 1995. Insolubility and redistribution of GPI-anchored proteins at the cell surface after detergent treatment, *Mol Biol Cell.* **6**: 929-944.
McLaughlin, S. and Aderem, A., 1995. The myristoyl-electrostatic switch: a modulator of reversible protein-membrane interactions, *Trends Biochem Sci.* **20**: 272-276.
McLaughlin, S., Wang, J., Gambhir, A. and Murray, D., 2002. PIP(2) and proteins: interactions, organization, and information flow, *Annu Rev Biophys Biomol Struct.* **31**: 151-175.
Meberg, P.J., Ono, S., Minamide, L.S., Takahashi, M. and Bamburg, J.R., 1998. Actin depolymerizing factor and cofilin phosphorylation dynamics: response to signals that regulate neurite extension, *Cell Motil Cytoskeleton.* **39**: 172-190.
Meiri, K.F., Bickerstaff, L.E. and Schwob, J.E., 1991. Monoclonal antibodies show that kinase C phosphorylation of GAP-43 during axonogenesis is both spatially and temporally restricted in vivo, *J Cell Biol.* **112**: 991-1005.
Meiri, K.F. and Burdick, D., 1991. Nerve growth factor stimulation of GAP-43 phosphorylation in intact isolated growth cones, *J Neurosci.* **11**: 3155-3164.
Meiri, K.F. and Gordon-Weeks, P.R., 1990. GAP-43 in growth cones is associated with areas of membrane that are tightly bound to substrate and is a component of a membrane skeleton subcellular fraction, *J Neurosci.* **10**: 256-266.
Meyer, G. and Feldman, E.L., 2002. Signaling mechanisms that regulate actin-based motility processes in the nervous system, *J Neurochem.* **83**: 490-503.

Miceli, M.C., Moran, M., Chung, C.D., Patel, V.P., Low, T. and Zinnanti, W., 2001. Co-stimulation and counter-stimulation: lipid raft clustering controls TCR signaling and functional outcomes, *Semin Immunol.* **13**: 115-128.

Michel, F., Mangino, G., Attal-Bonnefoy, G., Tuosto, L., Alcover, A., Roumier, A., Olive, D. and Acuto, O., 2000. CD28 utilizes Vav-1 to enhance TCR-proximal signaling and NF-AT activation, *J Immunol.* **165**: 3820-3829.

Mineo, C., James, G.L., Smart, E.J. and Anderson, R.G., 1996. Localization of epidermal growth factor-stimulated Ras/Raf-1 interaction to caveolae membrane, *J Biol Chem.* **271**: 11930-11935.

Molina, I.J., Sancho, J., Terhorst, C., Rosen, F.S. and Remold-O'Donnell, E., 1993. T cells of patients with the Wiskott-Aldrich syndrome have a restricted defect in proliferative responses, *J Immunol.* **151**: 4383-4390.

Montixi, C., Langlet, C., Bernard, A.M., Thimonier, J., Dubois, C., Wurbel, M.A., Chauvin, J.P., Pierres, M. and He, H.T., 1998. Engagement of T cell receptor triggers its recruitment to low-density detergent-insoluble membrane domains, *Embo J.* **17**: 5334-5348.

Moran, M. and Miceli, M.C., 1998. Engagement of GPI-linked CD48 contributes to TCR signals and cytoskeletal reorganization: a role for lipid rafts in T cell activation, *Immunity.* **9**: 787-796.

Mullins, R.D., Heuser, J.A. and Pollard, T.D., 1998. The interaction of Arp2/3 complex with actin: nucleation, high affinity pointed end capping, and formation of branching networks of filaments, *Proc Natl Acad Sci USA.* **95**: 6181-6186.

Mutoh, T., Hamano, T., Tokuda, A. and Kuriyama, M., 2000. Unglycosylated Trk protein does not co-localize nor associate with ganglioside GM1 in stable clone of PC12 cells overexpressing Trk (PCtrk cells), *Glycoconj J.* **17**: 233-237.

Narumiya, S., Ishizaki, T. and Watanabe, N., 1997. Rho effectors and reorganization of actin cytoskeleton, *FEBS Lett.* **410**: 68-72.

Neame, S.J., Uff, C.R., Sheikh, H., Wheatley, S.C. and Isacke, C.M., 1995. CD44 exhibits a cell type dependent interaction with triton X-100 insoluble, lipid rich, plasma membrane domains, *J Cell Sci.* **108** (Pt 9): 3127-3135.

Nebl, T., Pestonjamasp, K.N., Leszyk, J.D., Crowley, J.L., Oh, S.W. and Luna, E.J., 2002. Proteomic analysis of a detergent-resistant membrane skeleton from neutrophil plasma membranes, *J Biol Chem.* **277**: 43399-43409.

Newsome, T.P., Schmidt, S., Dietzl, G., Keleman, K., Asling, B., Debant, A. and Dickson, B.J., 2000. Trio combines with dock to regulate Pak activity during photoreceptor axon pathfinding in Drosophila, *Cell.* **101**: 283-294.

Niethammer, P., Delling, M., Sytnyk, V., Dityatev, A., Fukami, K. and Schachner, M., 2002. Cosignaling of NCAM via lipid rafts and the FGF receptor is required for neuritogenesis, *J Cell Biol.* **157**: 521-532.

Nobes, C.D. and Hall, A., 1995. Rho, rac and cdc42 GTPases: regulators of actin structures, cell adhesion and motility, *Biochem Soc Trans.* **23**: 456-459.

Ohashi, K., Hosoya, T., Takahashi, K., Hing, H. and Mizuno, K., 2000. A Drosophila homolog of LIM-kinase phosphorylates cofilin and induces actin cytoskeletal reorganization, *Biochem Biophys Res Commun.* **276**: 1178-1185.

Olofsson, B., 1999. Rho guanine dissociation inhibitors: pivotal molecules in cellular signalling, *Cell Signal.* **11**: 545-554.

Oude Weernink, P.A., Schulte, P., Guo, Y., Wetzel, J., Amano, M., Kaibuchi, K., Haverland, S., Voss, M., Schmidt, M., Mayr, G.W. and Jakobs, K.H., 2000. Stimulation of phosphatidylinositol-4-phosphate 5-kinase by Rho-kinase, *J Biol Chem.* **275**: 10168-10174.

Paratcha, G., Ledda, F., Baars, L., Coulpier, M., Besset, V., Anders, J., Scott, R. and Ibanez, C.F., 2001. Released GFRalpha1 potentiates downstream signaling, neuronal survival, and differentiation via a novel mechanism of recruitment of c-Ret to lipid rafts, *Neuron.* **29**: 171-184.

Patterson, S.I. and Skene, J.H., 1999. A shift in protein S-palmitoylation, with persistence of growth-associated substrates, marks a critical period for synaptic plasticity in developing brain, *J Neurobiol.* **39**: 423-437.

Porumb, T., Crivici, A., Blackshear, P.J. and Ikura, M., 1997. Calcium binding and conformational properties of calmodulin complexed with peptides derived from myristoylated alanine-rich C kinase substrate (MARCKS) and MARCKS-related protein (MRP), *Eur Biophys J.* **25**: 239-247.

Prehoda, K.E., Scott, J.A., Mullins, R.D. and Lim, W.A., 2000. Integration of multiple signals through cooperative regulation of the N-WASP-Arp2/3 complex, *Science.* **290**: 801-806.

Ramesh, N., Anton, I.M., Hartwig, J.H. and Geha, R.S., 1997. WIP, a protein associated with wiskott-aldrich syndrome protein, induces actin polymerization and redistribution in lymphoid cells, *Proc Natl Acad Sci USA.* **94**: 14671-14676.

Ridley, A.J. and Hall, A., 1992. Distinct patterns of actin organization regulated by the small GTP-binding proteins Rac and Rho, *Cold Spring Harb Symp Quant Biol.* **57**: 661-671.

Rodgers, W. and Zavzavadjian, J., 2001. Glycolipid-enriched membrane domains are assembled into membrane patches by associating with the actin cytoskeleton, *Exp Cell Res.* **267**: 173-183.

Rosenblatt, J., Agnew, B.J., Abe, H., Bamburg, J.R. and Mitchison, T.J., 1997. Xenopus actin depolymerizing factor/cofilin (XAC) is responsible for the turnover of actin filaments in Listeria monocytogenes tails, *J Cell Biol.* **136**: 1323-1332.

Rothberg, K.G., Heuser, J.E., Donzell, W.C., Ying, Y.S., Glenney, J.R. and Anderson, R.G., 1992. Caveolin, a protein component of caveolae membrane coats, *Cell.* **68**: 673-682.

Rozelle, A.L., Machesky, L.M., Yamamoto, M., Driessens, M.H., Insall, R.H., Roth, M.G., Luby-Phelps, K., Marriott, G., Hall, A. and Yin, H.L., 2000. Phosphatidylinositol 4,5-bisphosphate induces actin-based movement of raft-enriched vesicles through WASP-Arp2/3, *Curr Biol.* **10**: 311-320.

Salojin, K.V., Zhang, J. and Delovitch, T.L., 1999. TCR and CD28 are coupled via ZAP-70 to the activation of the Vav/Rac-1-/PAK-1/p38 MAPK signaling pathway, *J Immunol.* **163**: 844-853.

Sasaki, T. and Takai, Y., 1998. The Rho small G protein family-Rho GDI system as a temporal and spatial determinant for cytoskeletal control, *Biochem Biophys Res Commun.* **245**: 641-645.

Scheffzek, K., Ahmadian, M.R. and Wittinghofer, A., 1998. GTPase-activating proteins: helping hands to complement an active site, *Trends Biochem Sci.* **23**: 257-262.

Schlessinger, J., 2000. Cell signaling by receptor tyrosine kinases, *Cell.* **103**: 211-225.

Schnitzer, J.E., McIntosh, D.P., Dvorak, A.M., Liu, J. and Oh, P., 1995. Separation of caveolae from associated microdomains of GPI-anchored proteins, *Science.* **269**: 1435-1439.

Sechi, A.S. and Wehland, J., 2000. The actin cytoskeleton and plasma membrane connection: PtdIns(4,5)P(2) influences cytoskeletal protein activity at the plasma membrane, *J Cell Sci.* **113** (Pt 21): 3685-3695.

Shamah, S.M., Lin, M.Z., Goldberg, J.L., Estrach, S., Sahin, M., Hu, L., Bazalakova, M., Neve, R.L., Corfas, G., Debant, A. and Greenberg, M.E., 2001. EphA receptors regulate growth cone dynamics through the novel guanine nucleotide exchange factor ephexin, *Cell.* **105**: 233-244.

Shen, Y., Mani, S., Donovan, S.L., Schwob, J.E. and Meiri, K.F., 2002. Growth-associated protein-43 is required for commissural axon guidance in the developing vertebrate nervous system, *J Neurosci.* **22**: 239-247.

Simons, K. and Toomre, D., 2000. Lipid rafts and signal transduction, *Nat Rev Mol Cell Biol.* **1**: 31-39.

Smart, E.J., Graf, G.A., McNiven, M.A., Sessa, W.C., Engelman, J.A., Scherer, P.E., Okamoto, T. and Lisanti, M.P., 1999. Caveolins, liquid-ordered domains, and signal transduction, *Mol Cell Biol.* **19**: 7289-7304.

Snapper, S.B., Takeshima, F., Anton, I., Liu, C.H., Thomas, S.M., Nguyen, D., Dudley, D., Fraser, H., Purich, D., Lopez-Ilasaca, M., Klein, C., Davidson, L., Bronson, R., Mulligan, R.C., Southwick, F., Geha, R., Goldberg, M.B., Rosen, F.S., Hartwig, J.H. and Alt, F.W., 2001. N-WASP deficiency reveals distinct pathways for cell surface projections and microbial actin-based motility, *Nat Cell Biol.* **3**: 897-904.

Takenawa, T. and Itoh, T., 2001. Phosphoinositides, key molecules for regulation of actin cytoskeletal organization and membrane traffic from the plasma membrane, *Biochim Biophys Acta.* **1533**: 190-206.

Tanimura, N., Nagafuku, M., Minaki, Y., Umeda, Y., Hayashi, F., Sakakura, J., Kato, A., Liddicoat, D.R., Ogata, M., Hamaoka, T. and Kosugi, A., 2003. Dynamic changes in the mobility of LAT in aggregated lipid rafts upon T cell activation, *J Cell Biol.* **160**: 125-135.

Tansey, M.G., Baloh, R.H., Milbrandt, J. and Johnson, E.M., Jr., 2000. GFRalpha-mediated localization of RET to lipid rafts is required for effective downstream signaling, differentiation, and neuronal survival, *Neuron.* **25**: 611-623.

Threadgill, R., Bobb, K. and Ghosh, A., 1997. Regulation of dendritic growth and remodeling by Rho, Rac, and Cdc42, *Neuron.* **19**: 625-634.

Tilghman, R.W. and Hoover, R.L., 2002. E-selectin and ICAM-1 are incorporated into detergent-insoluble membrane domains following clustering in endothelial cells, *FEBS Lett.* **525**: 83-87.

Timasheff, S.N., 1995. Solvent stabilization of protein structure, *Methods Mol Biol.* **40**: 253-269.

Tolias, K. and Carpenter, C.L., 2000. In vitro interaction of phosphoinositide-4-phosphate 5-kinases with Rac, *Methods Enzymol.* **325**: 190-200.

Treanor, J.J., Goodman, L., de Sauvage, F., Stone, D.M., Poulsen, K.T., Beck, C.D., Gray, C., Armanini, M.P., Pollock, R.A., Hefti, F., Phillips, H.S., Goddard, A., Moore, M.W., Buj-Bello, A., Davies, A.M., Asai, N., Takahashi, M., Vandlen, R., Henderson, C.E. and Rosenthal, A., 1996. Characterization of a multicomponent receptor for GDNF, *Nature.* **382**: 80-83.

Uruno, T., Liu, J., Zhang, P., Fan, Y., Egile, C., Li, R., Mueller, S.C. and Zhan, X., 2001. Activation of Arp2/3 complex-mediated actin polymerization by cortactin, *Nat Cell Biol.* **3**: 259-266.

Valensin, S., Paccani, S.R., Ulivieri, C., Mercati, D., Pacini, S., Patrussi, L., Hirst, T., Lupetti, P. and Baldari, C.T., 2002. F-actin dynamics control segregation of the TCR signaling cascade to clustered lipid rafts, *Eur J Immunol.* **32**: 435-446.

van der Goot, F.G. and Harder, T., 2001. Raft membrane domains: from a liquid-ordered membrane phase to a site of pathogen attack, *Semin Immunol.* **13**: 89-97.

Veri, M.C., DeBell, K.E., Seminario, M.C., DiBaldassarre, A., Reischl, I., Rawat, R., Graham, L., Noviello, C., Rellahan, B.L., Miscia, S., Wange, R.L. and Bonvini, E., 2001. Membrane raft-dependent regulation of phospholipase Cgamma-1 activation in T lymphocytes, *Mol Cell Biol.* **21**: 6939-6950.

Vetter, I.R. and Wittinghofer, A., 2001. The guanine nucleotide-binding switch in three dimensions, *Science.* **294**: 1299-1304.

Villalba, M., Bi, K., Rodriguez, F., Tanaka, Y., Schoenberger, S. and Altman, A., 2001. Vav1/Rac-dependent actin cytoskeleton reorganization is required for lipid raft clustering in T cells, *J Cell Biol.* **155**: 331-338.

Viola, A., Schroeder, S., Sakakibara, Y. and Lanzavecchia, A., 1999. T lymphocyte costimulation mediated by reorganization of membrane microdomains, *Science*. **283**: 680-682.

Wahl, S., Barth, H., Ciossek, T., Aktories, K. and Mueller, B.K., 2000. Ephrin-A5 induces collapse of growth cones by activating Rho and Rho kinase, *J Cell Biol.* **149**: 263-270.

Walsh, F.S., Meiri, K. and Doherty, P., 1997. Cell signalling and CAM-mediated neurite outgrowth, *Soc Gen Physiol Ser.* **52**: 221-226.

Wange, R.L. and Samelson, L.E., 1996. Complex complexes: signaling at the TCR, *Immunity*. **5**: 197-205.

Weaver, A.M., Karginov, A.V., Kinley, A.W., Weed, S.A., Li, Y., Parsons, J.T. and Cooper, J.A., 2001. Cortactin promotes and stabilizes Arp2/3-induced actin filament network formation, *Curr Biol.* **11**: 370-374.

Welnhofer, E.A., Zhao, L. and Cohan, C.S., 1997. Actin dynamics and organization during growth cone morphogenesis in Helisoma neurons, *Cell Motil Cytoskeleton*. **37**: 54-71.

Witke, W., Sharpe, A.H., Hartwig, J.H., Azuma, T., Stossel, T.P. and Kwiatkowski, D.J., 1995. Hemostatic, inflammatory, and fibroblast responses are blunted in mice lacking gelsolin, *Cell.* **81**: 41-51.

Wong, K., Ren, X.R., Huang, Y.Z., Xie, Y., Liu, G., Saito, H., Tang, H., Wen, L., Brady-Kalnay, S.M., Mei, L., Wu, J.Y., Xiong, W.C. and Rao, Y., 2001. Signal transduction in neuronal migration: roles of GTPase activating proteins and the small GTPase Cdc42 in the Slit-Robo pathway, *Cell.* **107**: 209-221.

Wu, J., Motto, D.G., Koretzky, G.A. and Weiss, A., 1996. Vav and SLP-76 interact and functionally cooperate in IL-2 gene activation, *Immunity*. **4**: 593-602.

Wulfing, C., Chien, Y.H. and Davis, M.M., 1999. Visualizing lymphocyte recognition, *Immunol Cell Biol.* **77**: 186-187.

Xavier, R., Brennan, T., Li, Q., McCormack, C. and Seed, B., 1998. Membrane compartmentation is required for efficient T cell activation, *Immunity*. **8**: 723-732.

Yamamoto, M., Hilgemann, D.H., Feng, S., Bito, H., Ishihara, H., Shibasaki, Y. and Yin, H.L., 2001. Phosphatidylinositol 4,5-bisphosphate induces actin stress-fiber formation and inhibits membrane ruffling in CV1 cells, *J Cell Biol.* **152**: 867-876.

Yamamoto, M., Toya, Y., Schwencke, C., Lisanti, M.P., Myers, M.G., Jr. and Ishikawa, Y., 1998. Caveolin is an activator of insulin receptor signaling, *J Biol Chem.* **273**: 26962-26968.

Yang, N., Higuchi, O., Ohashi, K., Nagata, K., Wada, A., Kangawa, K., Nishida, E. and Mizuno, K., 1998. Cofilin phosphorylation by LIM-kinase 1 and its role in Rac-mediated actin reorganization, *Nature*. **393**: 809-812.

Yin, H.L. and Janmey, P.A., 2002. Phosphoinositide Regulation of the Actin Cytoskeleton, *Annu Rev Physiol.*

Yonemura, S., Hirao, M., Doi, Y., Takahashi, N., Kondo, T. and Tsukita, S., 1998. Ezrin/radixin/moesin (ERM) proteins bind to a positively charged amino acid cluster in the juxta-membrane cytoplasmic domain of CD44, CD43, and ICAM-2, *J Cell Biol.* **140**: 885-895.

Zallen, J.A., Cohen, Y., Hudson, A.M., Cooley, L., Wieschaus, E. and Schejter, E.D., 2002. SCAR is a primary regulator of Arp2/3-dependent morphological events in Drosophila, *J Cell Biol.* **156**: 689-701.

Zebda, N., Bernard, O., Bailly, M., Welti, S., Lawrence, D.S. and Condeelis, J.S., 2000. Phosphorylation of ADF/cofilin abolishes EGF-induced actin nucleation at the leading edge and subsequent lamellipod extension, *J Cell Biol.* **151**: 1119-1128.

Zhang, J., Shehabeldin, A., da Cruz, L.A., Butler, J., Somani, A.K., McGavin, M., Kozieradzki, I., dos Santos, A.O., Nagy, A., Grinstein, S., Penninger, J.M. and Siminovitch, K.A., 1999.

Antigen receptor-induced activation and cytoskeletal rearrangement are impaired in Wiskott-Aldrich syndrome protein-deficient lymphocytes, *J Exp Med.* **190**: 1329-1342.

Zhang, W., Sloan-Lancaster, J., Kitchen, J., Trible, R.P. and Samelson, L.E., 1998. LAT: the ZAP-70 tyrosine kinase substrate that links T cell receptor to cellular activation, *Cell.* **92**: 83-92.

Zhang, W., Trible, R.P. and Samelson, L.E., 1998. LAT palmitoylation: its essential role in membrane microdomain targeting and tyrosine phosphorylation during T cell activation, *Immunity.* **9**: 239-246.

Zhou, F.Q. and Cohan, C.S., 2001. Growth cone collapse through coincident loss of actin bundles and leading edge actin without actin depolymerization, *J Cell Biol.* **153**: 1071-1084.

Part IV

Regulation of Domain Formation

Chapter 9

GPI-anchored Protein Cleavage in the Regulation of Transmembrane Signals

Frances J. Sharom and Galina Radeva
Department of Chemistry and Biochemistry, University of Guelph, Guelph, Ontario N1G 2W1 Canada

Abstract
The structure of covalently-linked glycosylphosphatidylinositol (GPI) anchors of membrane proteins displayed on the cell surface is described. Evidence of how the GPI-anchors are sorted into membrane rafts in the plasma membrane is reviewed. Proteins are released by hydrolysis of the linkage to the GPI anchor and phospholipases from different sources involved in this process are characterised. The regulation of protein conformation and function resulting from phospholipase cleavage of the GPI anchor is discussed in the context of its role in signal transduction by insulin. In this signalling system, re-distribution of critical membrane components, including GPI-anchored proteins and non-receptor tyrosine kinases, between different raft domains appears to play a central role in the signal transduction pathway.

1. INTRODUCTION

A diverse group of proteins is anchored at the outer surface of eukaryotic plasma membranes via a molecule of covalently-linked glycosylphosphatidylinositol (GPI). The first description of this class of proteins began in the mid-1980's with the structural elucidation of the anchor of the variant surface glycoprotein (VSG) of the parasitic protozoan *Trypanosoma brucei* (Ferguson *et al.*, 1985). Since then, hundreds of GPI-anchored proteins have been

Subcellular Biochemistry, Volume 37: Membrane Dynamics and Domains
Edited by Peter J. Quinn. Kluwer Academic / Plenum Publishers, New York, 2004

identified in all major groups of eukaryotic organisms, including protozoans, yeast, plants, and animals (Eisenhaber *et al.*, 2001), and also reported in the archaebacterium, *Sulfolobus acidocaldarius* (Kobayashi *et al.*, 1997). GPI-anchored proteins are structurally and functionally diverse, and comprise ectoenzymes with hydrolytic activity at the external surface of the plasma membrane, such as 5′-nucleotidase (5′-NTase), alkaline phosphatase, and acetylcholinesterase; protozoan surface coats like trypanosomal VSG; adhesion proteins (e.g. lymphocyte-function-associated molecule, LFA-3); surface antigens, for example, Thy-1; various receptors, such as that for folate, and the prion protein implicated in spongiform encephalopathies. For a more complete list of eukaryotic GPI-anchored proteins, the reader is referred to a recent review article (Ikezawa, 2002).

Although GPI-anchored proteins do not apparently share common features, the presence of the anchor itself appears to confer some important functional and behavioural attributes on proteins to which it is attached. In particular, localization to lipid raft microdomains, and cleavage by endogenous and exogenous phospholipase enzymes, both of which depend on the presence of the GPI anchor, appear to play a major role in transduction of signals across the plasma membrane.

2. STRUCTURE OF THE GPI ANCHOR

The GPI anchor is covalently linked to the C-terminal amino acid of the protein. Figure 1 shows the structure of a typical protein GPI anchor. The protein is anchored in the outer leaflet of the plasma membrane by a moiety of phosphatidylinositol (PI), linked via a glucosamine residue to a central glycan core composed of mannose (and often other sugars), which is in turn coupled to the carboxylate group of the C-terminal amino acid of the protein via phosphoethanolamine. GPI anchors show microheterogeneity at the molecular level, with the existence of variants with different numbers and types of sugars in the glycan core and side-chains, and the presence of extra phosphoethanolamine residues. There are also variations in the lipid moiety, which may include both 1,2-diacyl and 1-alkyl-2-acyl substituents in a glycerol backbone, or a ceramide backbone in some protozoans. Mammalian GPI anchors commonly contain alkylacylglycerol structures. Additional fatty acylation of the *myo*-inositol ring in some proteins alters the susceptibility of the anchor to attack by phospholipase C (see below).

The GPI anchor is attached to proteins post-translationally in the endoplasmic reticulum (ER) of eukaryotic cells. The anchor is independently synthesized from PI, starting with the covalent linkage of glucosamine, followed by deacetylation, and the addition of various other components, including mannose

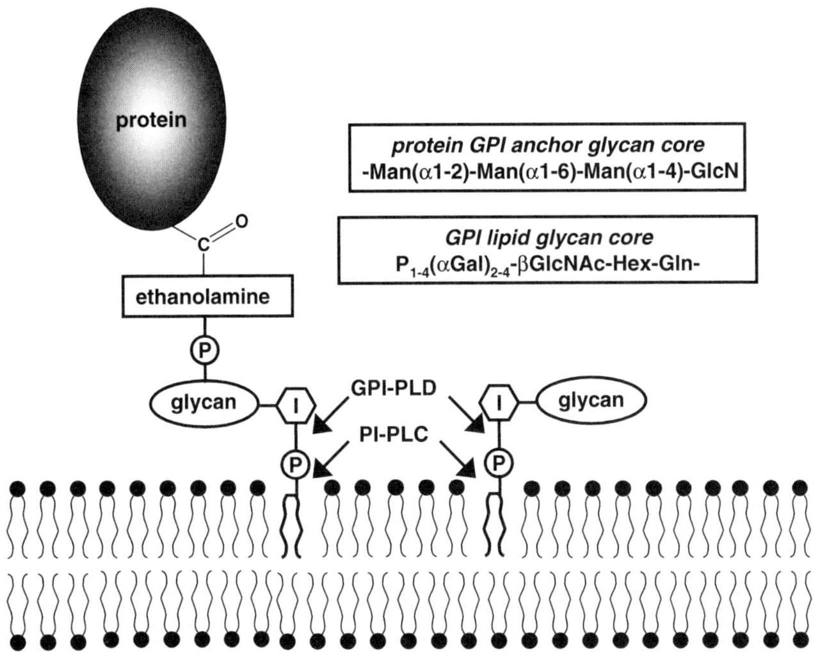

Figure 1. Structure of a typical protein GPI anchor and a free GPI lipid. The glycan core of the protein anchor, generally composed of Man(α1→2)Man(α1→6)Man(α1→4)GlcN, is glycosidically linked to the 6-hydroxyl group of phosphatidylinositol. The C-terminal residue of the protein is linked to the glycan core via a phosphoethanolamine bridge. Possible modifications of the anchor can occur at the core mannose residues, as well as additional acylation of the inositol ring. The proposed core of free GPI lipids is still not fully worked out, but is proposed to consist of phosphate, galactose, N-acetylglucosamine, glucosamine and an unidentified hexose residue (Jones and Varela-Nieto, 1998; Brodbeck, 1998). The GPI anchor and free GPI lipid are cleaved by PI-specific phospholipases C and D at the indicated sites.

and phosphoethanolamine. Once it is completed, the pre-formed anchor is transferred to a specific site upstream of the C-terminal end of the protein in the ER lumen by the action of a transamidase complex, which simultaneously cleaves off the remaining C-terminal peptide. The C-terminal sequence of the protein thus acts as a signal for GPI attachment. It is encoded in the sequences of genomic and cDNA, but does not appear in the final processed protein. Many studies have been carried out to define the characteristics of this sequence (reviewed by Ikezawa, 2002). Typically, a hydrophobic stretch of around 8–9 amino acids is required at the C-terminus, which acts to anchor the protein in the ER membrane prior to anchor attachment. Upstream, the attachment/ cleavage site is generally composed of three small amino acids, such as Ser,

Asn, Ala, Asp, Gly, and Cys. Recently, genome searches for potential cleavage/attachment sites have been used to identify candidates for GPI-anchored protein (Eisenhaber *et al.*, 2000; Eisenhaber *et al.*, 2001). Based on this analysis, about 0.5% of all proteins appear to be GPI-anchored. The GPI biosynthetic pathway has been extensively investigated in trypanosoma, yeast, and mammalian cells (Kinoshita *et al.*, 1997; Kinoshita and Inoue, 2000). Over 20 different proteins are involved, many in the synthesis of the anchor, and at least 6 in the transamidase complex itself (Ikezawa, 2002).

Free (unconjugated) glycosylphosphatidylinositol (GPI) lipids synthesized in the endoplasmic reticulum also make their way to the cell surface (Singh *et al.*, 1996). The glycan core of free GPI lipids appears to be different from that of the GPI anchors covalently linked to proteins (Varela-Nieto *et al.*, 1996; Jones and Varela-Nieto, 1998). Instead, the GPI lipid glycan core appears to consist of galactose, N-acetylgalactosamine, glucosamine, other hexose sugars and both *chiro* and *myo*-inositol, and unlike the common core of protein GPI anchors, it contains no mannose or ethanolamine (Figure 1). Multiple subtypes of these lipids exist in human, mouse, and rat liver, forming a family of closely-related structures (Jones *et al.*, 1997). Their cleavage by phospholipases may serve as a source of soluble inositolphosphoglycans (IPG), which are important second messengers in various hormone-stimulated transmembrane signalling events (see section 7.1).

3. LOCALIZATION OF GPI ANCHORED PROTEINS IN LIPID RAFT MICRODOMAINS

Over the last five years, substantial evidence has emerged suggesting that, rather than having a uniform distribution of lipid and proteins across its surface, the plasma membrane of eukaryotic cells contains specialized regions, or microdomains, with a different composition. The lipid raft hypothesis, which was proposed by Simons and Ikonen (1997), suggested that specialized microdomains composed of cholesterol and sphingolipids (both sphingomyelin and glycosphingolipids) exist in eukaryotic plasma membranes. These domains, which were named lipid rafts, have a more ordered structure, and likely exist in the liquid ordered phase as indicated by studies with lipid model systems (Ahmed *et al.*, 1997). The liquid ordered phase, l_o, has tightly packed and extended acyl chains, characteristic of the gel state, yet its constituent lipids have relatively high lateral mobility, as in the liquid crystalline state (Almeida *et al.*, 1992). The tighter packing of the l_o phase is believed to arise from specific interactions between cholesterol and the longer, more saturated acyl chains commonly found in sphingolipids, relative to the shorter unsaturated chains generally found in membrane glycerophospholipids. The rafts are envisaged as small ordered domains "floating" in a more fluid "sea" enriched in glycerophospholipids (see Figure 2). Thus, lipid immiscibility is proposed as the basis of lipid raft formation (Rietveld and Simons, 1998).

The increased ordering of lipid raft microdomains leads to poor solubility in cold nonionic detergents, such as Triton X-100 (London and Brown, 2000). Thus, one important characteristic of lipid rafts is their resistance to solubilization by such detergents at 4°C (Schroeder *et al.*, 1994; Hanada *et al.*, 1995). Treatment of intact cells with cold nonionic detergent leads to the isolation of what is termed detergent-resistant membranes (DRMs) (for example, see Melkonian *et al.*, 1995), which are believed to represent coalesced rafts domains.

Caveolae are small flask-shaped plasma membrane invaginations that appear to be a specialized type of lipid raft microdomain. They are characterized by the presence of a cytoplasmic coat composed of the caveolin proteins,

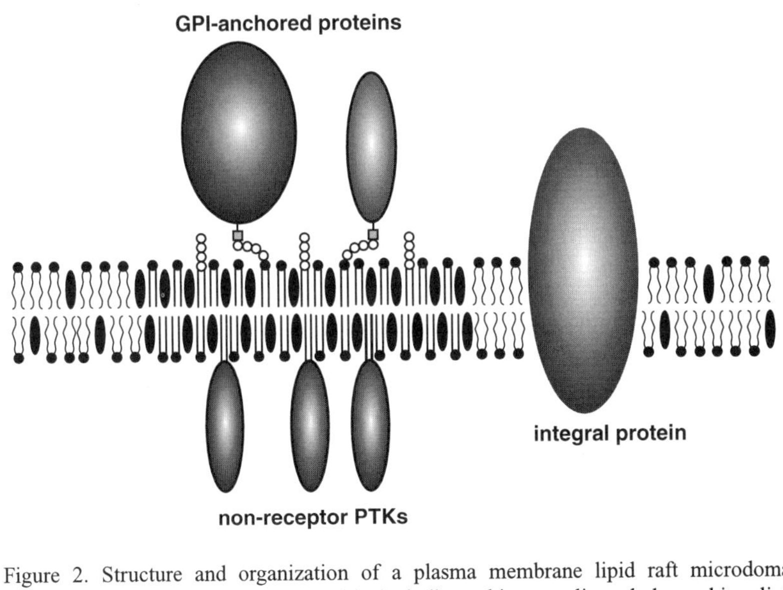

Figure 2. Structure and organization of a plasma membrane lipid raft microdomain. Cholesterol associates with sphingolipids, including sphingomyelin and glycosphingolipids, to form more tightly packed raft microdomains with increased order. The longer, saturated acyl chains of these species play an important role in stabilizing raft structure. Rafts are enriched in GPI-anchored proteins and acylated proteins, such as the Src family of protein tyrosine kinases ($p56^{lck}$, $p60^{fyn}$, $p53/p56^{lyn}$ etc). Most transmembrane integral proteins are excluded from raft microdomains. Rafts are surrounded by bulk lipids composed mainly of glycerophospholipids with shorter, unsaturated acyl chains, forming a fluid, more loosely packed matrix. The size and lifetime of lipid rafts in biological membranes are still controversial. Adapted from Sharom and Lehto (2002).

caveolin-1, -2, and -3, which bind cholesterol avidly. Caveolae appear to be involved in uptake of small molecules and ions via potocytosis, non-clathrin-mediated endocytosis, and cholesterol homeostasis, as well as sorting of membrane proteins. They are found in most terminally differentiated mammalian cell types and are particularly abundant in adipocytes, where they occupy more than 20% of the cell surface, and appear to play a major role in transmembrane signalling via insulin. However, some cells (such as lymphocytes) do not express caveolins, and are devoid of caveolae.

The presence of the GPI anchor, which is also characterized by a relatively long saturated acyl chain component, appears to direct proteins to localization within the external leaflet of lipid rafts and caveolae (Schroeder *et al.*, 1994; Benting *et al.*, 1999). The association of GPI-anchored proteins with lipid rafts appears to take place at the level of the Golgi apparatus during outward trafficking of the proteins, and helps to target them to the apical surface of polarized epithelial cells (Muñiz and Riezman, 2000). The cytoplasmic face of rafts is commonly enriched in acylated proteins , especially those containing two fatty acid chains, such as the non-receptor protein tyrosine kinases (PTKs) of the Src family, which include $p56^{lck}$, $p60^{fyn}$, and $p53/56^{lyn}$ (Cebecauer *et al.*, 1998). In contrast, prenylated proteins are reported to be excluded from lipid raft domains (Melkonian *et al.*, 1999). Most transmembrane integral proteins are not present in lipid rafts, although some exceptions have been noted (Salzer and Prohaska, 2001).

Several techniques have been used to demonstrate the existence of lipid raft domains in the plasma membranes of intact cells, including single particle tracking (Sheets *et al.*, 1997; Pralle *et al.*, 2000), fluorescence resonance energy transfer microscopy (Varma and Mayor, 1998), chemical crosslinking (Friedrichson and Kurzchalia, 1998), and atomic force microscopy (Giocondi *et al.*, 2000; Yuan *et al.*, 2002). However, there is still controversy regarding the size and lifetime of native lipid raft structures, as well as the fraction of the plasma membrane surface that they cover. Cholesterol appears to be important for raft formation, and its depletion, by agents such as methyl-β-cyclodextrin or saponin, has been an important tool in investigating the presence and functional significance of rafts in intact cells. However, not all raft microdomains appear to require high levels of cholesterol to stabilize their structure. The brush border membrane of enterocytes has been reported to contain low cholesterol rafts, which are not disrupted by cholesterol depletion (Hansen *et al.*, 2001). Using carbonate extraction, Müller *et al.* reported that rat adipocytes contained distinct species of detergent/carbonate-insoluble raft fractions, with higher buoyant density and lower amounts of cholesterol, caveolin, and GPI-anchored proteins (Müller *et al.*, 2002b). These atypical raft structures, which they named lcDIG (low cholesterol detergent-insoluble glycolipid-enriched raft domains), were distinct from high cholesterol rafts (hcDIG). Two separate populations of high-cholesterol and low-cholesterol lipid raft structures also appear to exist in 3T3-L1 fibroblasts and adipocytes.

Lipid rafts are now known to play a central role in various cellular signalling processes, primarily by organizing the spatial distribution of receptors and the proteins that regulate them (Cary and Cooper, 2000). Rafts also appear to be targets for invasion of eukaryotic cells by toxic and infectious agents, including bacterial toxins such as aerolysin, and pathogenic bacteria (Fantini *et al.*, 2000; Rosenberger *et al.*, 2000). Likewise, enveloped viruses appear to invade cells via rafts, and newly formed virus particles bud off from these regions of the membrane (Vincent *et al.*, 2000; Liao *et al.*, 2001).

4. CLEAVAGE OF THE GPI ANCHOR BY PI-SPECIFIC PHOSPHOLIPASES

4.1. Bacterial phospholipases

Several bacterial species secrete PI-specific type C phospholipases (PI-PLC) extracellularly, including *Bacillus cereus, Bacillus thuringiensis, Staphylococcus aureus, Listeria monocytogenes*, and *Clostridium novii*. Bacterial PI-PLCs are able to hydrolyse mammalian GPI anchors, and have been used extensively to both identify anchored proteins and aid in their structural characterization (Ikezawa, 1991). Expression of PI-PLC has been linked to pathogenicity in some bacterial species (Mengaud *et al.*, 1991; Daugherty and Low, 1993). PI-PLCs degrade PI by a two-step mechanism, first involving the production of diacylglycerol and *myo*-inositol-1,2-cyclic phosphate (cIP), which is subsequently ring-opened to produce *myo*-inositol-1-phosphate (I1P) (Figure 3). The cIP tends to accumulate, since the second hydrolysis step is slow. Acylation of the inositol ring in some species, often with palmitate, e.g. human acetylcholinesterase (Roberts *et al.*, 1988), blocks the action of bacterial PI-PLC, so that GPI anchors with this modification are resistant to its action (Rosenberry, 1991). Cleavage by bacterial PI-PLC produces diacylglycerol (DAG), and a soluble version of the protein with an inositolphosphoglycan (IPG) moiety linked at the C-terminus. The glycan portion of the anchor remaining on the cleaved protein contains cIP, forming an epitope known as the cross-reacting determinant or CRD. The CRD can be detected by specific antibodies (Broomfield and Hooper, 1993), and is the hallmark of cleavage by a PI-PLC enzyme, since it is not produced by phospholipase D.

High resolution X-ray crystal structures have been determined for the PI-PLC from *L. monocytogenes* (PDB 1AOD, Moser *et al.*, 1997) and *B. cereus* (Heinz *et al.*, 1995; Heinz *et al.*, 1996), both free (PDB 1PTD) and complexed with *myo*-inositol (PDB 1PTG) or glucosaminyl($\alpha 1 \rightarrow 6$)-D-*myo*-inositol (PDB

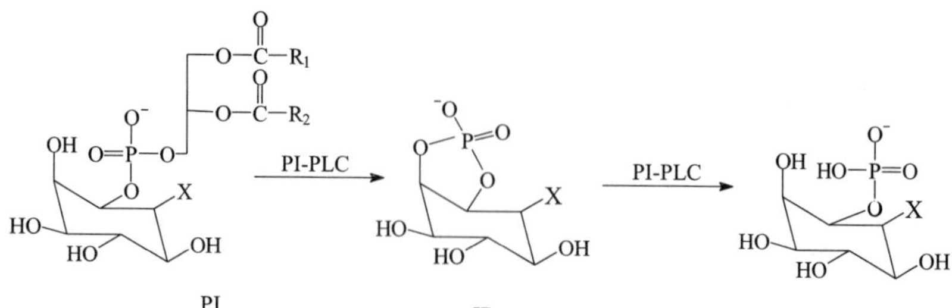

Figure 3. Cleavage of PI by bacterial and trypanosomal PI-PLC is proposed to take place via a two-step reaction sequence. The first step involves cleavage of PI to form a cyclic structure, *myo*-inositol-1,2-cyclic phosphate (cIP; the CRD), which is then hydrolysed to give *myo*-inositol-1-phosphate (I1P). The position of the glycan portion of the GPI anchor is indicated by an X.

1GYM). The *B. cereus* protein (*Bc*-PI-PLC) is made up of an αβ-barrel containing 8 β-strands, a structure which does not resemble that of any known eukaryotic phospholipases. The inositol moiety is involved in hydrogen bonding interactions with specific residues in a deep active site pocket, whereas the glucosamine residue lies at the entrance to the active site, exposed to solvent. Free *myo*-inositol occupies the same position as the inositol residue of glucosaminyl(α1→6)-D-*myo*-inositol, suggesting that the mechanism of cleavage of PI and GPI anchors is essentially the same. The phospholipase has almost no contacts with the glycan portion of the substrate, which likely accounts for the observation that it can cleave GPI anchors of varying structure. The *B. thuringiensis* enzyme (*Bt*-PI-PLC) differs from that of *B. cereus* at only 8 of 298 amino acids, and the two enzymes have almost identical properties (Griffith *et al.*, 1991), so they are likely to operate by the same mechanism. Bacterial phospholipases require interfacial activation by a membrane surface to display high levels of catalytic activity. They contact the membrane via an interfacial binding surface (IBS) located on a flat region surrounding the active site. Membrane fluidity greatly influenced the activity of *Bc*-PI-PLC on a protein GPI anchor, likely by affecting the ability of aromatic residues in the IBS to penetrate the bilayer and stabilize the enzyme-membrane complex (Lehto and Sharom, 2002a).

4.2. GPI-PLCs of parasitic protozoans

Several parasitic protozoans, for example, *T. brucei* and *Leishmania*, contain an endogenous GPI-specific PLC, which converts membrane-bound VSG (*T. brucei*) or gp63 (*Leishmania*) to hydrophilic soluble forms. Like the bacterial phospholipases, trypanosomal PI-PLC also proceeds via a cIP intermediate. The exact physiological role of the endogenous GPI-PLC of *Trypanosoma* remains unclear. Deletion of the gene encoding this phospholipase showed that it is not essential, but its loss compromises the ability of the parasite to grow in mammalian cells (Webb *et al.*, 1997), suggesting that it plays some role in infection and pathogenicity. It may be involved in removal of the surface coat of the parasite, thus aiding in the antigenic variation phenomenon that is critical for survival of the parasite in its host, or it may remove the coat proteins when the parasite enters the digestive tract of the tsetse fly, the insect host. However, the possibility remains that the GPI-PLC is involved in hydrolysing phosphatidylinositol-4,5-bisphosphate (PIP_2) to generate the second messengers *myo*-inositol-1,4,5-trisphosphate (IP_3) and *sn*-1,2-diacylglycerol, both of which function in signalling cascades.

4.3. Mammalian phospholipase D

Mammalian GPI-specific phospholipase D (GPI-PLD) is found in high concentrations in the serum (Low and Prasad, 1988; Hoener and Brodbeck, 1992; Huang *et al.*, 1990; Davitz *et al.*, 1987), although its source has been unclear until recently. This phospholipase is apparently secreted into the circulation, likely by pancreatic islet cells in response to stimulation by glucose and other insulin secretagogues. GPI-PLD co-localized with insulin within beta-TC3 cells (a mouse insulinoma line), and is likely secreted via the regulated pathway in these cells (Deeg and Verchere, 1997). GPI-PLD activity has also been reported in other mammalian tissues, including liver (Heller *et al.*, 1992). The physiological role and significance of this activity, which appeared to be membrane-bound, is not clear.

The serum enzyme has been purified, cloned, and characterized biochemically and enzymatically (Deeg and Davitz, 1995). Hydrolysis of GPI anchors by GPI-PLD gives rise to phosphatidic acid, leaving an inositol glycan tail on the formerly anchored protein. The enzyme appears to recognize glucosamine linked to PI, which is found in all GPI anchors as its minimal structural motif. Unlike bacterial and trypanosomal PI-PLC, GPI-PLD does not proceed through the cIP intermediate, so that the cleaved proteins do not cross-react with anti-CRD antibodies. GPI-PLD is active against GPI anchors containing an acylated inositol moiety, a modification which confers resistance to cleavage by PI-PLC.

Initial reports indicated that GPI-PLD was only active against GPI-anchored proteins in the presence of detergent, and was not able to cleave the anchors of proteins in a native membrane context (Low and Huang, 1991). The enzyme appeared to require activation, since trypsin cleaved it into products which remained associated with each other, and displayed increased activity towards GPI-anchored substrates. GPI-PLD was taken up from the medium into neuroblastoma cells, which processed it proteolytically to give catalytically active fragments (Hari *et al.*, 1997). Thus, it appears that GPI-PLD may act on its substrates intracellularly, after uptake from the serum and proteolytic activation. In support of this view, the enzyme is enriched in a lysosomal fraction from rat liver (Hari *et al.*, 1996). More recently, GPI-PLD has been reported to be post-translationally modified by glycosylation, leading to an increase in molecular mass from the immature form to a 115 kDa mature protein (Tsujioka *et al.*, 1998). GPI-PLD appears to undergo a conformational change upon processing, such that the mature form is capable of hydrolysing GPI anchors from intact membranes in the absence of detergent, whereas the immature form is not.

4.4. Mammalian phospholipase C

Since the first discovery of bacterial PI-PLC, endogenous mammalian GPI-specific PLCs have been postulated to serve as important regulatory factors, reducing surface expression of GPI-anchored proteins, while simultaneously increasing the levels of soluble protein. This hypothesis is supported by the discovery that soluble forms of several GPI-anchored proteins are indeed found in tissues (Piec and Le Hir, 1991; Vogel *et al.*, 1992), and body fluids such as synovial fluid (Huang *et al.*, 2001), urine (Park *et al.*, 2001), and cerebrospinal fluid (Almqvist and Carlsson, 1988). However, it has been hard to obtain conclusive proof of the existence of GPI-specific PLCs in mammalian cells. In particular, assaying GPI-specific phospholipase C in intact cells and tissues is difficult, since it is likely to be contaminated with serum-derived GPI-PLD (see section 4.3), which is abundant and highly active (Eliakim *et al.*, 1990). The product of GPI-PLD hydrolysis, phosphatidic acid, is rapidly converted to diacylglycerol by a phosphatase present in most tissues, thus giving the erroneous impression that a phospholipase C was responsible for the initial cleavage of the GPI anchor (Heller *et al.*, 1992).

More definitive proof of the existence of a GPI anchor-cleaving endogenous phospholipase C has been obtained by immunoblotting analysis of the soluble forms of putative cleaved proteins with anti-CRD antibodies that selectively recognise the cIP epitope formed on cleavage of the anchor by a type C, but not a type D, phospholipase (see sections 4.1 and 4.3). Using this approach, Vogel *et al.* (1992) showed that the soluble forms of 5′-NTase found in the electric organ of the electric-ray (*Torpedo marmorata*) and bovine cerebral cortex were generated by PI-PLC cleavage of the GPI-anchored membrane-bound form. More recently, Movahedi and Hooper (1997) found that membrane dipeptidase was released from intact 3T3-L1 adipocytes after treatment with insulin. The released enzyme was located in the aqueous phase after Triton X-114 partitioning, indicating that it was hydrophilic, and had lost its GPI anchor. The hydrophilic form of the enzyme was recognized by antibodies against the inositol-1,2-cyclic monophosphate CRD, showing that it had resulted from cleavage of the GPI anchor by an endogenous phospholipase C. Renal membrane dipeptidase (RDPase) is released spontaneously from porcine proximal tubules on incubation at 37°C (Park *et al.*, 2001). Again, the cleaved hydrophilic protein reacted with anti-CRD antibodies, strongly suggesting that the enzyme was released by a PI-PLC found endogenously in kidney tissue. RDPase is found in soluble form in the urine of several mammalian species. Western blotting of the protein isolated from human urine showed that it also reacted with anti-CRD antibodies, indicating that it was formed by the action of an endogenous PI-PLC (Park *et al.*, 2002a). This report represents the first evidence for an endogenous mammalian type C phospholipase that can cleave GPI anchors *in vivo*. The level of soluble RDPase in urine is decreased in acute renal failure, a pathological condition in which the level of nitric oxide, NO, is elevated. Park *et al.* recently discovered that NO donors were able to rapidly inhibit the release of RDPase from porcine kidney proximal tubules, likely by inhibiting the putative phospholipase enzyme (Park *et al.*, 2002b). They suggested that NO is involved as an inhibitory regulator of shedding of GPI-anchored RDPase in this disease state.

5. PHOSPHOLIPASE CLEAVAGE OF THE GPI ANCHOR CAN REGULATE PROTEIN CONFORMATION AND FUNCTION

The presence of the GPI anchor appears to affect the conformation of proteins to which it is linked and this may, in turn, affect protein function. Low and Finean (1978) first noted that incubation of porcine lymphocytes with PI-PLC from *S. aureus* resulted in substantially higher levels of catalytic activity for the ecto-enzyme, 5′-NTase. Further investigation showed that when PI-PLC from both *B. thuringiensis* and *S. aureus* cleaved the GPI anchor of this protein, it was catalytically activated (Sharom *et al.*, 1996). Purified 5′-NTase was reconstituted into lipid bilayers, and a detailed kinetic study of the effects of *Bt*-PI-PLC cleavage indicated that V_{max} was increased following release of the protein in soluble form (Figure 4B and C), whereas K_M remained essentially unchanged (Lehto and Sharom, 1998). The turnover number of membrane-bound 5′-NTase varied depending on the host lipid making up the bilayer, while that of the soluble enzyme appeared independent of the lipid originally used for reconstitution. Thus it was suggested that insertion of the

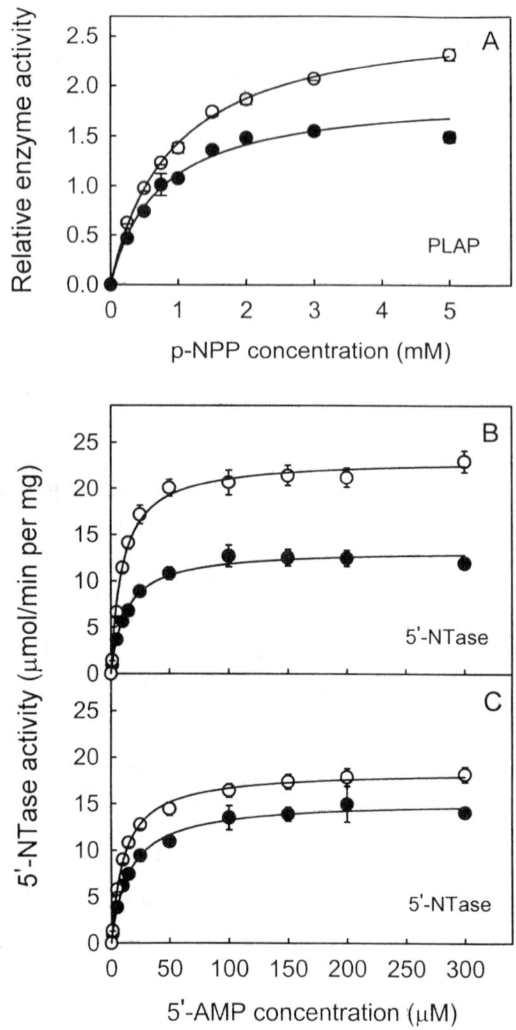

Figure 4. Catalytic activation of GPI-anchored ectoenzymes after PI-PLC cleavage. (A) Cleavage by *Bc*-PI-PLC of purified human placental alkaline phosphatase (PLAP) reconstituted into lipid bilayer vesicles. Enzyme activity was determined using hydrolysis of *p*-nitrophenylphosphate (*p*NPP) in a colorimetric assay. (B) and (C) Cleavage of purified porcine lymphocyte 5′-nucleotidase (5′-NTase) by *Bt*-PI-PLC following reconstitution into lipid bilayers of egg PE (B) and SCRL (C), a sphingolipid/cholesterol-rich lipid mixture known to form detergent-insoluble lipid rafts. Enzyme activities were measured before (•) and after (○) complete cleavage of the GPI anchor by the phospholipase. The solid lines represent the best fit of the kinetic data to the Michaelis-Menten equation.

GPI anchor into a membrane or lipid bilayer alters the protein to reduce its catalytic efficiency, perhaps by imposing conformational restraints on the active site, which are relaxed on hydrolysis of the anchor (Lehto and Sharom, 1998). For further discussion of this point, the reader is referred to a recent review (Sharom and Lehto, 2002). Several other GPI-anchored enzymes also display changes in catalytic activity upon anchor cleavage, including dipeptidase from the porcine kidney microvillar membrane (Brewis *et al.*, 1994), and the PH-20 hyaluronidase from guinea pig sperm (Gmachl *et al.*, 1993). We have recently determined that purified reconstituted human placental alkaline phosphatase (PLAP) is also catalytically activated following release from the lipid bilayer by bacterial PI-PLC (Figure 4A; T. To, J.W.K. Chu and F.J. Sharom, unpublished data).

It was recently reported that a GPI-anchored metalloproteinase from the non-pathogenic trypanosomatid, *Herpetomonas samuelpessoai*, was activated upon anchor hydrolysis by *Bt*-PI-PLC (Santos *et al.*, 2002). The 115 kDa protein appeared to be present at the cell surface in an inactive form, suggesting that it is activated by the action of PI-PLC. Thus, an endogenous PI-PLC produced by *H. samuelpessoai* may regulate the release and activation of the proteinase enzyme *in vivo*. Similarly, incubation of either intact merozoites or isolated schizont membranes from *Plasmodium falciparum* with PI-PLC resulted in release of a 76 kDa plasma membrane protein with proteolytic activity not seen in the membrane-bound form (Braun-Breton *et al.*, 1988).

GPI anchor cleavage may also affect the conformation of proteins with no catalytic activity. Barboni *et al.* were the first to note that removal of the GPI anchor from the Thy-1 antigen resulted in a major conformational change in the protein, which reduced binding of the OX-7 monoclonal antibody (Barboni *et al.*, 1995). More recently, Kukulansky and co-workers (1999) confirmed that several different monoclonal antibodies were completely unable to bind to Thy-1 after cleavage of its GPI anchor, and its reactivity with polyclonal antibodies was also greatly reduced. Treatment with PI-PLC was sufficient to release Thy-1 from pre-formed Thy-1-antibody complexes, indicating that a substantial conformational change takes place in the protein upon removal of the anchor. Carcinoembryonic antigen also shows decreased antibody binding after cleavage of the GPI anchor (Durbin *et al.*, 1994), suggesting that the anchor may have the general effect of altering the conformation of proteins to which it is linked. Alterations in GPI-anchored protein function following anchor removal have been reported for several other proteins, and may arise from conformational changes that alter their tertiary structure. A variant folate receptor with an acylated inositol ring displayed high folate binding affinity, which was reduced after removal of the anchor (Wang *et al.*, 1996), and two cAMP-binding proteins in yeast show a large decrease in the K_M for cAMP after PI-PLC or GPI-PLD-mediated cleavage of the GPI anchor (Müller and

Bandlow, 1994). The implications for transmembrane signalling are two-fold. GPI-anchored proteins released as a result of anchor cleavage mediated by a signalling process will likely have altered properties relative to the membrane-bound forms, such as enhanced catalytic activity. They will also be free to move to new locations via the circulation, and may exert that activity in new compartments, such as the serum, urine, or other body fluids.

The GPI anchor may mediate some of its effects on protein conformation and function via interaction of the protein with the membrane surface. Based on molecular dynamics simulations and 2-dimensional NMR experiments, Homans *et al.* (1988) suggested that the glycan moiety of trypanosomal VSG adopted an extended conformation on the surface of the membrane, where it occupied a large area. Later modelling studies of the GPI anchor of Thy-1, which is a small (28 kDa) species with a single immunoglobulin fold, suggested that the anchor lay within a carbohydrate-binding pocket within the protein (Rademacher *et al.*, 1991), or was tightly folded between the protein and the membrane surface (Barboni *et al.*, 1995). The lipid interactions of purified Thy-1 reconstituted into bilayer vesicles of defined phospholipids were studied using differential scanning calorimetry (Reid-Taylor *et al.*, 1999). Examination of the effects of Thy-1 on the lipid melting transition indicated that the protein perturbed about 53 membrane lipids. This is many more than would be predicted based on the size of the GPI anchor, which might be expected to interact with ~12 lipids in the outer leaflet of the membrane. Thus, it seems likely that the GPI anchor of Thy-1 is folded down onto the membrane surface, so that the protein itself interacts with the bilayer to perturb additional lipids (Figure 5). Examination of the interaction of membrane lipids with reconstituted Thy-1 suggested that the conformation of the protein may also change at high protein:lipid ratios in response to steric crowding, such that its interaction with the bilayer is reduced (Reid-Taylor *et al.*, 1999), perhaps by switching to a conformation in which the GPI anchor extends away from the membrane surface. Recent fluorescence studies using purified reconstituted PLAP also suggested that the protein is likely in contact with the lipid bilayer (Lehto and Sharom, 2002b).

Contact between GPI-anchored proteins and the membrane surface would allow transmission of structural changes from bilayer lipids to the protein, which might in turn modify their catalytic properties and behaviour. Biochemical studies of 5′-NTase, both in intact cells (Brasitus and Schachter, 1980) and reconstituted lipid bilayers (Sharom *et al.*, 1985), indicated that a change in the phase state of the lipid membrane was correlated with alterations in the activation energy. Modulation of the catalytic properties of 5′-NTase by the bilayer implies that the protein portion of the enzyme is in direct contact with the membrane surface. Recent experiments using purified reconstituted PLAP have shown that the activation energy of catalysis is also modulated by

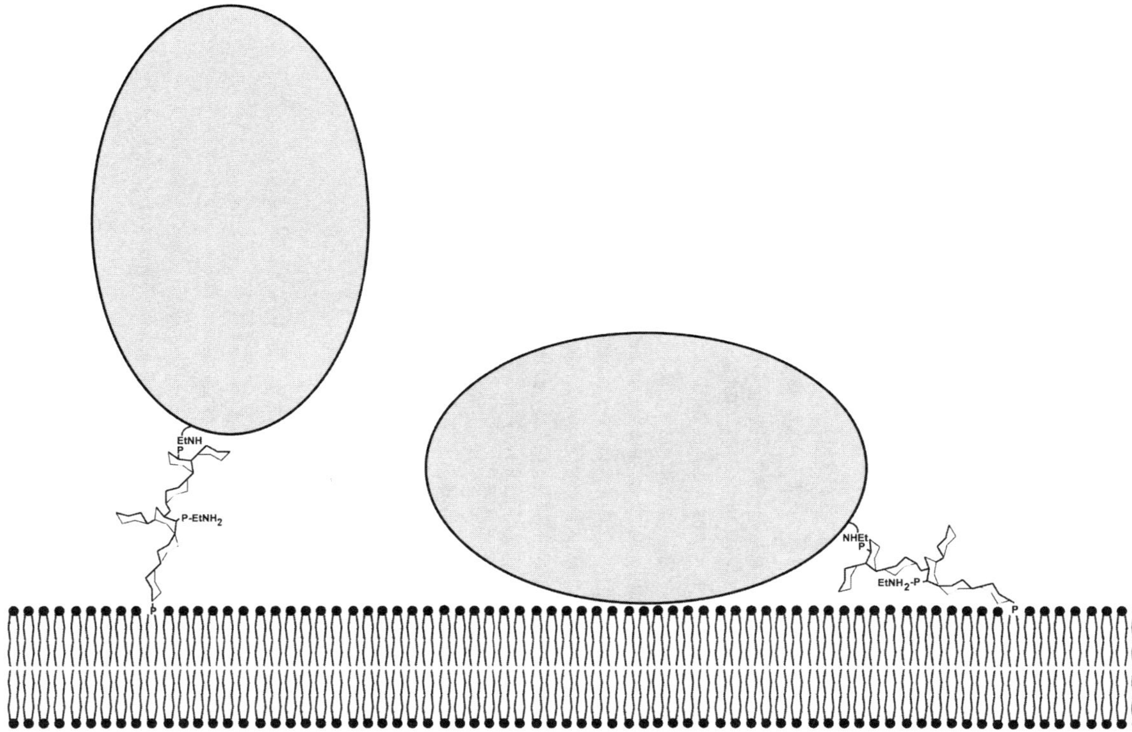

Figure 5. Model for potential interaction of GPI-anchored proteins with the membrane surface. GPI-anchored proteins may exist with the anchor in an extended conformation (left), projecting away from the bilayer surface, so that no contact exists between the protein and the membrane. However, for many proteins, the anchor may fold back to lie down on the membrane surface (right), so that the biophysical state of the lipid bilayer is observed to affect the catalytic properties and function of the protein. Surface crowding at high protein:lipid ratios may force GPI-anchored proteins to adopt the extended conformation.

bilayer phase transitions (T. To, J.W.K. Chu and F.J. Sharom, unpublished data). Thus the function of GPI-anchored proteins may, in general, be influenced by the biophysical properties of the membrane.

6. LIPID RAFTS AND GPI-ANCHORED PROTEINS CAN MEDIATE SIGNAL TRANSDUCTION

Lipid rafts are proposed to act as platforms or "hotspots" for signal transduction across the plasma membrane through a large number of cellular receptors (Langlet *et al.*, 2000; Cherukuri *et al.*, 2001; Viola, 2001). Their primary role appears to be spatial segregation of various receptors, and other agents that regulate receptor activity, such as kinases and phosphatases, into discrete microdomains within the plasma membrane. GPI-anchored proteins are able to mediate transduction of transmembrane signals in several cell types, most notably lymphocytes (the T- and B-cells of the immune system) and other hematopoietic cells. Crosslinking of GPI-anchored species via antibodies in these cells leads to protein phosphorylation, an increase in intracellular calcium concentration, and activation of transcription factors. Proteins that have been demonstrated to transduce signals when GPI-anchored include 5′-NTase, Thy-1 antigen, Ly-6/TAP, and decay accelerating factor (DAF). The signal itself appears to be transmitted by non-receptor PTKs of the Src family, such as $p56^{lck}$, $p60^{fyn}$ and $p53/56^{lyn}$, which are associated with the cytoplasmic face of lipid rafts via dual acylation (see Figure 2). It is thought that crosslinking of GPI-anchored species leads to raft coalescence, generating threshold levels of activation of these raft-associated PTKs.

7. CLEAVAGE OF PROTEIN GPI ANCHORS AND GPI PHOSPHOLIPIDS IN SIGNAL TRANSDUCTION

7.1. Cleavage of GPI lipids and production of IPG

A number of polypeptide growth factors, hormones and cytokines have been shown to stimulate hydrolysis of free GPI lipids as part of their mechanism of signal transduction. These include insulin, insulin-like growth factor-1 (IGF-1), the neurotrophins (such as nerve growth factor), transforming growth factor-β, epidermal growth factor, interleukins-1 and -2, chorionic gonadotropin, thyrotropin releasing hormone, and thyroid stimulating hormone (Jones and

Varela-Nieto, 1998; Petitfrere *et al.*, 1998). Hydrolysis of free GPI lipids leads to the production of a family of water-soluble oligosaccharide species termed inositolphosphoglycans (IPG) or phosphoinositolglycans (PIG). IPG compounds have been purified from mammalian organs (such as human and rat liver) and characterized structurally. They possess biological activity when added exogenously to cells, and mimic many of the biological effects of a variety of extracellular ligands, including hormones such as insulin, growth factors, and cytokines (Brodbeck, 1998). IPG modulate several key metabolic pathways controlled by insulin signalling; for example, they affect the activity of a number of enzymes, stimulate lipogenesis, and enhance glucose utilization. More recently, IPG with defined oligosaccharides have been chemically synthesized and also shown to be highly insulin mimetic (Frick *et al.*, 1998).

Jones *et al.* (1997) showed that GPI-PLD from mammalian serum was able to hydrolyse free GPI lipids to IPG. The IPG species generated by GPI-PLD cleavage displayed similar biological effects on intact cells to the compounds generated by PI-PLC, suggesting that phospholipase D might also play a role in signal transduction. When free GPI lipids were purified from rat liver and incorporated into large unilamellar vesicles of defined lipids, treatment with bacterial PI-PLC, trypanosomal PI-PLC, or serum GPI-PLD resulted in hydrolysis to IPG at a rate that depended on both the phospholipase and the bilayer composition (Villar *et al.*, 1998). Both exogenously added and endogenously expressed GPI-PLD were able to hydrolyse GPI anchors of proteins early in the secretory pathway (likely the ER), resulting in increased production of DAG, translocation of protein kinase C-α (PKCα) to the ER membrane, and phosphorylation of both PKCα and the proto-oncogenic protein, c-Fos (Tsujioka *et al.*, 1999).

7.2. Insulin signalling

7.2.1. Insulin-stimulated hydrolysis of GPI anchors

It was noted many years ago that treatment of myocytes or adipocytes with insulin resulted in the release of several GPI-anchored proteins into the extracellular medium. Chan and co-workers reported that lipoprotein lipase, which plays an important role in plasma lipoprotein metabolism, was released from insulin treated 3T3-L1 adipocytes by cleavage of its GPI anchor (Chan *et al.*, 1988), and they proposed that activation of an insulin-dependent PI-PLC was responsible (Figure 6). Another group showed that myocytes released GPI-anchored alkaline phosphatase upon insulin stimulation (Romero *et al.*, 1988), again suggesting the action of a phospholipase. Many other studies subsequently confirmed the release from myocytes or adipocytes of many different

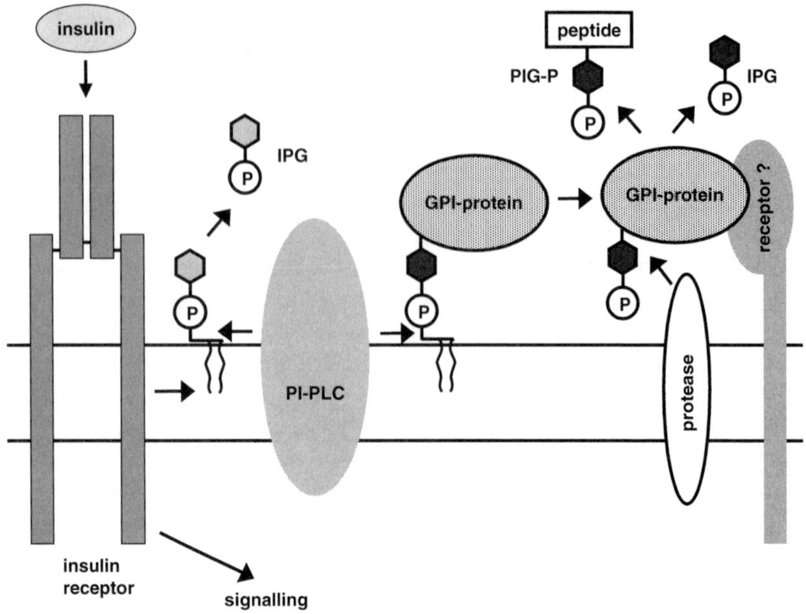

Figure 6. Cleavage of protein GPI anchors and free GPI lipids by signalling via the insulin receptor. Insulin binding to the receptor is assumed to activate an endogenous PI-specific phospholipase C enzyme, which cleaves free GPI lipids to form IPG compounds directly. In the case of GPI-anchored proteins, the released protein may be held at the plasma membrane via interaction with a specific receptor. Subsequent cleavage of the released protein by a surface protease can then generate IPG, or PIG-P compounds with a peptide still attached. The structures of the inositolglycan portions of IPG generated from GPI lipids and GPI protein anchors are expected to be different (see Figure 1). IPG/PIG-P formed from either GPI lipids or protein GPI anchors mimic many of the effects of insulin when added to intact adipocytes and cardiomyocytes.

GPI-anchored proteins in soluble form following treatment with insulin (Movahedi and Hooper, 1997; Klip *et al.*, 1988; Lisanti *et al.*, 1989). Insulin and EGF also released GPI-linked alkaline phosphatase from Hep-2 cells (Roberts *et al.*, 1990), again suggesting activation of an endogenous PI-specific PLC. Activation of this phospholipase appears to be controlled via protein phosphorylation by the activated insulin receptor tyrosine kinase (Villalba *et al.*, 1990).

The sulfonylurea drug glimepiride, is an insulin-mimetic drug that acts synergistically with insulin to stimulate glucose transport and metabolism in fat and muscle cells *in vitro*. Müller *et al.* (1994) were the first to show that the action of this drug is linked to GPI anchor hydrolysis. Glimepiride treatment of intact 3T3 adipocytes led to rapid release of several GPI-anchored proteins, including 5′-NTase, lipoprotein lipase, and a cAMP-binding protein (Gce1), from the plasma membrane into the extracellular medium. The released proteins were hydrophilic, retained inositol within their structure, and cross-reacted with anti-CRD antibodies, all of which suggested that they arose from PI-PLC cleavage. Thus, sulfonylureas may utilize part of the insulin signalling cascade, likely by activating an endogenous GPI-specific phospholipase C. Interestingly, the cleaved GPI-anchored proteins remained associated with the plasma membrane, likely as a result of interaction with a putative specific receptor, and could be released by competitors such as cIP.

7.2.2. Insulin-mimetic action of IPG

The GPI anchors of membrane-bound proteins also contain IPG as part of their core structure. Although some of the actions of insulin on cells are proposed to be mediated by IPG, it is not clear to what extent the IPG are derived from cleavage of GPI anchors of membrane proteins, or from hydrolysis of free GPI lipids. Early reports showing the release of GPI anchored proteins by insulin led to suggestions that IPG could be derived from membrane protein anchors (Chan *et al.*, 1988; Romero *et al.*, 1988; Lisanti *et al.*, 1989). Generation of IPG from protein GPI anchors would require the sequential action of both a phospholipase and a protease enzyme, whereas only a phospholipase would be needed if GPI lipid was the source (Figure 6). Since GPI lipids and protein GPI anchors appear to have dissimilar core structures (Figure 1), the structures of the two classes of IPG are also expected to be different.

Müller and co-workers isolated compounds, termed PIG-P, from the GPI-anchored protein, Gce1p, which is a cAMP-binding protein associated with the cell wall in *Saccharomyces cerevisiae*. A similar GPI-anchored cAMP-binding protein is present in rat and 3T3-L1 adipocytes, and its anchor is cleaved on stimulation with insulin (Müller *et al.*, 1993; Müller *et al.*, 1994). Sequential digestion of purified Gce1p with V8 protease and PI-PLC led to the large-scale production of C-terminal tripeptide fragments linked to the conserved glycan core of protein GPI anchors (Müller *et al.*, 1997). These PIG-P compounds were potent inducers of the metabolic actions of insulin in adipocytes, cardiomyocytes, and diaphragms, stimulating lipogenesis and glucose transport. In later work, Müller and co-workers chemically synthesized many defined IPG structures, and showed that they also were potent mimics of insulin action, stimulating up to 90% of the maximal insulin effect at concentrations of 2–10 μM (Frick *et al.*, 1998). However, Martin-Lomas *et al.* (2000) carried out chemical synthesis of IPG containing the conserved core of protein GPI anchors, and reported that they had no biological activity. They concluded that IPG second messengers are not, in fact, derived from GPI-anchored proteins. These discrepancies may be related to subtle structural differences among the various IPG compounds, and await a definitive resolution.

7.2.3. Involvement of IPG and lipid rafts in insulin-mimetic signalling

Lipid rafts have been implicated as playing a critical role in adipocytes in compartmentalizing insulin signalling (Bickel, 2002), which appears to be dependent on the presence of functional caveolae. Important components of the signal transduction machinery, including the insulin receptor itself, appear to be localized within caveolae or rafts. Lowering of the cholesterol content of adipocytes by treatment with methyl-β-cyclodextrin or cholesterol oxidation both disrupted caveolae and abrogated tyrosine phosphorylation via the insulin receptor (Gustavsson *et al.*, 1999; Parpal *et al.*, 2001). However, the insulin-mimetic activity of IPG, PIG-P, and glimepiride are all independent of surface expression of the insulin receptor. These agents appear to by-pass the insulin receptor tyrosine kinase, and act by triggering re-distribution of components of the signalling system from hcDIG to lcDIG (Müller, 2002). Interaction of PIG-P with the membrane surface, and its insulin-mimetic properties, were found to depend on the presence of a protein receptor. Treatment of the target cells with trypsin, NaCl, and N-ethylmaleimide resulted in a large reduction in the ability of PIG-P to mimic the actions of insulin, and a 115 kDa protein, p115, was identified as a putative receptor (Müller *et al.*, 1998d). The p115 receptor is located within hcDIGs, and may act as a physiological receptor for GPI lipids, or the GPI anchors of membrane proteins (Müller *et al.*, 2002a), sequestering these components into specific membrane domains and concentrating them. Addition of exogenous PIG-P is proposed to displace these natural ligands from the receptor. The action of PIG-P and IPG appears to also precipitate the redistribution of the dually acylated non-receptor PTKs, $pp59^{Lyn}$ and $pp125^{Fak}$, to lcDIG, disrupting their association with caveolin within hcDIG, which normally acts as an inhibitor to keep them in an inactive state (Müller *et al.*, 2002c). Subsequent incorporation of $pp59^{Lyn}$ and $pp125^{Fak}$ into lcDIG results in their activation. Some GPI-anchored proteins (Gce1 and 5′-NTase) are also re-distributed from hcDIG to lcDIG as a result of PIG-P binding (Müller *et al.*, 2001). Activation of the tyrosine kinases appears to set in motion further downstream signalling events, such as phosphorylation of the insulin receptor substrate proteins-1 and 2 (IRS-1/2), resulting in the complete array of metabolic events associated with the insulin cascade, including GLUT4 translocation to the plasma membrane. Depletion of cholesterol from adipocytes by methyl-β-cyclodextrin also led to redistribution of GPI-anchored proteins and Src-family tyrosine kinases from hcDIGs to lcDIGs, and their dissociation from caveolin complexes (Müller *et al.*, 2002b).

This resulted in a large inhibition of insulin-mimetic signalling mediated by IPG and glimepiride. The exact mechanism by which displacement of pp59^{Lyn} (which is located in the cytoplasmic leaflet of the plasma membrane) takes place is not yet known, but it may involve coupling via a "linker" or "bridge" protein to GPI-anchored proteins in the outer leaflet (Müller, 2002) (Figure 7).

The physiological significance of the IPG insulin-mimetic pathway, and its role in the insulin signalling cascade is still not clear. The signalling pathways for IPG and insulin appear to be divergent at the level of upstream events, but converge with the phosphorylation of IRS-1/2, and the activation of PI-3-kinase (Kessler *et al.*, 1998). It is not known whether IPG release in response to insulin signalling effects the same redistribution of lipid rafts components as

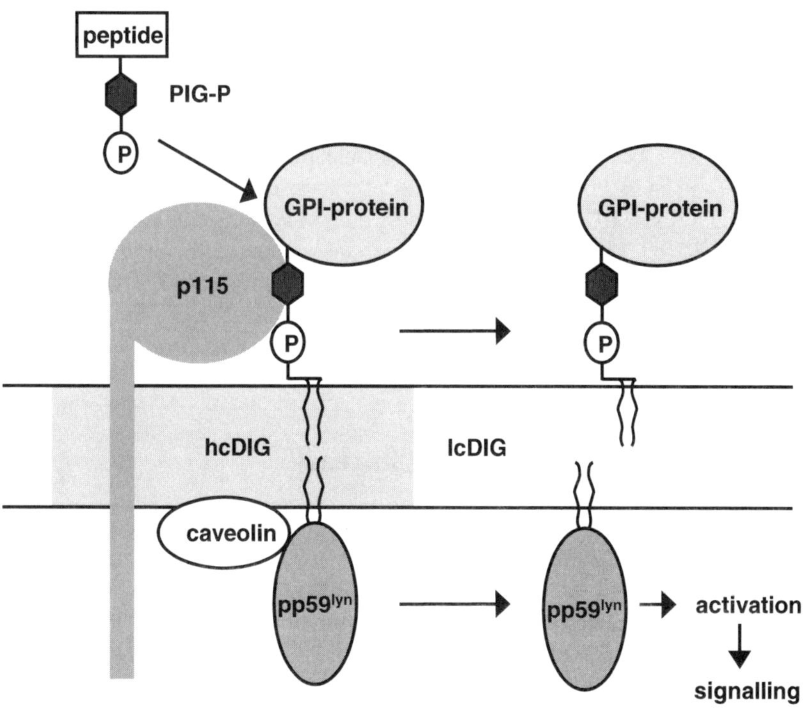

Figure 7. Proposed mechanism of insulin-mimetic signalling via PIG-P/IPG. PIG-P or IPG are proposed to displace GPI-anchored proteins from the receptor protein, p115, resulting in redistribution of the non-receptor PTK, pp59^{lyn}, from hcDIG raft domains to lcDIG domains. This movement of pp59^{lyn} results in disruption of the interaction between the protein and caveolin, which normally maintains the tyrosine kinase in an inactive state. Redistribution of pp59^{lyn} to lcDIG thus results in its activation, and initiation of downstream signalling events.

exogenously-added IPG. There is currently considerable interest in synthetic IPG as potential therapeutic agents for the treatment of insulin-resistant diabetes (Larner *et al.*, 1998).

7.2.4. Role of GPI anchor cleavage in GLUT4 translocation during insulin signalling

A critical downstream event in the insulin signalling cascade in fat and muscle is transfer of vesicles containing the glucose transporter GLUT4 from intracellular tubovesicular structures to the plasma membrane, where the large increase in transporter numbers causes an upsurge in cellular uptake of glucose. GLUT4 proteins translocated to the adipocyte plasma membrane in response to insulin are located predominantly within caveolar structures (Karlsson *et al.*, 2002). Further investigation showed that removal of GLUT4 from the plasma membrane of adipocytes by endocytosis appeared to occur via clathrin, but was also significantly influenced by plasma membrane caveolin (Shigematsu *et al.*, 2003). There has been a recent intriguing report that transfer of GLUT4 vesicles of rat skeletal muscle from large to small membrane structures in an *in vitro* assay for translocation was dependent on the activity of a cytosolic GPI-PLD enzyme (Kristiansen and Richter, 2002). Unexpectedly, highly purified GLUT4 protein (which is an integral protein with 12 membrane-spanning segments) was shown to contain the components of a GPI anchor, which presumably linked the transporter to the donor vesicles, and was subsequently cleaved by GPI-PLD action. Thus mobilization of GLUT4 in response to insulin in muscle may involve cleavage of a GPI anchor.

7.3. Insulin-like signalling in yeast

The yeast, *Saccharomyces cerevisiae*, appears to share some elements of the mammalian insulin signalling cascade involving glucose metabolism, when stimulated by either glucose or insulin itself. Transfer of yeast from lactate to glucose medium resulted in activation of an endogenous PI-PLC, and concomitant cleavage of the GPI anchor of the cAMP-binding protein, Gce1p (Müller and Bandlow, 1993). The cleaved protein remained associated with the plasma membrane surface, possibly as a result of interactions with a receptor protein. Cleavage of the anchor of Gce1p may thus play a role in signalling of nutritional status to the cell interior. Human insulin affected glucose metabolism in yeast by increasing glycogen synthase activity and decreasing glycogen phosphorylase activity, resulting in enhanced glycogen accumulation (Müller *et al.*, 1998a). This insulin response appeared to depend on the presence in the yeast plasma membrane of a 53 kDa transmembrane receptor, capable of binding human insulin with a K_d of 0.5 μM (Müller *et al.*, 1998c). Insulin stimulated Ser/Thr phosphorylation of the receptor and Tyr phosphorylation of another 70 kDa protein, and the two components appeared to form a large signalling complex (Müller *et al.*, 1998b).

7.4. Cleavage of GPI-anchored proteins may regulate other signalling processes

There have been a number of reports that stimulation of intact cells with a signalling ligand can result in cleavage of protein GPI anchors. Transforming growth factor-β (TGF-β) modulates development of the epidermis, and regulates keratinocyte function and migration during wound repair. A 150 kDa TGF-β accessory receptor, r150, was identified, which binds the growth factor with high affinity. An endogenous cell surface PI-PLC released the soluble form of r150, which still bound TGF-β (Tam *et al.*, 2001). When membrane-anchored, r150 may augment TGF-β signalling, while the soluble form may antagonize signalling. Thus, the mechanism of regulation of TGF-β action in keratinocytes may involve PI-PLC cleavage of the r150 protein.

Endocytosis from the apical surface of rat pancreatic acinar cells was activated by cleavage from the apical surface of GPI-anchored proteins, especially a protein known as GP2 (Freedman *et al.*, 1998). The mechanism of this effect is currently unknown.

Stimulation of human neutrophils by an inflammatory bacterial peptide (fMet-Leu-Phe) resulted in release of the GPI-anchored FcRIII (CD16) (Huizinga *et al.*, 1988), which acts as a receptor for the constant region (Fc) of immunoglobulin G. It is likely that this release occurs via activation of an endogenous PI-PLC. Shedding of FcRIII via PI-PLC release may be important in inflammatory reactions.

Signalling via the high density lipoprotein subfraction 3 (HDL3) in HepG2 cells also resulted in cleavage of protein-linked GPI anchors. HDL3 binds to the cell surfacc via a GPI-anchored HDL3 binding protein (HBP). This binding resulted in release of a 110 kDa protein into the supernatant, likely by activation of a phospholipase C enzyme (Nazih-Sanderson *et al.*, 1997a, b). Downstream signalling events included hydrolysis of PC, and generation of DAG. Antibodies to HBP also stimulated similar signalling events, which were blocked by treatment of the cells with bacterial PI-PLC. Further work implicated caveolae as the site where HDL3-binding proteins were localized (Nion *et al.*, 1997; Nazih-Sanderson *et al.*, 1997a).

The GPI-anchored form of CD58, an adhesion molecule capable of signal transduction, is expressed in the human B lymphoblastoid cell line JY in both transmembrane and GPI-anchored forms. Treatment of the cells with immobilized CD58-specific antibodies resulted in extensive tyrosine phosphorylation and secretion of tumour necrosis factor-α (TNF-α). Both forms of the CD58 protein appeared to be associated with protein kinases and protein kinase substrates (Itzhaky *et al.*, 1998b). GPI-anchored CD58 was released from the cell surface in soluble form following crosslinking by anti-CD58 antibodies, apparently mediated by a phospholipase (Itzhaky *et al.*, 1998a). The transmembrane form of CD58 was also released in parallel, although a protease appeared to be responsible in this case. Thus, the two isoforms of CD58 are both released upon antibody binding and transduction of a signal, but their release appears to be mediated by distinct mechanisms.

8. CONCLUDING REMARKS

Cleavage of the GPI-anchors of proteins and GPI lipids by an endogenous phospholipase C enzyme takes place during various signalling events, although the exact details of the processes involved at the molecular level remains to be worked out. Identification and biochemical characterization of the putative PI-PLC responsible for anchor cleavage in the signalling pathways for insulin and other agents will be important for further progress to be made in this area. In the case of insulin signalling, re-distribution of critical membrane components, including GPI-anchored proteins and non-receptor tyrosine kinases, between different raft domains appears to play a central role in the signal transduction pathway.

REFERENCES

Ahmed, S.N., Brown, D.A., and London, E., 1997, On the origin of sphingolipid/cholesterol-rich detergent-insoluble cell membranes: physiological concentrations of cholesterol and sphingolipid induce formation of a detergent-insoluble, liquid-ordered lipid phase in model membranes, *Biochemistry* **36**:10944-10953.

Almeida, P.F., Vaz, W.L., and Thompson, T.E., 1992, Lateral diffusion and percolation in two-phase, two-component lipid bilayers. Topology of the solid-phase domains in-plane and across the lipid bilayer, *Biochemistry* **31**:7198-7210.

Almqvist, P. and Carlsson, S.R., 1988, Characterization of a hydrophilic form of Thy-1 purified from human cerebrospinal fluid, *J. Biol. Chem.* **263**:12709-12715.

Barboni, E., Rivero, B.P., George, A.J., Martin, S.R., Renoup, D.V., Hounsell, E.F., Barber, P.C., and Morris, R.J., 1995, The glycophosphatidylinositol anchor affects the conformation of Thy-1 protein, *J. Cell Sci.* **108**:487-497.

Benting, J., Rietveld, A., Ansorge, I., and Simons, K., 1999, Acyl and alkyl chain length of GPI-anchors is critical for raft association in vitro, *FEBS Lett.* **462**:47-50.

Bickel, P.E., 2002, Lipid rafts and insulin signaling, *Am. J. Physiol. Endocrinol. Metab.* **282:** E1-E10.

Brasitus, T.A. and Schachter, D., 1980, Lipid dynamics and lipid-protein interactions in rat enterocyte basolateral and microvillus membranes, *Biochemistry* **19**:2763-2769.
Braun-Breton, C., Rosenberry, T.L., and da Silva, L.P., 1988, Induction of the proteolytic activity of a membrane protein in *Plasmodium falciparum* by phosphatidyl inositol-specific phospholipase C, *Nature* **332**:457-459.
Brewis, I.A., Turner, A.J., and Hooper, N.M., 1994, Activation of the glycosyl-phosphatidylinositol-anchored membrane dipeptidase upon release from pig kidney membranes by phospholipase C, *Biochem. J.* **303**:633-638.
Brodbeck, U., 1998, Signalling properties of glycosylphosphatidylinositols and their regulated release from membranes in the turnover of glycosylphosphatidylinositol-anchored proteins, *Biol. Chem. Hoppe Seyler* **379**:1041-1044.
Broomfield, S.J., and Hooper, N.M., 1993, Characterization of an antibody to the cross-reacting determinant of the glycosyl-phosphatidylinositol anchor of human membrane dipeptidase, *Biochim. Biophys. Acta* **1145**:212-218.
Cary, L.A., and Cooper, J.A., 2000, Signal transduction – Molecular switches in lipid rafts, *Nature* **404**:945-947.
Cebecauer, M., Cerny, J., and Horejsi, V., 1998, Incorporation of leucocyte GPI-anchored proteins and protein tyrosine kinases into lipid-rich membrane domains of COS-7 cells, *Biochem. Biophys. Res. Commun.* **243**:706-710.
Chan, B.L., Lisanti, M.P., Rodriguez-Boulan, E., and Saltiel, A.R., 1988, Insulin-stimulated release of lipoprotein lipase by metabolism of its phosphatidylinositol anchor, *Science* **241**:1670-1672.
Cherukuri, A., Dykstra, M., and Pierce, S.K., 2001, Floating the raft hypothesis: Lipid rafts play a role in immune cell activation, *Immunity* **14**:657-660.
Daugherty, S., and Low, M.G., 1993, Cloning, expression, and mutagenesis of phosphatidylinositol-specific phospholipase C from Staphylococcus aureus: a potential staphylococcal virulence factor, *Infect. Immun.* **61**:5078-5089.
Davitz, M.A., Hereld, D., Shak, S., Krakow, J., Englund, P.T., and Nussenzweig, V., 1987, A glycan-phosphatidylinositol-specific phospholipase D in human serum, *Science* **238**:81-84.
Deeg, M.A., and Davitz, M.A., 1995, Glycosylphosphatidylinositol-phospholipase D: a tool for glycosylphosphatidylinositol structural analysis, *Methods Enzymol.* **250**:630-640.
Deeg, M.A., and Verchere, C.B., 1997, Regulation of glycosylphosphatidylinositol-specific phospholipase D secretion from beta TC3 cells, *Endocrinology* **138**:819-826.
Durbin, H., Young, S., Stewart, L.M., Wrba, F., Rowan, A.J., Snary, D., and Bodmer, W.F., 1994, An epitope on carcinoembryonic antigen defined by the clinically relevant antibody PR1A3, *Proc. Natl. Acad. Sci. USA* **91**:4313-4317.
Eisenhaber, B., Bork, P., and Eisenhaber, F., 2001, Post-translational GPI lipid anchor modification of proteins in kingdoms of life: analysis of protein sequence data from complete genomes, *Protein Eng.* **14**:17-25.
Eisenhaber, B., Bork, P., Yuan, Y.P., Löffler, G., and Eisenhaber, F., 2000, Automated annotation of GPI anchor sites: case study *C. elegans*, *Trends Biochem. Sci.* **25**:340-341.
Eliakim, R., Becich, M.J., Green, K., and Alpers, D.H., 1990, Both tissue and serum phospholipases release rat intestinal alkaline phosphatase, *Am. J. Physiol.* **259:** G618-G625.
Fantini, J., Maresca, M., Hammache, D., Yahi, N., and Delézay, O., 2000, Glycosphingolipid (GSL) microdomains as attachment platforms for host pathogens and their toxins on intestinal epithelial cells: Activation of signal transduction pathways and perturbations of intestinal absorption and secretion, *Glycoconjugate J.* **17**:173-179.
Ferguson, M.A., Low, M.G., and Cross, G.A., 1985, Glycosyl-sn-1,2-dimyristylphosphatidylinositol is covalently linked to Trypanosoma brucei variant surface glycoprotein, *J. Biol. Chem.* **260**:14547-14555.

Freedman, S.D., Kern, H.F., and Scheele, G.A., 1998, Cleavage of GPI-anchored proteins from the plasma membrane activates apical endocytosis in pancreatic acinar cells, *Eur. J. Cell Biol.* **75**:163-173.

Frick, W., Bauer, A., Bauer, J., Wied, S., and Müller, G., 1998, Structure-activity relationship of synthetic phosphoinositolglycans mimicking metabolic insulin action, *Biochemistry* **37**:13421-13436.

Friedrichson, T., and Kurzchalia, T.V., 1998, Microdomains of GPI-anchored proteins in living cells revealed by crosslinking, *Nature* **394**:802-805.

Giocondi, M.C., Vié, V., Lesniewska, E., Goudonnet, J.P., and Le Grimellec, C., 2000, *In situ* imaging of detergent-resistant membranes by atomic force microscopy, *J. Struct. Biol.* **131**:38-43.

Gmachl, M., Sagan, S., Ketter, S., and Kreil, G., 1993, The human sperm protein PH-20 has hyaluronidase activity, *FEBS Lett.* **336**:545-548.

Griffith, O.H., Volwerk, J.J., and Kuppe, A., 1991, Phosphatidylinositol-specific phospholipases C from Bacillus cereus and Bacillus thuringiensis, *Methods Enzymol.* **197**:493-502.

Gustavsson, J., Parpal, S., Karlsson, M., Ramsing, C., Thorn, H., Borg, M., Lindroth, M., Peterson, K.H., Magnusson, K., and Strålfors, P., 1999, Localization of the insulin receptor in caveolae of adipocyte plasma membrane, *FASEB J.* **13**:1961-1971.

Hanada, K., Nishijima, M., Akamatsu, Y., and Pagano, R.E., 1995, Both sphingolipids and cholesterol participate in the detergent insolubility of alkaline phosphatase, a glycosylphosphatidylinositol-anchored protein, in mammalian membranes, *J. Biol. Chem.* **270**:6254-6260.

Hansen, G.H., Immerdal, L., Thorsen, E., Niels-Christiansen, L.L., Nystrom, B.T., Demant, E.J.F., and Danielsen, E.M., 2001, Lipid rafts exist as stable cholesterol-independent microdomains in the brush border membrane of enterocytes, *J. Biol. Chem.* **276**:32338-32344.

Hari, T., Butikofer, P., Wiesmann, U.N., and Brodbeck, U., 1997, Uptake and intracellular stability of glycosylphosphatidylinositol-specific phospholipase D in neuroblastoma cells, *Biochim. Biophys. Acta* **1355**:293-302.

Hari, T., Kunze, H., Bohn, E., Brodbeck, U., and Butikofer, P., 1996, Subcellular distribution of glycosylphosphatidylinositol-specific phospholipase D in rat liver, *Biochem. J.* **320**:315-319.

Heinz, D.W., Ryan, M., Bullock, T.L., and Griffith, O.H., 1995, Crystal structure of the phosphatidylinositol-specific phospholipase C from Bacillus cereus in complex with myo-inositol, *EMBO J.* **14**:3855-3863.

Heinz, D.W., Ryan, M., Smith, M.P., Weaver, L.H., Keana, J.F., and Griffith, O.H., 1996, Crystal structure of phosphatidylinositol-specific phospholipase C from Bacillus cereus in complex with glucosaminyl(alpha 1→6)-D-myo-inositol, an essential fragment of GPI anchors, *Biochemistry* **35**:9496-9504.

Heller, M., Bieri, S., and Brodbeck, U., 1992, A novel form of glycosylphosphatidylinositol-anchor converting activity with a specificity of a phospholipase D in mammalian liver membranes, *Biochim. Biophys. Acta* **1109**:109-116.

Hoener, M.C., and Brodbeck, U., 1992, Phosphatidylinositol-glycan-specific phospholipase D is an amphiphilic glycoprotein that in serum is associated with high-density lipoproteins, *Eur. J. Biochem.* **206**:747-757.

Homans, S.W., Ferguson, M.A., Dwek, R.A., Rademacher, T.W., Anand, R., and Williams, A.F., 1988, Complete structure of the glycosyl phosphatidylinositol membrane anchor of rat brain Thy-1 glycoprotein, *Nature* **333**:269-272.

Huang, J.B., Takeda, Y., Watanabe, T., and Sendo, F., 2001, A sandwich ELISA for detection of soluble GPI-80, a glycosylphosphatidyl-inositol (GPI)-anchored protein on human leukocytes involved in regulation of neutrophil adherence and migration – Its release from activated neutrophils and presence in synovial fluid of rheumatoid arthritis patients, *Microbiol. Immunol.* **45**:467-471.

Huang, K.S., Li, S., Fung, W.J., Hulmes, J.D., Reik, L., Pan, Y.C., and Low, M.G., 1990, Purification and characterization of glycosyl-phosphatidylinositol-specific phospholipase D, *J. Biol. Chem.* **265**:17738-17745.

Huizinga, T.W., van der Schoot, C.E., Jost, C., Klaassen, R., Kleijer, M., von dem, B., Roos, D., and Tetteroo, P.A., 1988, The PI-linked receptor FcRIII is released on stimulation of neutrophils, *Nature* **333**:667-669.

Ikezawa, H., 1991, Bacterial PIPLCs-unique properties and usefulness in studies on GPI anchors, *Cell Biol. Int. Rep.* **15**:1115-1131.

Ikezawa, H., 2002, Glycosylphosphatidylinositol (GPI)-anchored proteins, *Biol. Pharm. Bull.* **25**:409-417.

Itzhaky, D., Raz, N., and Hollander, N., 1998a, The glycosylphosphatidylinositol-anchored form and the transmembrane form of CD58 are released from the cell surface upon antibody binding, *Cell Immunol.* **187**:151-157.

Itzhaky, D., Raz, N., and Hollander, N., 1998b, The glycosylphosphatidylinositol-anchored form and the transmembrane form of CD58 associate with protein kinases, *J. Immunol.* **160**:4361-4366.

Jones, D.R., Avila, M.A., Sanz, C., and Varela-Nieto, I., 1997, Glycosyl-phosphatidylinositol-phospholipase type D: a possible candidate for the generation of second messengers, *Biochem. Biophys. Res. Commun.* **233**:432-437.

Jones, D.R., and Varela-Nieto, I., 1998, The role of glycosyl-phosphatidylinositol in signal transduction, *Int. J. Biochem. Cell Biol.* **30**:313-326.

Karlsson, M., Thorn, H., Parpal, S., Stralfors, P., and Gustavsson, J., 2002, Insulin induces translocation of glucose transporter GLUT4 to plasma membrane caveolae in adipocytes, *FASEB J.* **16**:249-251.

Kessler, A., Müller, G., Wied, S., Crecelius, A., and Eckel, J., 1998, Signalling pathways of an insulin-mimetic phosphoinositolglycan-peptide in muscle and adipose tissue, *Biochem. J.* **330**:277-286.

Kinoshita, T., and Inoue, N., 2000, Dissecting and manipulating the pathway for glycosylphosphatidylinositol-anchor biosynthesis, *Curr. Opin. Chem. Biol.* **4**:632-638.

Kinoshita, T., Ohishi, K., and Takeda, J., 1997, GPI-anchor synthesis in mammalian cells: genes, their products, and a deficiency, *J. Biochem. (Tokyo)* **122**:251-257.

Klip, A., Ramlal, T., Douen, A.G., Burdett, E., Young, D., Cartee, G.D., and Holloszy, J.O., 1988, Insulin-induced decrease in 5′-nucleotidase activity in skeletal muscle membranes, *FEBS Lett.* **238**:419-423.

Kobayashi, T., Nishizaki, R., and Ikezawa, H., 1997, The presence of GPI-linked protein(s) in an archaeobacterium, Sulfolobus acidocaldarius, closely related to eukaryotes, *Biochim. Biophys. Acta* **1334**:1-4.

Kristiansen, S., and Richter, E.A., 2002, GLUT4-containing vesicles are released from membranes by phospholipase D cleavage of a GPI anchor, *Am. J. Physiol. Endocrinol. Metab.* **283**:E374-E382.

Kukulansky, T., Abramovitch, S., and Hollander, N., 1999, Cleavage of the glycosylphosphatidylinositol anchor affects the reactivity of Thy-1 with antibodies, *J. Immunol.* **162**:5993-5997.

Langlet, C., Bernard, A.M., Drevot, P., and He, H.T., 2000, Membrane rafts and signaling by the multichain immune recognition receptors, *Curr. Opin. Immunol.* **12**:250-255.

Larner, J., Allan, G., Kessler, C., Reamer, P., Gunn, R., and Huang, L.C., 1998, Phosphoinositol glycan derived mediators and insulin resistance. Prospects for diagnosis and therapy, *J. Basic Clin. Physiol Pharmacol.* **9**:127-137.

Lehto, M.T., and Sharom, F.J., 1998, Release of the glycosylphosphatidylinositol-anchored enzyme ecto-5′-nucleotidase by phospholipase C: catalytic activation and modulation by the lipid bilayer, *Biochem. J.* **332**:101-109.

Lehto, M.T., and Sharom, F.J., 2002a, PI-specific phospholipase C cleavage of a reconstituted GPI-anchored protein: modulation by the lipid bilayer, *Biochemistry* **41**:1398-1408.

Lehto, M.T., and Sharom, F.J., 2002b, Proximity of the protein moiety of a GPI-anchored protein to the membrane surface: a FRET study, *Biochemistry* **41**:8368-8376.

Liao, Z.H., Cimakasky, L.M., Hampton, R., Nguyen, D.H., and Hildreth, J.E.K., 2001, Lipid rafts and HIV pathogenesis: Host membrane cholesterol is required for infection by HIV type 1, *AIDS Res. Hum. Retroviruses* **17**:1009-1019.

Lisanti, M.P., Darnell, J.C., Chan, B.L., Rodriguez-Boulan, E., and Saltiel, A.R., 1989, The distribution of glycosyl-phosphatidylinositol anchored proteins is differentially regulated by serum and insulin, *Biochem. Biophys. Res. Commun.* **164**:824-832.

London, E., and Brown, D.A., 2000, Insolubility of lipids in Triton X-100: physical origin and relationship to sphingolipid/cholesterol membrane domains (rafts), *Biochim. Biophys. Acta* **1508**:182-195.

Low, M.G., and Finean, J.B., 1978, Specific release of plasma membrane enzymes by a phosphatidylinositol-specific phospholipase C, *Biochim. Biophys. Acta* **508**:565-570.

Low, M.G., and Huang, K.S., 1991, Factors affecting the ability of glycosylphosphatidylinositol-specific phospholipase D to degrade the membrane anchors of cell surface proteins, *Biochem. J.* **279**:483-493.

Low, M.G., and Prasad, A.R., 1988, A phospholipase D specific for the phosphatidylinositol anchor of cell-surface proteins is abundant in plasma, *Proc. Natl. Acad. Sci. USA* **85**:980-984.

Martin-Lomas, M., Khiar, N., Garcia, S., Koessler, J.L., Nieto, P.M., and Rademacher, T.W., 2000, Inositolphosphoglycan mediators structurally related to glycosyl phosphatidylinositol anchors: synthesis, structure and biological activity, *Chemistry* **6**:3608-3621.

Melkonian, K.A., Chu, T., Tortorella, L.B., and Brown, D.A., 1995, Characterization of proteins in detergent-resistant membrane complexes from Madin-Darby canine kidney epithelial cells, *Biochemistry* **34**:16161-16170.

Melkonian, K.A., Ostermeyer, A.G., Chen, J.Z., Roth, M.G., and Brown, D.A., 1999, Role of lipid modifications in targeting proteins to detergent-resistant membrane rafts – Many raft proteins are acylated, while few are prenylated, *J. Biol. Chem.* **274**:3910-3917.

Mengaud, J., Braun-Breton, C., and Cossart, P., 1991, Identification of phosphatidylinositol-specific phospholipase C activity in Listeria monocytogenes: a novel type of virulence factor?, *Mol. Microbiol.* **5**:367-372.

Moser, J., Gerstel, B., Meyer, J.E., Chakraborty, T., Wehland, J., and Heinz, D.W., 1997, Crystal structure of the phosphatidylinositol-specific phospholipase C from the human pathogen Listeria monocytogenes, *J. Mol. Biol.* **273**:269-282.

Movahedi, S., and Hooper, N.M., 1997, Insulin stimulates the release of the glycosyl phosphatidylinositol-anchored membrane dipeptidase from 3T3-L1 adipocytes through the action of a phospholipase C, *Biochem. J.* **326**:531-537.

Müller, G., 2002, Dynamics of plasma membrane microdomains and cross-talk to the insulin signalling cascade, *FEBS Lett.* **531**:81-87.

Müller, G., and Bandlow, W., 1993, Glucose induces lipolytic cleavage of a glycolipidic plasma membrane anchor in yeast, *J. Cell Biol.* **122**:325-336.

Müller, G., and Bandlow, W., 1994, Lipolytic membrane release of two phosphatidylinositol-anchored cAMP receptor proteins in yeast alters their ligand-binding parameters, *Arch. Biochem. Biophys.* **308**:504-514.

Müller, G., Dearey, E.A., Korndorfer, A., and Bandlow, W., 1994, Stimulation of a glycosylphosphatidylinositol-specific phospholipase by insulin and the sulfonylurea, glimepiride, in rat adipocytes depends on increased glucose transport, *J. Cell Biol.* **126**:1267-1276.

Müller, G., Dearey, E.A., and Punter, J., 1993, The sulphonylurea drug, glimepiride, stimulates release of glycosylphosphatidylinositol-anchored plasma-membrane proteins from 3T3 adipocytes, *Biochem. J.* **289**:509-521.

Müller, G., Hanekop, N., Kramer, W., Bandlow, W., and Frick, W., 2002a, Interaction of phosphoinositolglycan(-peptides) with plasma membrane lipid rafts of rat adipocytes, *Arch. Biochem. Biophys.* **408**:17-32.

Müller, G., Hanekop, N., Wied, S., and Frick, W., 2002b, Cholesterol depletion blocks redistribution of lipid raft components and insulin-mimetic signaling by glimepiride and phosphoinositolglycans in rat adipocytes, *Mol. Med.* **8**:120-136.

Müller, G., Jung, C., Frick, W., Bandlow, W., and Kramer, W., 2002c, Interaction of phosphatidylinositolglycan(-peptides) with plasma membrane lipid rafts triggers insulin-mimetic signaling in rat adipocytes, *Arch. Biochem. Biophys.* **408**:7-16.

Müller, G., Jung, C., Wied, S., Welte, S., and Frick, W., 2001, Insulin-mimetic signaling by the sulfonylurea glimepiride and phosphoinositolglycans involves distinct mechanisms for redistribution of lipid raft components, *Biochemistry* **40**:14603-14620.

Müller, G., Rouveyre, N., Crecelius, A., and Bandlow, W., 1998a, Insulin signaling in the yeast *Saccharomyces cerevisiae*. 1. Stimulation of glucose metabolism and Snf1 kinase by human insulin, *Biochemistry* **37**:8683-8695.

Müller, G., Rouveyre, N., Upshon, C., and Bandlow, W., 1998b, Insulin signaling in the yeast *Saccharomyces cerevisiae*. 3. Induction of protein phosphorylation by human insulin, *Biochemistry* **37**:8705-8713.

Müller, G., Rouveyre, N., Upshon, C., Grobeta, E., and Bandlow, W., 1998c, Insulin signaling in the yeast *Saccharomyces cerevisiae*. 2. Interaction of human insulin with a putative binding protein, *Biochemistry* **37**:8696-8704.

Müller, G., Wied, S., Crecelius, A., Kessler, A., and Eckel, J., 1997, Phosphoinositolglycan-peptides from yeast potently induce metabolic insulin actions in isolated rat adipocytes, cardiomyocytes, and diaphragms, *Endocrinology* **138**:3459-3475.

Müller, G., Wied, S., Piossek, C., Bauer, A., Bauer, J., and Frick, W., 1998d, Convergence and divergence of the signaling pathways for insulin and phosphoinositolglycans, *Mol. Med.* **4**:299-323.

Muñiz, M., and Riezman, H., 2000, Intracellular transport of GPI-anchored proteins, *EMBO J.* **19**:10-15.

Nazih-Sanderson, F., Lestavel, S., Nion, S., Rouy, D., Denefle, P., Fruchart, J.C., Clavey, V., and Delbart, C., 1997a, HDL3 binds to glycosylphosphatidylinositol-anchored proteins to activate signalling pathways, *Biochim. Biophys. Acta* **1358**:103-112.

Nazih-Sanderson, F., Pinchon, G., Nion, S., Fruchart, J.C., and Delbart, C., 1997b, HDL3-signalling in HepG2 cells involves glycosyl-phosphatidylinositol-anchored proteins, *Biochim. Biophys. Acta* **1346**:45-60.

Nion, S., Briand, O., Lestavel, S., Torpier, G., Nazih, F., Delbart, C., Fruchart, J.C., and Clavey, V., 1997, High-density-lipoprotein subfraction 3 interaction with glycosylphosphatidylinositol-anchored proteins, *Biochem. J.* **328**:415-423.

Park, S.W., Choi, K., Kim, I.C., Lee, H.H., Hooper, N.M., and Park, H.S., 2001, Endogenous glycosylphosphatidylinositol-specific phospholipase C releases renal dipeptidase from kidney proximal tubules in vitro, *Biochem. J.* **353**:339-344.

Park, S.W., Choi, K., Lee, H.B., Park, S.K., Turner, A.J., Hooper, N.M., and Park, H.S., 2002a, Glycosyl-phosphatidylinositol (GPI)-anchored renal dipeptidase is released by a phospholipase C in vivo, *Kidney Blood Press Res.* **25**:7-12.

Park, S.W., Yoon, H.J., Lee, H.B., Hooper, N.M., and Park, H.S., 2002b, Nitric oxide inhibits the shedding of the glycosylphosphatidylinositol-anchored dipeptidase from porcine renal proximal tubules, *Biochem. J.* **364**:211-218.

Parpal, S., Karlsson, M., Thorn, H., and Stralfors, P., 2001, Cholesterol depletion disrupts caveolae and insulin receptor signaling for metabolic control via insulin receptor substrate-1, but not*et* for mitogen-activated protein kinase control, *J. Biol. Chem.* **276**:9670-9678.

Petitfrere, E., Sartelet, H., Vivien, D., Varela-Nieto, I., Elbtaouri, H., Martiny, L., and Haye, B., 1998, Glycosyl phosphatidylinositol (GPI)/inositolphosphate glycan (IPG): an intracellular signalling system involved in the control of thyroid cell proliferation, *Biochimie. J.* **80**:1063-1067.

Piec, G., and Le Hir, M., 1991, The soluble "low-Km" 5′-nucleotidase of rat kidney represents solubilized ecto-5'-nucleotidase, *Biochem. J.* **273**:409-413.

Pralle, A., Keller, P., Florin, E.L., Simons, K., and Hörber, J.K.H., 2000, Sphingolipid-cholesterol rafts diffuse as small entities in the plasma membrane of mammalian cells, *J. Cell Biol.* **148**:997-1007.

Rademacher, T.W., Edge, C.J., and Dwek, R.A., 1991, Dropping anchor with the lipophosphoglycans, *Glycobiology* **1**:41-42.

Reid-Taylor, K.L., Chu, J.W.K., and Sharom, F.J., 1999, Reconstitution of the glycosylphosphatidylinositol-anchored protein Thy-1: interaction with membrane phospholipids and galactosylceramide, *Biochem. Cell Biol.* **77**:189-200.

Rietveld, A., and Simons, K., 1998, The differential miscibility of lipids as the basis for the formation of functional membrane rafts, *Biochim. Biophys. Acta* **1376**:467-479.

Roberts, J.M., Kenton, P., and Johnson, P.M., 1990, Growth factor-induced release of a glycosyl-phosphatidylinositol (GPI)-linked protein from the HEp-2 human carcinoma cell line, *FEBS Lett.* **267**:213-216.

Roberts, W.L., Myher, J.J., Kuksis, A., Low, M.G., and Rosenberry, T.L., 1988, Lipid analysis of the glycoinositol phospholipid membrane anchor of human erythrocyte acetylcholinesterase. Palmitoylation of inositol results in resistance to phosphatidylinositol- specific phospholipase C, *J. Biol. Chem.* **263**:18766-18775.

Romero, G., Luttrell, L., Rogol, A., Zeller, K., Hewlett, E., and Larner, J., 1988, Phosphatidylinositol-glycan anchors of membrane proteins: potential precursors of insulin mediators, *Science* **240**:509-511.

Rosenberger, C.M., Brumell, J.H., and Finlay, B.B., 2000, Microbial pathogenesis: Lipid rafts as pathogen portals, *Curr. Biol.* **10:**R823-R825.

Rosenberry, T.L., 1991, A chemical modification that makes glycoinositol phospholipids resistant to phospholipase C cleavage: fatty acid acylation of inositol, *Cell Biol. Int. Rep.* **15**:1133-1150.

Salzer, U., and Prohaska, R., 2001, Stomatin, flotillin-1, and flotillin-2 are major integral proteins of erythrocyte lipid rafts, *Blood* **97**:1141-1143.

Santos, A.L., Abreu, C.M., Alviano, C.S., and Soares, R.M., 2002, Activation of the glycosylphosphatidylinositol-anchored membrane proteinase upon release from Herpetomonas samuelpessoai by phospholipase C, *Curr. Microbiol.* **45**:293-298.

Schroeder, R., London, E., and Brown, D.A., 1994, Interactions between saturated acyl chains confer detergent resistance on lipids and glycosylphosphatidylinositol (GPI)-anchored proteins: GPI- anchored proteins in liposomes and cells show similar behavior, *Proc. Natl. Acad. Sci. USA* **91**:12130-12134.

Sharom, F.J., and Lehto, M.T., 2002, GPI-anchored proteins: structure, function, and cleavage by PI-specific phospholipases, *Biochem. Cell Biol.* **80**:535-549.

Sharom, F.J., Lorimer, I., and Lamb, M.P., 1985, Reconstitution of lymphocyte 5′-nucleotidase in lipid bilayers: behaviour and interaction with concanavalin A, *Can. J. Biochem. Cell Biol.* **63**:1049-1057.

Sharom, F.J., McNeil, G.L., Glover, J.R., and Seier, S., 1996, Modulation of the cleavage of glycosylphosphatidylinositol-anchored proteins by specific bacterial phospholipases, *Biochem. Cell Biol.* **74**:701-713.

Sheets, E.D., Lee, G.M., Simson, R., and Jacobson, K., 1997, Transient confinement of a glycosylphosphatidylinositol-anchored protein in the plasma membrane, *Biochemistry* **36**:12449-12458.

Shigematsu, S., Watson, R.T., Khan, A.H., and Pessin, J.E., 2003, The adipocyte plasma membrane caveolin functional/structural organization is necessary for the efficient endocytosis of GLUT4, *J. Biol. Chem.* **278**:10683-10690.

Simons, K., and Ikonen, E., 1997, Functional rafts in cell membranes, *Nature* **387**:569-572.

Singh, N., Liang, L.N., Tykocinski, M.L., and Tartakoff, A.M., 1996, A novel class of cell surface glycolipids of mammalian cells. Free glycosyl phosphatidylinositols, *J. Biol. Chem.* **271**:12879-12884.

Tam, B.Y., Larouche, D., Germain, L., Hooper, N.M., and Philip, A., 2001, Characterization of a 150 kDa accessory receptor for TGF-beta 1 on keratinocytes: direct evidence for a GPI anchor and ligand binding of the released form, *J. Cell Biochem.* **83**:494-507.

Tsujioka, H., Misumi, Y., Takami, N., and Ikehara, Y., 1998, Posttranslational modification of glycosylphosphatidylinositol (GPI)-specific phospholipase D and its activity in cleavage of GPI anchors, *Biochem. Biophys. Res. Commun.* **251**:737-743.

Tsujioka, H., Takami, N., Misumi, Y., and Ikehara, Y., 1999, Intracellular cleavage of glycosylphosphatidylinositol by phospholipase D induces activation of protein kinase Calpha, *Biochem. J.* **342**:449-455.

Varela-Nieto, I., Leon, Y., and Caro, H.N., 1996, Cell signalling by inositol phosphoglycans from different species, *Comp. Biochem. Physiol. B* **115**:223-241.

Varma, R., and Mayor, S., 1998, GPI-anchored proteins are organized in submicron domains at the cell surface, *Nature* **394**:798-801.

Villalba, M., Alvarez, J.F., Russell, D.S., Mato, J.M., and Rosen, O.M., 1990, Hydrolysis of glycosyl-phosphatidylinositol in response to insulin is reduced in cells bearing kinase-deficient insulin receptors, *Growth Factors* **2**:91-97.

Villar, A.V., Goni, F.M., Alonso, A., Jones, D.R., Leon, Y., and Varela-Nieto, I., 1998, Phospholipase cleavage of glycosylphosphatidylinositol reconstituted in liposomal membranes, *FEBS Lett.* **432**:150-154.

Vincent, S., Gerlier, D., and Manié, S.N., 2000, Measles virus assembly within membrane rafts, *J. Virol.* **74**:9911-9915.

Viola, A., 2001, The amplification of TCR signaling by dynamic membrane microdomains, *Immunol. Today* **22**:322-327.

Vogel, M., Kowalewski, H., Zimmermann, H., Hooper, N.M., and Turner, A.J., 1992, Soluble low-Km 5′-nucleotidase from electric-ray (*Torpedo marmorata*) electric organ and bovine cerebral cortex is derived from the glycosyl-phosphatidylinositol-anchored ectoenzyme by phospholipase C cleavage, *Biochem. J.* **284**:621-624.

Wang, X., Jansen, G., Fan, J., Kohler, W.J., Ross, J.F., Schornagel, J., and Ratnam, M., 1996, Variant GPI structure in relation to membrane-associated functions of a murine folate receptor, *Biochemistry* **35**:16305-16312.

Webb, H., Carnall, N., Vanhamme, L., Rolin, S., Van Den, A.J., Welburn, S., Pays, E., and Carrington, M., 1997, The GPI-phospholipase C of Trypanosoma brucei is nonessential but influences parasitemia in mice, *J. Cell Biol.* **139**:103-114.

Yuan, C.B., Furlong, J., Burgos, P., and Johnston, L.J., 2002, The size of lipid rafts: An atomic force microscopy study of ganglioside GM1 domains in sphingomyelin/DOPC/cholesterol membranes, *Biophys. J.* **82**:2526-2535.

Chapter 10

Membrane Lipid Homeostasis

Claude Wolf[1] and Peter J. Quinn[2]
[1]Biochemistry Department, Mass Spectrometry Laboratory, INSERM U. 538, Faculté de Médecine Saint Antoine, Paris 75012, France; [2]Life Sciences Department, Biochemistry, King's College London, 150 Stamford Street, London SE1 9NN, UK

Abstract
The lipid matrix of biological membranes is composed of a complex mixture of polar lipids. It has been estimated that more than 600 distinct molecular species of lipid are constituents of biological membranes. This rather remarkable feature raises the questions of why such complexity is required when barrier properties and many protein functions can be reconstituted with relatively simple lipid systems. Secondly, the molecular species composition of morphologically distinct membranes appears to be preserved within fairly narrow limits. The biochemical mechanism(s) responsible for this homeostasis are not fully understood. This review examines the origin of membrane lipid complexity, the methods that are currently employed to measure and detect lipid molecular species and the biochemical reactions associated with the turnover of membrane lipids in resting and stimulated cells.

1. INTRODUCTION

The lipid bilayer of biological membranes was originally regarded as a simple fluid matrix with which the different intrinsic and extrinsic membrane proteins interact. Reconstitution experiments demonstrated that the function of membrane proteins often requires incorporation into a lipid bilayer but few specific requirements for particular lipids have been identified. The concept that the lipid matrix acted simply as a barrier to the movement of solutes was

Subcellular Biochemistry, Volume 37: Membrane Dynamics and Domains
Edited by Peter J. Quinn. Kluwer Academic / Plenum Publishers, New York, 2004

discarded when it became clear that the distribution of lipids on either side of the membrane was different and this asymmetry was maintained by an active process which, if disturbed, resulted in severe consequences (Quinn, 2002). More recently, it has become clear that the lateral distribution of lipid molecular species is also asymmetric leading to the formation of domains in the plane of the membrane (Edidin, 2001; London, 2002).

The creation of membrane domains is believed to be the result of specific interactions between particular membrane lipids located in one or other leaflet of the membrane lipid bilayer. As a consequence of domain formation the distribution of membrane proteins is altered and this alteration is associated with potentiation of signalling functions (Anderson and Jacobson, 2002; Kenworthy, 2002).

The modulation of protein function by domain creation must, in turn, be regulated by factors responsible for maintaining the molecular species composition of the membrane matrix and for their asymmetric distribution across the membrane. We will be concerned with an examination of where the complexity in lipid structure and composition arises, what purposes may be served by lipid heterogeneity and how biochemical processes may be exploited to maintain, on the one hand, and disturb lipid homeostasis in cells on the other.

2. MEMBRANE LIPID COMPLEXITY

2.1. The chemical complexity of membrane lipids

The complexity of membrane lipid composition extends to the number of chemically different molecular species of lipid that is associated with particular membranes as well as the relative proportions in which they are found. The variety of molecular species is comprised of lipids with different basic structures such as glycerolipids and sphingolipids and the substituents both polar and non-polar they possess. For example, the length, unsaturation and presence of branched groups on associated hydrocarbon chains and position and mode of attachment to the glycerol backbone. Sterols are an interesting case in that the type of sterol observed in membranes is characteristic of animals (cholesterol in higher organisms; ergosterol and 7-dehydrocholesterol in yeasts) and plants (camposterol, sitosterol, stigmasterol). The sterols have structures that are relatively similar indicating that the functions performed by sterols in biological membranes may be highly conserved. What these functions are remains unclear but a role in lateral domain creation is strongly inferred by recent evidence (see Chapter X). Certain functions identified with other membrane lipids are few. Specific glycolipids are known to be associated with the ABO blood group substances and phosphorylated molecular species of glycerophosphoinositides are involved in transmembrane signalling cascades.

Finally, the glycerophospholipids act as a reservoir of the metabolic precursor of eicosenoid biosynthesis, arachidonic acid.

In terms of representation, some molecular species may be present in relatively minor proportions and would not be expected to greatly influence the phase structure of the membrane lipid matrix. Others, for example, cholesterol and sphingomyelin in the plasma membrane, for example the membrane forming the myelin sheath, may dominate the lipid composition of the membrane and exert a major impact on its structure and properties.

The complexity of membrane lipid composition raises three major questions related to homeostasis:

(a) Why do (most) cells in a higher organism maintain such a complex lipid matrix despite being protected from variations in their environment by the homeostatic regulation of surrounding fluids (e.g. composition of the extravascular fluid and of serum)?
(b) Are the narrow ranges (homeostatic regulation) of any of the membrane lipids critical to the cell or are there cell functions specifically dependent on the presence of a particular native lipid which, if substituted results in perturbation of that function?
(c) Is lipid complexity the result of the complex assortment of membrane proteins or do lipids have their own rules for assembling which govern, in turn, those proteins "passively" inserted into the matrix?

Lipid homeostasis for a complex mixture represents a cost in terms of metabolic expense and gene diversity which encodes all enzymes responsible for catalysing the biosynthetic and degradation pathways. Multiple cross-regulation would be expected to achieve the characteristic composition of the membrane. The maintenance of a complex lipid composition requires regulated pathways to repair the alterations (literally "homeostasis") induced by the "activated enzymes" perturbing the cell membranes.

Methods of lipid analysis have currently achieved a level of sophistication such that a large number of molecular species can be identified accurately and quantitatively in a short period of time. Methods of establishing the physical consequences of the diversity of lipid composition have not kept pace with these developments. The reason for this is mainly that methods employed to characterise the conformation and structure of the membrane lipid matrix are averaging techniques such as spectroscopy, that are unable to distinguish subtle differences in local environment created by domains dependent on lipid complexity. As a result much information is indirect and based on the construction and examination of models to simulate the lipid matrix of biological membranes. Frequently such models diverge considerably from the real world because of the low energy barriers that separate conformational states in the complex mixtures which represent biological membranes that are less likely to occur in defined lipid mixtures.

2.2. The origin of membrane lipid complexity

The observed complexity of lipid composition particularly in the cell membranes of higher animals originates from endogenous biosynthesis of complex lipids from building blocks that are themselves products of the cell or imported into the cell from outside. Thus the consequence of the genetic diversity in evolving pathways of complex lipid biosynthesis is complemented by an external supply of membrane lipid building blocks from the diet. One of the more remarkable features of the process is that a number of the building blocks are essential nutrients in the diet of animals. In man these are vitamins and include fatty acids containing ω3 or ω6 unsaturation and fat-soluble vitamins like α-tocopherol.

The origin of membrane lipids has been investigated in a number of tissue culture cells *in vitro* (Sleight and Kent, 1983; Ladenson *et al.*, 1993). Fatty acid biosynthesis and *de novo* synthesis of polar lipids usually provide most of the lipids required for membrane biosynthesis. An interesting exception is the polyunsaturated fatty acids of membrane glycerophospholipids with their first double bond located at position ω3 and ω6 of the acyl chain (so called "essential" PUFA). Two families of essential PUFA are not able to be synthesized by higher animals. Poikilothermic animals possess the elongase and desaturase enzymes the activities of which provide the relevant end-products, arachidonic (AA, 20:4ω6) and docosahexaenoic (DHEA, 22:6ω3) acids from the respective precursor fatty acid, linoleic (LA, 18:2ω6) and eicosapentaenoic (EPA, 20:5ω3) acid supplied in the diet. In human the sources are plant lipids in the form of oil for LA (18:3ω3 is readily oxidized) and marine sources (EPA being slowly oxidized remains available for incorporation in complex lipids). While oxidative pathways in plants are the major source of essential fatty acids (Okuley *et al.*, 1994) they are not the sole source (Metz *et al.*, 2001).

2.3. The purpose of membrane lipid complexity

Membrane complexity allows flexibility in physiological functions in the face of varying environmental situations. While the detailed molecular species of membrane lipids is complex their physical properties are invariably that of weak surfactants with critical micelle concentrations in the nM concentration range. Most molecular species of lipid form either a bilayer or hexagonal-II phase when dispersed in dilute salt solutions at physiological temperatures. The exceptions are the sterols and heavily glycosylated lipids which are sparingly soluble and micellar-forming, respectively. The membrane lipid matrix is believed to be exclusively bilayer in arrangement so that the interaction of the non-bilayer forming lipids with other membrane constituents must constrain them into a bilayer configuration. This is consistent with structures observed when total polar lipid extracts of biological membranes are dispersed in aqueous systems which usually consist of phase separated bilayer and non-bilayer structures. Changes in temperature, salinity, etc. alter the stability of the matrix and this may be compensated to some extent by relocation of lipid and protein constituents into or out of the domains where bilayer stability may be altered in local regions of the membranes.

Adaptation of poikilothermic organisms to changes in environmental conditions involves changes in the lipid molecular species of the membrane matrix. Such changes may involve remodelling of molecular species such that a redistribution of the fatty acyl esters takes place within a particular class of lipid or between different lipid classes. For example, mixtures of phospholipids with saturated fatty acyl chains with molecular species acylated to unsaturated fatty acids results in phase separation of gel and fluid phases. A redistribution of the fatty acids between the two types of lipid would result in a homogeneous fluid bilayer at physiological temperatures. Remodelling without *de novo* fatty acid biosynthesis has been observed in adaptation of microorganisms such as protozoa (Dickens *et al.*, 1982) and algae (Sato and Murata, 1980). One of the most common strategies is to change the length and or extent of unsaturation of the acyl chains as the amphipathic balance is dependent on the hydrophobic-hydrophilic affinities within the complex lipid molecule (Cossins 1994). Such adaptations have been reported in yeast (Okuyama *et al.*, 1979) where a shift in growth temperature from 35° to 10°C resulted in a change in the ratio of C16:C18 fatty acids from 1.4 to 4.7 on a time-scale suggesting direct effects on the synthase rather than a control exerted at the level of genetic transcription. The most common method of adaptation, however, is an alteration in the extent of unsaturation of the fatty acyl residues. The insertion of a single cis double bond into a stearic acid molecule, for instance, reduces the melting point from 70° to 13°C due to the fact that the the chains are unable to pack into close alignment. Changes in desaturase activity can be bought about by direct modulation of the desaturase enzyme pathways (Kates, 1984), the rate of gene transcription (Wada *et al.*, 1990; Gombos *et al.*, 1996; Nakashima *et al.*, 1996) and by the altered availability of substrates including molecular oxygen the concentration of which is inversely proportional to temperature (Thomas *et al.*, 1998; Rutter *et al.*, 2002). While the lipid homeostatic process in homeothermic organisms is not required to cope with the changes that poikilothermic organisms are subjected to it is likely that the strategies employed to preserve the structure, stability and viscosity of the matrix are the same.

Changes in physiological state or activity can also be associated with changes in phospholipid molecular species in the membranes. Membrane phospholipid remodelling has been found to occur in tumor cells characterised by increased levels of fatty acid synthase. The products of the synthase are long-chain saturated fatty acids which incorporate into membrane phospholipids. The fate of these saturated molecular species has been traced to their incorporation into detergent resistant membranes leading to the suggestion that the synthase indirectly modulates transmembrane signalling processes mediated by membrane rafts (Swinnen *et al.*, 2003).

3. CHARACTERISATION OF MEMBRANE LIPID COMPLEXITY

A detailed analysis of the membrane lipids extracted from biological membranes was, until recently, a tedious procedure consisting of multiple steps. These included separation of the lipid classes according to the polarity by thin layer chromatography on a polar silica gel or by HPLC on an apolar grafted silica gel and subsequent separation of the molecular species for each of the lipid classes. The final step required a high degree of purity of each of the lipid classes to obtain a reliable pattern of fatty acids located at the *sn*-1 and *sn*-2 position of the glycerolipids. The profile of fatty acid methyl esters is obtained by gas-liquid chromatography after an additional step consisting of the position-selective enzymatic release by purified phospholipases A1 or A2. The direct assessment of molecular species for diglycerides (Gaskell and Brooks, 1977) and ceramides (Vieu *et al.*, 2002) can also be obtained by a one-step HPLC and GLC procedure. For diglycerides the derivatization of the primary hydroxyl group is required following the enzymatic cleavage of the polar head-group by phospholipase C from *Clostridiium welchii* or *Bacillus cereus* which shares a sphingomyelinase activity. For PI the use of a specific phospholipase C from *Staphylococcus aureus* is recommended. The pre-purification of ether PE lipids is required to prevent overlapping of the chromatographic patterns of diacyl-, alkyl- and alkenyl-compounds. The purification of the derivatized products either acetylated or trimethylsilylated is also advised before further analysis. Intramolecular acyl migration from position *sn*-2 to *sn*-3 is common and, although it does not change the fatty acid composition, the positional distribution is altered. A partial acyl migration which is facilitated by a pH shift results in appearance of 1,3-diacyl glycerol derivatives. Recently a single-step analysis of the molecular species of endogenous ceramides and of the ceramide moiety of sphingomyelins in biological samples, using gas-liquid chromatography (GLC) was introduced (Vieu *et al.*, 2002). Silylated sphingomyelins were quantitatively converted to monosilylated ceramide upon injection into the chromatography system, whereas the free ceramides were di-silylated on the primary and secondary alcohol function, as confirmed by mass spectrometry. The reproducible shift of the retention times between the mono- and di-silylated derivatives even without mass spectrometry data enables simultaneous quantification of the variety of sphingomyelin and ceramide molecular species. Overlapping diacylglycerols were first removed by a mild saponification of the lipid extract. Three major free ceramides (NC16:0, NC24:0, and NC24:1) were quantified in HEPG2 and Chinese hamster ovary (Lichtenbergova *et al.*,) cells. Upon induction of apoptosis in CHO cells by C6-ceramide, the disappearance of the C6-ceramide, its partial conversion to C6-sphingomyelin, and the prominent increase of NC16:0 ceramide are observed, which represents a unique procedure of simultaneous analysis of sphingomyelin and ceramide molecular species able to monitor the variation of the different pools in biological samples.

The straightforward procedure described above gives only the overall fatty acid composition of diglycerides or ceramides as total carbon number and double bond. Therefore the species C36-Δ4 is common to palmitoyl-arachidonyl (16 : 0–20:4) and dilinoleyl species (18 : 2–18 : 2) or other combinations. A rapid procedure of mass spectrometry has also been developed in which the direct introduction (DCI) of a TLC-purified phospholipid class yields the pattern of molecular species, after thermal desorption and soft ionization by chemical ionization with ammonia as the reagent gas, in a single step. The procedure was applied to screen the molecular species of sphingomyelin of red blood cell membranes from two geographic populations (Katsikas and Wolf, 1995). Spectra were obtained with three discrete areas corresponding to ions derived from sphingoid bases, fatty acids and ceramides, respectively. Eight sphingoid bases have been detected in both groups of red blood cells. Sphingosine was the major contributor. The rest of the sphingoid bases have 16 to 19 carbon atoms and 0 to 2 double bonds. Fourteen fatty acids have been detected with 16 to 26 carbon atoms and 0 to 2 double bonds. Palmitic acid was the most abundant fatty acid. Thirty eight ceramides have been identified. Statistically significant differences between some sphingoid bases, fatty acids and ceramides ($P < 0.01$) have been detected between the two populations which have been attributed to diet as well as other environmental and genetic factors.

Fast Atom Bombardment (FAB) has also been shown to resolve the lipid composition forming the annulus around intrinsic membrane proteins or the membrane anchor. An example of complete resolution of the structure is the mass spectrometric analysis of human acetylcholinesterase of the red blood cell membrane (Roberts *et al.*, 1988). The glycoinositol phospholipid membrane anchor was found to consist of a glycan linked through a glucosamine residue to an inositol phospholipid. Analysis by fast atom bombardment mass spectrometry with negative ion monitoring and by the complementary technique of collision-induced dissociation revealed molecular and daughter ions that indicated a plasmanylinositol with a palmitoyl group on an inositol hydroxyl residue.

The introduction of tandem MS-MS has considerably alleviated the analytical tasks. The procedure no longer requires the complete separation of every lipid class forming the matrix. The first stage in the mass spectrometer (quadrupole filter Q1) is used to trace the parent ion after a collision-induced fragmentation which induces a characteristic neutral loss detected in the second stage of the equipment (usually a quadrupole filter Q2). For example, the polar head group of phosphatidylcholine can be monitored by the loss of the phosphorylcholine group of 183 D by scanning each of the parent ions m/z = M in Q1 which have generated an ion mass m/z M-183 in Q2. The parent ions correspond in this instance to the various molecular species of phosphatidylcholine. The fragmentation between Q1 and Q2 in a gas collision chamber serves to produce the product ions identified in the second mass analyzer. The collision chamber is an intermediary quadrupole running in the transmission mode which can be filled with helium, argon or nitrogen at an appropriate pressure. A scan of the product ions allows a direct assignment of the fatty acids for a selected lipid. The whole procedure can be applied in the mode "selected reaction monitoring" (SRM) with a maximum selectivity and sensitivity for each lipid class. Because of the experimental simplicity of the method which requires no pre-purification of the lipid extract required the major advantages can be summarised:

(a) A definitive assessment of the fatty acid composition which hitherto could not be obtained from the overall carbon number-double bond content.
(b) The identification of minor molecular species among dominant species.
(c) An accurate quantification which cannot be obtained when the sample has to be pre-purified through multiple separation steps with varying yields.

This approach is essential if the significance of changes in minor species is to be fully appreciated in a particular cell type or as a result of cell activation while the dominant molecular species may be considered to form the common building-blocks of the matrix. In practice, the high sensitivity of the method allows analysis of small samples which greatly enhances utility of the method. This can be exploited so that analysis of membrane fractions from cultured cells, even if membrane sub-fractionation is required, for instance, for the preparation of the detergent-resistant membranes. An illustration of the application of the analytical procedure has been reported by Brugger *et al.* (1997).

Nano-electrospray ionization tandem mass spectrometry brings analysis of membrane lipids down to the picomole level. The soft ionization procedure applied in ESI-MS is appropriate for small volumes of chloroform:methanol extracts from less than 10^6 CHO cells (down to 1000). The resolution of the molecular species of lipid classes including plasmalogen analogues is obtained with negative ions [M−H]- mode for anionic lipids and positive mode [M + H] + for choline-containing phospholipids, PE being responsive in both modes. When operated in the single-stage MS, ESI mass spectra of the lipid extract shows almost exclusively molecular ion species without fragmentation. The assignments of the molecular ions can be obtained by recording the product ion spectra in Q2 after selection in Q1 and fragmentation in the collision cell. The polar head group detected in Q2 can be lost as a charged or a neutral fragment. The proportion of ion product (head group lost as a charged product) increases with the collision energy whereas the neutral loss is favored under 30 eV. In a precursor ion scan the second mass analyzer is set to transmit only ions from the parent ion selected by Q1 (for instance Q2 is set to detect m/z 184 corresponding to phosphorylcholine + H transferred to the headgroup of PC and SM) whereas in the neutral loss scan the analyzer Q2 is scanned in a synchronized fashion with Q1 but with an offset to lower m/z values equivalent to the neutral loss of interest (for instance, −183 to monitor PC).

Using the higher mass resolution of tandem Fourier transform ion cyclotron resonance mass spectrometry (Fridriksson *et al.*, 1999) over 90 different phospholipid species were resolved in detergent-resistant membranes prepared from mast cells. Coupling between the high affinity receptor for Ig E and tyrosine kinase, Lyn of the Src family anchored to DRM, is accompanied by a shift to polyunsaturated species of PC, SM, PE and PS found in this sub-fraction. Taking into account the critical role of fatty acids in the association of phospholipids with cholesterol a transition of the liquid-ordered to liquid-disordered fluid phase was inferred from this compositional shift (Biagi *et al.*, 1999). Subtle changes in compositon of discrete membrane domains of this order are not possible using conventional analytical methods. It may therefor be anticipated that in future identification of involvement of single molecular species with protein function may be realized.

In addition to monitoring changes in molecular species by chemical analysis methods have also been developed to monitor changes directly. One such method is to biosynthetically incorporate probes into membrane lipids in living cells and follow changes in the probes *in situ*. For example, parinaric acid when fed to cells in tissue culture becomes incorporated into polar membrane lipids as a substitute for the normal fatty acids (Shvedova *et al.*, 2001). Separation and analysis by conventional chromatography is simplified by following fluorescent analogues and changes as a consequence of oxidation, turnover etc can be detected at levels not yet accessible by direct chemical analysis (Koty *et al.*, 2002). No doubt the introduction of these sophisticated analytical techniques will underpin further studies of the response of the membrane matrix to activation and signalling processes.

4. DEPENDENCE OF MEMBRANE LIPID COMPOSITION ON EXTERNAL PROGENITORS

The interdependence of membrane lipid composition on domestic supply of building blocks for complex lipid biosynthesis compared from supply from external sources is best exemplified by cultured cells. When the growth media is supplemented by low percentages of serum a deficiency of essential fatty acid (EFA) can be demonstrated in cultured cells. The induction of EFA deficit can be used as a model of *in vitro* senescence for long term primary cultures of rat cardiomyocytes (Biagi *et al.*, 1999). Supplementation by ω6 and ω3 in the ratio 2 : 1 re-establishes the cell content of fatty acids without altering the composition appropriate for the balance of aggregation/anti-aggregation precursors. Similarly the epidermal linoleic acid (ω6 EFA) is essential for cutaneous barrier function. Cultured human keratinocytes, routinely used for studies of lipid metabolism, are grown in a serum-free medium (KSFM), under conditions that reveal EFA-deficient cells (Schurer *et al.*, 1999). The FA composition of the total cellular lipid and major lipid fractions of cells grown with supplements of 10% serum or 1% Ultroser balances their EFA-deficient state primarily with oleic acid but grown with supplements of primrose oil they are able to normalize the epidermal FA composition *in vivo* with raised linoleic and lower oleic acid contents. When cells are cultured longer than 48 h with this supplement a decreased cell growth is observed. The results show also that the uptake of linoleic, but not that of oleic acid, is influenced by the extracellular fatty acid composition, and that EFA-supplemented keratinocytes compared to EFA-deficient cells might serve as a better *in vitro* model. Strikingly, not only the overall fatty acid composition but also the proportion of lipid classes represented in cell membranes can be perturbed by EFA-deficiency in cultured cells. EFA deficiency in hepatoma HepG2 cells caused more linoleic acid given as a tracer to be converted to eicosanoid precursors by Δ6- and Δ5-desaturases and to distribute in neutral fats and PE and less to cardiolipin than control did (Melin and Nilsson, 1997). It is expected that the depletion of polyunsaturated species of particular pools of cell phospholipids has consequences on the matrix physical state.

4.1. Essential fatty acids

Large shifts of the fatty acid composition of the membrane lipids reported above in cultured cells illustrates an artefact resulting from a shortage of EFA in the growth medium. With a perspective of homeostatic regulation on membrane lipids the depletion of essential fatty acids in cultured cells may introduce a bias in studies conducted *in vitro*.

Indeed, *in vivo*, the culmination in healthy living animals of the action of pancreatic enzyme digestion and absorption in the gut followed by lipoprotein packaging and fatty acid interconversion in the liver, and their distribution *via* the serum, provides the peripheral tissues with an abundant supply of EFA. The relevance of such abundance of EFA has probably been underestimated. Recently, dietary recommendations have been revised to incorporate an increased supplementation in ω3 and ω6 EFA in milk formula for newborns. Vegetable-oil additives appropriately complement the relatively low levels of EFA in cows milk to the levels found in human breast milk. The recommendations given as AI (Adequate Intake) (Simopoulos *et al.*, 2000) are values expected to meet or exceed the amounts needed to maintain a defined nutritional state in all members of a healthy population. Because the infant formula should also cover the requirements for a large number of premature infants and the needs of the sick infant there has been considerable discussion on daily recommended dietary intakes of ω-6 and ω-3 EFA. With the objective of achieving a standard brain and visual development AI are considered to be; linoleic acid (ω6 18:2), 10% (2% of the energy/24 hours); linolenic acid (ω3 18:3), 1.5%; arachidonic acid (ω6 20:4); DHA (ω3 22:6), 0.35%; and EPA (ω3 20:5),$<0.1\%$. These values are similar to the values found in human breast milk in Western countries but much higher (2-folds) for linoleic acid than found in Japan and Mediterranean countries. It is noteworthy that a consensus has been reached that ω-6 EFA should be reduced in the diet of newborns to reduce the adverse effects of excessive arachidonic and linoleic acids generating inappropriate levels of eicosanoïds. It has also been found that linoleic acid competitively inhibits the $\Delta 6$ desaturation of α-linolenic acid, the precursor of ω-3 EFAs. The role of ω-3 EFA (especially DHA) incorporated into membrane phospholipids surrounding the Na^+ and Ca^{2+} channels was demonstrated in cultures of neural heart cells. Dietary supplementation changed the fatty acid composition of the membrane lipids and this was reflected in data recorded by electroretinograms. Deprivation in ω3-22:6 resulted in the accumulation in ω6-22:5, a compensatory mechanism to maintain the proportion of C22 chains in particular neurosensorial cells (Bourre *et al.*, 1989a, 1989b; Bui *et al.*, 1998; Weisinger *et al.*, 1998; Weisinger *et al.*, 1999). Such a compensatory increase is amongst the most convincing evidence of the homeostasis of the membrane lipid composition *in vivo*.

4.2. Essential fatty acid alterations in pathological conditions

Deficiency in membrane lipid EFAs occurs in pathological conditions associated with disorders like cystic fibrosis or when lipids digested and absorbed normally but are "leaking" via lymphangectasia. Deviation in the red cell membrane phospholipid fatty acid composition can be used to detect a long-term deficiency with greater reliability than the pattern of circulating lipid fatty acids. Figure 1 shows the compensatory mechanisms triggered in infants when nutritional sources of ω6-18:2 has been decreased. ω6 C18:2 can be decreased variably under the pathological conditions cited above, by – 48% in a case of severe cystic fibrosis and by – 15% in moderate lymphangectasia. The proportions of non-essential fatty acids which are produced in the liver by an active Δ9-desaturase, such as ω7-16:1 (not shown), ω7-18:1 and ω9-20:3 are found to be inversely related to the level of ω6-18:2. In adults with a liver disease this compensatory reaction can be absent.

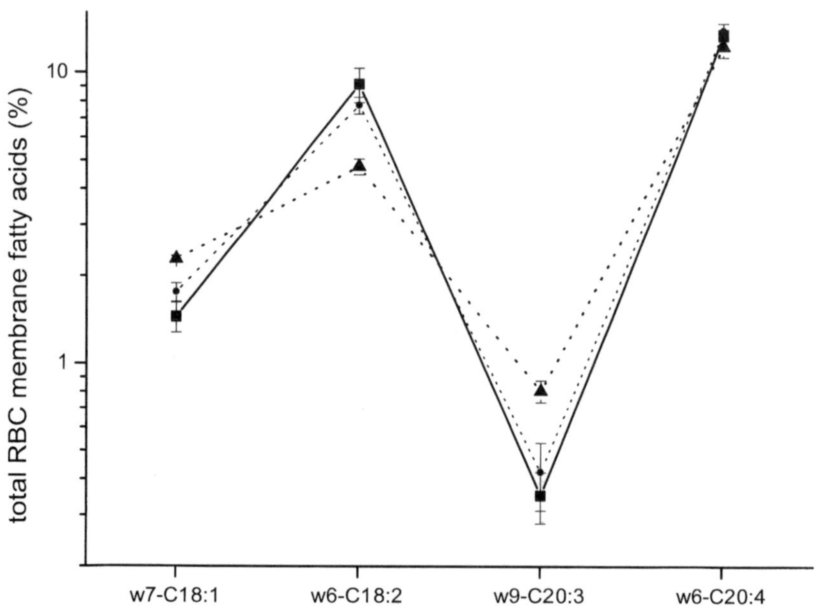

Figure 1. Fatty acid profiles of red blood cell phospholipid membrane illustrate the homeostatic compensatory reaction for their overall double bond content. As a result of the extensive ω6-EFAs deperdition through lymphangectasia (dotted line) or of the poor digestion of the ω6-rich dietary lipids of plant origin due to pancreatic juice deficit (dashed line) the content of ω6 linoleic acid is decreased relative to normal profile (full line). ω6-18:2 decrease used as a sensitive marker has a strenghened diagnostic value if associated to the ω7 and ω9-desaturase "compensatory" reaction which results in the increase of non essential polyunsaturated fatty acids such as ω7- and ω9-20:3 produced from the ω7- and ω9-monounsaturated precursors. Particular derivatives such as ω7 and ω9 20:3 are "non physiological" components of the erythrocyte phospholipids (normal < 0.5%) and their accumulation is highly significant of EFA long-term deficiency.symbol values; full square, normal; full circle, lymphangectasia; full triangle, cystic fibrosis.

The effect of cystic fibrosis on membrane phospholipid molecular species has been the subject of a number of studies. In a knockout mice model of cystic fibrosis, *cftr*$^{-/-}$, it has been found that phospholipids of lung, pancreas and ileum are enriched with arachidonic acid at the expense of docosahexenoic acid. The characteristic balance of these fatty acids could be restored to proportions observed in the wild-type animals by dietary supplementation with docosahexenoic acid (Freedman *et al.*, 1999). The consequence of fatty acid imbalance in the knockout mice was that when the lungs were challenged with *Pseudomonas* lipopolysaccharide the production of eicosenoids such as $PGF_{1\alpha}$, $PGF_{2\alpha}$, PGE_2 and thromboxane B_2 were all significantly elevated above that of the control mice (Freedman *et al.*, 2002). One of the most striking examples of homeostatic regulation is the increased conversion of ω6-18:2 to ω6-20:4 which serves to maintain a normal or sub-normal level of arachidonic acid in the membrane lipids. For instance, arachidonic acid is decreased by only 9% whilst ω6-18:2 is reduced by 48% in infants affected by severe cystic fibrosis (Figure 1).

4.3. Choline

Choline is designated as a lipotropic factor in the diet preventing liver steatosis and its shortage can induce apoptosis (Yen *et al.*, 2002). When auxotroph PC12 cells were cultivated in a choline-free medium, apoptosis was induced (27.4% of cells apoptotic at 72 h as compared to 4.4% in control medium). In choline-free medium at 72 h, there was a 49% decrease in phosphatidylcholine concentration ($P < 0.01$) and a 34% decrease in sphingomyelin concentration ($P < 0.01$); however, there was no change in phosphatidylethanolamine concentration. Before detecting increased apoptosis in choline-deficient cells, a significant increase in ceramide (218% control) and diacyglycerol (155% control) concentrations was measured. Choline deficiency-induced apoptosis was prevented completely by replacing choline or adding a caspase inhibitor into the medium within 48 h of initial choline deprivation. In those cells rescued by replacing choline at 36 h, the concentrations of phosphatidylcholine, sphingomyelin, ceramide, and diacyglycerol returned to levels of control cells. According to this model the signaling activity of diacylglycerol is considered negligible in comparison to ceramides.

4.4. Cholesterol homeostasis

Cholesterol is a significant lipid component of plasma membranes and its role in membrane domain formation is discussed in detail elsewhere in this volume. The source of cholesterol in membranes is from both the diet and endogenous biosynthesis. A reduction of cholesterol levels in the body can be induced by the statin group of drugs which modulate the activity of the key enzyme in cholesterol biosynthesis, HMG-CoA reductase (Miyazaki *et al.*, 2002). The ultimate aim of such therapies is prevent plaque formation by reducing blood cholesterol not to modulate membrane cholesterol although perturbations in plasma membrane cholesterol levels has been reported as a consequence of treatment (Kirsch *et al.*, 2003).

A blockade of the ultimate step in cholesterol biosynthesis due to a genetic defect in the enzyme catalysing the reaction, 7-dehydrocholesterol reductase (Tint, 1993), or following the administration of a potent inhibitor of the enzyme (Roux and Aubry, 1966) produces a phenotype similar to Shh−/−, the defect associated with the patterning protein, Sonic Hedgehog (Porter *et al.*, 1996). A possible explanation the phenotype is that the malformation in the brain (agenesis of the pituitary gland and corpus callosum associated with a variable degree of holoprosencephaly) and limb malformations (frequent ectrosyndactyly) (Gofflot *et al.*, 2003) are caused by a deficiency of cholesterol which normally assists the autoprocessing of Sonic Hedgehog in the embryo (Mann and Beachy, 2000). This, however, appears to be inconsistent with observations of *in vitro* processing where 7-dehydrocholesterol, the precursor accumulated, is able to substitute for cholesterol with full activity (Cooper *et al.*, 1998). An alternative possibility is that "rafts" are required to serve as a platform for Sonic Hedgehog maturation (autocleavage and cholesterol addition) (Rietveld *et al.*, 1999) and 7-dehydrocholesterol is unable to substitute for cholesterol in raft function. According to this hypothesis, embryonic requirements for cholesterol are quantitative and cannot depend entirely on the maternal influx when the embryonic synthesis is suppressed. The relatively high requirements for cholesterol have been assessed with embryo cultures in which endogenous biosynthesis is blocked and the supply of cholesterol can be varied in the growth medium (Llirbat *et al.*, 1997). Extremely high rates of cholesterol accumulation were found between gestational day 10 and 14 (mid-pregnancy in the rat species). The rates were such that it would be unlikely that the cholesterol requirements could be met by a normal maternal influx alone. *In vivo* only pregnant dams rendered hypercholesterolemic by a diet highly enriched with cholesterol can prevent the malformations in the offspring (Roux *et al.*, 1980; Barbu *et al.*, 1988). The dependence of the embryo on external sources can be also demonstrated when the influx is perturbed in mouse mutants lacking the endocytic receptor megalin (Willnow *et al.*, 1996) or lacking apolipoprotein B affecting the cholesterol transport via the yolk sac (Farese *et al.*, 1996). Under these circumstances the embryos can display either malformations or a much higher sensitivity to cholesterol inhibitors. The experimental data support the view that an embryonic dependence exists for cholesterol imported from the dam and despite the fact that cholesterol is synthesized by embryonic cells the quantitative requirements exceeds the endogenous supply during the early pregnancy. This is consistent with the involvement of cholesterol in the phase separation of "rafts" in the lipid matrix at a point when cholesterol reaches a critical level in the membrane matrix. This level would be reached when cholesterol begins to associate with saturated molecular species of sphingomyelin to form the liquid-ordered phase (Ahmed *et al.*, 1997).

5. ENZYMES ALTERING MEMBRANE LIPIDS

The predominant lipids of most cell membranes are the phospholipids. These lipids consist of four chemically-distinct components: fatty acids, glycerol, phosphate and an amide-containing alcohol or *myo*-inositol. Each is linked by an ester bond with the notable exception of the vinyl ether or ether bond of the plasmalogens or glyceryl ether phospholipids, respectively. This means that the acyl phospholipids are synthesized and degraded by condensation and hydrolytic reactions. Some of the more important of these reactions are shown in Figure 2.

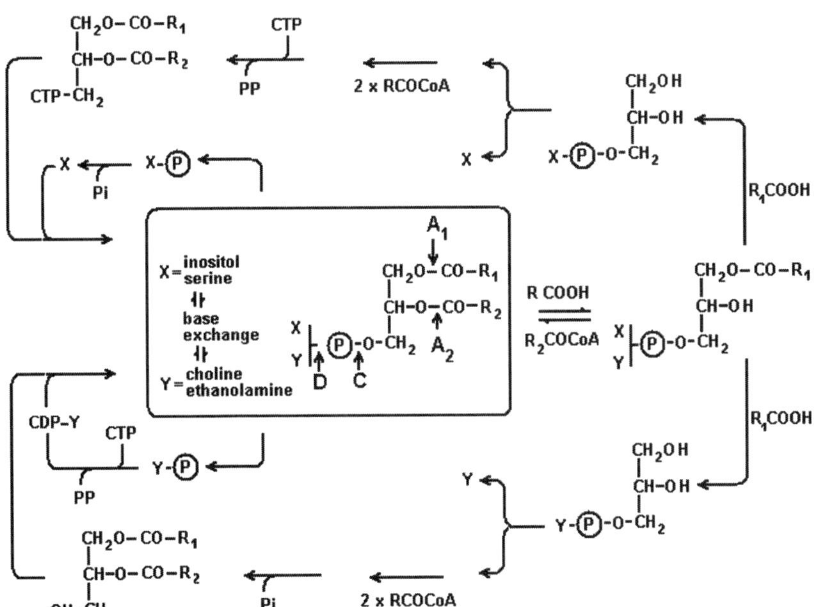

Figure 2. Major pathways of phospholipid turnover in mammalian cells.The arrows in the central box indicate the sites of hydrolase action by phospholipases. The combined action of phospholipases A_1 and A_2 results in the pathways depicted as B.

It can be seen from the figure that phospholipid synthesis and degradation can be divided into distinct categories. In A the essential amphipathic character of the lipid is preserved as a consequence of the reactions that occur since hydrolysis of only one acyl ester bond is involved followed by recondensation with another. This cycle represents a partial turnover of the molecule as the lysophosphatide intermediate retains its identity as a phospholipid, albeit with physical properties that are not conducive to a stable bilayer arrangement. In B and C, degradation is more extensive and the yields are water-soluble products such as glyceryl and phosphoryl esters. The difference between pathways B and C is in the particular group of the phospholipids that is hydrolysed first. Pathway B is initiated by removal of one acyl chain then by the other followed by degradation of the resultant water-soluble products. Degradation by pathway C begins with removal of the phosphoryl base or sugar. In no case does synthesis proceed by a reversal of degradation. Biosynthesis is by an entirely different pathway and involves condensation of an unsubstituted diglyceride with a CDP-derivative or vice versa, condensation of CDP-diglyceride with an alcohol.

5.1. Phospholipase A_2

A number of early studies showed evidence of phospholipase A activity hydrolysing fatty acyl residues at the C-2 position of the glycerol moiety of phosphoglycerides (PLA_2) by enzymes that were bound to cell membranes and acting on endogenous substrates. The enzyme activity was shown to change the level of cyclic AMP in fat cells (de Cingolani *et al.,* 1972) and to be influenced by insulin/glucagon ratio (Polonovski *et al.*, 1974; Wolf *et al.*, 1977). The possibility that the enzyme retro-regulates membrane fluidity was also implied in these early reports (Petkova *et al.*, 1987). The notion that these enzymes are involved in signal transduction by mediation of metabolic turnover of membrane phospholipids has been less favoured by the discovery of cytosolic species of PLA_2 whose action appears more appropriate to this function (Clark *et al.*, 1991). It was also realized that phospholipids of the membrane matrix were relatively resistant to enzyme hydrolysis. A number of mechanisms are responsible for the protection of phospholipids against PLA_2 attack. These include:

(a) Limitation of the penetration of the enzyme into the substrate due to the tight packing of the lipid molecules.
(b) Restricted access of the enzyme to its prefered substrate. For instance, exogenous secretory type II PLA_2 outside cells cannot gain access to phosphatidylethanolamine substrate located in the inner leaflet of plasma membranes.

(c) Dilution of susceptible substrates within non-substrate membrane lipids such as sphingolipids and cholesterol.

These limitations serve to restrict enzyme activity to a relatively small proportion of the lipids forming the membrane lipid matrix leading to the conclusion that the hydrolysis of such a small number of molecules would be unlikely to result in any physiological consequences.

Detection of phospholipid turnover resulting from the action of endogenous PLA_2 acting on the membrane lipid matrix requires pre-labeling of phospholipids by biosynthetic incorporation of radio-labeled precursors, either ^{32}P- or ^{14}C-labelled fatty acids, administered beforehand to the organism. Furthermore, the release of lysoderivatives and fatty acids can only be related to PLA_2 under particular experimental conditions in which subsequent enzyme action on the products (lysophospholipases, transacylase, oxygenation of PUFA, etc) are negligible. These conditions are not easily achieved and assays of enzyme activity against exogenous substrate, i.e. dispersions of radio-labeled or fluorescent analogues of phospholipids, are assumed to reflect the membrane bound enzyme activities. Despite such problems it has been shown that activities against exogenous and endogenous substrates vary according to the fluidity of the membrane matrix (Momchilova *et al.*, 1985; Momchilova *et al.*,1986). In these studies the enrichment of the membrane in saturated molecular species of phosphatidylcholine resulted in an increase in PLA_2 activity against exogenous substrate but a decrease against endogenous membrane phospholipids. This was interpreted as an effect on vertical displacement of the phospholipid due to the particular packing arrangements of the moleculaes in the bilayer.

The demonstration that $cPLA_2$ is the particular phospholipase involved in the release of arachidonic acid from the *sn*-2 position of phosphatidylcholine under conditions where the cascade of reactions leading to eisosanoid biosynthesis is triggered has been fundamental to understanding the role of PLA_2 in this process (Clark *et al.*, 1991). In this work it was shown that cloning and expression of a cDNA encoding a high molecular weight (85.2 kd) $cPLA_2$, assigned to type IV PLA_2, has no detectable sequence homology with the non-pancreatic secreted forms of PLA_2 (type II) described previously. Whereas type II PLA_2 represented non-specific activities observed earlier, $cPLA_2$ selectively cleaved arachidonic acid from microsomes of intact biological membranes. It was demonstrated that $cPLA_2$ translocated to membrane vesicles in response to physiologically relevant changes in free calcium. By contrast, secretory type II PLA_2 is known to be calcium-dependent with an optimum concentration greater than 1 mM consistent with its preferential extra-cellular activity. Moreover, an amino-terminal 140 amino acid fragment was identified in $cPLA_2$ which translocated to natural membrane vesicles in a Ca^{2+}-dependent fashion. Maximal activation required phosphorylation of Ser-505 via the

MAP kinase pathway (Lin *et al.*, 1993). Treatment of cells with agents that stimulated the release of arachidonic acid cause increased serine phosphorylation and activation. The site of $cPLA_2$ phosphorylation by MAP kinase, Ser-505, was identical to the major site of $cPLA_2$ phosphorylation observed in phorbol ester-treated cells. Replacement of Ser-505 with Ala resulted in a mutant $cPLA_2$ that caused little or no enhanced agonist-stimulated arachidonate release from intact cells.

The delineation of the two functionally distinct domains of $cPLA_2$ has been obtained experimentally (Nalefski *et al.*, 1994). Isolation and sequence analysis of $cPLA_2$ cDNA clones from four different species revealed several highly conserved regions. The N-terminal conserved region is homologous to a number of other Ca^{2+}-dependent lipid-binding proteins. The first 178 residues of $cPLA_2$, containing the homologous Ca^{2+}-dependent lipid-binding (CaLB) motif, and another recombinant protein containing the $cPLA_2$(1–178) fragment placed at the C-terminus of the maltose-binding protein (MBP-CaLB) associated with membranes in a Ca^{2+}-dependent manner. $cPLA_2$ and MBP-CaLB also bind to synthetic liposomes at physiological Ca^{2+} concentrations, demonstrating that accessory proteins are not required for this process. In contrast, ΔC2, a truncated $cPLA_2$ lacking the CaLB domain, failed to associate with membranes and failed to hydrolyze liposomal substrates but hydrolyzed monomeric 1-palmitoyl-2-lysophosphatidylcholine at identical rates to that of $cPLA_2$ in a Ca^{2+}-independent fashion. On the basis on the requirement for calcium, the Ca^{2+}-independent catalytic domain, and the regulatory CaLB domain were clearly distinguished. In Chinese Hamster Ovary (Lichtenbergova *et al.*, 1998) cells the binding to membrane of $cPLA_2$ can be modulated by the ratio SM/cholesterol (Klapisz *et al.*, 2000), which in turn, modulates membrane susceptibility. Cholesterol depletion of CHO-2B cells by treatment with methyl-β-cyclodextrin resulted in inhibition of the release of arachidonic acid whereas the restoration of cholesterol by incubation with cholesterol-loaded cyclodextrin relieved the inhibition. Conversion of CHO-2B cellular sphingomyelin to ceramides by *Staphylococcus aureus* sphingomyelinase enhanced endogenous $cPLA_2$ activation by calcium ionophore and epinephrine as well as uptake by cells of C2- and C6-ceramide analogs. These results were confirmed *in vitro* with purified human recombinant $cPLA_2$ acting on a model phospholipid substrate. The enzyme activity was inhibited by sphingomyelin but reactivated by ceramides as well as by cholesterol added to glycerophospholipid liposomal substrates containing sphingomyelin. These results regarding $cPLA_2$ are precisely in line with earlier observations of secretory type II PLA_2 (Koumanov *et al.*, 1998). Secretory type II phospholipase A2 ($sPLA_2$) is inhibited by sphingomyelin. Cholesterol, either mixed with the model glycerophospholipid substrate or added to the assay medium in separate liposomes, effectively counteracts this inhibition.

The inhibition of fatty acid release, assayed by quantitative gas chromatography-MS, is observed when sphingomyelin is added to erythrocyte membranes as substrate. Hydrolysis of sphingomyelin by bacterial sphingomyelinase suppresses its inhibitory potency. The addition of cholesterol to sphingomyelin liposomes with a 1:1 stoichiometry completely relieves the inhibition of $sPLA_2$. An enzyme binding assay showed that $sPLA_2$ binds with relatively high affinity to the sphingomyelin, after either phase segregation at the assay temperature or on the pure sphingomyelin liposomes added to the incubation medium. Cholesterol was shown to suppress the binding of the enzyme to the sphingomyelin interface. Interestingly, it turns out that the specificity for the release of polyunsaturated fatty acids, mostly ω6-C20:4 for $cPLA_2$, is a property that can also be acquired by other types of PLA_2 when the ratio sphingomyelin/cholesterol is manipulated. When the ratio is decreased from 10 to 1 in the lipid mixture serving as the substrate the release by $sPLA_2$ of ω6-C20:4 relative to ω9-C18:1 increases from 1.5 to 2.06. This supports the notion that enzymes active on lipids have a specificity that not only depends on the particular configuration of the catalytic site but also on the form of presentation of the substrate.

5.2. Acyltransferases

The activity of acyltransferase able to reverse the action of PLA_2 was initially described in membranes of the endoplasmic reticulum (Magargal *et al.*, 1978). The subcellular distribution of oleoyl-CoA:1-acyl-sn-glycero-3-phosphocholine acyltransferase (EC 2.3.1.23) has been examined in cultured swine aorta endothelial cells and smooth muscle cells. The distribution of acyltransferase activity was found to be similar to that of enzymes of the endoplasmic reticulum. Treatment of microsomal membranes with digitonin caused a specific increase in density of plasma membranes allowing their convenient separation from endoplasmic reticulum vesicles. The plasma membranes were shown to be free of acyltransferase activity. In contrast to the activity in prokaryotes such as *Escherichia coli*, where modulation occurs by the lipid environment of the substrate (Scheideler and Bell 1986), the only phospholipid regulating activity in eucaryote cells was sphingomyelin (Momchilova *et al.*, 1999).

The influence of membrane lipids on arachidonoyl-CoA:lysophosphatidylcholine acyltransferase has been investigated in microsomal membranes from control and ras-transformed NIH 3T3 fibroblasts. Of all the phospholipids examined only sphingomyelin induced activation of acyltransferase in membranes from ras-transformed cells. No effect on the acyltransferase was observed in microsomal membranes from control fibroblasts. Diacylglycerol was found to inhibit acyltransferase in both cell lines, whereas ceramide accumulation induced inhibition only in membranes from the transformed cells.

These results appeared to contradict previous experiments where the influence of different phospholipases C on acyltransferase activity of membranes of the endoplasmic reticulum was tested (Sonoki and Ikezawa, 1976). Three kinds of phospholipase C were used to selectively hydrolyze phospholipids in rat liver microsomes, and their effects on the acyl-CoA:glycerophosphate and acyl-CoA:lysophospholipids acyltransferase systems were examined. The glycerophosphate acyltransferase [EC 2.3.1.15] system was rapidly inactivated by treatment with phospholipase C of *Pseudomonas aureofaciens* or *Bacillus cereus* and the loss of activity paralleled the degradation of phosphatidylcholine and phosphatidylethanolamine. The 1-acylglycerylphosphorylcholine acyltransferase [EC 2.3.1.23] system was only partially inactivated under the same conditions, whereas the 1-acylglycerophosphate acyltransferase [EC 2.3.1.51] system retained most of its activity even when more than 95% of phosphatidylcholine and phosphatidylethanolamine had been hydrolyzed. The results demonstrate the heterogeneity of acyltransferase systems with respect to their dependence on the intact membrane phospholipids. Hydrolysis of more than 80% of phosphatidylinositol by phosphoinositidase of *B. cereus* did not significantly affect these acyltransferase systems. These results suggest that 1-acylglycerophosphate and 1-acylglycerylphosphorylcholine acyl-transferase systems do not require specific phospholipids such as phosphatidylcholine, phosphatidylethanolamine, and phosphatidylinositol for their catalytic activities, but the integrity of these phospholipids is necessary for the proper functioning and stability of the enzymes.

A purification scheme has been now proposed for the enzyme (Kerkhoff *et al.*, 2000). The defined experimental conditions that allowed the extraction of the integral membrane protein lysophospholipid : acyl-CoA acyltransferase [EC 2.3.1.23] from membranes while maintaining the full enzyme activity using the nonionic detergent n-octyl glucopyranoside (OGP) and solutions of high ionic strength have been given. Detailed lipid analysis of the different protein/lipid/detergent mixed micelles showed that the protein/lipid/OGP mixed micelles were relatively enriched with sphingomyelin compared to protein/lipid/Chaps mixed micelles, indicating that the differences in the solubilization efficiency may be due to the ability to extract more sphingomyelin from membranes.

The activity of acyltranferase in the plasma membrane of liver cells has been suggested to play a major role in the secretion of bile lecithins (Colard *et al.*, 1980). Phospholipid acyltransferase activities of plasma membranes have been investigated with various acyl-CoA thioesters (palmitoyl, stearoyl, oleoyl, linoleoyl and arachidonoyl) with and without added lysoderivatives. Different patterns of incorporation were observed for each acyl-CoA into endogenous phosphatidylcholine and phosphatidylethanolamine. The turnover rates calculated with tracer amounts (10 microM) of acyl-CoA thioesters were five times faster for the polyunsaturated than for the saturated acyl molecular species of phosphatidylethanolamine and phosphatidylcholine. Arachidonoyl-CoA was the most efficient acyl donor at low concentrations and maximal turnover rate was observed at about 25 microM. No saturation was observed at concentrations at up to 100 microM linoleoyl-CoA. Linoleoyl-CoA transacylase acylated the lyso- compounds in the following order: lysophosphatidylcholine > lysophosphatidylserine = lysophosphatidylinositol. Lysophosphatidylethanolamine was found to inhibit linoleate incorporation into the phosphatidylethanolamine. No satisfactory explanation was given for this effect so it remains uncertain as to whether it is connected with an interference of translocators for lysoderivatives and phosphatidylethanolamine. Linoleoyl-CoA transacylation was not affected by the fatty acyl moiety at the 1-position of the lysophosphatidylcholine.

5.3. Acylation–deacylation cycle

Few details have been published on the relative activities of PLA_2 and acyltransferase in relation to the homeostatic regulation of membrane lipids. Regulation of phosphorylation/dephosphorylation was suggested to be a key factor from studies of the incorporation of ^{14}C-palmitoyl CoA into membrane phospholipids via the deacylation/acylation cycle conducted in rat liver microsomes (Hutson and Higgins, 1987). This activity was reversibly inactivated by treatment of the microsomes with 105,000 g supernatant in either the presence or absence of ATP and $MgCl_2$. Evidence was obtained that the acylation cycle is controlled by a mechanism involving phosphorylation/dephosphorylation. Because the pool of lysolecithin in the membranes was not altered by conditions increasing incorporation of palmitoyl CoA into phospholipid, it is likely that the site of regulation of deacylation/acylation is the acyltransferase reaction rather than the phospholipase without accumulation of lysophospholipid intermediate. This conclusion was reached on the basis of studies of resting tissue, however, results obtained on activation of granulocytes provided a different perspective (Tou, 1981). The effect of phorbol myristate acetate, known to reproduce the stimulated oxidative activities characteristic of phagocytosis, was examined on the metabolism of the fatty acyl groups of granulocyte phospholipids and compared with that of phagocytic stimuli. Phorbol myristate acetate was found to stimulate the labeling of phosphatidylethanolamine, phosphatidylcholine and phosphatidylinositol by1-^{14}C palmitic acid but not by U-^{14}C glycerol. Challenge of the cells with starch granules, by contrast, selectively increased the labeling of phosphatidylinositol by both radioactive tracers. Labeled palmitic acid was found at both *sn*-1- and *sn*-2-positions of phospholipids and more radioactivity was recovered from the 2-position.

The radioactivity at both positions was enhanced in stimulated cells. These data suggest that phorbol myristate acetate increased palmitic acid incorporation into glycerophospholipids by increasing the acylation of the lyso derivatives and that starch granules enhanced the formation of phosphatidylinositol via both *de novo* synthesis and acylation of the lyso derivative. Both phorbol myristate acetate and starch granules selectively augmented the incorporation of [1-^{14}C]arachidonic acid into phosphatidylinositol which exhibited the highest specific radioactivity among the phospholipids in control and in stimulated cells. The significance of the increased incorporation of arachidonic acid into phosphatidylinositol is thought to reflect the degradation that produces the eicosanoid precursor. This result can be understood in terms of an immediate reacylation of lysoderivatives by acylCoA conjugates of saturated or unsaturated fatty acids to prevent detergent-like effects followed subsequently by a specific incorporation of polyunsaturated acyl group at position *sn*-2. Only the second step represents a remodelling of the membrane phospholipids and takes place as a homeostatic mechanism.

The high specificity of the recipient lysoderivative has been examined in platelet, another activable circulating cell (McKean and Silver 1985). The transfer of unsaturated fatty acyl groups to 1-alkyl-*sn*-glycero-3-phosphocholine is many-fold slower than to 1-acyl-*sn*-glycero-3-phosphocholine. The CoA esters of linoleate and arachidonate, two unsaturated fatty acyl groups commonly found in platelet phospholipids, are the preferred fatty acyl group donors. In macrophages, 3 mechanisms of reacylation participating in the remodeling of membrane phospholipids were examined. Intact alveolar macrophages were found to acylate alkyl- and acyllysophospholipids with a high selectivity for arachidonate. A specific mechanism appears responsible for the incorporation of arachidonate into lysophospholipids in intact cells since the kinetic pattern for the formation of the C20:4 species was different from all other species. This specificity was investigated in more detail by examining the enzymatic acylation of 1-alkyl-2-lyso-*sn*-glycero-3-phosphocholine by membrane fractions; in the absence of CoA, ATP, and Mg^{2+}, this lysophospholipid was still re-acylated with a high preference for arachidonate that was independent of added free fatty acids. The addition of CoA alone increased the rate of acylation of 1-alkyl-2-lyso-*sn*-glycero-3-phosphocholine, mainly due to an increase in the formation of species other than those containing arachidonate. When CoA, ATP, and Mg^{2+} were present, the macrophage membranes catalyzed the acylation of 1-alkyl-2-lyso-*sn*-glycero-3-phosphocholine without preference for arachidonate. The acylation of alkyl- and acyllysophospholipids by rabbit alveolar macrophages appears to take place by three distinct mechanisms: a CoA-independent transacylation, a CoA-dependent transacylation (reverse reaction catalyzed by acyl-CoA acyltransferase), and an acyl-CoA-dependent acylation. The CoA-independent transacylation reaction is unique in that it is specific for arachidonate and accounts for the selective acylation of alkyl- and acyllysophospholipids by arachidonate in membrane preparations of alveolar macrophages. This reaction appears to be extremely important in the remodeling of phospholipid molecular species and the mobilization of arachidonate into ether-linked lipids. The transfer of arachidonate to 1-alkyl-2-lyso-*sn*-glycero-3-phosphocholine also is of importance in the termination step for platelet activating factor (1-alkyl-2-acetyl-*sn*-glycero-3-phosphocholine) activity, whereby 1-alkyl-2-arachidonoyl- *sn*-glycerol-3-phosphocholine (a stored precursor of both platelet activating factor and arachidonic acid metabolites) is restored. Different reacylation rates of ether and diacylphospholipids has also been reported in immuno-competent cells (Le Gouvello *et al.*, 1992). CD2 triggering stimulates the formation of platelet-activating factor-acether from akyl-arachidonoyl-glycerophosphocholine in a human $CD4^+$ T-lymphocyte clone (Masliah *et al.*, 1988). In rabbit alveolar macrophage the transacylation system was shown to exhibit a complex selectivity according to distinct donor and acceptor and CoA dependency (Sugiura *et al.*, 1987). Microsomes were found to acylate 1-^{3}H-alkyl-glycero-3-phosphocholine (GPC) (lyso platelet-activating factor) in the absence of any cofactors, indicating the presence of transacylation activity. The transacylation activity was comparable to the activity of acyl-CoA:1-alkyl-GPC acyltransferase. The fatty acyl moieties introduced into 1-alkyl-GPC from membrane lipids by microsomes were mainly C20:4ω6. ^{14}C-labeled C20:4ω6, C20:5ω3, C22:4ω6, and C22:6ω3 were transferred efficiently from diacyl-GPC to 1-alkyl-GPC in a CoA-independent manner. The transfer rates for C16:0, C18:0, and C18:1 from diacyl-GPC to 1-alkyl-GPC were relatively low in the presence and absence of CoA. On the other hand, the transfer of C20:4 from diacyl-glycero-3-phosphoethanolamine (GPE) or diacyl-glycero-3-phosphoinositol (GPI) to 1-alkyl-GPC or 1-acyl-GPC was markedly increased by the addition of CoA. This observation confirmed that acylation can be a specific and active pathway for polyunsaturated fatty acids cleaved from the *sn*-2 position of phospholipids serving as substrate during cell activation. In the activated human neutrophil the circulation of arachidonate between alkyl and alkenyl-derivatives participate in the generation of the lyso-derivative precursor of platelet activating factor (Sugiura *et al.*, 1987). These studies indicate that lyso-PAF is formed by the transfer of arachidonate from 1-O-alkyl-2-arachidonoyl-GPC to the alkenyl-lyso-GPE by a CoA-independent transacylase reaction.

Mass measurements revealed a rapid loss of arachidonate from 1-radyl-2-acyl-GPE and a concomitant increase in alkenyl-lyso-GPE upon stimulation of the neutrophils by addition of ionophore A23187. The circulation of polyunsaturated fatty acids between alkylPC and lysoPAF via alkenyPE was demonstrated in a variety of tissues (Blank *et al.*, 1995; Nixon *et al.*, 1996).

Microsomal membranes from six different rat tissues (spleen, lung, kidney, brain, testis, and liver) were found to possess CoA-independent transacylase activity that could both acylate lyso-PAF (1-hexadecyl-2-lyso-*sn*-glycero-3-phosphocholine) and then deacylate the 1-hexadecyl-2-acyl-*sn*-glycero-3-phosphocholine product *via* the transacylation of added exogenous 1-alk-1′-enyl-2-lyso-*sn*-glycero-3-phosphoethanolamine. Analysis of molecular species of 1-hexadecyl-2-acyl-*sn*-glycero-3-phosphocholine before and after addition of 1-alk-1′-enyl-2-lyso-*sn*-glycero-3-phosphoethanolamine as the acyl acceptor demonstrated a high selectivity for polyunsaturated fatty acids (>3 double bonds/acyl group) in both the acylation and deacylation processes that occurred in testicular microsomal membranes. The transfer of acyl groups by the transacylase appeared to be equally effective for either arachidonic or docosapentaenoic (ω6) fatty acids, whereas linoleic and oleic acids were not transferred. The results indicate the PAF-related transacylase is widely distributed among tissues and, although highly selective for polyunsaturated acyl groups, does not discriminate selectively among the polyunsaturates.

A very particular role was ascribed to the deacylation/reacylation of lysoderivatives on the inner monolayer of red blood cell membrane. This was the maintenance of the highly asymmetric distribution of phosphatidylethanolamine (PE) in ruminant membrane and the relatively low content of phosphatidylcholine (PC) (Florin-Christensen *et al.*, 2001). Ruminant erythrocytes are remarkable for their choline-phospholipid anomalies; namely, low or absent PC along with high sphingomyelin levels. Another anomaly in bovine erythrocytes affects aminophospholipids: PE shows almost absolute asymmetry, with only 2% of the total present in the outer leaflet. Furthermore, phospholipase A_2, an enzyme located on the external surface of the erythrocytes, shows 3-fold greater activity against PC than against PE. Because acylation of PE is by far the most important biosynthetic event in this cell following deacylation by phospholipase A_2 to generate lysocompounds, the selective reacylation of lyso-PE on the inner side, can account for the asymmetry of PE distribution and the departure from the outer side of lyso-PC extracted by serum albumin for the low content of PC in bovine erythrocyte membranes.

5.4. Phospholipases C

Phospholipases C are enzymes that hydrolyse diacylglycerophospholipids to diacylglycerols and a water-soluble phosphorylated product. Both lipophilic and water-soluble products are involved in the transduction of membrane signals and act as modulators of enzymes or ion channels. There are four main families of mammalian PLC. These include PLCβ, −γ, −δ and −ε all of which are characterised by their structural organization and mechanism of regulation (Rebecchi and Pentyala, 2000; Rhee *et al.*, 2001). All families are regulated by numerous cell regulators but some are specific to particular phospholipases. The PLCβ family are uniquely regulated by heterotrimeric G-proteins; PLCγ enzymes are regulated by both receptor and non-receptor tyrosine kinases; PLCδ are modulated by agents which include RhoGAP and α_H; PLCε enzymes contain a GTP exchange factor and are regulated by RAS that interacts via two RAS-binding domains (Kelley *et al.*, 2001; Lopez *et al.*, 2001).

Phospholipase Cβ are enzymes that specifically hydrolyse phosphatidylinositol 4,5-bisphosphate (PIP_2) to form diacylglycerol and inositol-1,4,5-trisphosphate. The enzymes are activated to differing extents by G-protein αq subunits and by G-protein βγ dimers (Rhee 2001). It is known that the C-terminal region of PLCβ is essential for stimulation by αq but the phospholipase can still be activated by Rho GTPases and G-protein βγ subunits that bind to different regions of the enzyme (Illenberger *et al.*, 2003). Thus the catalytic subdomains of PLCβ2 are all that is required for efficient stimulation by βγ dimers whereas additionally the putative pleckstrin homology domain is required for stimulation by Rho GTPases. Amongst the Rho GTPases, Rac 1 and Rac 2 were found to be more important stimulators than Cdc42 and all are implicated in receptor-mediated stimulation of PLCβ2 activity. The molecular mechanisms of stimulation by either heterotrimeric G-proteins or Rho GTPases is presently unknown but targetting of the enzyme to the substrate and allosteric regulation of the enzyme are both factors (McLaughlin *et al.*, 2002). The role of pleckstrin homology domains serve to modulate the activity of their catalytic sites upon either interaction with the substrate or G-protein activators (Philip *et al.*, 2002)

The PLCγ family, like PLCβ acts on PIP_2. Two isoforms have been identified, PLCγ1 which is widely distributed in mammalian tissues, and PLCγ2 the expression of which is restricted mainly to haematopoietic cells (Wilde and Watson 2001). PLCγ1 has a function essential to growth and development and cells deficient in the enzyme fail to mobilize Ca^{2+} in response to epidermal growth factor but proliferate normally. Animals defective in PLCγ2 survive but affected mice show decreased B-cell number, failure of platelets to aggregate in response to challenge with collagen and a failure to respond to Fc receptor-mediated processes (Wang *et al.*, 2000). The molecular features that distinguish PLCγ from other PLC isotypes is the presence of two Src homology domains located within a pleckstrin homology domain which are responsible for localization at the membrane substrate and an activation by phosphorylation of multiple tyrosyl residues on the enzyme mediated by tyrosine kinases at sites that are system-dependent.

The regulatory process controlling activation of PLCε is mediated by two Ras/Rap-1 associating domains located at the C-terminus of the molecule and a CDC25 homology domain near the N-terminus. These sites suggest that modulation is achieved by binding of Ras family GTPases to these sites (Kelley *et al.*, 2001). The molecular basis of activation of PLCε by Ras and Rap-1 provoked by platelet-derived growth factor stimulation of cells has been investigated by Song *et al.* (2002). It was found that binding platelet-derived growth factor to its receptor site on the cell surface induces persistent activation of PLCε. An initial rapid phase of activation of the enzyme is mediated by Ras and the prolonged activation is sustained by Rap-1. The CDC25 homology domain, which displays a guanine nucleotide exchange factor activity for Rap-1, but not ras, is a requirement for the prolonged activation of the enzyme.

5.5. Phospholipases D

Phospholipase D hydrolyses phospholipids yielding phosphatidic acid and water-soluble base such as choline (Singer *et al.*, 1997). The role of phosphatidic acid generation in membranes and its role in signalling remains conjectural. The membrane-bound product does not occur in appreciable proportions although locally it may be generated in amounts that may have consequential effects on membrane structure and stability. Efforts to implicate phospholipase D in signalling have therefore concentrated on how and where the enzyme is activated on a subcellular level.

Two genes coding for phospholipase D have been identified in mammals, PLD1 and PLD2. Both genes have been cloned and overexpressed in different cell lines (Frohman *et al.*, 1999). PLD1 is located in association with intracellular membranes and is known to be active in living cells. PLD2, by contrast, is associated with plasma membranes but has low resting activity that can be activated by a variety of factors including protein kinase C, family members of ADP-ribosylation factor and Rho and the lipid PIP_2 (Singer *et al.*, 1997; Exton 2002).

Targetting of PLD1 to membranes may involve palmitoylation of cysteine residues at a domain near the C-terminus of the protein (Sugars *et al.*, 2002). The importance of the domain in the vicinity of the palmitoylated cysteinyl residues suggested that not only did palmitoylation target the enzyme to the membrane but there was specificity in the membrane site of interaction.

How phosphatidic acid may act as a signalling device at the sites of enzyme activity has been inferred from evidence that the phospholipid binds to specific proteins. Candidates for phosphatidic acid binding include Raf-1, a serine/threonine kinase (Rizzo *et al.*, 2000), c-AMP-specific phosphodiesterase

(Grange *et al.*, 2000), the mammalian target of rapamycin, mTOR (Fang *et al.*, 2001), protein phosphatase-1 (Jones and Hannun 2002) and Src homology 2-domain containing protein tyrosine phosphatase (Frank *et al.*, 1999). Recently, a solid-phase adsorption system has been described to identify traffic-related phosphatidic acid binding proteins and proteins that may be implicated in phospholipase D dependent pathways and a number of specific proteins have been characterised (Ktistakis *et al.*, 2003).

5.6. Sphingomyelinases

Sphingomyelinases are enzymes that cleave the phosphorylcholine moiety from sphingomyelin to yield lipophilic ceramide. The significance of sphingomyelinases has been recognised by the discovery that sphingomyelin is a substrate progenitor for initiation of signal transduction pathways (Kolesnick, 1991) in which ceramide acts as a second messenger in functions like cell differentiation, proliferation and apoptosis (Gomez-Munoz, 1998; Venkataraman and Futerman, 2000; Ohanian and Ohanian, 2001; Hannun and Obeid, 2002). Five different categories of sphingomyelinase have been characterised on the basis of their optimal conditions for catalytic activity (Samet and Barenholz, 1999) although a different sphingomyelinase has been identified in bacteria (Goni and Alonso, 2002). In the sphingomyelin cycle, a number of extracellular agents and insults such as tumor necrosis factor, Fas ligands, and chemotherapeutic agents cause activation of sphingomyelinases, which act on membrane sphingomyelin and release ceramide. The amount of accumulating ceramide may reach up to 10% of membrane lipids (Hannun, 1996). It is believed that the ceramides accumulate as the product of hydrolysis of sphingomyelin by a neutral cytosol sphingomyelinase acting on caveolae-like structures (Levade and Jaffrezou, 1999; Veldman *et al.*, 2001). The origin of accumulating ceramides is conjectural since an increased *de novo* synthesis can also be detected under abnormal circumstances which trigger apoptosis (Yen *et al.*, 1999). This example is relevant to membrane lipid homeostasis perturbed in mutant cells. It is not often appreciated that nutritional status can modulate apoptosis, a process that eliminates unwanted or damaged cells.

Evidence from model membrane studies has indicated that the action of sphingomyelinase in generating ceramide from sphingomyelin can induce lamellar to non-lamellar phase transitions leading to membrane fusion (Holopainen *et al.*, 2000) and to lateral phase separations of ceramide-enriched domains which exist in gel phase in fluid bilayers of phospholipid (Huang *et al.*, 1996). Ceramide, and to a much lesser extent dihydroceramide, are also able to induce the formation of pores in phospholipid bilayers and mitochondrial membranes, a property that has been attributed to the extensive hydrogen-bonding capacity of ceramide (Siskind *et al.*, 2002). Ceramides in the presence of high concentrations of cholesterol such as in caveolae from which they are formed have been recently shown to cause formation of crystal-like materials embedded in the membrane plane (McIntosh *et al.*, 2003).

There is evidence that the action of neutral sphingomyelinase causes domain formation of phases enriched in ceramide and a consequent clustering of cell surface receptors. One example is the clustering of L-selectin in lymphocytes (Junge *et al.*, 1999). The action of a Zn-dependent acid sphingomyelinase in response to interleukin-1β treatment of human fibroblasts has been found to be associated with depletion of sphingomyelin and a corresponding increase of ceramide in the caveolae compartment (Liu and Anderson 1995). The interleukin-1β treatment was also shown to be associated with decreased platelet-derived growth factor-induced thymidine uptake by these cells suggesting that perturbation of trans-membrane signalling processes had occurred. Similar studies of the action of nerve growth factor signalling have emphasized the importance of intact caveolae for sphingomyelinase action (Dobrowsky *et al.* 2000).

Ceramide generation and creation of rafts has been shown to be essential for optimal Fas signalling and induction of apoptosis in lymphocytes and in host defense mechanisms associated with infection (Kirschnek *et al.*, 2000; Grassme *et al.*, 2002; 2003). On the basis of these studies a model has been proposed for the action of ceramide in signalling processes associated with apoptosis (Cremesti *et al.* 2002). Essentially, the engagement of Fas triggers translocation of acid sphingomyelinase to the plasma membrane where it acts on its substrate segregated into sphingomyelin-cholesterol rafts. The formation of ceramide induces coalescence of the rafts into large domains in which oligomerisation of downstream effectors such as FADD/MORT-1 and pro-caspase-8 can take place leading to the Fas death signal.

5.7. Phospholipid turnover and membrane lipid disposition

The particular molecular species composition and arrangement of lipids within the structure is influenced by metabolic turnover and lipid translocases embedded in the structure. The primary level of homeostatic regulation is the maintenance of particular proportions of each of the major lipid classes characterised by their polar constituent. There is some latitude for flexibility in the case of the choline phospholipids of which phosphatidylcholine and sphingomyelin are the predominant members with choline plasmalogen and glycerylether lipids representing less conspicuous amounts and possibily fulfilling more specific functions. These two classes of choline phospholipids in erythrocyte membranes of different animal species appear to be interchangable.

Choline-containing phospholipids amount to 46.3–60.8 mole% in the red cell membranes of most mammalian species (Virtanen *et al.*, 1998). Detailed analysis of ruminant erythrocytes membranes, however, have shown that phosphatidylcholine, the most abundant lipid in human erythrocytes (approximately 56 mole%) can be substituted by sphingomyelin in ruminants such as goat and sheep (50 and 68%, respectively). Therefore phosphatidylcholine of the outer and the inner leaflet of the human erythrocyte membrane (44.8 and 14%, respectively) must be substituted by sphingomyelin. Sphingomyelin is known to be located predominantly in the outer leaflet of human red blood cells (van Deenen and de Gier, 1974). Because the sphingophospholipid is resistant to phospholipases A and C hydrolysis it is often inferred that the alteration of ratio sphingomyelin : phosphatidylcholine in ruminants results from the degradation of phosphatidylcholine by high levels of circulating enzymes. According to this view the homeostasis of membrane phospholipid composition represents the remnants of phosphatidylcholine degradation throughout the life of the cell. Intermediate steps of the membrane degradation between bone marrow erythropoiesis and the circulating cells remain to be characterized.

The influence of secretory phospholipases on membrane phospholipid composition has been reported in a number of cell types and different physiological and pathological states. An increased synthesis of secretory PLA_2 and resultant activity in serum is associated with inflammatory responses triggered by cytokines (IL-1 and TNFα) in humans. Regulation of the activity of the phospholipase appears to be exerted at the substrate interface by alteration of the sphingomyelin:glycerophospholipid ratio. A limit to the extent of degradation of glycerophospholipid located on the outer leaflet of the plasma membrane was found which represents only one third of the rate of secretory PLA_2 acting on inside-out vesicles (Koumanov, K. *et al.*, 1997). The observed resistance to hydrolysis was assumed to result of the high sphingomyelin content on the outer leaflet of the human red blood cell membrane (42.1 vs 9.1 mole% on the inner side). On the right-side out oriented membrane it was found that annexin VI brought about a perturbation of the phospholipid sideness (Burgoyne and Geisow, 1989) which resulted in the activation of PLA_2. By contrast, annexin VI acted as an inhibitor of PLA_2 on randomly oriented dispersions of membrane lipid extracts, in agreement with the "substrate depletion" model proposed by Comera (Comera *et al.*, 1990). Modulation of the regulation by sphingomyelin was demonstrated when it was found that cholesterol relieved the inhibition of PLA_2 (Koumanov *et al.*, 1998). The sequestration of the hydrolytic enzyme on sphingomyelin-enriched interfaces was demonstrated by binding assays, and the enzyme was released by the addition of cholesterol to the sphingomyelin. This observation suggests a dual role for sphingomyelin and cholesterol, not only as constituents on the physical structure of the membrane but also on modulating the activity of enzymes acting to alter the chemical composition. Indeed, cells are normally exposed to an environment replete with enzymes capable of acting on their constituent membranes. sphingomyelin has been recognized as a protector toward many circulating enzyme activities in addition to PLA_2 type IIa such as pancreatic type I and *Naja* venon type II PLA_2, lipoprotein lipase (Lobo and Wilton, 1997), phospholipase C (Scarlata *et al.*, 1996). Sphingomyelin has been demonstrated to efficiently protect the membrane of activated cells from the attack of diglyceride-triggered intracellular phospholipases (Dawson *et al.*, 1984, 1985). The relevance of this observation can be correlated with the presence of sphingomyelin in the cytoplasmic leaflet of the plasma membrane of activated cells.

In contrast with the protective role played by sphingomyelin, ceramides produced by sphingomyelinase increase the susceptibility of lipids to hydrolytic attack. Indeed, *Staphylococcus aureus* sphingomyelinase reverses the inhibition of type IIa PLA_2 brought about by sphingomyelin (Koumanov *et al.*, 1998). The bacterial enzyme is also able to increase 3-fold the release of arachidonic acid by cytoplasmic type IV PLA_2 from Chinese Hamster Ovary-2B cells activated by calcium-ionophore and epinephrine (Klapisz *et al.*, 2000). The degradation of sphingomyelin on the external leaflet of the plasma membrane by the exogenous sphingomyelinase followed by the activation of an intracellular enzyme acting on the inner leaflet is an intriguing observation. It raises the question of how sphingomyelin can control the inhibition of enzymes acting on the outer leaflet but ceramides can trigger hydrolysis on both leaflets. Similarly with the ^{31}P-NMR data collected for lipid mixtures enriched with diacylglycerol (Dawson *et al.*, 1984) it is suspected that ceramides remain inside the bilayer but phase separate to form a non-bilayer arrangement from the low concentration of 5 mole% (Holopainen *et al.*, 1998). These microdomains could induce the structural defects on both sides of the membrane which expose the matrix lipids to enzyme penetration (Lichtenbergova *et al.*, 1998). X-ray diffraction studies of codispersions of phosphatidylethanolamine and phosphatidyserine containing ceramide confirm that ceramides at 10 mole % favors the lamellar/non lamellar phase separation of the phospholipids at high temperatures. At low temperatures ceramides act like sphingomyelin to oppose separation in this lipid binary mixture selected to mimic the composition of the cytoplasmic leaflet of plasma membrane. Such contrasting behavior between sphingomyelin and ceramide may explain how sphingomyelinase activity is activated during apoptosis. It is worth remembering that the control of the cascade of events leading to apoptosis is integral to the homeostasis of the membrane lipid matrix.

The anionic headgroups of phosphatidylserine and phosphatidylinositol (and pluri-phosphorylated derivatives) are considered to be stringently located on the inner leaflet of the plasma membrane in resting cells. The early observations conducted by van Deenen *et al.*, (Verkleij *et al.*, 1973) have been confirmed after the isolation of proteins with an activity of translocator, either a scramblase that randomises the distribution by accelerating the bidirectional lipid transport between the 2 leaflets or a translocase which establishes the asymmetric distribution (Quinn, 2002). Scramblase is an energy-independent translocator (Williamson *et al.*, 2001) which reduces the time for transverse motion of amphipathic lipids across bilayers from hours to a matter of minutes for lysophosphatidylcholine (Mohandas *et al.*, 1982), phosphatidylcholine and sphingomyelin (Bai and Pagano 1997). The appearance of phosphatidylserine on the outer leaflet of the plasma membrane under the influence of an activated scramblase is responsible for triggering a number of physiological events such as the stimulation of blood clotting on activated platelets (Williamson *et al.*, 1995; Comfurius *et al.*, 1996) or the binding of annexin V in the presence of calcium (Willems *et al.*, 2000). These irreversible alterations point to the critical role of the maintenance of the bilayer asymmetry. The cost of maintenance of this distribution is a continuous energy-dependent process involving the aminophospholipid inward translocase (Seigneuret and Devaux 1984; Devaux *et al.*, 1986). ATP consumption under the usual blood bank conditions can explain the formation of echinocytes when the level of aminophospholipidids on the outer leaflet is increased. The translocator activity has been tentatively assigned to a P-type ATPase II (Tang *et al.*, 1996). The bilayer asymmetry relies also on the outward translocation (floppase) of phosphatidylcholine by the ATP-binding cassette transporter Mdr2 (ABCB4) (Smit *et al.*, 1993) and of glucosylceramide and possibly other glycolipids by MDR1 (ABCB1). The crossed-regulation of these continuous inward and outward translocators prevents the triggering of apoptosis induced by the exposure of sphingomyelin to intracellular sphingomyelinase or the cell engulfment by macrophages triggered by the presence of phosphatidylserine on the extracellular surface of the plasma membrane.

A high degree of stringency can be assessed in the less reactive cellular types but it is difficult to evaluate in activable cells. For instance, a significant proportion of phosphatidylserine and phosphatidylethanolamine has been observed on the outer leaflet in rat platelets before activation (Joo *et al.*, 1993). Interestingly, the molecular species of these two phospholipids were the most saturated of the molecular species in the membrane indicating that an additional degree of asymmetry can be superimposed upon the asymmetric distribution of the lipid class itself. The sorting of lipids according to the acyl chain length by MDR1 (ABCB1) has been reported (van Helvoort *et al.*, 1996; Raggers *et al.*, 2001).

The distribution of anionic membrane phospholipids is a sentitive index of the homeostatic regulation of the phospholipid distribution. The binding of the annexin V on anionic headgroups of membrane phospholipids acting as a receptor can be measured with high sensitivity using radio-labeled derivatives (Schutte *et al.*, 1998; Shounan *et al.*, 1998). The measurement promises an interesting medical application to monitor *in vivo* perturbation of the distribution of membrane phospholipids in tissues affected by apoptosis after ischemia. The marker should distinguish necrotic from apoptotic cells that can be subsequently assessed by other methods to locate recovery of tissues following reperfusion.

6. CONCLUSIONS

The perception of the structure of biological membranes consisting of a lipid matrix comprised of a bilayer of fluid lipids serving to support the various associated proteins has dramatically changed over the past decade or so. The change has been largely driven by the recognition that the lipids are not randomly distributed either laterally in the plane of the membrane or between the two leaflets of the bilayer.

Integral to these discoveries has been the question not only of how interactions between lipids can drive lateral phase separations and create domains within the structure but what are the homeostatic processes that underly the regulation of membrane lipid composition. Considerable progress has been made in understanding processes that trigger hydrolysis of membrane lipids in response to specific receptor-mediated events but less is known about the regulation of enzymes that are responsible for membrane lipid turnover in the preservation of the molecular species composition. Doubtless this will be one of the major research challenges in the future.

REFERENCES

Ahmed, S. N., Brown, D. A. and London, E. (1997) On the origin of sphingolipid/cholesterol-rich detergent-insoluble cell membranes: physiological concentrations of cholesterol and sphingolipid induce formation of a detergent-insoluble, liquid-ordered lipid phase in model membranes. *Biochemistry*, **36**, 10944-10953.

Anderson, R. G. and Jacobson, K. (2002) A role for lipid shells in targeting proteins to caveolae, rafts, and other lipid domains. *Science*, **296**, 1821-1825.

Bai, J. and Pagano, R. E. (1997) Measurement of spontaneous transfer and transbilayer movement of BODIPY-labeled lipids in lipid vesicles. *Biochemistry*, **36**, 8840-8848.

Barbu, V., Roux, C., Lambert, D., Dupuis, R., Gardette, J., Maziere, J. C., Maziere, C., Elefant, E. and Polonovski, J. (1988) Cholesterol prevents the teratogenic action of AY 9944: importance of the timing of cholesterol supplementation to rats. *J Nutr*, **118**, 774-779.

Biagi, P. L., Bordoni, A., Lorenzini, A., Horrobin, D. F. and Hrelia, S. (1999) Essential fatty acid metabolism in long term primary cultures of rat cardiomyocytes: a beneficial effect of n-6:n-3 fatty acids supplementation. *Mech Ageing Dev*, **107**, 181-195.

Blank, M. L., Smith, Z. L., Fitzgerald, V. and Snyder, F. (1995) The CoA-independent transacylase in PAF biosynthesis: tissue distribution and molecular species selectivity. *Biochim Biophys Acta*, **1254**, 295-301.

Bourre, J. M., Durand, G., Pascal, G. and Youyou, A. (1989*a*) Brain cell and tissue recovery in rats made deficient in n-3 fatty acids by alteration of dietary fat. *J Nutr*, **119**, 15-22.

Bourre, J. M., Francois, M., Youyou, A., Dumont, O., Piciotti, M., Pascal, G. and Durand, G. (1989*b*) The effects of dietary alpha-linolenic acid on the composition of nerve membranes, enzymatic activity, amplitude of electrophysiological parameters, resistance to poisons and performance of learning tasks in rats. *J Nutr*, **119**, 1880-1892.

Brugger, B., Erben, G., Sandhoff, R., Wieland, F. T. and Lehmann, W. D. (1997) Quantitative analysis of biological membrane lipids at the low picomole level by nano-electrospray ionization tandem mass spectrometry. *Proc Natl Acad Sci USA*, **94**, 2339-2344.

Bui, B. V., Weisinger, H. S., Sinclair, A. J. and Vingrys, A. J. (1998) Comparison of guinea pig electroretinograms measured with bipolar corneal and unipolar intravitreal electrodes. *Doc Ophthalmol*, **95**, 15-34.

Burgoyne, R. D. and Geisow, M. J. (1989) The annexin family of calcium-binding proteins. Review article. *Cell Calcium*, **10**, 1-10.

Clark, J. D., Lin, L. L., Kriz, R. W., Ramesha, C. S., Sultzman, L. A., Lin, A. Y., Milona, N. and Knopf, J. L. (1991) A novel arachidonic acid-selective cytosolic PLA2 contains a Ca(2+)-dependent translocation domain with homology to PKC and GAP. *Cell*, **65**, 1043-1051.

Colard, O., Bard, D., Bereziat, G. and Polonovski, J. (1980) Acylation of endogenous phospholipids and added lysoderivatives by rat liver plasma membranes. *Biochim Biophys Acta*, **618**, 88-97.

Comera, C., Rothhut, B. and Russo-Marie, F. (1990) Identification and characterization of phospholipase A2 inhibitory proteins in human mononuclear cells. *Eur J Biochem*, **188**, 139-146.

Comfurius, P., Williamson, P., Smeets, E. F., Schlegel, R. A., Bevers, E. M. and Zwaal, R. F. (1996) Reconstitution of phospholipid scramblase activity from human blood platelets. *Biochemistry*, **35**, 7631-7634.

Cooper, M. K., Porter, J. A., Young, K. E. and Beachy, P. A. (1998) Teratogen-mediated inhibition of target tissue response to Shh signaling. *Science*, **280**, 1603-1607.

Cossins, A. R. (1994). *Temperature adaptation of Biological Membranes*. London, Portland Press.

Cremesti, A. E., Goni, F. M. and Kolesnick, R. (2002) Role of sphingomyelinase and ceramide in modulating rafts: do biophysical properties determine biologic outcome? *FEBS Lett*, **531**, 47-53.

Dawson, R. M., Hemington, N. and Irvine, R. F. (1985) The inhibition of diacylglycerol-stimulated intracellular phospholipases by phospholipids with a phosphocholine-containing polar group. A possible physiological role for sphingomyelin. *Biochem J*, **230**, 61-68.

Dawson, R. M., Irvine, R. F., Bray, J. and Quinn, P. J. (1984) Long-chain unsaturated diacylglycerols cause a perturbation in the structure of phospholipid bilayers rendering them susceptible to phospholipase attack. *Biochem Biophys Res Commun*, **125**, 836-842.

de Cingolani, G. E., van den Bosch, H. and van Deenen, L. L. (1972) Phospholipase A and lysophospholipase activities in isolated fat cells: effect of cyclic 3′,5′-AMP. *Biochim Biophys Acta*, **260**, 387-392.

Devaux, P. F., Zachowski, A., Favre, E., Fellmann, P., Cribier, S., Geldwerth, D., Herve, P. and Seigneuret, M. (1986) [Energy-dependent translocation of amino-phospholipids in the erythrocyte membrane]. *Biochimie*, **68**, 383-393.

Dickens, B. F., Ramesha, C. S. and Thompson, G. A., Jr. (1982) Quantification of phospholipid molecular species by coupled gas chromatography-mass spectrometry of deuterated samples. *Anal Biochem*, **127**, 37-48.

Dobrowsky, R. T. (2000) Sphingolipid signalling domains floating on rafts or buried in caves? *Cell Signal*, **12**, 81-90.

Edidin, M. (2001) Shrinking patches and slippery rafts: scales of domains in the plasma membrane. *Trends Cell Biol*, **11**, 492-496.

Exton, J. H. (2002) Regulation of phospholipase D. *FEBS Lett*, **531**, 58-61.

Fang, Y., Vilella-Bach, M., Bachmann, R., Flanigan, A. and Chen, J. (2001) Phosphatidic acid-mediated mitogenic activation of mTOR signaling. *Science*, **294**, 1942-1945.

Farese, R. V., Jr., Cases, S., Ruland, S. L., Kayden, H. J., Wong, J. S., Young, S. G. and Hamilton, R. L. (1996) A novel function for apolipoprotein B: lipoprotein synthesis in the yolk sac is critical for maternal-fetal lipid transport in mice. *J Lipid Res*, **37**, 347-360.

Florin-Christensen, J., Suarez, C. E., Florin-Christensen, M., Wainszelbaum, M., Brown, W. C., McElwain, T. F. and Palmer, G. H. (2001) A unique phospholipid organization in bovine erythrocyte membranes. *Proc Natl Acad Sci USA*, **98**, 7736-7741.

Frank, C., Keilhack, H., Opitz, F., Zschornig, O. and Bohmer, F. D. (1999) Binding of phosphatidic acid to the protein-tyrosine phosphatase SHP-1 as a basis for activity modulation. *Biochemistry*, **38**, 11993-12002.

Freedman, S. D., Katz, M. H., Parker, E. M., Laposata, M., Urman, M. Y. and Alvarez, J. G. (1999) A membrane lipid imbalance plays a role in the phenotypic expression of cystic fibrosis in cftr(-/-) mice. *Proc Natl Acad Sci USA*, **96**, 13995-14000.

Freedman, S. D., Weinstein, D., Blanco, P. G., Martinez-Clark, P., Urman, S., Zaman, M., Morrow, J. D. and Alvarez, J. G. (2002) Characterization of LPS-induced lung inflammation in cftr-/- mice and the effect of docosahexaenoic acid. *J Appl Physiol*, **92**, 2169-2176.

Fridriksson, E. K., Shipkova, P. A., Sheets, E. D., Holowka, D., Baird, B. and McLafferty, F. W. (1999) Quantitative analysis of phospholipids in functionally important membrane domains from RBL-2H3 mast cells using tandem high-resolution mass spectrometry. *Biochemistry*, **38**, 8056-8063.

Frohman, M. A., Sung, T. C. and Morris, A. J. (1999) Mammalian phospholipase D structure and regulation. *Biochim Biophys Acta*, **1439**, 175-186.

Gaskell, S. J. and Brooks, C. J. (1977) Gas-liquid chromatography-mass spectrometry of phospholipid mixtures after enzymic hydrolysis. *J Chromatogr*, **142**, 469-480.

Gofflot, F., Hars, C., Illien, F., Chevy, F., Wolf, C., Picard, J. J. and Roux, C. (2003) Molecular mechanisms underlying limb anomalies associated with cholesterol deficiency during gestation: implications of Hedgehog signaling. *Hum Mol Genet*, **12**, 1187-1198.

Gombos, Z., Wada, H., Varkonyi, Z., Los, D. A. and Murata, N. (1996) Characterization of the Fad12 mutant of Synechocystis that is defective in delta 12 acyl-lipid desaturase activity. *Biochim Biophys Acta*, **1299**, 117-123.

Gomez-Munoz, A. (1998) Addendum to 'Modulation of cell signalling by ceramides'. *Biochim Biophys Acta*, **1394**, 261.

Gomez-Munoz, A. (1998) Modulation of cell signalling by ceramides. *Biochim Biophys Acta*, **1391**, 92-109.

Goni, F. M. and Alonso, A. (2002) Sphingomyelinases: enzymology and membrane activity. *FEBS Lett*, **531**, 38-46.

Grange, M., Sette, C., Cuomo, M., Conti, M., Lagarde, M., Prigent, A. F. and Nemoz, G. (2000) The cAMP-specific phosphodiesterase PDE4D3 is regulated by phosphatidic acid binding. Consequences for cAMP signaling pathway and characterization of a phosphatidic acid binding site. *J Biol Chem*, **275**, 33379-33387.

Grassme, H., Jendrossek, V., Bock, J., Riehle, A. and Gulbins, E. (2002) Ceramide-rich membrane rafts mediate CD40 clustering. *J Immunol*, **168**, 298-307.

Grassme, H., Jendrossek, V., Riehle, A., von Kurthy, G., Berger, J., Schwarz, H., Weller, M., Kolesnick, R. and Gulbins, E. (2003) Host defense against Pseudomonas aeruginosa requires ceramide-rich membrane rafts. *Nat Med*, **9**, 322-330.

Hannun, Y. A. (1996) Functions of ceramide in coordinating cellular responses to stress. *Science*, **274**, 1855-1859.

Hannun, Y. A. and Obeid, L. M. (2002) The Ceramide-centric universe of lipid-mediated cell regulation: stress encounters of the lipid kind. *J Biol Chem*, **277**, 25847-25850.

Holopainen, J. M., Angelova, M. I. and Kinnunen, P. K. (2000) Vectorial budding of vesicles by asymmetrical enzymatic formation of ceramide in giant liposomes. *Biophys J*, **78**, 830-838.

Holopainen, J. M., Subramanian, M. and Kinnunen, P. K. (1998) Sphingomyelinase induces lipid microdomain formation in a fluid phosphatidylcholine/sphingomyelin membrane. *Biochemistry*, **37**, 17562-17570.

Huang, H. W., Goldberg, E. M. and Zidovetzki, R. (1996) Ceramide induces structural defects into phosphatidylcholine bilayers and activates phospholipase A2. *Biochem Biophys Res Commun*, **220**, 834-838.

Hutson, J. L. and Higgins, J. A. (1987) Reversible activation/inactivation of the deacylation/acylation cycle in rat liver microsomes. *Biosci Rep*, **7**, 73-80.

Illenberger, D., Walliser, C., Nurnberg, B., Diaz Lorente, M. and Gierschik, P. (2003) Specificity and structural requirements of phospholipase C-beta stimulation by Rho GTPases versus G protein beta gamma dimers. *J Biol Chem*, **278**, 3006-3014.

Jones, J. A. and Hannun, Y. A. (2002) Tight binding inhibition of protein phosphatase-1 by phosphatidic acid. Specificity of inhibition by the phospholipid. *J Biol Chem*, **277**, 15530-15538.

Joo, F., Chevy, F., Colard, O. and Wolf, C. (1993) The activation of rat platelets increases the exposure of polyunsaturated fatty acid enriched phospholipids on the external leaflet of the plasma membrane. *Biochim Biophys Acta*, **1149**, 231-240.

Junge, S., Brenner, B., Lepple-Wienhues, A., Nilius, B., Lang, F., Linderkamp, O. and Gulbins, E. (1999) Intracellular mechanisms of L-selectin induced capping. *Cell Signal*, **11**, 301-308.

Kates, M., Pugh, E.L. and Ferrante, G. (1984) Influence of temperature and growth phase on desaturase activity of the mesophilic yeast Candida lipolytica. *Biomembranes*, **12**, 379-395.

Katsikas, H. and Wolf, C. (1995) Blood sphingomyelins from two European countries. *Biochim Biophys Acta*, **1258**, 95-100.

Kelley, G. G., Reks, S. E., Ondrako, J. M. and Smrcka, A. V. (2001) Phospholipase C (epsilon): a novel Ras effector. *Embo J*, **20**, 743-754.

Kenworthy, A. (2002) Peering inside lipid rafts and caveolae. *Trends Biochem Sci*, **27**, 435-437.

Kerkhoff, C., Trumbach, B., Gehring, L., Habben, K., Schmitz, G. and Kaever, V. (2000) Solubilization, partial purification and photolabeling of the integral membrane protein lysophospholipid:acyl-CoA acyltransferase (LAT). *Eur J Biochem*, **267**, 6339-6345.

Kirsch, C., Eckert, G. P. and Mueller, W. E. (2003) Statin effects on cholesterol microdomains in brain plasma membranes. *Biochem Pharmacol*, **65**, 843-856.

Kirschnek, S., Paris, F., Weller, M., Grassme, H., Ferlinz, K., Riehle, A., Fuks, Z., Kolesnick, R. and Gulbins, E. (2000) CD95-mediated apoptosis in vivo involves acid sphingomyelinase. *J Biol Chem*, **275**, 27316-27323.

Klapisz, E., Masliah, J., Bereziat, G., Wolf, C. and Koumanov, K. S. (2000) Sphingolipids and cholesterol modulate membrane susceptibility to cytosolic phospholipase A(2). *J Lipid Res*, **41**, 1680-1688.

Kolesnick, R. N. (1991) Sphingomyelin and derivatives as cellular signals. *Prog Lipid Res*, **30**, 1-38.

Koty, P. P., Tyurina, Y. Y., Tyurin, V. A., Li, S. X. and Kagan, V. E. (2002) Depletion of Bcl-2 by an antisense oligonucleotide induces apoptosis accompanied by oxidation and externalization of phosphatidylserine in NCI-H226 lung carcinoma cells. *Mol Cell Biochem*, **234–235**, 125-133.

Koumanov, K., Wolf, C. and Bereziat, G. (1997) Modulation of human type II secretory phospholipase A2 by sphingomyelin and annexin VI. *Biochem J*, **326**, 227-233.

Koumanov, K. S., Quinn, P. J., Bereziat, G. and Wolf, C. (1998) Cholesterol relieves the inhibitory effect of sphingomyelin on type II secretory phospholipase A2. *Biochem J*, **336**, 625-630.

Ktistakis, N. T., Delon, C., Manifava, M., Wood, E., Ganley, I. and Sugars, J. M. (2003) Phospholipase D1 and potential targets of its hydrolysis product, phosphatidic acid. *Biochem Soc Trans*, **31**, 94-97.

Ladenson, R. C., Monsey, J. D., Allin, J. and Silbert, D. F. (1993) Utilization of exogenously supplied sphingosine analogues for sphingolipid biosynthesis in Chinese hamster ovary and mouse LM cell fibroblasts. *J Biol Chem*, **268**, 7650-7659.

Le Gouvello, S., Vivier, E., Debre, P., Thomas, Y. and Colard, O. (1992) CD2 triggering stimulates the formation of platelet-activating factor-acether from alkyl-arachidonoyl-glycerophosphocholine in a human CD4 + T lymphocyte clone. *J Immunol*, **149**, 1289-1293.

Levade, T. and Jaffrezou, J. P. (1999) Signalling sphingomyelinases: which, where, how and why? *Biochim Biophys Acta*, **1438**, 1-17.

Lichtenbergova, L., Yoon, E. T. and Cho, W. (1998) Membrane penetration of cytosolic phospholipase A2 is necessary for its interfacial catalysis and arachidonate specificity. *Biochemistry*, **37**, 14128-14136.

Lin, L. L., Wartmann, M., Lin, A. Y., Knopf, J. L., Seth, A. and Davis, R. J. (1993) cPLA2 is phosphorylated and activated by MAP kinase. *Cell*, **72**, 269-278.

Liu, P. and Anderson, R. G. (1995) Compartmentalized production of ceramide at the cell surface. *J Biol Chem*, **270**, 27179-27185.

Llirbat, B., Wolf, C., Chevy, F., Citadelle, D., Bereziat, G. and Roux, C. (1997) Normal and inhibited cholesterol synthesis in the cultured rat embryo. *J Lipid Res*, **38**, 22-34.

Lobo, L. I. and Wilton, D. C. (1997) Combined effects of sphingomyelin and cholesterol on the hydrolysis of emulsion particle triolein by lipoprotein lipase. *Biochim Biophys Acta*, **1349**, 122-130.

London, E. (2002) Insights into lipid raft structure and formation from experiments in model membranes. *Curr Opin Struct Biol*, **12**, 480-486.

Lopez, I., Mak, E. C., Ding, J., Hamm, H. E. and Lomasney, J. W. (2001) A novel bifunctional phospholipase c that is regulated by Galpha 12 and stimulates the Ras/mitogen-activated protein kinase pathway. *J Biol Chem*, **276**, 2758-2765.

Magargal, W. W., Dickinson, E. S. and Slakey, L. L. (1978) Distribution of membrane marker enzymes in cultured arterial endothelial and smooth muscle cells. The subcellular location of oleoyl-CoA:1-acyl-sn-glycero-3-phosphocholine acyltransferase. *J Biol Chem*, **253**, 8311-8318.

Mann, R. K. and Beachy, P. A. (2000) Cholesterol modification of proteins. *Biochim Biophys Acta*, **1529**, 188-202.

Masliah, J., Bachelet, M., Colard, O., Bereziat, G. and Vargaftig, B. B. (1988) Mobilization of arachidonic acid from diacyl and ether-linked phospholipids in FMLP stimulated alveolar macrophages. *Biochem Pharmacol*, **37**, 547-550.

McIntosh, A. L., Gallegos, A. M., Atshaves, B. P., Storey, S. M., Kannoju, D. and Schroeder, F. (2003) Fluorescence and multiphoton imaging resolve unique structural forms of sterol in membranes of living cells. *J Biol Chem*, **278**, 6384-6403.

McKean, M. L. and Silver, M. J. (1985) Phospholipid biosynthesis in human platelets. The acylation of lyso- platelet-activating factor. *Biochem J*, **225**, 723-729.

McLaughlin, S., Wang, J., Gambhir, A. and Murray, D. (2002) PIP(2) and proteins: interactions, organization, and information flow. *Annu Rev Biophys Biomol Struct*, **31**, 151-175.

Melin, T. and Nilsson, A. (1997) Delta-6-desaturase and delta-5-desaturase in human Hep G2 cells are both fatty acid interconversion rate limiting and are upregulated under essential fatty acid deficient conditions. *Prostaglandins Leukot Essent Fatty Acids*, **56**, 437-442.

Metz, J. G., Roessler, P., Facciotti, D., Levering, C., Dittrich, F., Lassner, M., Valentine, R., Lardizabal, K., Domergue, F., Yamada, A. *et al.,* (2001) Production of polyunsaturated fatty acids by polyketide synthases in both prokaryotes and eukaryotes. *Science*, **293**, 290-293.

Miyazaki, A., Koieyama, T., Shimada, Y., Kikuchi, T., Nezu, H., Ito, K., Kasanuki, N. and Koga, T. (2002) Effects of pravastatin sodium on mevalonate metabolism in common marmosets. *J Biochem (Tokyo)*, **132**, 395-400.

Mohandas, N., Wyatt, J., Mel, S. F., Rossi, M. E. and Shohet, S. B. (1982) Lipid translocation across the human erythrocyte membrane. Regulatory factors. *J Biol Chem*, **257**, 6537-6543.

Momchilova, A., Markovska, T. and Pankov, R. (1999) Arachidonoyl-CoA:lysophosphatidylcholine acyltransferase activity in ras-transformed NIH 3T3 fibroblasts depends on the membrane composition. *Biochem Mol Biol Int*, **47**, 555-561.

Momchilova, A., Petkova, D. and Koumanov, K. (1986) Rat liver microsomal phospholipase A2 and membrane fluidity. *Int J Biochem*, **18**, 659-663.

Momchilova, A., Petkova, D., Mechev, I., Dimitrov, G. and Koumanov, K. (1985) Sensitivity of 5′-nucleotidase and phospholipase A2 towards liver plasma membranes modifications. *Int J Biochem*, **17**, 787-792.

Nakashima, S., Zhao, Y. and Nozawa, Y. (1996) Molecular cloning of delta 9 fatty acid desaturase from the protozoan Tetrahymena thermophila and its mRNA expression during thermal membrane adaptation. *Biochem J*, **317 (Pt 1)**, 29-34.

Nalefski, E. A., Sultzman, L. A., Martin, D. M., Kriz, R. W., Towler, P. S., Knopf, J. L. and Clark, J. D. (1994) Delineation of two functionally distinct domains of cytosolic phospholipase A2, a regulatory Ca(2+)-dependent lipid-binding domain and a Ca(2+)-independent catalytic domain. *J Biol Chem*, **269**, 18239-18249.

Nixon, A. B., Greene, D. G. and Wykle, R. L. (1996) Comparison of acceptor and donor substrates in the CoA-independent transacylase reaction in human neutrophils. *Biochim Biophys Acta*, **1300**, 187-196.

Ohanian, J. and Ohanian, V. (2001) Sphingolipids in mammalian cell signalling. *Cell Mol Life Sci*, **58**, 2053-2068.

Okuley, J., Lightner, J., Feldmann, K., Yadav, N., Lark, E. and Browse, J. (1994) Arabidopsis FAD2 gene encodes the enzyme that is essential for polyunsaturated lipid synthesis. *Plant Cell*, **6**, 147-158.

Okuyama, H., Saito, M., Joshi, V. C., Gunsberg, S. and Wakil, S. J. (1979) Regulation by temperature of the chain length of fatty acids in yeast. *J Biol Chem*, **254**, 12281-12284.

Petkova, D. H., Momchilova-Pankova, A. B. and Koumanov, K. S. (1987) Effect of liver plasma membrane fluidity on endogenous phospholipase A2 activity. *Biochimie*, **69**, 1251-1255.

Philip, F., Guo, Y. and Scarlata, S. (2002) Multiple roles of pleckstrin homology domains in phospholipase Cbeta function. *FEBS Lett*, **531**, 28-32.

Polonovski, J., Bard, D., Colard, O. and Bereziat, G. (1974) [Change in phospholipase activity in plasma membranes of rat hepatocytes under influence of insulin and glucagon]. *C R Seances Soc Biol Fil*, **168**, 1211-1215.

Porter, J. A., Young, K. E. and Beachy, P. A. (1996) Cholesterol modification of hedgehog signaling proteins in animal development. *Science*, **274**, 255-259.

Quinn, P. J. (2002) Plasma membrane phospholipid asymmetry. *Subcell Biochem*, **36**, 39-60.

Raggers, R. J., Vogels, I. and van Meer, G. (2001) Multidrug-resistance P-glycoprotein (MDR1) secretes platelet-activating factor. *Biochem J*, **357**, 859-865.

Rebecchi, M. J. and Pentyala, S. N. (2000) Structure, function, and control of phosphoinositide-specific phospholipase C. *Physiol Rev*, **80**, 1291-1335.

Rhee, S. G. (2001) Regulation of phosphoinositide-specific phospholipase C. *Annu Rev Biochem*, **70**, 281-312.

Rhee, S. G., Illenberger, D., Walliser, C., Nurnberg, B., Diaz Lorente, M., Gierschik, P., McLaughlin, S., Wang, J., Gambhir, A. and Murray, D. (2001) Regulation of phosphoinositide-specific phospholipase C

Specificity and structural requirements of phospholipase C-beta stimulation by Rho GTPases versus G protein beta gamma dimers

PIP(2) and proteins: interactions, organization, and information flow. *Annu Rev Biochem*, **70**, 281-312.

Rietveld, A., Neutz, S., Simons, K. and Eaton, S. (1999) Association of sterol- and glycosylphosphatidylinositol-linked proteins with Drosophila raft lipid microdomains. *J Biol Chem*, **274**, 12049-12054.

Rizzo, M. A., Shome, K., Watkins, S. C. and Romero, G. (2000) The recruitment of Raf-1 to membranes is mediated by direct interaction with phosphatidic acid and is independent of association with Ras. *J Biol Chem*, **275**, 23911-23918.

Roberts, W. L., Santikarn, S., Reinhold, V. N. and Rosenberry, T. L. (1988) Structural characterization of the glycoinositol phospholipid membrane anchor of human erythrocyte acetylcholinesterase by fast atom bombardment mass spectrometry. *J Biol Chem*, **263**, 18776-18784.

Roux, C. and Aubry, M. (1966) [Teratogenic action in the rat of an inhibitor of cholesterol synthesis, AY 9944]. *C R Seances Soc Biol Fil*, **160**, 1353-1357.

Roux, C., Dupuis, R., Horvath, C. and Talbot, J. N. (1980) Teratogenic effect of an inhibitor of cholesterol synthesis (AY 9944) in rats: correlation with maternal cholesterolemia. *J Nutr*, **110**, 2310-2312.

Rutter, A. J., Thomas, K. L., Herbert, D., Henderson, R. J., Lloyd, D. and Harwood, J. L. (2002) Oxygen induction of a novel fatty acid n-6 desaturase in the soil protozoon, Acanthamoeba castellanii. *Biochem J*, **368**, 57-67.

Samet, D. and Barenholz, Y. (1999) Characterization of acidic and neutral sphingomyelinase activities in crude extracts of HL-60 cells. *Chem Phys Lipids*, **102**, 65-77.

Sato, N. and Murata, N. (1980) Temperature shift-induced responses in lipids in the blue-green alga, Anabaena variabilis: the central role of diacylmonogalactosylglycerol in thermo-adaptation. *Biochim Biophys Acta*, **619**, 353-366.

Scarlata, S., Gupta, R., Garcia, P., Keach, H., Shah, S., Kasireddy, C. R., Bittman, R. and Rebecchi, M. J. (1996) Inhibition of phospholipase C-delta 1 catalytic activity by sphingomyelin. *Biochemistry*, **35**, 14882-14888.

Scheideler, M. A. and Bell, R. M. (1986) Efficiency of reconstitution of the membrane-associated sn-glycerol 3-phosphate acyltransferase of Escherichia coli. *J Biol Chem*, **261**, 10990-10995.

Schurer, N. Y., Rippke, F., Vogelsang, K., Schliep, V. and Ruzicka, T. (1999) Fatty acid uptake by cultured human keratinocytes grown in medium deficient in or supplemented with essential fatty acids. *Arch Dermatol Res*, **291**, 47-53.

Schutte, B., Nuydens, R., Geerts, H. and Ramaekers, F. (1998) Annexin V binding assay as a tool to measure apoptosis in differentiated neuronal cells. *J Neurosci Methods*, **86**, 63-69.

Seigneuret, M. and Devaux, P. F. (1984) ATP-dependent asymmetric distribution of spin-labeled phospholipids in the erythrocyte membrane: relation to shape changes. *Proc Natl Acad Sci USA*, **81**, 3751-3755.

Shounan, Y., Feng, X. and O'Connell, P. J. (1998) Apoptosis detection by annexin V binding: a novel method for the quantitation of cell-mediated cytotoxicity. *J Immunol Methods*, **217**, 61-70.

Shvedova, A. A., Tyurina, Y. Y., Tyurin, V. A., Kikuchi, Y., Kagan, V. E. and Quinn, P. J. (2001) Quantitative analysis of phospholipid peroxidation and antioxidant protection in live human epidermal keratinocytes. *Biosci Rep*, **21**, 33-43.

Simopoulos, A. P., Leaf, A. and Salem, N., Jr. (2000) Workshop statement on the essentiality of and recommended dietary intakes for Omega-6 and Omega-3 fatty acids. *Prostaglandins Leukot Essent Fatty Acids*, **63**, 119-121.

Singer, W. D., Brown, H. A. and Sternweis, P. C. (1997) Regulation of eukaryotic phosphatidylinositol-specific phospholipase C and phospholipase D. *Annu Rev Biochem*, **66**, 475-509.

Siskind, L. J., Kolesnick, R. N. and Colombini, M. (2002) Ceramide channels increase the permeability of the mitochondrial outer membrane to small proteins. *J Biol Chem*, **277**, 26796-26803.

Sleight, R. and Kent, C. (1983) Regulation of phosphatidylcholine biosynthesis in mammalian cells. II. Effects of phospholipase C treatment on the activity and subcellular distribution of CTP:phosphocholine cytidylyltransferase in Chinese hamster ovary and LM cell lines. *J Biol Chem*, **258**, 831-835.

Smit, J. J., Schinkel, A. H., Oude Elferink, R. P., Groen, A. K., Wagenaar, E., van Deemter, L., Mol, C. A., Ottenhoff, R., van der Lugt, N. M., van Roon, M. A. *et al.* (1993) Homozygous disruption of the murine mdr2 P-glycoprotein gene leads to a complete absence of phospholipid from bile and to liver disease. *Cell*, **75**, 451-462.

Song, C., Satoh, T., Edamatsu, H., Wu, D., Tadano, M., Gao, X. and Kataoka, T. (2002) Differential roles of Ras and Rap1 in growth factor-dependent activation of phospholipase C epsilon. *Oncogene*, **21**, 8105-8113.

Sonoki, S. and Ikezawa, H. (1976) Effect of phospholipase C hydrolysis of membrane phospholipids on acyltransferase systems in rat liver microsomes. *J Biochem (Tokyo)*, **80**, 1233-1239.

Sugars, J. M., Cellek, S., Manifava, M., Coadwell, J. and Ktistakis, N. T. (2002) Hierarchy of membrane-targeting signals of phospholipase D1 involving lipid modification of a pleckstrin homology domain. *J Biol Chem*, **277**, 29152-29161.

Sugiura, T., Masuzawa, Y., Nakagawa, Y. and Waku, K. (1987) Transacylation of lyso platelet-activating factor and other lysophospholipids by macrophage microsomes. Distinct donor and acceptor selectivities. *J Biol Chem*, **262**, 1199-1205.

Swinnen, J. V., Van Veldhoven, P. P., Timmermans, L., De Schrijver, E., Brusselmans, K., Vanderhoydonc, F., Van de Sande, T., Heemers, H., Heyns, W. and Verhoeven, G. (2003) Fatty acid synthase drives the synthesis of phospholipids partitioning into detergent-resistant membrane microdomains. *Biochem Biophys Res Commun*, **302**, 898-903.

Tang, X., Halleck, M. S., Schlegel, R. A. and Williamson, P. (1996) A subfamily of P-type ATPases with aminophospholipid transporting activity. *Science*, **272**, 1495-1497.

Thomas, K., Rutter, A., Suller, M., Harwood, J. and Lloyd, D. (1998) Oxygen induces fatty acid (n-6)-desaturation independently of temperature in Acanthamoeba castellanii. *FEBS Lett*, **425**, 171-174.

Tint, G. S. (1993) Cholesterol defect in Smith-Lemli-Opitz syndrome. *Am J Med Genet*, **47**, 573-574.

Tou, J. S. (1981) Activation of the metabolism of the fatty acyl group in granulocyte phospholipids by phorbol myristate acetate. *Biochim Biophys Acta*, **665**, 491-497.

van Deenen, L. L. M. and de Gier, J. (1974). *The red blood cell*. New York, academic press.

van Helvoort, A., Smith, A. J., Sprong, H., Fritzsche, I., Schinkel, A. H., Borst, P. and van Meer, G. (1996) MDR1 P-glycoprotein is a lipid translocase of broad specificity, while MDR3 P-glycoprotein specifically translocates phosphatidylcholine. *Cell*, **87**, 507-517.

Veldman, R. J., Maestre, N., Aduib, O. M., Medin, J. A., Salvayre, R. and Levade, T. (2001) A neutral sphingomyelinase resides in sphingolipid-enriched microdomains and is inhibited by the caveolin-scaffolding domain: potential implications in tumour necrosis factor signalling. *Biochem J*, **355**, 859-868.

Venkataraman, K. and Futerman, A. H. (2000) Ceramide as a second messenger: sticky solutions to sticky problems. *Trends Cell Biol*, **10**, 408-412.

Verkleij, A. J., Zwaal, R. F., Roelofsen, B., Comfurius, P., Kastelijn, D. and van Deenen, L. L. (1973) The asymmetric distribution of phospholipids in the human red cell membrane. A combined study using phospholipases and freeze-etch electron microscopy. *Biochim Biophys Acta*, **323**, 178-193.

Vieu, C., Terce, F., Chevy, F., Rolland, C., Barbaras, R., Chap, H., Wolf, C., Perret, B. and Collet, X. (2002) Coupled assay of sphingomyelin and ceramide molecular species by gas liquid chromatography. *J Lipid Res*, **43**, 510-522.

Virtanen, J. A., Cheng, K. H. and Somerharju, P. (1998) Phospholipid composition of the mammalian red cell membrane can be rationalized by a superlattice model. *Proc Natl Acad Sci USA*, **95**, 4964-4969.

Wada, H., Gombos, Z. and Murata, N. (1990) Enhancement of chilling tolerance of a cyanobacterium by genetic manipulation of fatty acid desaturation. *Nature*, **347**, 200-203.

Wang, D., Feng, J., Wen, R., Marine, J. C., Sangster, M. Y., Parganas, E., Hoffmeyer, A., Jackson, C. W., Cleveland, J. L., Murray, P. J. *et al.* (2000) Phospholipase Cgamma2 is essential in the functions of B cell and several Fc receptors. *Immunity*, **13**, 25-35.

Weisinger, H. S., Vingrys, A. J., Abedin, L. and Sinclair, A. J. (1998) Effect of diet on the rate of depletion of n-3 fatty acids in the retina of the guinea pig. *J Lipid Res*, **39**, 1274-1279.

Weisinger, H. S., Vingrys, A. J., Bui, B. V. and Sinclair, A. J. (1999) Effects of dietary n-3 fatty acid deficiency and repletion in the guinea pig retina. *Invest Ophthalmol Vis Sci*, **40**, 327-338.

Wilde, J. I. and Watson, S. P. (2001) Regulation of phospholipase C gamma isoforms in haematopoietic cells: why one, not the other? *Cell Signal*, **13**, 691-701.

Willems, G. M., Janssen, M. P., Comfurius, P., Galli, M., Zwaal, R. F. and Bevers, E. M. (2000) Competition of annexin V and anticardiolipin antibodies for binding to phosphatidylserine containing membranes. *Biochemistry*, **39**, 1982-1989.

Williamson, P., Bevers, E. M., Smeets, E. F., Comfurius, P., Schlegel, R. A. and Zwaal, R. F. (1995) Continuous analysis of the mechanism of activated transbilayer lipid movement in platelets. *Biochemistry*, **34**, 10448-10455.

Williamson, P., Christie, A., Kohlin, T., Schlegel, R. A., Comfurius, P., Harmsma, M., Zwaal, R. F. and Bevers, E. M. (2001) Phospholipid scramblase activation pathways in lymphocytes. *Biochemistry*, **40**, 8065-8072.

Willnow, T. E., Hilpert, J., Armstrong, S. A., Rohlmann, A., Hammer, R. E., Burns, D. K. and Herz, J. (1996) Defective forebrain development in mice lacking gp330/megalin. *Proc Natl Acad Sci USA*, **93**, 8460-8464.

Wolf, C., Colard-Torquebiau, O., Bereziat, G. and Polonovski, J. (1977) [Phospholipase A2 of the adipocyte cytoplasmic membrane]. *Biochimie*, **59**, 115-118.

Yen, C. L., Mar, M. H., Craciunescu, C. N., Edwards, L. J. and Zeisel, S. H. (2002) Deficiency in methionine, tryptophan, isoleucine, or choline induces apoptosis in cultured cells. *J Nutr*, **132**, 1840-1847.

Yen, C. L., Mar, M. H. and Zeisel, S. H. (1999) Choline deficiency-induced apoptosis in PC12 cells is associated with diminished membrane phosphatidylcholine and sphingomyelin, accumulation of ceramide and diacylglycerol, and activation of a caspase. *Faseb J*, **13**, 135-142.

Chapter 11

Phospholipid Metabolism in Lung Surfactant

Ruud Veldhuizen[1] and Fred Possmayer[2]
[1]Departments of Medicine and Physiology and Pharmacology, Lawson Health Research Institute, 268 Grosvenor Street, London, ON, N6A 4V2, Canada; [2]Departments of Obstetrics & Gynaecology and Biochemistry, CIHR Group in Fetal and Neonatal Health and Development, The University of Western Ontario, London, ON N6A 5A5 Canada

Abstract
Pulmonary surfactant is a mixture of lipids, mostly phospholipids, and proteins that allows for breathing with minimal effort. The current chapter discusses the metabolism of the phospholipids of this material. Surfactant phospholipids are synthesized in the type II epithelial cells of the lung. The lipids and surfactant proteins are assembled in intracellular storage organelles, called lamellar bodies, and are subsequently secreted into the alveolar space. Within this extracellular space surfactant undergoes several transformations. First the lamellar bodies unravel to form a highly organized lattice-like lipid : protein structure tubular myelin. Second, the organized structures, in particular tubular myelin, adsorb to form a lipid at the air-liquid interface of the alveoli. It is, in fact, this surface tension reducing film that is responsible for the physiological role of surfactant, to prevent lung collapse and allow ease of inflation. Third, the surface film is converted to a small vesicular form. Finally, these small vesicles are taken-up by the type II cells for recycling and degradation and by alveolar macrophages for degradation.

1. INTRODUCTION

Pulmonary surfactant promotes normal respiration by facilitating lung expansion during inhalation and preventing alveolar collapse during exhalation to low lung volumes. These essential functions of lung surfactant are readily apparent from the severe lung dysfunction that occurs in premature infants deficient in surfactant and the marked and immediate improvement observed with these babies following surfactant supplementation (Avery and Mead, 1959; Avery, 2000; Jobe and Ikegami, 2000; Yost and Soll, 2000). Surfactant stabilizes the terminal airspaces through the formation of a phospholipid rich film that reduces surface tension at the alveolar air-liquid interface (Goerke, 1998). Surfactant synthesis and secretion, the formation of the surface film and the removal of surfactant remnants from the airspaces represent a series of dynamic metabolic interactions, which are the focus of the current review. More specifically, we will emphasize aspects of phospholipid metabolism that are unique to surfactant rather than describing metabolic aspects of phospholipids in general. As such we will only briefly discuss the synthetic pathways involved in surfactant phospholipid synthesis and will concentrate on the alveolar metabolism of surfactant. In order to put these various metabolic issues into context, we will first describe surfactant composition and function. For more detailed information about those areas that are only briefly covered by the current chapter, the reader will be referred to other reviews.

1.1. Surfactant composition

Pulmonary surfactant functions at the air:fluid interfaces of the lung. Surfactant can be readily recovered from the lung in a relatively pure state through bronchoalveolar lavage with saline. As will be elaborated below, centrifugation of recovered surfactant generates the large aggregates (LA) fraction, leaving small aggregates (SA) in the supernatant. For the most part, LA have been used for analyses. Although minor differences are present, surfactant composition appears to be fairly consistent among different mammalian species (Shelley *et al.*, 1984; Cockshutt and Possmayer, 1992). Surfactant is composed of about 85% phospholipids, 5–8% neutral lipids and 5–10% surfactant associated proteins (Possmayer, 2003) (Figure 1). The neutral lipid fraction of surfactant consists mostly of cholesterol.

The surfactant-associated proteins (SP-) consist of 2 hydrophobic proteins designated SP-B and SP-C and 2 large hydrophilic proteins, SP-A and SP-D (Possmayer, 1988; Persson *et al.*, 1989). The structures and properties of these proteins have been described in detail elsewhere (Crouch, 1998; Hawgood *et al.*, 1998; Johansson, 1998; Haagsman and Diemel, 2001; McCormack and Whitsett, 2002). The current review will refer to these proteins only in terms of their roles in influencing the metabolism of the surfactant phospholipids.

Phosphatidylcholine (PC) is the most abundant phospholipid (60–80% of total phospholipid) and phosphatidylglycerol (PG) is also present in a relative high abundance at 7–15% of total phospholipids (Shelley *et al.*, 1984;

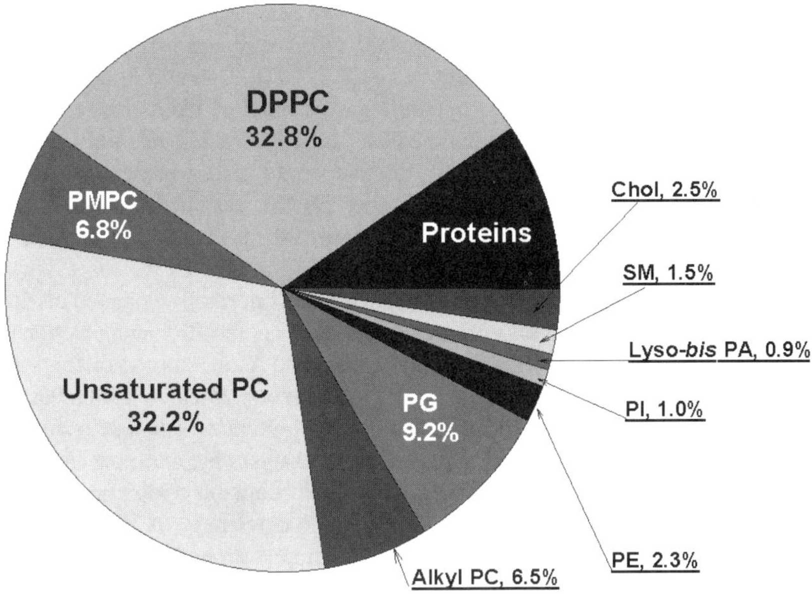

Figure 1. The composition of bovine pulmonary surfactant. Abbreviations: Alkyl-PC, alkyl, acyl phosphatidylcholine; Chol, cholesterol; PMPC, palmitoyl, myristoylphosphatidylcholine; DPPC, dipalmitoylphosphatidylcholine; PA, phosphatidic acid; PC, phosphatidylcholine; PE, phosphatidylethanolamine; PG, phosphatidylglycerol; PI, phosphatidylinositol; SM, sphingomyelin. (From Possmayer, 2003, the authors thank Dr. K. Rodriguez Capote for unpublished data.)

Cockshutt and Possmayer, 1992). Other phospholipids, such as phosphatidylethanolamine (PE), sphingomyelin (SM), phosphatidylinositol (PI), and phosphatidylserine, comprise the remaining phospholipid fraction. Although these lipid values represent a generic mammalian surfactant composition, differences among species occur. For example, the rhesus monkey contains a relatively low amount of PG but has a high level of another acidic surfactant phospholipid, PI. (Egberts *et al.*, 1987). PI tends to be high in surfactant from human fetuses and premature infants, and its replacement by PG represents an indication of lung maturation, particularly with diabetic mothers (Hallman *et al.*, 1977). The levels of cholesterol have also been reported to be variable among species and can change with physiologic state (Doyle *et al.*, 2000; Orgeig and Daniels, 2001).

Analysis of the acyl chains of the phospholipid species has focused mostly on PC and PG. Initial investigations employed a variety of techniques involving derrivatization and chromatography to establish surfactant phospholipid acyl chain composition and, in some cases, molecular species distribution

(Kahn *et al.*, 1995; Bernhard *et al.*, 2001). More recently, electrospray ionization mass spectrometry has been applied effectively toward phospholipid molecular species determination (Postle *et al.*, 2001). It should be noted that mass spectrometry separates compounds on the basis of mass/charge (m/z). Thus, dipalmitoyl phosphatidylcholine (16:0/16:0-PC) at 734 m/z units or the commonly used sodiated adduct at 756 m/z could also represent stearoyl, palmitoyl-PC (18:0/16:0-PC). Furthermore, mass spectrometry analysis as normally conducted does not reveal distribution of the acyl chains at the 1- and 2-sn-position of the glycerol moieties (i.e., 1–16:0, 2–18:1-PC versus 1–18:1, 2–16:0-PC). Consequently, acyl chain distribution is normally assigned on the basis of new or prior analyses employing fatty acyl composition and positional determination through hydrolysis by phospholipase A_2 or a lipase or through fragmentation analysis by tandem mass spectrometry (Healey 2000, Postle *et al.*, 2001). Thus 16:0/18:1-PC indicates a peak at 796 m/z (sodiated) primarily composed of 1–16:0, 2–18:1-PC. Phospholipid analyses by a number of these techniques consistently confirm that surfactant PC contains a high percentage of the disaturated species 16:0:16:0-PC. The high percentage of 16:0/16:0-PC (normally 30–60% of total PCs) has been accepted as an important characteristic of pulmonary surfactant and in many studies measurements of disaturated PC are used as an indication of the total surfactant levels. It should be noted however that the most widely-used method to estimate disaturated PC, based on the reaction of osmium tetroxide with unsaturated acyl chains (Mason *et al.*, 1976), results in an overestimation of disaturated PC due to the inclusion of some unreacted monounsaturated PC (Chan *et al.,* 1983; Holm *et al.*, 1996). Nevertheless, 16:0/16:0-PC is the most abundant PC species in mammal surfactants reported to date, with a single exception. Recent mass spectrometric analyses on the fat-tailed dunnart, *Sminthopsis crassicaudata*, reveal surfactant 16:0/16:0-PC at ~15%, with 16:0/16:1-PC at ~30% being the major molecular species (unpublished observations). The significance of this interesting observation must still be investigated. In addition to 16:0/16:0-PC, mammalian surfactants contain ~10% of the disaturated PCs 16:/14:0-PC (Postle *et al.*, 2001). As suggested by Postle *et al.*, this unusual PC species can be considered even more characteristic of surfactant phospholipids than 16:0/16:0-PC (Heeley *et al.*, 2000; Postle *et al.*, 2001). In addition, surfactant also contains relatively high levels of 16:0/16:1-PC, rising in the fat-tailed dunnart to the noted ~30%. Other relatively abundant PC species are 16:0/18:2 and 16:0/18:1, present in 5–15% range, depending on the source.

The acyl composition of PG is more variable among different species (Postle *et al.*, 2001). For example, 16:0/18:1-PG is the major PG species in rabbit and human surfactant but 16:0/16:0-PG is the most abundant PG species in rat (Cockshutt and Possmayer, 1992). A number of other molecular PG species are present in amounts between 1% and 10% of total PG.

1.2. Surfactant function

As mentioned in the introduction, the primary role of pulmonary surfactant is to ensure alveolar stability of the lung by reducing the surface tension at the air-liquid interface (Goerke, 1998). The importance of surface tension reduction in maintaining alveolar stability can be explained by considering the law of Young and Laplace. This law states that the pressure differential across a bubble (or an alveolus) is 2 times the surface tension divided by the radius. Thus, at similar surface tensions a smaller bubble has a higher pressure gradient across the bubble than a larger bubble (i.e. the small bubble would tend to collapse into the more stable larger bubble. The presence of pulmonary surfactant in the alveoli, however, reduces the surface tension so that pressure differences are minimized. In addition, surface tension varies with alveolar size such that as alveolar radius decreases, there is a parallel decrease in surface tension so that the pressure gradient does not cause alveolar collapse.

The effects of the law of Young and Laplace can be observed in whole lung by examining the relationship between lung volume and pressure. In a normal lung, inflation does not require a large pressure gradient and the lung remains stable at low lung volumes. In contrast, due to the need to overcome the large surface tension forces, a surfactant deficient lung will require large pressures to inflate and the lung will easily collapse during deflation. This surfactant deficient scenario defines the underlying pathophysiology for Neonatal Respiratory Distress Syndrome (NRDS) which preterm infants can develop due to surfactant deficiency (Avery and Mead, 1959; Avery, 2000). This critical role of surfactant is further emphasized by the fact that surfactant supplementation of neonates with NRDS results in marked improvements in lung function (Jobe and Ikegami, 2000; Yost and Soll, 2000). Surfactant treatment, currently a standard therapy for NRDS, facilitates lung expansion and improves gaseous exchange in these premature infants, thereby buying time until the infant can generate sufficient surfactant of its own. Improvements in the care of premature infants, including surfactant supplementation, have resulted in a significant reduction in mortality associated with this syndrome (Jobe and Ikegami, 2000). Surfactant remains essential for lung expansion and stability throughout life.

2. OVERVIEW OF SURFACTANT METABOLISM

As indicated above, surfactant functions extracellularly. In order to generate and maintain the critical surface tension reducing film at the air-liquid interface of the lung, surfactant is synthesized within the alveolar type II cell and stored intracellularly as lamellar bodies until it is secreted into the alveolar space, where it subsequently forms the surface film (Figure 2) (Wright and

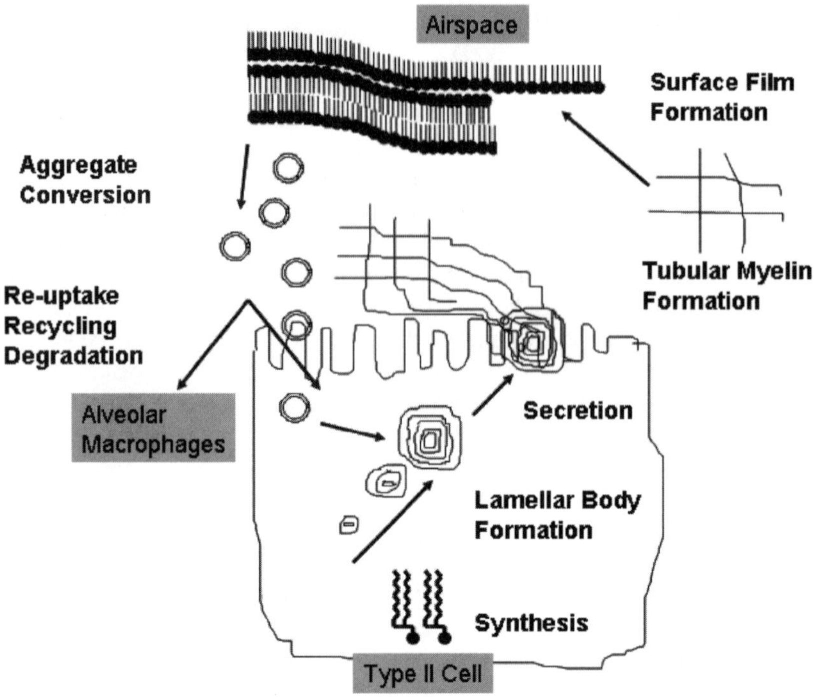

Figure 2. Schematic overview of surfactant metabolism. Pulmonary surfactant is synthesized in the alveolar type II cell; it is stored intracellularly as lamellar bodies which are, upon stimulation, secreted into the alveolar space. Within the alveolar space surfactant forms highly organized lipid protein structures which subsequently form the surface tension reducing surface film. Small surfactant vesicles are formed during respiration which are taken up for recycling or degradation.

Dobbs, 1991; Goerke, 1998). Furthermore, remnants from the surface film must be cleared from the airspace (Wright, 1990; Gross, 1995). In the subsequent section we will briefly cover the intracellular metabolism of surfactant phospholipids. The remainder of the chapter will then focus on the alveolar metabolism of surfactant.

3. INTRACELLULAR SURFACTANT PHOSPHOLIPID METABOLISM

3.1. Synthesis

Surfactant is synthesized, assembled, and secreted by the alveolar type II cell (Batenburg and Haagsman, 1998). The primary focus of many studies on surfactant phospholipid synthesis has been on the manner in which type II cells produce the specific, conserved phospholipid composition of surfactant. Specific discussion will be directed toward the type II cell's ability to produce a high proportion of the disaturated PC species 16:0/16:0-PC compared to other organs.

The experimental approaches taken to study surfactant phospholipid synthesis include incorporation studies of radioactive precursors in whole animals, studies on isolated type II cells from neonatal and adult animals, and investigations into the specificity and kinetics of enzymes involved in phospholipid biosynthesis (Post *et al.*, 1984; Batenburg *et al.*, 1986; Jobe *et al.*, 1989; Cockshutt and Possmayer, 1992). More recently, investigators have employed genetic approaches to study enzymes involved in biosynthesis, for example by overexpressing phosphocholine : CTP cytidylyltransferase in type II cells *in vitro* and *in vivo* (Spragg and Li, 2000; Li *et al.*, 2002). In addition, clinical studies have utilized stable isotopes to assess incorporation of surfactant precursors in neonates (Cogo *et al.*, 1999; Li *et al.*, 2002). These extensive studies, using various approaches, have provided a reasonably clear, overview of surfactant phospholipid synthesis.

The main substrate for synthesis of the glycerol backbone of surfactant lipids is glucose supplied by the circulation (Cockshutt and Possmayer, 1992; Batenburg and Haagsman, 1998). The fatty acids and choline required for lipid synthesis can also be taken up from the circulation. However, for the fatty acids, degradation of phospholipids (for example those recovered from the alveolar space) and *de novo* synthesis by type II cells represent important additional sources. These substrates are utilized within the type II cell to produce phospholipids via the classical pathway first described by Kennedy (Kennedy, 1961; Rooney, 1985; Batenburg and Haagsman, 1998) (Figure 3). The first steps of this pathway involve the enzymatically induced formation of phosphatidic acid (PA) from glycerol-3-phosphate or dihydroxyacetonephosphate. PA marks an important branchpoint in phospholipid synthesis since it can be utilized for the formation of CDP-diacylglycerol or for the formation of diacylglycerol. CDP-diacylglycerol represents an intermediate for the formation of PG and PI whereas diacylglycerol can react with CDP-choline to form PC and with CDP-ethanolamine to form PE. Formation of CDP-choline for PC synthesis involves choline kinase to generate cholinephosphate and phosphocholine : CTP cytidylyltransferase. Considerable evidence has accumulated showing that the reaction catalysed by phosphocholine cytidylyltransferase is both rate-limiting and regulatory in lung and type II cells (Post *et al.*, 1984; Zimmermann *et al.*, 1993) as well as in other organs, such as liver (Pritchard and Vance, 1981). This is somewhat unusual in that the cytidylyltransferase reaction is reversible. Nevertheless, it is evident that flux through this reaction increases under the developmental and hormonal regulation associated with pulmonary maturation, resulting

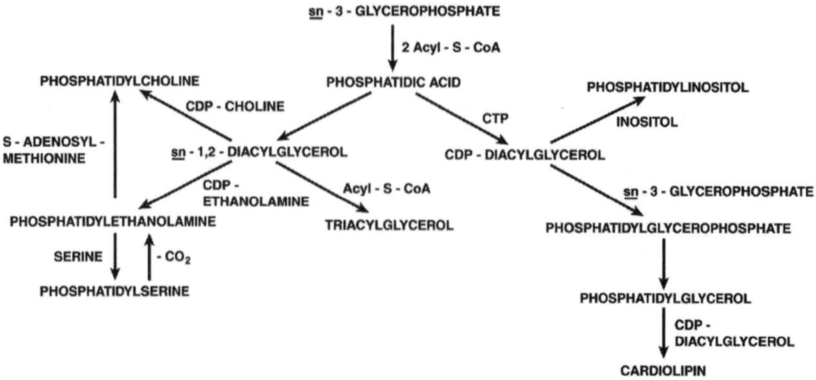

Figure 3. The Kennedy Pathway for phospholipid and triacylglycerol biosynthesis in animal tissues. RCO-ScoA, fatty acyl CoA. (From Nanjundan and Possmayer 2003, as modified from Kennedy E., 1961.)

in increased PC synthesis for surfactant (Rooney, 1985; Cockshutt and Possmayer, 1992; Batenburg and Haagsman, 1998). Increased PC synthesis requires a supply of diacylglycerol to accept phosphocholine from CDP-diacylglycerol. Relatively little is known about the regulation of the lipid branch of the PC synthetic pathway. However, it is evident that glycerolphosphate/dihydroxyacetonephosphate acyltransferaseactivity in type II cells is affected by redox potential and is decreased with reduction in levels of reduced glutathione, which occurs during hyperoxia and ozone exposure (Holm *et al.*, 1985; Van Bree *et al.*, 1988).

Control of the synthesis of the other surfactant phospholipids such as PG has not been studied as intensively. As stated above, the proportion of PG in most animals and in human surfactant increases near term. The increase in PG synthesis for surfactant appears related to a developmental decrease in the inositol/glycerophosphate ratio in type II cells (Hallman *et al.*, 1976; 1977). Although PG is generated in mitochondria as an intermediate in cardiolipin synthesis, developmental studies have shown PG for surfactant is synthesized in the endoplasmic reticulum (Hallman and Gluck, 1975).

The basis for the ability of lung to generated high levels of 16:0/16:0-PC relative to other tissues has not been completely resolved. It is clear that the enzymes involved in the *de novo* pathway are not selective towards DPPC formation. The acyl composition of PC formed by the *de novo* pathway is influenced by the availability of the fatty acid substrates and a high percentage of palmitoyl-CoA is found in isolated Type II cells (Post *et al.*, 1983; Haagsman and van Golde, 1991). This suggests that these cells have the potential to produce a relatively high proportion of this disaturated PC by *de novo* synthesis. However, although 16:0/16:0-PG represents ~10% of surfactant PGs, there is no evidence for enrichment in disaturated PI or PE. The available evidence would suggest that *de novo* synthesis provides only a portion of the DPPC. Remodelling of PCs containing unsaturated fatty acids at the 2-position is thought to be responsible for the formation of additional 16:0/16:0-PC. The major remodelling pathway is the Land's deacylation-reacylation cycle, in which phospholipase A_2 removes a fatty acid at the 2-position and the resulting lyso-PC is then reacylated with palmitoyl-CoA (Rooney, 1985; Cockshutt and Possmayer, 1992; Batenburg and Haagsman, 1998).

3.2. Lamellar body formation

Following synthesis in the endoplasmatic reticulum, surfactant components are assembled in intracellular storage organelles, the lamellar bodies (Rooney *et al.*, 1975; Haagsman and van Golde, 1991; Schmitz and Muller, 1991; Weaver *et al.*, 2002). Evidence for this function comes from radioactive precursor studies and from the fact that the lipid composition of isolated lamellar bodies is similar to that of isolated surfactant from the air space (Rooney *et al.*, 1975; Post *et al.*, 1982). Structurally lamellar bodies can be recognized within the type II cell as densely packed concentric bilayers of approximately 1–2 μm in size (Williams, 1982; Schmitz and Muller, 1991). Generation of lamellar bodies has recently been reviewed by Weaver *et al.* and, as pointed out by that review, progress in our understanding of lamellar body formation is surprisingly limited (Weaver *et al.*, 2002). It appears that newly synthesized phospholipids are transported to lamellar bodies directly from the endoplasmic reticulum. This contrasts with the surfactant proteins which are processed, at least in part, in the Golgi apparatus and routed through multivesicular bodies. Surfactant proteins (SP-) SP-B and SP-C are processed through the Golgi and then are proteolytically processed within the lamellar bodies (Voorhout *et al.*, 1992; Weaver, 1998). SP-A and SP-D (large, oligomeric glycoproteins) are processed by the Golgi apparatus and secreted by the constitutive pathway independent of surfactant phospholipids (Crouch, 1998; McCormack, 1998). In addition to *de novo* synthesized phospholipids, surfactant material taken up from the alveolar space by type II cells can also be directed toward lamellar body formation (Hallman *et al.*, 1981; Kalina and Socher, 1990; Kalina *et al.*, 1993). Recycling of surfactant material taken up from the alveoli contributes to lamellar body formation (Weaver *et al.*, 2002).

3.3. Secretion

Secretion of lamellar bodies occurs by exocytosis (Rooney, 1985; 2001; Mason and Voelker, 1998). Briefly, following a signal for secretion, there is directed movement of the lamellar body towards the apical plasma membrane of the type II cell. Little is known about the mechanisms for this intracellular transport. Electron microscopic studies have documented that the outer membrane of the lamellar bodies fuses with the plasma membrane (Chander and Wu, 1991; Haller *et al.*, 2001). This fusion likely involves GTP binding proteins of the RAB family and is followed by exocytosis of the lamellar body contents into the alveolar space (Rubins *et al.*, 1992; Liu, 1999).

A number of factors and situations can either stimulate or inhibit secretion (Mason and Voelker, 1998). For example, stretching of the type II cell, as well as beta-adrenergic agonists, diacylglycerol analogues such as phorbol esters, ATP and numerous other bioactive compounds can stimulate surfactant secretion from isolated type II cells (Mason and Voelker, 1998; Rooney, 2001). Due to technical limitations in studying secretion *in vivo*, the majority of the studies studying secretion have utilized isolated type II cells. The *in vivo* studies available have both complemented and contradicted the results of the *in vitro* culture experiments. For example, studies with hyperventilation would appear to support the concept that lung stretch stimulates secretion (Massaro and Massaro, 1983; Nicholas and Barr, 1983; Veldhuizen *et al.*, 2000).

It has often been reported that SP-A inhibits surfactant secretion from isolated type II cells (Kuroki *et al.*, 1988; Wright and Dobbs, 1991). The physiological significance of this role for SP-A under normal circumstances has been challenged since SP-A deficient mice exhibit surfactant metabolism similar to wildtype controls (Korfhagen *et al.*, 1996; 1998). However, recent evidence using SP-A deficient mice suggests that SP-A does play a role in surfactant uptake by type II cells under stress conditions such as hyperventilation and lung injury (Quintero *et al.*, 2002; Jain *et al.*, 2003). In conclusion, most of the information available on surfactant secretion is obtained from *in vitro* experiments; mechanisms of secretion as it occurs *in vivo* are less well-defined.

4. EXTRACELLULAR SURFACTANT PHOSPHOLIPID METABOLISM

Surfactant metabolism within the airspaces has been studied extensively since this is where it functions (Gross, 1995; Goerke, 1998; Rooney, 2001). The secreted lamellar bodies undergo transformation to a more complex structure, known as tubular myelin, which is thought to be the direct precursor of the surface tension-reducing surface film (Sanders *et al.*, 1980; Williams, 1982). During respiration, likely due to repeated compression:expansion cycles, small vesicles are formed which are taken up by macrophages and type II cells for degradation or recycling (Magoon *et al.*, 1983; Baritussio *et al.*, 1984; Stevens *et al.*, 1987). These small, mainly unilamellar, vesicles comprise the small aggregate (SA) fraction. The following discussion of the alveolar metabolism of surfactant phospholipids will highlight the following aspects: (1) formation of tubular myelin, a unique structure within the LA fraction, (2) formation of the surface tension reducing film, (3) the conversion of the LA into the small vesicular (SA) forms, and (4) SA degradation and recycling.

4.1. Formation of tubular myelin

Following secretion the lamellar bodies combine with SP-A, which is secreted separately, to form the highly organized lipid-protein structure known as tubular myelin (Froh *et al.*, 1990). Tubular myelin is one of the specific and unique structures isolated with the LA fraction. This structure (Figure 4) consists of a lattice like structure of elongated lipid bilayers (Williams, 1982; Young *et al.*, 1992). Not surprisingly this unique structure has raised numerous questions regarding its formation and function and electron microscopy has been utilized extensively to study tubular myelin. Electron microscopic pictures showing bilayers from lamellar bodies connected to tubular myelin structures strongly suggest the unraveling of one and sometimes more lamellar bodies leads to the formation of tubular myelin. This concept is also supported by the observation that isolated lamellar bodies can be transformed into tubular myelin-like structures *in vitro* in the presence of exogenous SP-A and calcium (Sanders *et al.*, 1980; Froh *et al.*, 1990).

Unfortunately, it has not yet been possible to isolate pure tubular myelin from lavage material, and this has hampered determination of its precise composition. It is known however, that SP-A is a critical component of tubular myelin. This is based on three lines of evidence: (1) Using gold- labeled anti SP-A antibodies, Voorhout *et al.* showed that SP-A was present in tubular myelin and appeared to be located near the corners of this structure (Voorhout *et al.*, 1991). (2) tubular myelin can be formed from purified surfactant components or purified lamellar bodies; SP-A is essential in this process (Suzuki *et al.*, 1989; Williams *et al.*, 1991; Poulain *et al.*, 1992) , and (3) mice deficient in SP-A do not contain tubular myelin and antisense inhibition of SP-A formation in human lung explants decrease tubular myelin formation (Korfhagen *et al.*, 1996; 1998; Klein *et al.*, 2002). The other components that are known to be important for tubular myelin structures are 16:0/16:0-PC PG, SP-B and calcium since all these components are required for tubular myelin formation

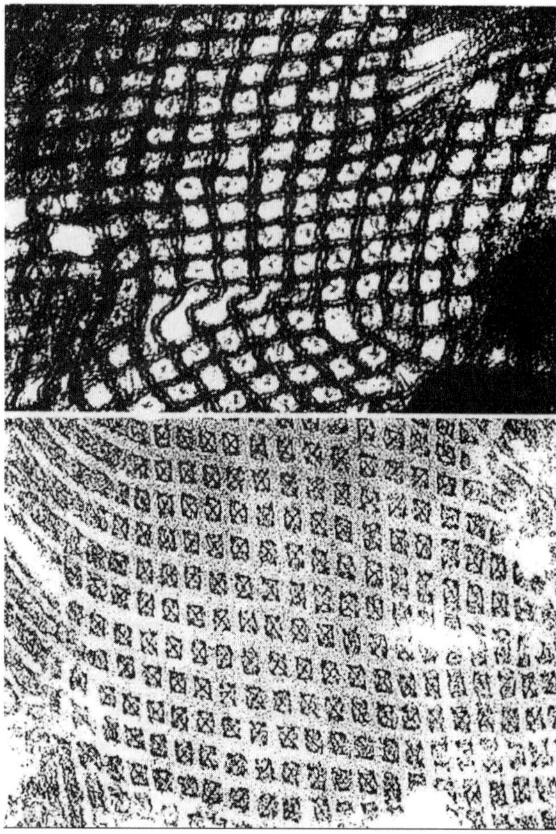

Figure 4. Tubular Myelin. Top: Electron micrograph showing cross-section of tubular myelin from bovine surfactant. Tubular myelin consists of fused tubes which here extend perpendicular to the plane of the image. The squares are ~50 μm across. Bottom: Structures remaining after acetone washing to remove phospholipids. The X-shaped structures are thought to be SP-As which interact with the phospholipids via their carbohydrate recognition domains. (From Nag *et al.*, 1999.)

in vitro (Benson *et al.*, 1984; Suzuki *et al.*, 1989; Williams *et al.*, 1991; Poulain *et al.*, 1992).

The inability to obtain pure tubular myelin has also limited the ability to assign a specific role for this structure. Furthermore, the lack of tubular myelin in healthy SP-A knockout mice demonstrates that this structure is not absolutely essential for lung function under normal conditions (Korfhagen *et al.*, 1998). Nevertheless, tubular myelin, and other related structures, likely play a role in forming the surface film and in providing stability to the LA structures in normal lung as will be discussed below.

4.2. Formation of the surface tension reducing film

As described above, the main function of surfactant is to form and maintain a surface tension reducing film at the air:liquid interface. In addition to the physiological and clinical importance of this process, the mechanisms involved in this aspect of surfactant phospholipid metabolism have prompted a significant amount of novel investigation in this area. With the availability of new analytical techniques developed for studying surface films during the last few years, theoretical models of how the surface film is formed and organized are continuously being restructured and updated. We will discuss the critical process of surfactant surface tension reduction by first describing the "classical model" of surfactant function, which has directed research in this area for over three decades. Subsequently we will present more recent experimental findings which argue against this classical model thereby illustrating the need for further conceptual alterations.

The scientific basis of the classical model for surfactant function arose from surface balance studies, almost exclusively with the modified Langmuir-Wilhelmy apparatus introduced by Dr. John Clements in 1957 (Clements, 1957). Such studies demonstrated that solvent-spread monolayers of 16:0/16:0-PC could withstand sufficient surface pressure to drive surface tension (γ) at 37°C to low values near 0 mN/m required to stabilize the lung (Goerke, 1998). This stability can contrast with the behaviour of films of unsaturated phospholipids, such as 18:0/18:1-PC (stearoyl/oleoyl-phosphatidylcholine), which are not able to maintain γ below the equilibrium γ of ~22 mN/m (Figure 5). Equilibrium γ is the lowest, which can be achieved either by spreading phospholipid at an air-water interface or through adsorption of phospholipid vesicles (or surfactant) from the subphase. The equilibrium γ of 22 mN/m is not sufficiently low to stabilize the alveoli at low lung volumes. Values below equilibrium can be achieved through lateral compression of surfactant films through surface area reduction on a surface balance (Figure 5), in a pulsating bubble surfactometer, or in the alveolus during exhalation. It must be stressed that compressed 16:0/16:0-PC monolayers near 0 mN/m are metastable by definition: they are below equilibrium γ. However, such films are quite stable kinetically and remain at low γs for much longer time periods than are required to stabilize the lungs.

The ability of gel phase phospholipids such as 16:0/16:0-PC to remain at γ near 0 mN/m at 37°C for sufficient periods to stabilize the alveoli is related to the phase transition (T_m = 41°C) for 16:0/16:0-PC bilayers. Compression of 18:0/18:1-PC monolayers below that phospholipid's T_m of ~5°C also results in γs near 0 mN/m. (Note T_m is given for bilayers because there is no direct equivalent for monolayers. Thus DPPC slightly above 41°C can be compressed to γs near 0 mN/m simply by applying more lateral pressure. The extent to which temperature can be increased above 41°C while maintaining low γs

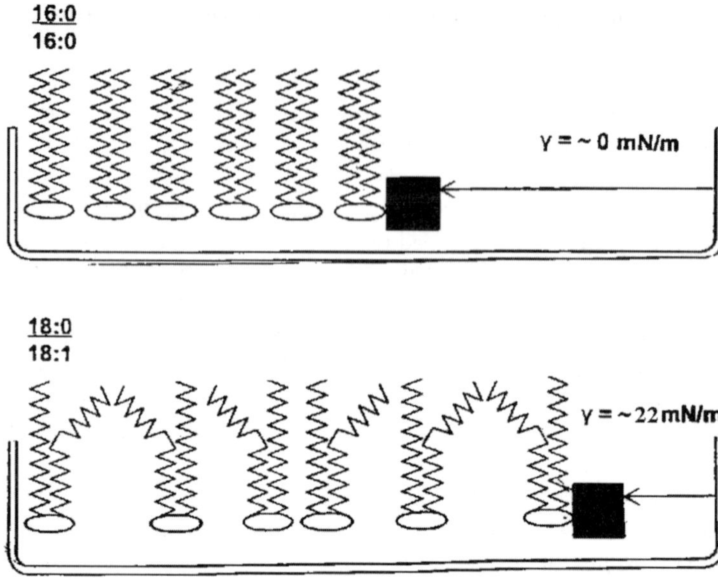

Figure 5. Effect of fatty acid composition of phosphatidylcholines on the minimum surface tension that can be achieved on a Langmuir-Wilhelmy surface balance. The upper monolayer composed of 16:0/16:0-PC can withstand sufficient surface pressure to lower surface tension to near 0 mN/m. The lower monolayer composed of a PC containing an unsaturated fatty acid (stearoyl, oleoyl phosphatidylcholine) cannot be compressed to low tensions. Further compression results in expulsion of PC from the monolayer. (From Possmayer *et al.*, 2001.)

depends on the nature of the surface balance and, as will be explained below, the manner in which the film is compressed.)

The observation that 16:0/16:0-PC was the only significant component of pulmonary surfactant capable of withstanding sufficient surface pressure on the Langmuir-Wilhelmy surface balance to reduce γ to near 0 mN/m led to the apparently logical conclusion that a monolayer highly enriched in this disaturated phospholipid was required to stabilize the terminal airways (Watkins, 1968; Clements, 1977; Bangham *et al.*, 1979). It was further hypothesized that monolayers enriched in 16:0/16:0-PC could arise through a "squeeze-out" mechanism, whereby the less stable fluid phospholipids were expelled during monolayer compression. This process would allow uptake of fresh surfactant during film expansion. Repeated compression:expansion events occurring during respiration would result in progressive increases in monolayer 16:0/16:0-PC content. This concept was reinforced by observations showing that repeated compression:expansion cycling of adsorbed films resulted in progressive lowering of γ (Enhorning, 1977; Schűrch *et al.*, 1989).

More recent studies have indicated that, in addition to fluid phospholipid squeeze-out, monolayer enrichment in 16:0/16:0-PC might also arise through selective DPPC adsorption (Schűrch *et al.*, 1992; 1998). These studies, which employed a greatly improved surface balance, the captive bubble tensiometer, demonstrated that surface area reduction required to achieve $\gamma \approx 0$ mN/m during the first compression of adsorbed surfactant films was significantly smaller than would be predicted on the basis of 16:0/16:0-PC content. For example, 16:0/16:0-PC:PG:SP-B (70:30:1) films required ~20% surface area reduction and BLES (~35% 16:0/16:0-PC) films required only ~30% surface area reduction to attain $\gamma \approx 0$ mN/m during the initial compression after adsorption (Possmayer *et al.*, 2001). Taken together, these accumulated surface balance studies were consistent with the hypothesis that surfactant monolayers were enriched in 16:0/16:0-PC content during adsorption. Furthermore, results implied that the surface concentration of 16:0/16:0-PC increased during compression:expansion cycling due to the squeeze-out mechanism, so that the surface monolayer became composed of almost pure 16:0/16:0-PC.

It has recently become apparent that the classical model can no longer adequately explain surfactant function. Four independent approaches using newly-developed techniques will be briefly described.

1. Captive bubble tensiometer and electron microscopic studies promote strong evidence indicating that adsorbed surfactant films act as multilayers, not monolayers (Schűrch *et al.*, 1995; 1998; Possmayer *et al.*, 2001)
2. Attempts to demonstrate directly selective 16:0/16:0-PC adsorption by non-surface tension related approaches have proven unsuccessful. For example, studies have been conducted in which the surface monolayer is sampled by allowing the film to spread from an adsorbing dish over a filter paper-supported wet bridge to a clean surface on a spreading dish. Experiments using surfactant labeled with trace amounts of [^{3}H] 16:0/16:0-PC and [^{14}C]-labeled unsaturated PC, 16:0/16:0-PG, unsaturated PG or cholesterol have revealed no difference in ^{3}H/^{14}C ratio between the surfactant injected into or recovered from the bulk phase of the adsorbing dish and the monolayer recovered from the spreading dish under a wide range of conditions (Yu and Possmayer, 2003). These results provide strong evidence counter to 16:0/16:0-PC enrichment during adsorption.
3. Fluorescent microscopic studies use fluorescent phospholipid probes (which selectively dissolve into fluid phases) to monitor fluid phase to gel phase interconversion during compression. Such studies demonstrate spread monolayers of calf lung surfactant phospholipids or BLES (bovine lipid extract surfactant) compressed to $\gamma \approx 0$ retain >60% of the surface monolayer in the fluid phase (Piknova *et al.*, 2001; 2002). The proportion of the compressed monolayers present as gel lipid-condensed phase approximates

surfactant's content of 16:0/16:0-PC + 16:0/16:0-PG. This observation revealed for the first time that surfactant phospholipids can attain surface tensions near zero without fluid phospholipid squeeze-out.

4. Captive bubble tensiometer studies show that compression of 14:0/14:0-PC (dimyristoylphosphatidylcholine, $T_m = 24.5°C$) or 16:0/18:1-PC ($T_m = -6°C$) at rapid rates (>30% of initial surface area/second) permits such films to attain $\gamma = 2$ mN/m. Like 16:0/16:0-PC below the equilibrium γ of 22 mN/m, such films are metastable but like 16:0/16:0-PC can maintain low γs for many minutes and are thus capable of stabilizing the alveoli at low lung volumes. The observation that fluid phospholipids compressed rapidly can attain low tensions was quite unexpected.

These four lines of investigation, in particular the observation that fluid phase phospholipids can attain low tensions when compressed rapidly, provide seemingly irrefutable evidence to counter the classical model and signal the need for a paradigm shift to explain surfactant function, at least *in vitro*. As emphasized by Piknova and associates, whether the lung is stabilized by a saturated/unsaturated supercompressed surfactant film, or a film highly enriched in 16:0/16:0-PC remains unknown (Piknova *et al*., 2002). It appears likely that further technical advances will be required for fruitful investigation in this area. Piknova and associates have suggested generation of low γs during rapid compression occurs because the monolayer is compressed so quickly that it does not have sufficient time to collapse. Consequently, the film forms an amorphous solid which can remain kinetically trapped in this state for remarkably long periods. Although no direct evidence is available, it can be speculated that the presence of some gel phase phospholipids such as 16:0/16:0-PC and possibly the hydrophobic surfactant proteins SP-B and SP-C may promote formation of this newly described amorphous solid phase.

4.3. Conversion of LA into SA

Whereas the above discussion focused on the formation of specific structural forms of surfactant, we will now consider the role of alveolar surfactant metabolism in regulating surfactant subtype pool sizes within the airspace (Gross, 1995). In addition to the previously mentioned synthesis and secretion, the amount of surface active material within the alveoli is influenced by the process of aggregate subtype conversion. As indicated earlier, surfactant obtained by saline lavage can be separated into two distinct fractions via differential or density centrifugation, the large aggregates (LA) and the small aggregates (SA) (Magoon *et al*., 1983; Lewis *et al*., 1990). The LA, also called the heavy subtype, consist of organized lipid-protein structures such as tubular myelin, lamellar bodies and multi-lamellar structures, as determined by electron microscopy. LA contain all the surfactant lipids as well as SP-A, SP-B and SP-C. Isolated LA material can reduce the surface tension when tested *in vitro* on a pulsating or captive bubble surfactometer (Putz *et al.*, 1994; Brackenbury *et al.*, 2002). Purified LA can also restore lung function in surfactant deficient rabbits and rats (Yamada *et al.*, 1990; Brackenbury *et al.*, 2002). Thus the LA represent the surface active subfraction of surfactant. The SA, or light subfraction, consists of small unilamellar vesicles. The lipid composition of SA is very similar to LA, however, SA contain very low amounts of the surfactant proteins. As a consequence of the limited amounts of surfactant proteins, SA are not capable to reduce the surface tension to low values (Putz *et al.*, 1994; Brackenbury *et al.*, 2002).

Pulse chase studies with radio-labeled surfactant precursors have demonstrated that radioactivity is first observed in the LA fraction and subsequently in the SA. The level of the SA fraction increases markedly after birth, when air-breathing is established. These data strongly suggest that SA are the metabolic products of the LA. It is important to note that studies on surfactant aggregates use centrifugation to separate lavage surfactant into subtypes. The fate of the surface film during lavage and centrifugation is unknown and may form some of the SA during the lavaging process. However, electron microscopy studies have observed SA in the hypophase of the alveoli of fixed lung tissue demonstrating that SA are present *in vivo* (Savov *et al.*, 1999; Ochs *et al.*, 2000). Furthermore, the relative amounts of SA determined via morphological methods in normal rats was very similar to what has been determined via the biochemical methodology (Malloy *et al.*, 1997; Savov *et al.*, 1999; Veldhuizen *et al.*, 2002), apparently because the surface monolayer accounts for only ~5% of total phospholipid in the airspaces.

The mechanisms involved in the conversion of LA into SA can be studied with an *in vitro* system, called surface area cycling, developed by Dr Nicholas Gross and associates (Gross and Narine, 1989). This technique involves rotating small plastic test tubes containing LA suspensions end over end. Rotation at 40RPM results in a continuous change in the air-liquid interface of the LA suspension (~8-fold). LA are converted into SA during surface area cycling. The SA generated in this *in vitro* system are morphologically and functionally similar to SA obtained from lung lavage material (Gross and Narine, 1989; Veldhuizen *et al.*, 1993A; Gunther *et al.*, 1999). An obvious disadvantage is this technique requires large changes in surface area that are not reflective of the alterations occurring *in vivo*. Furthermore, as will be discussed below, several of the mechanisms described for the *in vitro* situation are based on adding or removing specific components to or from the LA surfactant suspension and these mechanisms may be difficult to test *in vivo*. Nevertheless, surface area cycling has revealed some interesting properties of surfactant and has helped elucidate some of the mechanisms involved in the conversion process.

Early studies on surface area cycling demonstrated LA isolated from a number of mammalian species convert to SA (Gross and Narine, 1989; Veldhuizen *et al.*, 1993A). Conversion is dependent on continuously changing the surface area. Little conversion occurs in completely filled tubes or tubes which are not rotated. The extent of subtype conversion varies with time, surface area, and temperature, with 37°C being optimal (Gross and Narine, 1989; Veldhuizen *et al.*, 1999).

In a series of papers, Gross and colleagues demonstrated that serine protease inhibitors inhibited SA formation (Gross and Schultz, 1990; 1992; Gross, 1995). The active-serine enzyme involved was initially termed "convertase" and identified as a 73kD protein (Gross and Schultz, 1992). Immunoreactive SP-B "disappeared" from surfactant samples during surface area cycling (Veldhuizen *et al.*, 1993A; Gunther *et al.*, 1999). These observations suggested that a specific protease was involved in the conversion process, likely through the degradation of SP-B. However, subsequent studies identified convertase as a carboxy-esterase rather than a protease (Krishnasamy *et al.*, 1997; Barr *et al.*, 1998). Furthermore, SP-B appears to dissociate from LA lipids but is not degraded since full size, bioactive, SP-B can be recovered from the cycling tubes (Inchley *et al.*, 1999). Although addition of the carboxy-esterase to surfactant increases conversion *in vitro*, it is currently unclear if this or any other enzymatic activity is necessary for aggregate conversion *in vivo* (Krishnasamy *et al.*, 1997).

Other surface area cycling studies have investigated the surfactant properties involved in the conversion process (Ueda *et al.*, 1994; Veldhuizen *et al.*, 1996A; 1999; Dhand *et al.*, 1998). Since a changing surface area is required, it seemed obvious that surfactant adsorption was important for aggregate conversion. BLES, which consists of only surfactant phospholipids and SP-B and SP-C, converts rapidly (note: BLES does not have all the structural and compositional features of native LA. Therefore, LA and SA in these experiments were experimentally defined as the pellet (LA) and supernatant (SA) of a 15 min 40,000 g centrifugation). Adsorption inhibitors, such as albumin, reduce BLES conversion to SA (Veldhuizen *et al.*, 1999). Only LA samples that can rapidly adsorb to an air-liquid interface convert to SA (Ueda *et al.*, 1994; Veldhuizen *et al.*, 1999), consistent with the concept that adsorption of LA is important for SA formation *in vitro*. However, rapid adsorption is not sufficient for rapid conversion of LA to SA. For example, the addition of SP-A to BLES increases its rate of adsorption but limits the rate of conversion. The role of SP-A in maintaining LA integrity was also confirmed by experiments in which SP-A was degraded using proteases and by cycling experiments using LA from SP-A knockout mice (Veldhuizen *et al.*, 1994; 1996A; Ikegami *et al.*, 1998). Collectively, studies on conversion using surface area cycling have led to the following model for *in vitro* aggregate conversion. During the increase in surface area, LA adsorb to the air liquid interface, and during the subsequent decrease in surface area LA can either be reformed via a process requiring SP-A or else SA are formed. The balance between LA reformation and formation of SA is influenced by the composition of the LA and by "convertase" activity. Whether similar processes occur *in vivo* is unknown. Nevertheless, surface area cycling of samples from *in vivo* situations, such as lung injury, has proven useful in characterizing surfactant properties (Higuchi *et al.*, 1992; Veldhuizen *et al.*, 1997). However, considering the current state of our knowledge of these mechanisms, the *in vivo* significance of the results can be difficult to establish.

Based on the *in vitro* observations, several studies have extended investigation of aggregate conversion to the *in vivo* situation (Veldhuizen *et al.*, 1996B; 1997). *In vivo* surface area varies during respiration, depending on tidal volume. Mechanically ventilated rabbits have been used to demonstrate that trace doses of radio-labelled LA convert to SA *in vivo* and that the rate depends on tidal volume (i.e. surface area) during ventilation (Veldhuizen *et al.*, 1996B). This result, combined with the observation that SA formation initiates with the onset of air breathing after birth (Stevens *et al.*, 1987), provides strong evidence for the concept that, similar to *in vitro*, a changing surface area of the air liquid interface of the alveoli is required for aggregate conversion *in vivo*. There is also evidence that SP-A plays a role in maintaining the integrity of LA *in vivo*. SP-A deficient mice have a lower percentage of LA than their wildtype counterparts (Korfhagen *et al.*, 1998; Ikegami *et al.*, 1998). It should be noted however that these differences in pool sizes are relatively minor and further, more definitive evidence for the role of SP-A in this process is required. As mentioned above, there is no direct *in vivo* evidence arguing either in favour of or against the involvement of a serine-active enzyme in the conversion process.

4.4. Re-uptake and degradative pathways

Surfactant uptake from the airspace is accomplished predominantly by two cell types, the alveolar macrophages and the type II cell (Wright, 1990; Wright and Dobbs, 1991). Studies with radio-labelled and fluorescence-labelled lipids suggest that most of the surfactant is taken up by type II cells and is both recycled into lamellar bodies and degraded by lysosomes (~50% each). Surfactant recycling into lamellar bodies has been studied with isolated type II cells as well as whole lung. The pathways involved in the uptake of surfactant lipids by type II cells include both clathrin mediated and actin mediated pathways (Tsilibary and Williams, 1983; Rose *et al.*, 1999; Ruckert *et al.*, 2003).

The role of the macrophage in surfactant uptake is, in terms of amounts, relatively minor. Rider and colleagues estimated that, in rabbits, 20% of DPPC was cleared from the lung through uptake by macrophages (Rider *et al.*, 1992). Although the percentage is low, recent studies indicate that this clearance of surfactant is important metabolically. The most convincing evidence for the importance of macrophages in surfactant metabolism comes, somewhat surprisingly, from the studies on the knockout mouse for granulocyte macrophage colony stimulating factor (GM-CSF) (Dranoff *et al.*, 1994). The phenotype of this null mutation shows a progressive accumulation of surfactant in the airspaces which has subsequently been related to impaired surfactant clearance by macrophages (Dranoff *et al.*, 1994). SP-D KO mice also accumulate surfactant in the alveoli and this may also be related to macrophage function (Botas *et al.*, 1998; Ikegami *et al.*, 2000; Wert *et al.*, 2000) Thus although the relative contribution of macrophages is small, this clearance pathway appears important in maintaining normal pool sizes of surfactant lipids in the alveolar space.

5. CLINICAL SITUATIONS IN WHICH SURFACTANT METABOLISM IS ALTERED

Considering the importance of surfactant for normal lung function, it should not be surprising that alterations in surfactant may lead to, and are certainly associated with, a number of clinical conditions (Griese, 1999). The prototypic clinical condition involving the surfactant system is NRDS (Avery and Mead 1959; Avery, 2000; Jobe and Ikegami, 2000). The fundamental basis for this syndrome is premature birth with surfactant deficiency of the immature lung accounting for mechanism of lung dysfunction. It should be noted that, with NRDS, surfactant metabolism is inadequate, rather than dysfunctional. The obvious requirement for more surfactant has prompted investigation of potential mechanisms for stimulating endogenous surfactant phospholipid synthesis prior to the premature delivery. Maternal administration of glucocorticoids 24–48 h prior to pending parturition has proven an effective prophylactic treatment for NRDS. The synthetic glucocorticoids used pass through the placenta and act on the developing fetal lung. Animal studies indicate two potential mechanisms: increased phosphocholine cytidylyltransferase activity to produce CDP-choline and enhanced fatty acid synthetase activity to provide palmitic acid (Rooney, 1989). While the overall mechanisms remain unclear, fatty acid stimulation of cytidylyltransferase through promoting translocation from cytosol to the endoplasmic reticulum may provide part of the stimulus for increased PC synthesis (Cockshutt and Possmayer, 1992; Batenburg and Haagsman, 1998). It should also be mentioned that glucocorticoids enhance maturation of the whole lung and that steroid effects on collagen and/or elastin synthesis and on generation of the surfactant apoproteins clearly contribute to the overall beneficial effects (Mendelson and Boggaram, 1991). As indicated earlier, surfactant supplementation of infants suffering from NRDS is also very effective therapy until endogenous surfactant synthetic pathways become established (Jobe and Ikegami, 2000; Yost and Soll, 2000).

The second common clinical scenario involving an altered surfactant system is acute lung injury (ALI), in combination with its more severe form, the acute respiratory distress syndrome (ARDS) (Lewis and Jobe, 1993; Lewis and Veldhuizen, 2003). These conditions, which are defined by a number of physiological criteria including poor blood oxygenation, can be caused by one or several initiating events such as infection, near drowning, smoke inhalation and trauma (Petty, 1990; Livingston *et al.*, 1995). With these patients, the surfactant system is both significantly altered in terms of composition and functionally impaired (Frerking *et al.*, 2001; Lewis and Veldhuizen, 2003). Compared to normal controls, surfactant from ALI/ARDS patients has decreased relative amounts of 16:0/16:0-PC and PG, with increased levels of lyso-PC, PC, SM, and sometimes PI and other lipids. Patients also exhibit relatively higher levels of SA compared to LA. In addition, patients with ARDS have decreases in surfactant-associated proteins, particularly SP-B, whereas non-surfactant proteins, mostly serum proteins, are markedly increased. The injury-induced changes in the metabolic processes causing these surfactant alterations have been extensively studied and appear to be multi-factorial. For example, changes in lipid composition may be the result of altered type II cell metabolism, as it has been shown that inflammatory mediators such as tumor necrosis factor can affect the synthesis of surfactant (Arias-Diaz *et al.*, 1994; Pryhuber *et al.*, 1998). In addition, surfactant phospholipid composition may be affected by phospholipases in the alveolar space (Hite *et al.*, 1998). The decrease in the percentage LA with ALI/ARDS patients may be the result of reduced levels of SP-A, since this protein is important for LA integrity (Veldhuizen *et al.*, 1994; 1996A). More importantly, it is likely that mechanical ventilation, which is a necessary supportive therapy for these patients, generates an increased change in surface area that may promote the conversion of LA into SA. Evidence for this mechanism for the altered aggregate levels in ALI arises from animal studies in which spontaneously breathing animals had a normal percentage LA but as soon as animal were mechanically ventilated this percentage dropped (Veldhuizen *et al.*, 1996B). It was hypothesized that the spontaneously breathing animals breathed with a lower tidal volume to preserve their LA. Consistent with these animal studies, a large clinical trial in patients with ARDS demonstrated that low tidal volume ventilation resulted in a significantly lower mortality compared to traditional volumes (Brower *et al.*, 2000). Unfortunately, surfactant aggregates were not analyzed in these patients.

A relatively rare but interesting disorder with altered surfactant metabolism is alveolar proteinosis (Summers, 1966; Maygarden *et al.*, 2001). With respect to surfactant metabolism, alveolar proteinosis is defined by the accumulation of large amounts of surfactant lipids and proteins within the airspaces. Conventional therapy for alveolar proteinosis is periodic lung lavage to remove excess material from the airspace. The discovery that GM-CSF deficient mice develop a murine version of alveolar proteinosis has led to elucidation of some of the mechanisms involved in this disease (Dranoff *et al.*, 1994). Specifically, neutralizing anti-bodies against GM-CSF have been detected in some of these patients (Thomassen *et al.*, 2000). The lack of normal surfactant uptake by alveolar macrophages, as described for the GM-CSF deficient mouse model may ultimately account for the surfactant accumulation (Yoshida *et al.*, 2001).

Several other clinical conditions, such as asthma, lung transplantation, congenital SP-B deficiency and pneumonia, may involve altered surfactant metabolism that is either a consequence of the disease or a contributing component of the pathophysiology (Veldhuizen *et al.*, 1993B; Nogee *et al.*, 1994; Griese, 1999; Hohlfeld *et al.*, 1999; Ochs *et al.*, 2000; Nogee *et al.*, 2002).

6. CONCLUDING REMARKS

Since surfactant performs a critical physiological role in lung function it is important to understand the mechanisms involved in the metabolism of surfactant. Since the discovery of pulmonary surfactant in the 1950's tremendous progress has been made in surfactant research, with the highlight being the establishment of surfactant therapy for premature infants. Still, numerous questions about surfactant metabolism, structure and function remain. The exciting progress of the last several years has been largely propagated through insightful use of various technological advances. The development of transgenic and knock-out animals has led to new insights into surfactant metabolism and, more often than not, phenotypic abnormalities of a transgenic animal have been surprisingly different than those anticipated based on previous studies. Novel biophysical techniques to investigate surface film formation have altered our thinking regarding surfactant function. New analytical techniques have provided more detailed compositional and structural information on surfactant. Further understanding of surfactant metabolism will no doubt further depend on these, and other, technological advancements. But, even more importantly, studies on the surfactant system encompass such a variety of scientific disciplines, including biophysics, biochemistry, physiology, medicine, endocrinology and molecular biology, that ultimately our understanding of surfactant metabolism will further develop based on cross disciplinary approaches.

REFERENCES

Arias-Diaz, J., Vara, E., Garcia, C., and Balibrea, J. L. 1994. Tumor Necrosis factor-alpha-induced inhibition of phosphatidylcholine synthesis by human type 2 Pneumocytes is partially mediated by Prostaglandins. *J. Clin. Invest.* **94**:244-250.

Avery, M. E. and Mead, J. 1959. Surface properties in relation to atelectasis and hyaline membrane disease. *Am. J. Dis. Child* **97**:517-523.

Avery, M. E. 2000. Surfactant deficiency in hyaline membrane disease: the story of discovery. *Am. J. Respir. Crit. Care Med.* **161**:1074-1075

Bangham, A. D., Morley, C. J. and Phillips, M. C. 1979. The physical properties of an effective lung surfactant. *Biochim. Biophys. Acta* **573**:552-556.

Baritussio, A., Bellina, L., Carraro, R., Rossi, A., Enzi, G., Magoon, M. W., and Mussini, I. 1984. Heterogeneity of alveolar surfactant in the rabbit: composition, morphology, and labelling of subfractions isolated by centrifugation of lung lavage. *Eur. J. Clin. Invest.* **14**:24-29.

Barr, F., Clark, H., and Hawgood, S. 1998. Identification of a putative surfactant convertase in rat lung as a secreted serine carboxylesterase. *Am. J. Physiol* **274**:(Pt 1):L404-10.

Batenburg, J. J., den Breejen, J. N., Yost, R. W., Haagsman, H. P., and Van Golde, L. M. 1986. Glycerol 3-phosphate acylation in microsomes of type II cells isolated from adult rat lung. *Biochim. Biophys. Acta* **878**:301-309.

Batenburg, J. J. and Haagsman, H. P. 1998. The lipids of pulmonary surfactant: dynamics and interactions with proteins. *Prog. Lipid Res.* **37**:235-276.

Benson, B. J., Hawgood, S., and Williams, M. C. 1984. Role of apoprotein and calcium ions in surfactant function. *Exp. Lung Res.* **6**:223-236.

Bernhard, W., Postle, A. D., Rau, G. A., and Freihorst, J. 2001. Pulmonary and gastric surfactants. A comparison of the effect of surface requirements on function and phospholipid composition. *Comp. Biochem. Physiol. A Mol. Integr .Physiol.* **129**:173-182.

Botas, C., Poulain, F., Akiyama, J., Brown, C., Allen, L., Goerke, J., Clements, J., Carlson, E., Gillespie, A. M., Epstein, C., and Hawgood, S. 1998. Altered surfactant homeostasis and alveolar type II cell morphology in mice lacking surfactant protein d. *Proc. Natl. Acad. Sci. USA* **95**:11869-11874.

Brackenbury, A. M., Malloy, J. L., McCaig, L. A., Yao, L. J., Veldhuizen, R. A., and Lewis, J. F. 2002. Evaluation of alveolar surfactant aggregates in vitro and in vivo. *Eur. Respir. J* **19**:41-46.

Brower, R. G., Matthay, M. A., Morris, A., Schoenfeld, D., Thompson, B. T., and Wheeler, A. 2000. Ventilation with lower tidal volumes as compared with traditional tidal volumes for acute lung injury and the acute respiratory distress syndrome. The Acute Respiratory Distress Syndrome Network. *N. Engl. J. Med.* **342**:1301-1308.

Chan, F., Harding, P. G. R., Wong, T., Fellows, G. F., and Possmayer, F. 1983. Cellular distribution of enzymes involved in phosphatidylcholine synthesis in developing rat lung. *Can. J. Biochem.* **61**:107-114.

Chander, A. and Wu, R. D. 1991. In vitro fusion of lung lamellar bodies and plasma membrane is augmented by lung synexin. *Biochim. Biophys. Acta* **1086**:157-166.

Clements, J. A. 1957. Surface tension of lung extracts. *Proc. Soc. Exp. Biol. Med.* **95**:170-172.

Clements, J. A. 1977. Functions of the alveolar lining. *Am. Rev. Resp. Dis.* **115**:67-71.

Cockshutt, A. M. and Possmayer, F. 1992. Metabolism of surfactant lipids and proteins in developing lung. In *Pulmonary Surfactant: from Molecular Biology to Clinical Practice* (Robertson, B., Van Golde, L. M. G., and Batenburg, J. J., eds.), Elsevier Science Publishers, Amsterdam, pp. 339-377.

Cogo, P. E., Carnielli, V. P., Bunt, J. E., Badon, T., Giordano, G., Zacchello, F., Sauer, P. J., and Zimmermann, L. J. 1999. Endogenous surfactant metabolism in critically ill infants measured with stable isotope labeled fatty acids. *Pediatr. Res.* **45**:242-246.

Crouch, E. C. 1998. Structure, biologic properties, and expression of surfactant protein D (SP-D). *Biochim. Biophys. Acta* **1408**:278-289.

Dhand, R., Sharma, V. K., Teng, A. L., Krishnasamy, S., and Gross, N. J. 1998. Protein-lipid interactions and enzyme requirements for light subtype generation on cycling reconstituted surfactant. *Biochem. Biophys. Res. Commun.* **244**:712-719.

Doyle, I. R., Morton, S., Crockett, A. J., Barr, H. A., Davidson, K. G., Jones, M. J., Jones, M. E., and Nicholas, T. E. 2000. Composition of alveolar surfactant changes with training in humans. *Respirology*. **5**:211-220.

Dranoff, G., Crawford, A. D., Sadelain, M., Ream, B., Rashid, A., Bronson, R. T., Dickersin, G. R., Bachurski, C. J., Mark, E. L., and Whitsett, J. A. 1994. Involvement of granulocyte-macrophage colony-stimulating factor in pulmonary homeostasis. *Science* **264**:713-716.

Egberts, J., Beintema-Dubbeldam, A., and de Boers, A. 1987. Phosphatidylinositol and not phosphatidylglycerol is the important minor phospholipid in rhesus-monkey surfactant. *Biochim.Biophys.Acta* **919**:90-92.

Enhorning, G. 1977. Pulsating bubble technique for evaluating pulmonary surfactant. *J. Appl. Physiol.* **43**:198-203.

Frerking, I., Gunther, A., Seeger, W., and Pison, U. 2001. Pulmonary surfactant: functions, abnormalities and therapeutic options. *Intensive Care Med.* **27**:1699-1717.

Froh, D., Ballard, P. L., Williams, M. C., Gonzales, J., Goerke, J., Odom, M. W., and Gonzales, L. W. 1990. Lamellar bodies of cultured human fetal lung: content of surfactant protein A (SP-A), surface film formation and structural transformation in vitro. *Biochim. Biophys. Acta* **1052**:78-89.

Goerke, J. 1998. Pulmonary surfactant: functions and molecular composition. *Biochim. Biophys. Acta* **1408**:79-89.

Griese, M. 1999. Pulmonary surfactant in health and human lung diseases: state of the art. *Eur. Respir. J.* **13**:1455-1476.

Gross, N. J. and Narine, K. R. 1989. Surfactant subtypes of mice: metabolic relationships and conversion in vitro. *J. Appl. Physiol.* **67**:414-421.

Gross, N. J. and Schultz, R. M. 1990. Serine proteinase requirement for the extra-cellular metabolism of pulmonary surfactant. *Biochim. Biophys. Acta* **1044**:222-230.

Gross, N. J. and Schultz, R. M. 1992. Requirements for extracellular metabolism of pulmonary surfactant: tentative identification of serine protease. *Am. J. Physiol.* **262**: L446-L453.

Gross, N. J. 1995. Extracellular metabolism of pulmonary surfactant: the role of a new serine protease. *Annu. Rev. Physiol.* **57**:135-150.

Gunther, A., Schmidt, R., Feustel, A., Meier, U., Pucker, C., Ermert, M., and Seeger, W. 1999. Surfactant subtype conversion is related to loss of surfactant apoprotein B and surface activity in large surfactant aggregates. Experimental and clinical studies. *Am. J. Respir. Crit. Care. Med.* **159**:244-251.

Haagsman, H. P. and Van Golde, L. M. 1991. Synthesis and assembly of lung surfactant. *Annu. Rev. Physiol.* **53**:441-464.

Haagsman, H. P. and Diemel, R. V. 2001. Surfactant-associated proteins: functions and structural variation. *Comp. Biochem. Physiol. A Mol. Integr. Physiol.* **129**:91-108.

Haller, T., Pfaller, K., and Dietl, P. 2001. The conception of fusion pores as rate-limiting structures for surfactant secretion. *Comp. Biochem. Physiol. A Mol.Integr. Physiol.* **129**:227-231.

Hallman, M. and Gluck, L. 1975. Phosphatidylglycerol in lung surfactant. II. Subcellular distribution and mechanism of biosynthesis in vitro. *Biochim. Biophys. Acta* **409**:172-191.

Hallman, M., Kulovich, M., Kirkpatrick, E., Sugarman, R. G., and Gluck, L. 1976. Phosphatidylinositol and phosphatidylglycerol in amniotic fluid: indices of lung maturity. *Am. J. Obstet. Gynecol.* **125**:613-617.

Hallman, M., Feldman, B. H., Kirkpatrick, E., and Gluck, L. 1977. Absence of phosphatidylglycerol (PG) in respiratory distress syndrome in the newborn. Study of the minor surfactant phospholipids in newborns. *Pediatr. Res.* **11**:714-720.

Hallman, M., Epstein, B. L., and Gluck, L. 1981. Analysis of labeling and clearance of lung surfactant phospholipids in rabbit. Evidence of bidirectional surfactant flux between lamellar bodies and alveolar lavage. *J. Clin. Invest.* **68**:742-751.

Hawgood, S., Derrick, M., and Poulain, F. 1998. Structure and properties of surfactant protein B. *Biochim. Biophys. Acta.* **1408**:150-160.

Heeley, E. L., Hohlfeld, J. M., Krug, N., and Postle, A. D. 2000. Phospholipid molecular species of bronchoalveolar lavage fluid after local allergen challenge in asthma. *Am. J. Physiol. Lung Cell. Mol. Physiol.* **278**:L305-L311.

Higuchi, R., Lewis, J., and Ikegami, M. 1992. In vitro conversion of surfactant subtypes is altered in alveolar surfactant isolated from injured lungs. *Am. Rev. Respir. Dis.* **145**:1416-1420.

Hite, R. D., Seeds, M. C., Jacinto, R. B., Balasubramanian, R., Waite, M., and Bass, D. 1998. Hydrolysis of surfactant-associated phosphatidylcholine by mammalian secretory phospholipases A_2. *Am. J. Physiol.* **275**:L740-L747.

Hohlfeld, J. M., Ahlf, K., Enhorning, G., Balke, K., Erpenbeck, V. J., Petschallies, J., Hoymann, H. G., Fabel, H., and Krug, N. 1999. Dysfunction of pulmonary surfactant in asthmatics after segmental allergen challenge. *Am. J. Respir. Crit. Care Med.* **159**:1803-1809.

Holm, B. A., Notter, R. H., Siegle, J., and Matalon, S. 1985. Pulmonary physiological and surfactant changes during injury and recovery from hyperoxia. *J. Appl. Physiol.* **59**:1402-1409.

Holm, B. A., Wang, Z., Egan, E. A., and Notter, R. H. 1996. Content of dipalmitoyl phosphatidylcholine in lung surfactant: ramifications for surface activity. *Pediatr. Res.* **39**:805-811.

Ikegami, M., Korfhagen, T. R., Whitsett, J. A., Bruno, M. D., Wert, S. E., Wada, K., and Jobe, A. H. 1998. Characteristics of surfactant from SP-A-deficient mice. *Am. J. Physiol.* **275**:L247-L254.

Ikegami, M., Whitsett, J. A., Jobe, A., Ross, G., Fisher, J., and Korfhagen, T. 2000. Surfactant metabolism in SP-D gene-targeted mice. *Am. J. Physiol. Lung Cell. Mol. Physiol.* **279**:L468-L476.

Inchley, K., Cockshutt, A., Veldhuizen, R., and Possmayer, F. 1999 Dissociation of surfactant protein B from canine surfactant large aggregates during formation of small surfactant aggregates by in vitro surface area cycling. *Biochem. Biophys. Acta.* **1440**:49-58.

Jain, D., Dodia, C., Bates, S. R., Hawgood, S., Poulain, F. R., and Fisher, A. B. 2003. SP-A is necessary for increased clearance of alveolar DPPC with hyperventilation or secretagogues. *Am. J. Physiol. Lung Cell Mol. Physiol.* **284**:L759-L765.

Jobe, A. H., Ikegami, M., Seidner, S. R., Pettenazzo, A., and Ruffini, L. 1989. Surfactant phosphatidylcholine metabolism and surfactant function in preterm, ventilated lambs. *Am. Rev. Respir. Dis.* **139**:352-359.

Jobe, A. H. and Ikegami, M. 2000. Lung development and function in preterm infants in the surfactant treatment era. *Annu. Rev. Physiol.* **62**:825-846.

Johansson, J. 1998. Structure and properties of surfactant protein C. *Biochim. Biophys. Acta* **1408**:161-172.

Kahn, M. C., Anderson, G. J., Anyan, W. R., and Hall, S. B. 1995. Phosphatidylcholine molecular species of calf lung surfactant. *Am. J. Physiol.* **269**:L567-L573.

Kalina, M. and Socher, R. 1990. Internalization of pulmonary surfactant into lamellar bodies of cultured rat pulmonarytype II cells. *J. Histochem. Cytochem.* **38**:483-492.

Kalina, M., McCormack, F. X., Crowley, H., Voelker, D. R., and Mason, R. J. 1993. Internalization of surfactant protein A (SP-A) into lamellar bodies of rat alveolar type II cells in vitro. *J. Histochem. Cytochem.* **41**:57-70.

Kennedy, E. P. 1961. Biosynthesis of complex lipids. *Fed. Proc.* **20**:934-940.

Klein, J. M., McCarthy, T. A., Dagle, J. M., and Snyder, J. M. 2002. Antisense inhibition of surfactant protein A decreases tubular myelin formation in human fetal lung in vitro. *Am. J. Physiol. Lung Cell Mol. Physiol.* **282**:L386-L393.

Korfhagen, T. R., Bruno, M. D., Ross, G. F., Huelsman, K. M., Ikegami, M., Jobe, A. H., Wert, S. E., Stripp, B. R., Morris, R. E., Glasser, S. W., Bachurski, C. J., Iwamoto, H. S., and Whitsett, J. A. 1996. Altered surfactant function and structure in SP-A gene targeted mice. *Proc. Natl. Acad. Sci. USA* **93**:9594-9599.

Korfhagen, T. R., LeVine, A. M., and Whitsett, J. A. 1998. Surfactant protein A (SP-A) gene targeted mice. *Biochim. Biophys. Acta* **1408**:296-302.

Krishnasamy, S., Gross, N. J., Teng, A. L., Schultz, R. M., and Dhand, R. 1997. Lung "surfactant convertase" is a member of the carboxylesterase family. *Biochem. Biophys. Res. Commun.* **235**:180-4.

Kuroki, Y., Mason, R. J., and Voelker, D. R. 1988. Pulmonary surfactant apoprotein A structure and modulation of surfactant secretion by rat alveolar type II cells. *J. Biol. Chem.* **263**:3388-3394.

Lewis, J. F., Ikegami, M., and Jobe, A. H. 1990. Altered surfactant function and metabolism in rabbits with acute lung injury. *J. Appl. Physiol.* **69**:2303-2310.

Lewis, J. F. and Jobe, A. H. 1993. Surfactant and the adult respiratory distress syndrome. *Am. Rev. Respir. Dis.* **147**:218-233.

Lewis, J. F. and Veldhuizen, R. 2003. The role of exogenous surfactant in the treatment of acute lung injury. *Annu. Rev. Physiol.* **65**:613-642.

Li, J., Marsh, J. J., and Spragg, R. G. 2002. Effect of CTP:phosphocholine cytidylyltransferase overexpression on the mouse lung surfactant system. *Am. J. Respir. Cell Mol. Biol.* **26**:709-715.

Liu, L. 1999. Calcium-dependent self-association of annexin II: a possible implication in exocytosis. *Cell Signal* **11**:317-324.

Livingston, D. H., Mosenthal, A. C., and Deitch, E. A. 1995. Sepsis and multiple organ dysfunction syndrome: a clinical- mechanistic overview. *New Horiz.* **3**:257-264.

Magoon, M. W., Wright, J. R., Baritussio, A., Williams, M. C., Goerke, J., Benson, B. J., Hamilton, R. L., and Clements, J. A. 1983. Subfractionation of lung surfactant. Implications for metabolism and surface activity. *Biochim. Biophys. Acta.* **750**:18-31.

Malloy, J., McCaig, L., Veldhuizen, R., Yao, L. J., Joseph, M., Whitsett, J., and Lewis, J. 1997. Alterations of the endogenous surfactant system in septic adult rats. *Am. J. Respir. Crit. Care Med.* **156**:617-623.

Mason, R. J., Nellenbogen, J., and Clements, J. A. 1976. Isolation of disaturated phosphatidylcholine with osmium tetroxide. *J. Lipid Res.* **17**:281-284.

Mason, R. J. and Voelker, D. R. 1998. Regulatory mechanisms of surfactant secretion. *Biochim. Biophys. Acta.* **1408**:226-240.

Massaro, G. D. and Massaro, D. 1983. Morphologic evidence that large inflations of the lung stimulate secretion of surfactant. *Am. Rev. Respir. Dis.* **127**:235-236.

Maygarden, S. J., Iacocca, M. V., Funkhouser, W. K., and Novotny, D. B. 2001. Pulmonary alveolar proteinosis: a spectrum of cytologic, histochemical, and ultrastructural findings in bronchoalveolar lavage fluid. *Diagn. Cytopathol.* **24**:389-395.

McCormack, F. X. 1998. Structure, processing and properties of surfactant protein A. *Biochim. Biophys. Acta* **1408**:109-131.

McCormack, F. X. and Whitsett, J. A. 2002. The pulmonary collectins, SP-A and SP-D, orchestrate innate immunity in the lung. *J. Clin. Invest.* **109**:707-712.

Mendelson, C. R. and Boggaram, V. 1991. Hormonal control of the surfactant system in fetal lung. *Annu. Rev. Physiol.* **53**:415-440.

Nag, K., Munro, J. G., Hearn, S. A., Rasmusson, J., Petersen, N. O., and Possmayer, F. 1999. Correlated atomic force and transmission electron microscopy of nanotubular structures in pulmonary surfactant. *J. Struct. Biol.* **126**:1-15.

Nanjundan, M. and Possmayer, F. 2003. Phosphatidic acid phosphatase and lipid phosphate phosphohydrolase. *Am. J. Physiol. Lung Cell Mol. Physiol.* **284**:L1-L23.

Nicholas, T. E. and Barr, H. A. 1983. The release of surfactant in rat lung by brief periods of hyperventilation. *Respir. Physiol.* **52**:69-83.

Nogee, L. M., Garnier, G., Dietz, H. C., Singer, L., Murphy, A. M., deMello, D. E., and Colten, H. R. 1994. A mutation in the surfactant protein B gene responsible for fatal neonatal respiratory disease in multiple kindreds. *J. Clin. Invest.* **93**:1860-1863.

Nogee, L. M., Dunbar, A. E., III, Wert, S., Askin, F., Hamvas, A., and Whitsett, J. A. 2002. Mutations in the surfactant protein C gene associated with interstitial lung disease. *Chest* **121**:20S-21S.

Ochs, M., Fehrenbach, H., Nenadic, I., Bando, T., Fehrenbach, A., Schepelmann, D., Albes, J. M., Wahlers, T., and Richter, J. 2000. Preservation of intraalveolar surfactant in a rat lung ischaemia/reperfusion injury model. *Eur. Respir. J.* **15**:526-531.

Orgeig, S. and Daniels, C. B. 2001. The roles of cholesterol in pulmonary surfactant: insights from comparative and evolutionary studies. *Comp. Biochem. Physiol. A. Mol. Integr. Physiol.* **129**:75-89.

Persson, A., Chang, D., Rust, K., Moxley, M., Longmore, W., and Crouch, E. 1989. Purification and biochemical characterization of CP4 (SP-D), a collagenous surfactant-associated protein. *Biochemistry* **28**:6361-6367.

Petty, T. L. 1990. Acute respiratory distress syndrome (ARDS). *Dis.Mon.* **36**:1-58.

Piknova, B., Schief, W. R., Vogel, V., Discher, B. M., and Hall, S. B. 2001. Discrepancy between phase behavior of lung surfactant phospholipids and the classical model of surfactant function. *Biophys. J.* **81**:2172-2180.

Piknova, B., Schram, V., and Hall, S. B. 2002. Pulmonary surfactant: phase behavior and function. *Curr. Opin. Struct. Biol.* **12**:487-494.

Possmayer, F. 1988. A proposed nomenclature for pulmonary surfactant-associated proteins. *Am. Rev. Respir. Dis.* **138**:990-998.

Possmayer, F., Nag, K., Rodriguez, K., Qanbar, R., and Schűrch, S. 2001. Surface activity in vitro: role of surfactant proteins. *Comp. Biochem. Physiol. A Mol. Integr. Physiol.* **129**:209-220.

Possmayer, F. 2004. Physicochemical aspects of pulmonary surfactant. In *Fetal and Neonatal Physiology* (3rd edn.), (Polin, R. A., Fox, W. W., and Abman, S. H., eds.), W.B. Saunders Company, Philadelphia, pp. 1014-1034.

Post, M., Batenburg, J. J., Schuurmans, E. A., Laros, C. D., and Van Golde, L. M. 1982. Lamellar bodies isolated from adult human lung tissue. *Exp. Lung Res.* **3**:17-28.

Post, M., Schuurmans, E. A., Batenburg, J. J., and Van Golde, L. M. 1983. Mechanisms involved in the synthesis of disaturated phosphatidylcholine by alveolar type II cells isolated from adult rat lung. *Biochim. Biophys. Acta* **750**:68-77.

Post, M., Batenburg, J. J., Van Golde, L. M., and Smith, B. T. 1984. The rate-limiting reaction in phosphatidylcholine synthesis by alveolar type II cells isolated from fetal rat lung. *Biochim. Biophys. Acta.* **795**:558-563.

Postle, A. D., Heeley, E. L., and Wilton, D. C. 2001. A comparison of the molecular species compositions of mammalian lung surfactant phospholipids. *Comp. Biochem. Physiol. A Mol. Integr. Physiol.* **129**:65-73.

Poulain, F. R., Allen, L., Williams, M. C., Hamilton, R. L., and Hawgood, S. 1992. Effects of surfactant apolipoproteins on liposome structure: implications for tubular myelin formation. *Am. J. Physiol.* **262**:L730-L739.

Pritchard, P. H. and Vance, D. E. 1981. Choline metabolism and phosphatidylcholine biosynthesis in cultured rat hepatocytes. *Biochem. J.* **196**:261-267.

Pryhuber, G. S., Bachurski, C., Hirsh, R., Bacon, A., and Whitsett, J. A. 1998. Tumor necrosis factor-alpha decreases surfactant protein B mRNA in murine lung. *Am. J. Physiol.* **270**:L714-L721.

Putz, G., Goerke, J., and Clements, J. A. 1994. Surface activity of rabbit pulmonary surfactant subfractions at different concentrations in a captive bubble. *J. Appl. Physiol.* **77**:597-605.

Quintero, O. A., Korfhagen, T. R., and Wright, J. R. 2002. Surfactant protein A regulates surfactant phospholipid clearance after LPS-induced injury in vivo. *Am. J. Physiol. Lung Cell Mol. Physiol.* **283**:L76-L85.

Rider, E. D., Ikegami, M., and Jobe, A. H. 1992. Localization of alveolar surfactant clearance in rabbit lung cells. *Am. J. Physiol.* **263**:L201-L209.

Rooney, S. A., Page-Roberts, B. A., and Motoyama, E. K. 1975. Role of lamellar inclusions in surfactant production: studies on phospholipid composition and biosynthesis in rat and rabbit lung subcellular fractions. *J. Lipid. Res.* **16**:418-425.

Rooney, S. A. 1985. The surfactant system and lung phospholipid biochemistry. *Am. Rev. Respir. Dis.* **131**:439-460.

Rooney, S.A. 1989. Fatty acid biosynthesis in the fetal lung *Am. J. Physiol.* **257**:L195-201.

Rooney, S. A. 2001. Regulation of surfactant secretion. *Comp. Biochem. Physiol. A Mol. Integr. Physiol.* **129**:233-243.

Rose, F., Kurth-Landwehr, C., Sibelius, U., Reuner, K. H., Aktories, K., Seeger, W., and Grimminger, F. 1999. Role of actin depolymerization in the surfactant secretory response of alveolar epithelial type II cells. *Am. J. Respir. Crit. Care Med.* **159**:206-212.

Rubins, J. B., Panchenko, M., Shannon, T. M., and Dickey, B. F. 1992. Identification of ras and ras-related low-molecular-mass GTP-binding proteins associated with rat lung lamellar bodies. *Am. J. Respir. Cell Mol. Biol.* **6**:253-259.

Ruckert, P., Bates, S. R., and Fisher, A. B. 2003. Role Of clathrin- and actin-dependent endocytotic pathways in lung phospholipid uptake. *Am. J. Physiol. Lung Cell Mol. Physiol.* **284**:L981-L989.

Sanders, R. L., Hassett, R. J., and Vatter, A. E. 1980. Isolation of lung lamellar bodies and their conversion to tubular myelin figures in vitro. *Anat. Rec.* **198**:485-501.

Savov, J., Silbajoris, R., and Young, S. L. 1999. Mechanical ventilation of rat lung: effect on surfactant forms. *Am. J. Physiol.* **277**:L320-L326.

Schmitz, G. and Muller, G. 1991. Structure and function of lamellar bodies, lipid-protein complexes involved in storage and secretion of cellular lipids. *J. Lipid. Res.* **32**:1539-1570.

Schűrch, S., Bachofen, H., Goerke, J., and Possmayer, F. 1989. A captive bubble method reproduces the in situ behavior of lung surfactant monolayers. *J. Appl. Physiol.* **67**:2389-2396.

Schűrch, S., Possmayer, F., Cheng, S., and Cockshutt, A. M. 1992. Pulmonary SP-A enhances adsorption and appears to induce surface sorting of lipid extract surfactant. *Am. J. Physiol.* **263**:L210-L218.

Schűrch, S., Qanbar, R., Bachofen, H., and Possmayer, F. 1995. The surface-associated surfactant reservoir in the alveolar lining. *Biol. Neonate.* **67 Suppl 1**:61-76.

Schűrch, S., Green, F. H., and Bachofen, H. 1998. Formation and structure of surface films: captive bubble surfactometry. *Biochim. Biophys.Acta.* **1408**:180-202.

Shelley, S. A., Paciga, J. E., and Balis, J. U. 1984. Lung surfactant phospholipids in different animal species. *Lipids* **19**:857-862.

Spragg, R. G. and Li, J. 2000. Effect of phosphocholine cytidylyltransferase overexpression on phosphatidylcholine synthesis in alveolar type II cells and related cell lines. *Am. J. Respir. Cell Mol. Biol* **22**:116-124.

Stevens, P. A., Wright, J. R., and Clements, J. A. 1987. Changes in quantity, composition, and surface activity of alveolar surfactant at birth. *J. Appl. Physiol.* **63**:1049-1057.

Summers, J. E. 1966. Pulmonary alveolar proteinosis. Review of the literature with follow-up studies and report of two new cases. *Calif. Med.* **104**:428-436.

Suzuki, Y., Fujita, Y., and Kogishi, K. 1989. Reconstitution of tubular myelin from synthetic lipids and proteins associated with pig pulmonary surfactant. *Am. Rev. Respir. Dis.* **140**:75-81.

Thomassen, M. J., Yi, T., Raychaudhuri, B., Malur, A., and Kavuru, M. S. 2000. Pulmonary alveolar proteinosis is a disease of decreased availability of GM-CSF rather than an intrinsic cellular defect. *Clin. Immunol.* **95**:85-92.

Tsilibary, E. C. and Williams, M. C. 1983. Actin and secretion of surfactant. *J. Histochem. Cytochem.* **31**:1298-1304.

Ueda, T., Ikegami, M., and Jobe, A. 1994. Surfactant subtypes: In vitro conversion, in vivo function, and effects of serum proteins. *Am. J. Respir. Crit. Care Med.* **149**:1254-1259.

Van Bree, L., Haagsman, H. P., Van Golde, L. M., and Rombout, P. J. 1988. Phosphatidylcholine synthesis in isolated type II pneumocytes from ozone-exposed rats. *Arch. Toxicol.* **61**:224-228.

Veldhuizen, R. A., Inchley, K., Hearn, S. A., Lewis, J. F., and Possmayer, F. 1993A. Degradation of surfactant-associated protein B (SP-B) during in vitro conversion of large to small surfactant aggregates. *Biochem. J.* **295**:141-147.

Veldhuizen, R. A., Lee, J., Sandler, D., Hull, W., Whitsett, J. A., Lewis, J., Possmayer, F., and Novick, R. J. 1993B. Alterations in pulmonary surfactant composition and activity after experimental lung transplantation. *Am. Rev. Respir. Dis.* **148**:208-215.

Veldhuizen, R. A. W., Hearn, S. A., Lewis, J. F., and Possmayer, F. 1994. Surface-area cycling of different surfactant preparations: SP-A and SP-B are essential for large-aggregate integrity. *Biochem. J.* **300**:519-524.

Veldhuizen, R. A. W., Yao, L.-J., Hearn, S. A., Possmayer, F., and Lewis, J. F. 1996A. Surfactant-associated protein A is important for maintaining large aggregate forms during surface area cycling. *Biochem. J.* **313**:835-840.

Veldhuizen, R. A. W., Marcou, J., Yao, L.-J., McCaig, L., Ito, Y., and Lewis, J. F. 1996B. Alveolar surfactant aggregate conversion in ventilated normal and injured rabbits. *Am. J. Physiol.* **270**:L152-L158.

Veldhuizen, R. A. W., Ito, Y., Marcou, J., Yao, L. J., McCaig, L., and Lewis, J. F. 1997. Effects of lung injury on pulmonary surfactant aggregate conversion in vivo and in vitro. *Am. J. Physiol.* **16**:L 872-L 878.

Veldhuizen, R. A., Yao, L., and Lewis, J. F. 1999. An examination of the different variables affecting surfactant aggregate conversion in vitro. *Exp. Lung Res.* **25**:127-141.

Veldhuizen, R. A. W., Tremblay, L. N., Govindarajan, A., Rozendaal, B. A. W. M., Haagsman, H. P., and Slutsky, A. S. 2000. Pulmonary surfactant is altered during mechanical ventilation of isolated rat lung. *Crit. Care Med.* **28**:2545-2551.

Veldhuizen, R. A., Welk, B., Harbottle, R., Hearn, S., Nag, K., Petersen, N., and Possmayer, F. 2002. Mechanical ventilation of isolated rat lungs changes the structure and biophysical properties of surfactant. *J. Appl. Physiol.* **92**:1169-1175.

Voorhout, W. F., Veenendaal, T., Haagsman, H. P., Verkleij, A. J., Van Golde, L. M., and Geuze, H. J. 1991. Surfactant protein A is localized at the corners of the pulmonary tubular myelin lattice. *J. Histochem. Cytochem.* **39**:1331-1336.

Voorhout, W. F., Veenendaal, T., Haagsman, H. P., Weaver, T. E., Whitsett, J. A., and Geuze, H. J. 1992. Intracellular processing of pulmonary surfactant protein B in an endosomal/lysosomal compartment. *Am. J. Physiol.* **263**:L479-L486.

Watkins, J. C. 1968. The surface properties of pure phospholipid in relation to those of lung extracts. *Biochim. Biophys. Acta.* **152**:293-306.

Weaver, T. E. 1998. Synthesis, processing and secretion of surfactant proteins B and C. *Biochim. Biophys. Acta* **1408**:173-179.

Weaver, T. E., Na, C. L., and Stahlman, M. 2002. Biogenesis of lamellar bodies, lysosome-related organelles involved in storage and secretion of pulmonary surfactant. *Semin. Cell Dev. Biol.* **13**:263-270.

Wert, S., Jones, T., Korfhagen, T., Fisher, J., and Whitsett, J. 2000. Spontaneous emphysema in surfactant protein D gene-targeted mice. *Chest.* **117**:248S.

Williams, M. C. 1982. Ultrastructure of tubular myelin and lamellar bodies in fast-frozen adult rat lung. *Exp. Lung Res.* **4**:37-46.

Williams, M. C., Hawgood, S., and Hamilton, R. L. 1991. Changes in lipid structure produced by surfactant proteins SP-A, SP-B, and SP-C. *Am. J. Resp. Cell. Mol. Biol.* **5**:41-50.

Wright, J. R. 1990. Clearance and recycling of pulmonary surfactant. *Am. J. Physiol.* **259**:L1-12.

Wright, J. R. and Dobbs, L. G. 1991. Regulation of pulmonary surfactant secretion and clearance. *Annu. Rev. Physiol.* **53**:395-414.

Yamada, T., Ikegami, M., and Jobe, A. H. 1990. Effects of surfactant subfractions on preterm rabbit lung function. *Pediatr. Res.* **27**:592-598.

Yoshida, M., Ikegami, M., Reed, J. A., Chroneos, Z. C., and Whitsett, J. A. 2001. GM-CSF regulates protein and lipid catabolism by alveolar macrophages. *Am. J. Physiol. Lung Cell Mol. Physiol.* **280**:L379-L386.

Yost, C. C. and Soll, R. F. 2000. Early versus delayed selective surfactant treatment for neonatal respiratory distress syndrome. *Cochrane. Database. Syst. Rev.* CD001456.

Young, S. L., Fram, E. K., and Larson, E. W. 1992. Three-dimensional reconstruction of tubular myelin. *Exp. Lung Res.* **18**:497-504.

Yu, S. H. and Possmayer, F. 2003. Lipid compositional analysis of pulmonary surfactant monolayers and monolayer-associated reservoirs. *J. Lipid. Res.* **44**:621-629.

Zimmermann, L. J., Hogan, M., Carlson, K. S., Smith, B. T., and Post, M. 1993. Regulation of phosphatidylcholine synthesis in fetal type II cells by CTP:phosphocholine cytidylyl-transferase. *Am. J. Physiol.* **264**:L575-L580.

Part V

Signal Transduction Processes

Chapter 12

Membrane Targeting in Secretion

Michael Schrader
Department of Cell Biology and Cell Pathology, University of Marburg
Robert Koch Str. 6, 35037 Marburg, Germany

Abstract

Regulated secretion and exocytosis require the selective packaging of regulated secretory proteins in secretory storage organelles and the controlled docking and fusion of these organelles with the plasma membrane. Secretory granule biogenesis involves sorting of secretory proteins and membrane components both at the level of the *trans*-Golgi network and the immature secretory granule. Sorting is thought to be mediated by selective protein aggregation and the interaction of these proteins with specific membrane domains. There is now considerable interest in the understanding of the complex lipid–protein and protein–protein interactions at the *trans*-Golgi network and the granule membrane. A role for lipid microdomains and associated sorting receptors in membrane targeting and granule formation is vividly discussed for (neuro)endocrine cells. In exocrine cells, however, little has been known of granule membrane composition and membrane protein function. With the cloning and characterization of granule membrane proteins and their interactions at the inner leaflet of zymogen granules of pancreatic acinar cells,

Abbreviations: CgA, B, chromogranin A, B; CPE, carboxypeptidase E; CV, condensing vacuole; GAG, glucosaminoglycan; GP, glycoprotein; GPI, glycosylphosphatidylinositol; ISGs, immature secretory granules; MβCD, methyl-β-cyclodextrin; MSGs, mature secretory granules; NSF, *N*-ethyl-maleimide-sensitive factor; PC, prohormone convertase; PI-PLC, phosphatidylinositol-specific phospholipase C; POMC, pro-opiomelanocortin; PRPs, proline-rich proteins; RSPs, regulated secretory proteins; SH3, src homology 3 domain binding motif; SNAP, soluble *N*-ethyl-maleimide-sensitive factor attachment protein; SNAREs, soluble *N*-ethyl-maleimide-sensitive factor attachment protein receptors; TGN, *trans*-Golgi network; THP, Tamm-Horsfall protein; ZGM, zymogen granule membrane.

Subcellular Biochemistry, Volume 37: Membrane Dynamics and Domains
Edited by Peter J. Quinn. Kluwer Academic / Plenum Publishers, New York, 2004

it is now possible to elucidate their function in membrane targeting and sorting of zymogens at the molecular level.

1. INTRODUCTION

1.1. The biogenesis of secretory granules

Regulated secretion of hormones, peptides or proteins occurs from specialized cell types such as endocrine, neuroendocrine and exocrine cells via the regulated secretory pathway. Regulated secretory proteins (RSPs) are packaged and stored in unique intracellular organelles, the secretory granules. When the specialized cells receive an external stimulus, the secretory granules undergo fusion with the plasma membrane and release their content into the extracellular environment. Secretory granule formation takes place at the *trans*-Golgi network (TGN) and involves two major sorting steps through which the RSPs are separated from constitutively released (or lysosomal) proteins and are packaged in a concentrated form within secretion granules. The initial step is a pH-dependent selective aggregation of RSPs in the slightly acidic, high-Ca^{2+} milieu of the TGN. The second sorting event is mediated by the interaction of the aggregates with specific membrane domains of the TGN, which then pinch off as immature secretory granules (ISGs). After various subsequent changes, for example in granule size or modification/processing of RSPs, which depend on the cell type and the RSPs, ISGs are converted to mature secretory granules (MSGs). Secretory granules can be actively transported via cytoskeletal elements to distinct domains of the plasma membrane (Rudolf *et al.*, 2001) where they are stored until membrane fusion and the release of RSPs is triggered by an external signal.

In an active model for sorting ('sorting by entry'), membrane binding of secretory proteins is assumed to depend on a 'sorting receptor' within the TGN (e.g., a transmembrane protein), and entry into forming granules is restricted to receptor-mediated trafficking. In an alternative, passive model ('sorting by retention'), aggregation of RSPs is the primary sorting event. In this model, a receptor is not required, and entry into the forming granule is not solely restricted to RSPs. The non-secretory proteins fail to aggregate and are removed from the maturing granules in a clathrin-dependent process (reviewed in Arvan and Castle, 1992; Arvan and Castle, 1998). In this case the ISG can also be a site for sorting. However, the retention of the granule-specific membrane components is likely to involve their interaction with RSPs. The extent of sorting at the level of the TGN and the ISG seems to differ between the various regulated secretory cell types.

1.2. Interaction of secretory proteins with the *trans*-Golgi/granule membrane

Secretory granule formation at the TGN has not yet been demonstrated to require cytoplasmic coat and adaptor proteins (Tooze, 1998; Thiele and Huttner, 1998b), which usually mediate clustering of receptor-ligand complexes and membrane budding. In the case of secretory granules, membrane deformation at the TGN is thought to occur by the adhesion of the membrane to an already curved particle, the condensed secretory cargo. Therefore, the initiation of RSP aggregation, the proper sorting of RSP and their tight association with the lipid bilayer are crucial for the generation of ISGs.

Most of the RSPs in endocrine and exocrine cells have been well characterized at the molecular level, and the selective aggregation of secretory proteins has been well documented (Chanat and Huttner, 1991; Leblond *et al.*, 1993; Rindler, 1998). In contrast, little information is available on the components of the granule membrane (Wagner and Williams, 1994; Wasmeier *et al.*, 2002). Furthermore, the specific interaction of RSPs and their aggregates with the TGN membrane and their ultimate sorting into ISGs is poorly understood, especially in exocrine systems. A key issue for many investigations nowadays is the unraveling of the molecular machinery required to form and mature secretory granules. Many investigators have therefore focused on the understanding of complex lipid-protein and protein-protein interactions at the TGN/granule membrane. Reviews on the aggregation, sorting, processing and maturation of RSPs have been published (Tooze, 1998; Arvan and Castle, 1998; Thiele and Huttner, 1998b; Glombik and Gerdes, 2000; Tooze *et al.*, 2001). This chapter is concerned with the process of membrane interaction/targeting of RSPs in endocrine and exocrine cells. A major part will focus on membrane targeting in the acinar cells of the exocrine pancreas.

2. SORTING IN ENDOCRINE CELLS

2.1. Biogenesis of secretory granules in (neuro)endocrine cells

In (neuro)endocrine cells evidence has been presented that many RSPs exist in two forms: a soluble subpopulation, which can associate to create aggregates, and a tightly membrane-associated form (Fricker *et al.*, 1990; Pimplikar and Huttner, 1992; Yoo, 1993). Based on these findings, RSPs are thought to associate with a specialized region of the TGN by assembling into oligomers with each other and with the membrane-associated subpopulation of RSPs (Thiele and

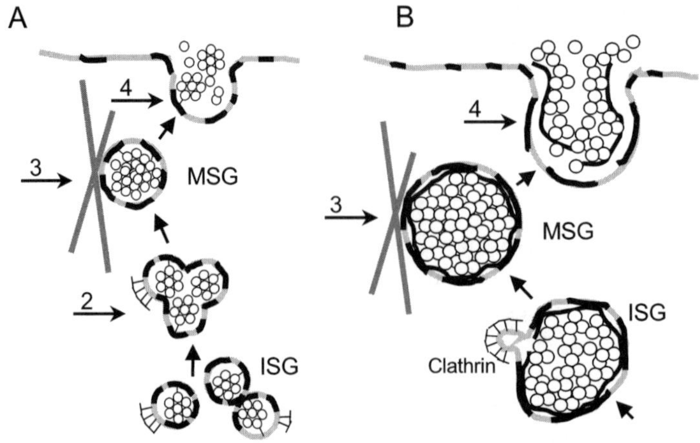

Figure 1. Biogenesis of secretory granules in neuroendocrine and exocrine cells. (A) Formation, homotypic fusion, maturation/remodeling and exocytosis of secretory granules in neuroendocrine cells (for details, see 2.). (B) Formation, maturation/remodeling and exocytosis of secretory granules in the acinar cells of the exocrine pancreas (for details, see 3.). The potential roles of lipid rafts (black lines) in secretory granule biogenesis are illustrated. Lipid rafts may mediate the binding/sorting of RSPs (white circles) to the TGN/ granule membrane (1). In zymogen granules of the exocrine pancreas, rafts may be involved in the attachment of a submembranous protein matrix (striped line) to the granule membrane (B). The granule matrix is supposed to mediate sorting (see 3.2.2). Lipid rafts may operate as platforms for the clustering and organization of the docking and fusion machinery required for homotypic fusion of ISG (2) or regulated exocytosis of MSG (4). Furthermore, rafts may mediate membrane interactions with the cytoskeleton/motor protein machinery (3). White circles in (B) represent protein complexes of RSPs. CV, condensing vacuole; ISG, immature secretory granule; MSG, mature secretory granule; TGN, *trans*-Golgi network.

Huttner, 1998b; Tooze *et al.*, 2001) (Figure 1A). Afterwards, ISGs form from the specialized regions of the TGN. The ISGs undergo homotypic fusion mediated by SNAREs (see 4), and non-secretory granule proteins, usually soluble proteins, peptides and membrane proteins, can be removed from the maturing granules in a clathrin-dependent process (Figure 1A). Since the soluble proteins are often secreted in a non-regulated manner, this process is described as 'constitutive-like secretion' (Arvan *et al.*, 1991). Further processing of the proforms of RSPs by prohormone convertases (PCs) (Zhou *et al.*, 1999) results in the formation of highly ordered structures within the mature granules and guarantees retention of RSPs in the regulated secretory pathway.

2.2. The role of the granins

The granins represent a family of RSPs that are ubiquitously distributed in amine- and peptide-containing secretory granules of endocrine, neuroendocrine and neuronal cells. Due to their abundance and widespread occurence, granins have often been used to study the mechanisms of protein targeting into secretory granules. The members of this well-characterized protein family (e.g., chromogranins A and B, secretogranins II, III, IV) form aggregates in a pH- and calcium-dependent manner (Yoo and Albancsi, 1990; Chanat and Huttner, 1991; Huttner *et al.*, 1991; Yoo, 1995; Ozawa and Takata, 1995), and have been proposed to act as accessory or 'helper' proteins which support aggregate formation and assist sorting of prohormones (Huttner and Natori, 1995). Since granins exist as membrane-bound forms (Fricker *et al.*, 1990; Yoo, 1993), it has been suggested that these forms interact in a homo- or heterophilic manner with soluble secretory proteins and might function as 'nucleation receptors' during the process of aggregation (Pimplikar and Huttner, 1992; Tooze *et al.*, 2001). The assembly into oligomers and multimers and thus the potentiation of sorting signals is supposed to enhance the interaction of RSPs with proteins and lipids of the TGN/granule membrane. Recently, evidence has been presented that secretory granule biogenesis is also dependent on chromogranin A (CgA). Granule formation in endocrine cells could be induced or inhibited through the expression level of CgA. These data indicate that CgA is acting as an on/off switch and controls granule biogenesis (Kim *et al.*, 2001).

2.3. Loops and membrane receptors

For granins and some other secretory proteins, for example proopiomelanocortin (POMC), a disulfide-bonded loop at the N-terminus has been identified as a signal for sorting into ISGs (Chanat *et al.*, 1993; Cool *et al.*, 1995; Krömer *et al.*, 1998). Disruption of its loop structure by dithiothreitol (DTT) resulted in the mistargeting of chromogranin B (CgB) into the constitutive pathway (Chanat *et al.*, 1993). Krömer *et al.* (1998) were the first to demonstrate that the N-terminal loop of CgB was essential and sufficient to sort it into ISGs in PC12 cells in the absence of any other RSPs. The C-terminal part allowed sorting by aggregation only in the presence of other RSPs. Furthermore, a role of the loop structure in multimerization and membrane binding has been confirmed (Thiele and Huttner, 1998a; Glombik *et al.*, 1999). In elegant experiments, Glombik *et al.* (1999) have demonstrated that the loop in CgB is sufficient to target a constitutively secreted protein, α1-antitrypsin, into ISGs. It is supposed that the loop mediates interactions between monomers of RSPs and/or binding to the membrane. In earlier studies, a pH-dependent binding of CgA and CgB to the granule membrane has been reported (Pimplikar and Huttner, 1992; Yoo, 1993). Whether the granin-loop interacts directly with the lipid leaflet of the TGN/granule membrane or is bound via interaction with yet unidentified membrane receptors is presently unknown.

The loop structure of POMC, the precursor of adrenocorticotrophic hormone (ACTH), has recently been demonstrated to bind to carboxypeptidase E (CPE), and a role for CPE as a sorting receptor has been postulated (Cool *et al.*, 1997; Normant and Loh, 1998). CPE is a protease involved in the generation of mature hormones. It binds to granule membranes in a pH-dependent manner via a hydrophobic region at the C-terminus (Fricker *et al.*, 1990; Mitra *et al.*, 1994; Dhanvantari *et al.*, 2002), and has been shown to form aggregates with other RSPs (Palmer and Christie, 1992; Rindler, 1998). The POMC-loop binding site of CPE was shown to be distinct from the substrate binding site, indicating that RSP sorting into granules is entirely independent of the enzymatic activity of CPE (Zhang *et al.*, 1999). In mice with defective CPE (Cpe^{fat}/Cpe^{fat}), missorting to the constitutive pathway occurs for several hormones and neuropeptides (Cool *et al.*, 1997). However, whether this effect is due to loss of CPE itself or is secondary is not clear. These mice also have decreased expression of PCs, which are required for the processing of hormone precursors, and this deficiency might affect storage in secretory granules. Therefore, the possible role of CPE as a universal sorting receptor is controversial (Irminger *et al.*, 1997; Thiele *et al.*, 1997; Varlamov *et al.*, 1997; Arvan and Castle, 1998).

Specific sorting domains for ISGs have also been identified for other RSPs, such as the pro-peptide of somatostatin (Stoller and Shields, 1989), or the C-terminal regions of prohormone convertase 2 (PC2) (Creemers *et al.*, 1996), and PC5-A (De Bie *et al.*, 1996), but no common structural targeting motif emerged from these studies. In addition, evidence for sorting signals on granule membrane proteins has been presented, for example for P-selectin, an adhesion receptor that is concentrated in secretory granules of endothelial cells and platelets (Disdier *et al.*, 1992; Koedam *et al.*, 1992), or for PAM (peptidylglycine α-amidating mono-oxygenase), which is widely distributed in regulated secretory cells (Kolhekar *et al.*, 1997; Milgram *et al.*, 1997).

2.4. Lipid microdomains and regulated secretion in (neuro)endocrine cells

It has been postulated that lipid microdomains mediate the tight association of RSPs with the TGN/granule membrane by protein–lipid interactions. Aggregation of the soluble and membrane-bound RSPs (see above) and further clustering within the membrane might promote the assembly of specialized subdomains in the TGN and thus the budding and formation of an ISG (Thiele and Huttner, 1998b; Tooze *et al.*, 2001) (Figure 1A). The significance of lipid microdomains for membrane curvature, budding and fission has been discussed recently (Huttner and Zimmerberg, 2001). Furthermore, their importance for apical targeting in epithelial cells has been clearly demonstrated

(Ikonen and Simons, 1998). There is now growing evidence that not only constitutive but also regulated secretion utilizes raft-dependent interactions (reviewed in Ikonen, 2001). To investigate the role of lipid microdomains in membrane targeting and granule formation, lipid levels in different cellular model systems have been manipulated by pharmacological compounds, and the protein association with rafts has been probed by detergent-treatment and flotation assays (Brown and Rose, 1992). Wang *et al.* (2000) were the first to demonstrate that cholesterol depletion with lovastatin inhibited the formation of secretory granules containing POMC as well as the formation of constitutive vesicles from the TGN of neuroendocrine AtT20 cells. Granule as well as vesicle formation could be restored by the addition of cholesterol, whereas excess cholesterol was shown to promote ISG generation. Blazquez *et al.* (2000) showed that sphingolipid depletion with fumonisin resulted in the missorting of PC2 in AtT20 cells. The authors presented further evidence for the raft-dependent binding of PC2 to the membranes of chromaffin granules, which was mediated by the pro-region of PC2. PC2 was found to associate with detergent-insoluble rafts together with the granule protein GPIII (Palmer and Christie, 1992), whereas CgA was removed. In pituitary cells, the membrane-associated CPE, the potential sorting receptor for RSPs (see 2.3), was also found in detergent-insoluble rafts isolated from ISG and TGN membranes, whereas CgA was again absent (Dhanvantari and Loh, 2000). Cholesterol depletion with methyl-β-cyclodextrin (MβCD) resulted in the dissociation of CPE and a reduction of the binding of POMC to the granule membranes. A domain within the C-terminal 25 residues of CPE is supposed to function as a signal for both raft association and sorting to the regulated secretory pathway (Zhang *et al.*, 2003). Finally, thyroglobulin, a RSP of the thyroid, has been demonstrated to enter lipid rafts en route to the apical surface (Martin-Belmonte *et al.*, 2000).

It is not yet clear if single or multiple types of lipid rafts, which are supposed to form specialized membrane domains by clustering, exist in the membranes of secretory granules (Tooze *et al.*, 2001). Furthermore, the molecular basis of the interaction between RSPs and lipids during granule formation is currently unknown.

3. SORTING IN EXOCRINE CELLS

3.1. Biogenesis of secretory granules in the acinar cells of the exocrine pancreas

The primary function of pancreatic acinar cells is the synthesis, packaging and regulated secretion of a variety of digestive enzymes, which are finally released into the duodenum via an elaborate pancreatic duct system. The exocrine pancreas served as the model that first established the role of intracellular compartments in the secretory pathway (Palade, 1975). Along with hepatocytes, the pancreatic acinar cells exhibit the highest rate of protein synthesis among cells in higher organisms. More than 90% of the newly synthesized proteins are targeted to the secretory pathway (Scheele *et al.*, 1978) and are packaged into large secretory granules, also designated as zymogen granules (for a comprehensive review see Scheele and Kern, 1993). The secretory products (RSPs) are comprised of five functional groups of hydrolytic enzymes and pro-enzymes (endo- and exoproteases, lipases, glycosidases, and RNAse). Most pancreatic enzymes are synthesized as inactive precursors (zymogens) which are only activated through limited proteolysis within the intestinal lumen. The most popular example of a digestive enzyme is trypsinogen, which is converted into trypsin by an intestinal enteropeptidase, and thus activated. In contrast to (neuro)endocrine cells, a quite complex mixture of 15–20 different enzyme and isoenzyme proteins has to be sorted and packaged within the TGN of the acinar cells (Figure 1B). Some of the enzymes form protein complexes already within the lumen of the ER (Kleene *et al.*, 1999b; Tooze *et al.*, 1989). These complexes exhibit distinct protein compositions and contain certain zymogens in association with non-enzyme proteins, which have been proposed to act as 'helper' proteins in complex formation, aggregation and/or sorting to the membrane (Kleene *et al.*, 1999b; Kleene *et al.*, 2000). The protein complexes progressively aggregate in a Ca^{2+}- and pH-dependent manner and form dense core aggregates which interact with the TGN membrane (Leblond *et al.*, 1993; Freedman and Scheele, 1993b; Dartsch *et al.*, 1998). Ultrastructural studies indicate that parts of the Golgi cisternae become dilated and filled with electron-opaque material. These condensing vacuoles (CVs), the initial stage of zymogen granules, then pinch off as ISGs, which mature into zymogen granules by the removal of non-secretory granule proteins via 'constitutive-like secretion' and further condensation of the aggregated RSPs (Figure 1B). In comparison with the relatively small secretory granules of (neuro)endocrine cells, zymogen granules are huge, with a diameter of $\sim$1 μm. Upon nervous and/or hormonal stimulation (for review see Williams, 2001), zymogen granules move towards the apical plasma membrane in a microtubule- and actin-dependent manner (Kraemer *et al.*, 1999; Ueda *et al.*, 2000; Valentijn *et al.*, 2000) and intermittently fuse with it to release their content via exocytosis into the lumen of the pancreatic duct system. Secretagogue stimulation of the cells causes an elevation of the intracellular Ca^{2+} concentration, which in turn triggers granule fusion. Despite extensive study, the mechanism underlying zymogen granule exocytosis is still far from clear; however, some information is beginning to emerge about the proteins that mediate the exocytotic membrane fusion event (see 4).

The membrane material of the granule which is transiently incorporated into the apical plasma membrane is removed and recycled via endocytosis by clathrin-coated vesicles. Upon exocytosis of the zymogen granule, the digestive enzymes are exposed to the alkaline pH of the acinar/ductal lumen, and the proteins are solubilized for their transit to the duodenum (Freedman and Scheele, 1994).

3.2. Membrane proteins of zymogen granules

The selective aggregation of pancreatic secretory proteins has been well documented (Leblond *et al.*, 1993; Freedman and Scheele, 1993b; Dartsch *et al.*, 1998), but little information is available on both the function of individual membrane proteins and their specific interaction with the RSPs of the granule content (Wagner and Williams, 1994). Membrane proteins involved in the sorting and packaging of zymogens are presumed to have their functional domains exposed on the luminal side of the granule membrane, whereas membrane proteins involved in trafficking and exocytosis are likely to be exposed on the cytosolic side. However, the list of known membrane proteins still remains short. Only a small subset of zymogen granule membrane (ZGM) proteins that have been identified so far seems to be restricted to pancreatic acinar cells or other regulated secretory cells. A predominant subset of ZGM proteins is also identified in other kinds of membranes, for example proteins of the trafficking and fusion machinery (see 4). The following section will focus on ZGM proteins bound to the inner leaflet of ZGM, which are expected to be involved in membrane binding/sorting of zymogens.

3.2.1. The role of GP-2, the major glycoprotein of zymogen granule membranes

It has long been hypothesized that the most abundant membrane protein of zymogen granules, a glycoprotein of about 75 kDa referred to as GP-2, might act as a sortase for aggregated secretory proteins (Jacob *et al.*, 1992). GP-2 is present as membrane-bound and soluble forms, which are encoded by a single gene (Fukuoka *et al.*, 1991). The membrane-bound form is attached to the luminal leaflet of ZGMs via a glycosylphosphatidylinositol (GPI) anchor (LeBel and Beattie, 1988; Scheele *et al.*, 1994) and represents up to 40% of the total membrane proteins within the ZGM (Ronzio *et al.*, 1978). The amino acid sequence of GP-2 is similar to that of Tamm-Horsfall protein (THP), a GPI-anchored protein of the kidney (Hoops and Rindler, 1991). It has been suggested that GP-2 and THP constitute a family of homologous genes (GP-2/THP gene family) (Fukuoka *et al.*, 1992). When expressed in endocrine cells, pancreatic GP-2 is not sorted into endogenous granules but accumulates in multivesicular bodies and at the cell surface instead (Hoops *et al.*, 1993; Colomer *et al.*, 1994). This effect indicates potential differences in targeting signals for RSPs in endocrine and exocrine secretory granules. Interestingly, GP-2 can form complexes with pancreatic zymogens, which are stable in mildly acidic media that mimic the TGN interior, but not in alkaline media that mimic ductal fluid (Leblond *et al.*, 1993; Colomer *et al.*, 1996). An endogenous phosphatidylinositol-specific phospholipase C (PI-PLC) is believed to be responsible for the release of GP-2 both into the granule content and into the apical medium during exocytosis (Freedman *et al.*, 1998a,b). However, several studies demonstrated that zymogen granules can form in the complete absence of GP-2, for example during embryonic development (Dittie and Kern, 1992), and in partially differentiated acinar carcinoma cell lines (Hansen *et al.*, 1983; Hoops *et al.*, 1993; Laine *et al.*, 1996). This argues against a direct role of GP-2 as specific sortase, but does not generally exclude its involvement in granule formation and in apical sorting (Scheele *et al.*, 1994; Schmidt *et al.*, 2001), at least in the adult exocrine pancreas. Since hormone-regulated secretion is absent in prenatal pancreatic cells (Chang and Jamieson, 1986) and largely muted in pancreatic tumor cells (Iwanij and Jamieson, 1982; Logsdon *et al.*, 1985), GP-2 may play an important role in the regulation of secretion/signal transduction (Kalus *et al.*, 2002). In support of this idea, membrane-bound GP-2 has been recently reported to be associated with detergent-insoluble lipid microdomains isolated from zymogen granule membranes (Schmidt *et al.*, 2001) (see 3.3.1). Furthermore, an *in vitro* interaction with syncollin, a novel membrane-bound protein of zymogen granules (see 3.2.5), has been described (Kalus *et al.*, 2002).

3.2.2. The submembranous granule matrix

Previous studies have shown that GP-2 is associated with fibrillar material in lysates of zymogen granules, in the acinar lumen and in pancreatic juice (Beaudoin *et al.*, 1991; Grondin *et al.*, 1992). Freedman and Scheele (1993a) have demonstrated that globular GP-2 detached from the membrane at slightly acidic pH forms tetrameric complexes which might be associated with sulfated proteoglycans. Sulfated proteoglycans have previously been identified as constituents of pancreatic zymogen granules and in the fluid of the pancreatic ductal system (Berg and Young, 1971; Tartakoff *et al.*, 1975; Reggio and Palade, 1978). Furthermore, Scheele *et al.* (1994) reported that in acinar cells of the rat more than 90% of sulfate-labeled proteoglycans were associated with the ZGM, and could be released by treatment of ZGM with sodium carbonate at pH 11.2. In addition, a reticular network on the inner surface of ZGM had been observed by freeze-fracture and deep-etching studies several years ago

(Cabana *et al.*, 1981). Based on these findings, Scheele and co-workers postulated the existence of a submembranous granule matrix at the luminal surface of ZGM which at least in part is composed of GP-2 and sulfated proteoglycans attached to it. They presented further evidence that during granule exocytosis the release of the putative matrix is mediated by the alkaline pH present in the acinar lumen and the activation of PI-PLC (Freedman *et al.*, 1998a). The release of the hypothetical matrix is supposed to be a prerequisite for the efficient recycling of granule membranes via endocytosis (Freedman *et al.*, 1998b). Furthermore, a function of submembranous proteoglycans and the secretory lectin ZG16p (Cronshagen *et al.*, 1994) in the binding of aggregated zymogens to the granule membrane has been suggested, based on *in vitro* studies (Kleene *et al.*, 1999a) using a recently developed condensation-sorting assay (Dartsch *et al.*, 1998). Our laboratory has undertaken a first approach to characterize the components of the putative submembranous granule matrix in more detail (Schmidt *et al.*, 2000). Analysis of the composition of a carbonate extract from ZGM revealed the presence of the lectin ZG16p, GP-2, the serpin ZG46p (Chen *et al.*, 1997) and a group of so far unknown acidic as well as basic proteins. These proteins turned out to be methanol-soluble proteoglycans and glycine-rich glycoproteins. In addition, the liberated matrix components enhanced the binding of aggregated zymogens to the granule membrane in an *in vitro* assay (Schmidt *et al.*, 2000). In a recent publication, we have found sulfated proteoglycans to be associated with lipid microdomains in the ZGM (Schmidt *et al.*, 2001; Kalus *et al.*, 2002) (see 3.3.1). These lipid microdomains and their associated proteins (e.g., GP-2) were involved in the attachment of the matrix proteoglycans to the granule membrane (Schmidt *et al.*, 2001).

3.2.3. The role of sulfated proteoglycans and glycoproteins

Sulfated proteoglycans are present in multiple intracellular compartments, such as endosomes, lysosomes, the nucleus, and especially in storage granules of immunosecretory and neurosecretory cells (Gallagher *et al.*, 1986; Gallagher, 1989; Kolset and Gallagher, 1990; Prydz and Dalen, 2000). The negatively charged proteoglycans in secretory granules of haemopoietic cells and mast cells are engaged in the binding of small positively charged molecules, such as histamine (Castle and Castle, 1998a; Brion *et al.*, 1992; Grimes and Kelly, 1992) and proteases (Huang *et al.*, 1998; Lutzelschwab *et al.*, 1997), and have therefore been considered to promote the efficient packaging and concentration of secretory products (Matsumoto *et al.*, 1995). In support of this hypothesis, it has been shown recently that mast cells from mice that lack the enzyme *N*-deacetylase/*N*-sulfotransferase-2 (NDST-2), which is required for heparin sulfate synthesis, fail to store histamine and several proteases that are normally bound to heparin sulfate in secretory granules. Mast cells from these knockout mice contained smaller granules and large empty vacuoles, which indicates that heparin glucosaminoglycans (GAGs) are required for normal granule formation (Forsberg *et al.*, 1999; Humphries *et al.*, 1999a,b; Forsberg and Kjellen, 2001).

As discussed above, sulfated proteoglycans within zymogen granules are supposed to interact electrostatically and through specific protein-protein and carbohydrate-protein binding domains with the RSPs of the granule content. In the model of the submembranous granule matrix, a proteoglycan scaffold underneath the granule membrane serves as the backbone of a complex membrane matrix, and a bridge between the granule membrane and the zymogens. Recent ultrastructural data from our laboratory revealed a submembranous localization of sulfated proteoglycans within zymogen granules using Cupromeronic Blue cytochemistry (Schmidt *et al.*, manuscript submitted), further supporting the existence of a proteoglycan-based submembranous matrix. Removal of the matrix components by carbonate treatment of isolated granule membranes led to a loss of membrane curvature. The inhibition of proteoglycan synthesis by β-D xyloside resulted in the constitutive secretion of zymogens in pancreatic acinar cells, and impaired dexamethasone-induced granule formation in AR42J cells. We conclude from these findings that a submembranous proteoglycan matrix is required for proper granule formation and secretion of zymogens, as well as for granule shape and stability in acinar cells of the pancreas.

Sulfated membrane-associated glycoproteins have also been demonstrated to contribute to secretory granule formation. SV2, a keratan sulfate-containing integral membrane protein has been proposed to promote the concentration of neurotransmitters in neuroendocrine cells (Scranton *et al.*, 1993). Furthermore, the sulfated glycoprotein chromogranin A is necessary for granule formation in neuroendocrine cells (Kim *et al.*, 2001) (see 2.2). For the mouse exocrine pancreas, a major membrane-associated sulfated glycoprotein, Pro-Muclin/GP-300, has been proposed to mediate interactions between the granule content and the ZGM (De Lisle, 1994). GP-300/Muclin shares homologies with Ebnerin, a von Ebner's gland protein (Li and Snyder, 1995), and with the glycoprotein GP-340, which is synthesized in lung, trachea, salivary gland, small intestine and stomach (Holmskov *et al.*, 1997). Pro-Muclin is a type I membrane protein that is presumably cleaved in the ISG to yield Muclin and the membrane protein p80 (De Lisle and Ziemer, 2000). Pro-Muclin acquires fixed negative charges via sulfation of its O-linked sugars in the TGN, where it is supposed to interact ionically with the zymogens. Sulfation and O-linked glycosylation turned out to be required for the normal formation of zymogen granules (De Lisle, 2002). In rat parotid acinar cells, a chondroitin sulfate-containing proline-rich proteoglycan has been shown to be the major sulfated secretory protein among the salivary products (e.g., amylase, DNase, RNase, peroxidase, proline-rich proteins, histatins) (Blair *et al.*, 1991; Castle and Castle, 1993). It was demonstrated that acidic salivary proline-rich proteins (PRPs) are the substrates for GAG addition (Castle and Castle, 1993). PRPs comprisc a family of proteins that are unique secretory products of salivary glands. Certain PRPs can become N-glycosylated in the ER. In humans they represent more than two-thirds of the secretory content, and it has been suggested that sulfation improves their sorting and storage in secretory granules (Castle and Castle, 1998b). However, the mechanism by which PRPs are anchored to the membrane has not been identified. In this context, our laboratory reported the cloning and expression of ZG29p, a novel pancreas-specific zymogen granule protein, which has been proposed to act as a 'helper' protein in granule formation (Kleene *et al.*, 1999c). ZG29p possesses proline-rich motifs which match the consensus sequences of SH3 (src homology 3 domain binding motif) domain binding sites. Furthermore, it is associated with an amylase-containing protein complex. Cross-linking and competition experiments indicate that a direct interaction of ZG29p with amylase is mediated by its SH3 binding domains, and that these domains are involved in the sorting of amylase (and other granule proteins) to the granule membrane (Kleene *et al.*, 2000). When attached to the membrane (or granule matrix), ZG29p-bound amylase could function as a 'nucleation site' for further association and aggregation of amylase molecules.

3.2.4. The secretory lectin ZG16p

There is also evidence that lectin-like proteins are involved in the association of granule content proteins with the granule membrane. Membrane-association is supposed to be mediated by the binding of distinct carbohydrates on secretory proteins and/or by specific lectin–protein interactions. From studies of polarized protein trafficking performed mainly with MDCK cells, it is known that lectins are involved in protein sorting to the apical domain of the plasma membrane (Fiedler *et al.*, 1994; Ikonen and Simons, 1998). One interesting candidate identified in the acinar cells of the exocrine pancreas is ZG16p, which is also expressed in the duodenum and colon (Cronshagen *et al.*, 1994). ZG16p shows homologies to two secretory proteins from other exocrine glands, common salivary protein 1 and prostatic spermine binding protein. According to its sequence homology with the plant lectin jacalin, ZG16p was considered a secretory lectin. It is predominantly associated with the luminal surface of ZGM, but can be removed in conjunction with proteoglycans by carbonate- or chondroitinase-treatment (Kleene *et al.*, 1999a). It is likely that ZG16p represents a component of the submembranous granule matrix, where it interacts with sulfated proteoglycans (see 3.2.2). Furthermore, a function as a linker/helper protein in the binding of aggregated zymogens to the granule membrane has been proposed from *in vitro* studies (Kleene *et al.*, 1999a). Upon exocytosis of the zymogen granules, ZG16p is found in the acinar/ductal lumen, indicating that it is removed from the membrane, presumably in a pH-dependent manner. Interestingly, ZG16p has also been found to associate with lipid microdomains of the ZGM together with GP-2 and sulfated proteoglycans (Schmidt *et al.*, 2001) (see 3.3.1). In addition, a tightly membrane-associated form of about 32 kDa, presumably a dimer, has been identified (Kalus *et al.*, 2002). Thévenod and co-workers recently proposed an additional function of the protein in the regulation of a K^+ conductance in zymogen granules (Braun and Thévenod, 2000). In an approach to identify a ZGM protein involved in the regulation of an ATP-sensitive K^+ and Cl^- conductance they used a dihydropyridine derivative in photoaffinity labeling experiments. To their surprise, they labeled and purified ZG16p as a high-affinity dihydropyridine binding protein of rat ZGM. They suggest a regulatory role for ZG16p in the direct coupling between granule fusion to the plasma membrane and the activation of channels in the ZGM (Thévenod, 2002). This intriguing possibility remains to be established. If ZG16p is indeed a regulatory membrane protein, it might regulate granule biogenesis/sorting and/or exocytosis via lipid microdomains.

3.2.5. Syncollin

Syncollin, a novel zymogen granule protein, was originally isolated through its ability to bind to syntaxin in a Ca^{2+}-sensitive manner (Edwardson *et al.*, 1997). This property, together with the observation that recombinant syncollin was able to inhibit Ca^{2+}-dependent fusion *in vitro* between zymogen granules and pancreatic plasma membranes, led to the proposal that syncollin was involved in the control of exocytotic membrane fusion in the pancreatic acinar cell. However, it has recently been shown that syncollin posesses an N-terminal signal sequence that directs it into the lumen of the endoplasmic reticulum (An *et al.*, 2000; Hodel and Edwardson, 2000). In acinar cells, syncollin is found tightly attached to the luminal surface of the ZGM. It resists salt-washing of the granule membranes, but is removed by treatment with sodium carbonate. These results are difficult to reconcile with a role of syncollin in membrane trafficking. However, the levels of syncollin are modulated in response to changes in feeding behaviour (Tan and Hooi, 2000), suggesting that it is involved in the secretion of digestive enzymes. Syncollin is strongly expressed in the acinar cells of the pancreas, but has also been found in the rat duodenum, colon, spleen and parotid gland (Tan and Hooi, 2000; Imai *et al.*, 2001). More recently, evidence has been presented that there are two pools of syncollin in pancreatic acinar cells: one free in the granule lumen and the other tightly bound to the granule membrane (Hodel *et al.*, 2001; Kalus *et al.*, 2002). The nature of the interaction of syncollin with the membrane has been studied in detail. Membrane-associated syncollin was found to form homo-oligomeric structures (Hodel *et al.*, 2001). Cross-linking studies demonstrated that membrane-bound syncollin interacts with itself and with the glycoprotein GP-2, which is linked to the luminal surface of the membrane via a GPI-anchor (see 3.2.1) (Kalus *et al.*, 2002). Cleavage-experiments with PI-PLC indicate that the membrane association is not mediated exclusively by GP-2. Interestingly, syncollin was identified as a component of cholesterol-rich microdomains in ZGM, together with GP-2, ZG16p, SNARE proteins and sulfated proteoglycans (Kalus *et al.*, 2002) (see 3.3.1). Extraction of cholesterol from ZGM by treatment with MβCD caused the detachment of syncollin (and GP-2). Furthermore, purified syncollin was able to bind to liposomes in a pH- and cholesterol-dependent manner, and cholesterol could be co-immunoprecipitated with syncollin (Hodel *et al.*, 2001). These findings indicate that membrane-bound syncollin is a component of lipid rafts, where it interacts with both GP-2 and membrane lipids, and inserts into the membrane in a cholesterol-dependent manner.

The notion that syncollin exists as a homo-oligomer, in conjunction with its lipid interaction, has led to the suggestion that syncollin might be a pore-forming protein. In support of this idea, the oligomer has been shown to form a doughnut-shaped, pore-like structure, both when bound to a solid support, and in association with a lipid bilayer. Furthermore, addition of syncollin to liposomes rendered them permeable to the water-soluble fluorescent probe 5(6)-carboxyfluorescein (Geisse *et al.*, 2002). In fact, the existence of pores in zymogen granule membranes has been reported (Cabana *et al.*, 1981; Goncz and Rothman, 1992, 1995), although their function remains unclear. If syncollin is indeed a pore-forming protein, it might fulfil important functions in signal transduction, and might regulate granule biogenesis and/or exocytosis of secretory proteins via lipid microdomains.

To analyze the function of syncollin in more detail, Jahn and co-workers have generated syncollin-deficient mice (Antonin *et al.*, 2002). These mice are viable and show no obvious changes in pancreatic morphology, stimulated release of zymogens, or protein content of zymogen granules. However, syncollin-deficient mice developed a more severe pancreatitis after hormonal hyperstimulation with caerulein. Furthermore, the authors reported that the rates of both protein synthesis and intracellular transport of secretory proteins were reduced.

Although the precise role of syncollin is still unclear, the current findings point to an involvement of syncollin in the formation and/or maturation of zymogen granules rather than in membrane fusion. Since syncollin can be defined as a further component of the submembranous granule matrix (see 3.2.2), it might form a partially redundant network with other granule membrane proteins (e.g., GP-2), that is required for sorting/packaging of zymogens as well as for granule formation. It might even be involved in signal transduction processes across the granule membrane, and thus in the regulation of granule maturation (e.g., via membrane pores) and/or exocytosis of secretory proteins via lipid microdomains.

3.3. Lipid microdomains and regulated secretion in acinar cells of the pancreas

The acinar cells of the pancreas are highly polarized, with regulated secretion directed to the apical domain surrounding the acinar lumen. From studies of polarized protein trafficking, performed mainly with MDCK cells, it is known that protein sorting to the apical domain of the plasma membrane is controlled by N- and/or O-glycosylation of protein ectodomains (Fiedler and Simons, 1995; Rodriguez-Boulan and Gonzalez, 1999; Alfalah *et al.*, 1999), by apical sorting determinants present in the cytoplasmic tails of seven-transmembrane proteins (Chuang and Sung, 1998; Sun *et al.*, 1998), or by the incorporation of apically sorted proteins into lipid microdomains in the Golgi complex, that have been proposed to act as sorting platforms for the apical delivery of plasma membrane proteins (Brown and Rose, 1992; Harder and Simons, 1997; Simons and Ikonen, 1997). Inclusion into rafts and subsequent apical sorting has been shown for GPI-anchored proteins at the plasma membrane (Harder and Simons, 1997) and for some transmembrane proteins (Kundu *et al.*, 1996; Lin *et al.*, 1998).

3.3.1. Lipid microdomains in zymogen granule membranes

Early studies addressing the lipid composition of zymogen granules showed that ZGM are characterized by a higher phospholipid: protein ratio, and by a higher concentration of sphingomyelin and of cholesterol, when compared to microsomal membranes (Meldolesi *et al.*, 1971). Our laboratory obtained the first evidence for the presence of lipid microdomains in the membranes of zymogen granules of the exocrine pancreas (Schmidt *et al.*, 2001). These microdomains showed an enrichment in cholesterol and sphingomyelin, and formed Triton X-100-insoluble glycolipid-enriched complexes which floated to the lighter fractions of sucrose density gradients. Furthermore, the complexes were found to contain the GPI-anchored glycoprotein GP-2 (see 3.2.1), the secretory lectin ZG16p (see 3.2.4) and sulfated proteoglycans (see 3.2.3). Similar observations have recently been made for syncollin (Kalus *et al.*, 2002) (see 3.2.5): it was found to be associated with detergent-insoluble cholesterol-enriched complexes in conjunction with GP-2, ZG16p and proteoglycans (Figure 2). The association of syncollin with the membrane was more sensitive to Triton X-100 or octyl glucoside than to Lubrol. As Triton X-100 treatment has been shown to interfere with the oligomeric structure of syncollin (An *et al.*, 2000), one might speculate that an oligomeric form is important for its membrane interaction. In addition, syntaxin 3 and synaptobrevin 2, two SNAREs of the ZGM which mediate granule fusion (Hansen *et al.*, 1999), have been found in lipid microdomains after Lubrol treatment (Kalus *et al.*, 2002). While the association of GP-2 with lipid microdomains is obviously mediated by its GPI-anchor, the association of ZG16p is less clear. It is likely that the interaction with the membrane is mediated by its lectin domain, possibly by binding to microdomain-associated glycolipids or glycoproteins. Another mode of interaction may be the association of ZG16p with chondroitin sulfate moieties of proteoglycans which are themselves attached to lipid microdomains (Kleene *et al.*, 1999a) (Figure 2). The association of syncollin with lipid microdomains is presumably mediated by its interaction with cholesterol.

It is not yet clear if single or multiple types of lipid rafts, which are able to form specialized membrane domains by aggregation, exist in the membranes of secretory granules (Tooze *et al.*, 2001). It is also unknown whether the composition of these rafts is changed after granule fusion with the apical plasma membrane. Recently (Röper *et al.*, 2000), Lubrol treatment was used to define cholesterol-based lipid rafts which differ from those isolated by Triton X-100. In the ZGM, the presence of syntaxin 3, synaptobrevin 2 and syncollin in detergent-insoluble complexes produced by extraction with Lubrol, but not

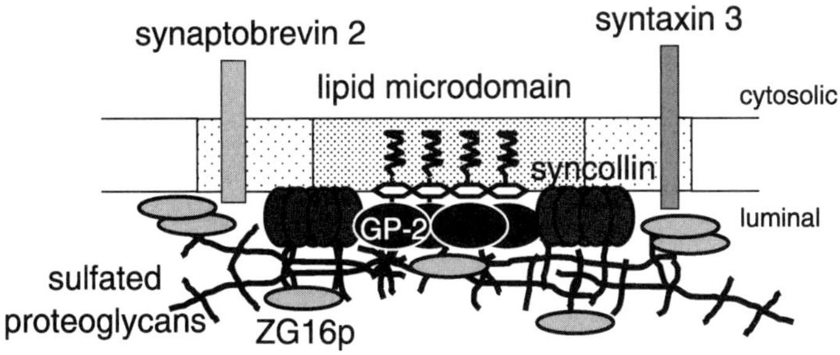

Figure 2. Schematic model of lipid microdomains and asscociated components in zymogen granule membranes. The GPI-anchored glycoprotein GP-2 is thought to exist as a tetramer and may be involved in the binding of a submembranous protein matrix (see 3.2.2) composed of proteoglycans, glycoproteins and lectins (e.g., ZG16p). Additional components are the SNARE proteins syntaxin 3 and synaptobrevin 2. Syncollin binds to membranes in a cholesterol-dependent manner and forms homo-oligomers. A hypothetical 'core' domain containing highly-ordered lipids is indicated. For further details, see 3.3.1.

with Triton X-100, might support the existence of separate or structurally different microdomains. Interestingly, Triton X-100 extraction of granule membranes produced a detergent-resistant complex of greater buoyant density than that observed after Lubrol treatment (Kalus *et al.*, 2002). It is possible that the rafts consist of highly-ordered (and Triton X-100-resistant) 'core' domains, containing GP-2, ZG16p and matrix proteoglycans, surrounded by semi-ordered (and Triton X-100-sensitive, but Lubrol-resistant) structures containing syncollin, syntaxin 3 and synaptobrevin 2 (Figure 2). A hypothetical function of such a marginal zone of semi-ordered lipids could be to enhance interactions between GPI-anchored proteins (e.g., GP-2) and transmembrane or peripheral membrane proteins that are essential for protein function (e.g., during exocytosis) (Madore *et al.*, 1999).

3.3.2. Lipid microdomains and regulated secretion of zymogens

Our laboratory demonstrated that after raft-disruption by cholesterol depletion or inhibition of glycosphingolipid synthesis by fumonisin B1 the formation of zymogen granules was reduced. Moreover, in cholesterol-depleted cells amylase secretion was increased in both the presence and absence of a hormonal stimulus, indicating missorting of secretory proteins to the constitutive pathway (Schmidt *et al.*, 2001). These findings underline the importance of lipid rafts for the regulated secretion of zymogens (Figure 1B).

3.3.3. Lipid microdomains and anchorage of the submembranous matrix

In recent studies (Kleene *et al.*, 1999a; Schmidt *et al.*, 2000; 2001), we have identified the above mentioned raft-associated proteins to be components of a submembranous proteoglycan-based matrix at the inner surface of zymogen granules, which is supposed to play a key role in membrane sorting of zymogens as well as in granule formation and stability (Figure 1B). The raft-associated GP-2 is a likely (but not the only) candidate for tethering the submembranous proteoglycan matrix to the lipid bilayer (Scheele *et al.*, 1994; Freedman *et al.*, 1998b) (Figure 2). This notion is supported by our observation that inhibition of GPI-anchor biosynthesis by mannosamine or YW 3548 interferes with protein trafficking and granule formation. Furthermore, the attachment of proteoglycans to Golgi membranes was also diminished by these manipulations (Schmidt *et al.*, 2001). It is likely that the anchorage of the matrix to the membrane (and to rafts) is a prerequisite for proper packaging of zymogens (Kleene *et al.*, 1999a). In addition, lipid microdomains and associated proteoglycans might act together to provide the driving force for membrane deformation and granule formation at the TGN (Huttner and Zimmerberg, 2001). During granule biogenesis, raft-associated proteoglycans might therefore provide both a mechanical device for granule formation at the TGN (that substitutes for a cytoplasmic coat), and a submembranous scaffold for the attachment of zymogen complexes (see 3.2.2).

3.3.4. Lipid microdomains, zymogen granule fusion and signal transduction

Recently, the cholesterol-dependent clustering of SNARE proteins has been reported in PC12 cell membranes, although there is disagreement about whether the SNAREs are present in detergent-insoluble complexes (Lang *et al.*, 2001; Chamberlain *et al.*, 2001). However, in the pancreatic acinus cell syntaxin 3 and synaptobrevin 2, two SNAREs which mediate granule fusion (Hansen *et al.*, 1999), as well as syncollin, which might be involved in exocytosis, have been found in lipid microdomains of ZGM (Figure 2). It remains to be elucidated whether lipid rafts of the ZGM are involved in the assembly and clustering of the docking and fusion machinery required for regulated secretion (Figure 1B).

Consequently, syncollin and GP-2, which show direct interaction with each other, might have a regulatory function in granule biogenesis and secretion in conjunction with other raft-associated proteins. It has been reported that after granule fusion the cleavage of GP-2 from the apical membrane is required to activate endocytosis (Freedman *et al.*, 1998b), which is regulated by tyrosine kinases (Freedman *et al.*, 1999). Freedman and co-workers suggested that GP-2 may exist in a complex with src kinases and caveolin to regulate endocytosis at the apical plasma membrane (Parker *et al.*, 2000). As GPI-anchored proteins have been shown to exist in a complex with proteins that are important in signal transduction (Stefanova *et al.*, 1991; Sargiacomo *et al.*, 1993; Rodgers *et al.*, 1994), the interaction between GP-2 and syncollin might have a related function.

3.3.5. Assembly of lipid microdomains and their role in zymogen granule formation

An analysis of the association of GP-2 with Golgi membranes and ZGM allowed us to define distinct pools of GP-2 which undergo changes during granule formation and may represent different stages of lipid raft assembly (Schmidt *et al.*, 2001). ZGM contained a characteristic pool of GP-2 which was inaccessible to PI-PLC treatment and insoluble in Triton X-100 even at 37°C. This pool was also found to be insoluble in detergents like octyl glucoside or deoxycholate, which are reported to dissolve rafts even at low temperature. It might represent a population of GP-2 present in highly ordered lipid microdomains with compact packaging and strong interactions between the components. It is supposed that GP-2 forms tetramers in the membranes of zymogen granules (Scheele *et al.*, 1994), which might account for its inaccessibility. Furthermore, both syncollin and ZG16p are observed to form multimeric structures which are tightly associated with the granule membrane (Hodel *et al.*, 2001; Kalus *et al.*, 2002). These observations suggest that the assembly of distinct protein and lipid components into lipid rafts is a progressive process involving different assembly steps.

Assembly of lipid rafts is initiated in the Golgi complex when lipids are glycosylated. In the acinar cells of the exocrine pancreas, glycosylation and sulfation of proteins are thought to represent a prerequisite for the succeeding assembly steps. In a recently proposed model (Kleene *et al.*, 1999a; Schmidt *et al.*, 2001; Kalus *et al.*, 2002), regulated granule formation at the TGN is initiated by lateral interactions of membrane components, such as glycolipids, integral and/or GPI-anchored glycoproteins (e.g., GP-2) and sulfated proteoglycans. These macromolecules interact via their carbohydrates and through specific protein–protein, protein–lipid and protein–carbohydrate binding domains, leading to the formation of lipid/glycoprotein/proteoglycan patches (Figure 2). In the course of granule formation lectins may stabilize these structures and/or mediate the binding of other components, resulting in tightly packaged lipid/protein platforms in the membrane. A submembranous protein matrix composed of accessory glycoproteins, lectins and proteoglycans found attached to the platforms has the role of a 'sorting receptor' in the recruitment of aggregated zymogens to the membrane (Figure 1B). These components interact electrostatically and through specific protein-protein and protein-carbohydrate binding domains with the regulated secretory proteins. Furthermore, the granule matrix may represent a mechanical device for granule formation and a guarantor of granule shape and stability. Finally, granule microdomains and associated components may be retained in CVs and in mature zymogen granules during the process of maturation and membrane remodeling, whereas non-associated components are removed in a clathrin-dependent process (Figure 1B). Since in exocrine cells neither a common sorting signal nor a sorting receptor have been found so far, the above model represents an interesting approach to explain the efficient packaging and sorting of a complex mixture of secretory proteins in an exocrine system.

4. SNAREs, RAFTS AND GRANULE FUSION

During exocytosis, membrane-bound SNARE proteins (SNAP receptors) of secretory vesicles/granules (v-SNAREs) have to interact with the corresponding SNARE proteins (t-SNAREs) on the target membrane (the plasma membrane) to initiate membrane fusion. The SNAREs are thought to form high-affinity complexes that bridge the two membranes and thus support bilayer mixing (reviewed in Jahn and Südhof, 1999; Lin and Scheller, 2000).

Furthermore, membrane fusion *in vivo* is regulated by a large number of accessory proteins that need to interact with individual SNAREs in a controlled manner. Many of the proteins of the fusion machinery are highly conserved in eukaryotic cells and function in a wide range of fusion events. For example, cytosolic components involved in all membrane fusion events are NSF (*N*-ethyl-maleimide-sensitive fusion protein) and α-SNAP (soluble NSF attachment protein).

Recent evidence suggests that lipid microdomains may help to concentrate SNAREs and other components of the membrane docking and fusion machinery at specialized sites of the plasma membrane and thereby promote efficient fusion. Although there is some disagreement about whether the SNAREs are present in detergent-insoluble complexes, their association with cholesterol-dependent structures is likely to be important for exocytosis. In support of this suggestion, the rate of regulated exocytosis in PC12 cells was reduced after cholesterol depletion (Chamberlain *et al.*, 2001; Lang *et al.*, 2001).

Well-characterized SNAREs involved in secretory granule exocytosis are VAMP 1/2 (a v-SNARE on the granule membrane) and syntaxin 1 and SNAP-25 (both t-SNAREs at the plasma membrane). These SNAREs are known to interact with Ca^{2+}-binding regulatory proteins, the synaptotagmins (Davis *et al.*, 1999; Gerona *et al.*, 2000), and are found on both ISGs and MSGs in (neuro)endocrine cells. Recently, additional SNAREs have been found on ISGs only (e.g., syntaxin 6, VAMP 4) (Klumperman *et al.*, 1998; Eaton *et al.*, 2000). These SNAREs are likely to be involved in the homotypic fusion and maturation of ISG in (neuro)endocrine cells (Wendler *et al.*, 2001).

Regulated exocytosis in the pancreatic acinar cell involves the fusion of the zymogen granules with the apical domain of the plasma membrane. Granule fusion is triggered by secretagogue stimulation of the cells which causes an elevation of the intracellular Ca^{2+} concentration. Apical exocytosis is an important and critical step, because fusion with the basolateral membrane would result in the release of digestive enzymes into the extracellular space and may lead to pancreatic injury. Despite extensive study, the mechanism underlying zymogen granule exocytosis is still far from clear; however, some information is beginning to emerge about the proteins that mediate the exocytotic membrane fusion event. For instance, it is known that the SNARE protein syntaxin 2, present on the apical plasma membrane, is essential for fusion between the zymogen granules and the plasma membrane (Gaisano *et al.*, 1996a), while syntaxin 3, on the granule membrane, mediates homotypic fusion between granules (Hansen *et al.*, 1999). Other granule membrane proteins that may contribute to the secretion of zymogens from the acinar cells are synaptobrevin 2 (Gaisano *et al.*, 1996b), the Rab proteins Rab3D (Ohnishi *et al.*, 1996), Rab4 (Ohnishi *et al.*, 1999), and Rab5 (Wagner *et al.*, 1994), and secretory carrier membrane proteins (SCAMP) (Brand *et al.*, 1991). However, synaptobrevin 2 appears to play only a minor role in membrane fusion (Gaisano *et al.*, 1994; Hansen *et al.*, 1999), and Rab3D has recently been suggested to function in granule maturation but not in exocrine exocytosis (Riedel *et al.*, 2002). Furthermore, it is still unknown how the components are regulated and which proteins are involved in transmitting the Ca^{2+} signal to the exocytic machinery.

5. CONCLUDING REMARKS

For RSPs several possibilities exist to explain their stable association with the TGN/granule membrane during granule formation: (1) they bind to a subpopulation of RSPs that are already tightly associated with the membrane; (2) they interact with a membrane protein; or (3) they interact with membrane lipids. Although the importance of these interactions might vary depending on the cellular system and the secretory protein studied, several lines of evidence suggest that lipid microdomains may represent unifying key structures in membrane targeting and granule formation in (neuro)endocrine and exocrine cells. Cholesterol, a crucial component of lipid microdomains, is an abundant constituent of secretory granules, and manipulation of the cholesterol and lipid levels has significant effects on granule formation and regulated secretion in (neuro)endocrine and exocrine cells. Whereas in (neuro)endocrine cells membrane-associated forms of RSPs are thought to couple the formation of lipid microdomains to the aggregation/sorting of regulated secretory cargo, specific accessory proteins (e.g., sulfated proteoglycans, glycoproteins, lectins) may be required in exocrine cells, at least in the exocrine pancreas. It is an intriguing possibility that lipid microdomains support the assembly, clustering and interaction of granule membrane components, which themselves have important functions in sorting, packaging, trafficking and release of secretory proteins. Defects in assembly may even be responsible for the development of human diseases (e.g., acute pancreatitis). For further understanding of these processes, our knowledge about granule membrane components, their precise interactions and the regulation of these interactions has to be increased.

ACKNOWLEDGEMENTS

I would like to thank Katja Schmidt, Ralf Kleene, Barbara Wäsle, J. Michael Edwardson and Horst F. Kern for stimulating discussions, and I apologize to those whose work has not been cited due to space limitations. This work was supported by the Deutsche Forschungsgemeinschaft, Bonn-Bad Godesberg, Germany (SFB 286, TP B2). This work was supported by the Deutsche Forschungsgemeinschaft, Bonn-Bad Godesberg, Germany (SFB 286, TP B2) and by the P. E. Kempkes foundation, Marburg, Germany.

REFERENCES

Alfalah, M., Jacob, R., Preuss, U., Zimmer, K. P., Naim, H. and Naim, H. Y. (1999). O-linked glycans mediate apical sorting of human intestinal sucrase-isomaltase through association with lipid rafts. *Curr Biol* **9**, 593-6.

An, S. J., Hansen, N. J., Hodel, A., Jahn, R. and Edwardson, J. M. (2000). Analysis of the association of syncollin with the membrane of the pancreatic zymogen granule. *J Biol Chem* **275**, 11306-11.

Antonin, W., Wagner, M., Riedel, D., Brose, N. and Jahn, R. (2002). Loss of the zymogen granule protein syncollin affects pancreatic protein synthesis and transport but not secretion. *Mol Cell Biol* **22**, 1545-54.

Arvan, P. and Castle, D. (1992). Protein sorting and secretion granule formation in regulated secretory cells. *Trends Cell Biology* **2**, 327-31.

Arvan, P. and Castle, D. (1998). Sorting and storage during secretory granule biogenesis: looking backward and looking forward. *Biochem J* **332**, 593-610.

Arvan, P., Kuliawat, R., Prabakaran, D., Zavacki, A. M., Elahi, D., Wang, S. and Pilkey, D. (1991). Protein discharge from immature secretory granules displays both regulated and constitutive characteristics. *J Biol Chem* **266**, 14171-4.

Beaudoin, A. R., St-Jean, P. and Grondin, G. (1991). Ultrastructural localization of GP2 in acinar cells of pancreas: presence of GP2 in endocytic and exocytic compartments. *J Histochem Cytochem* **39**, 575-88.

Berg, N. B. and Young, R. W. (1971). Sulfate metabolism in pancreatic acinar cells. *J Cell Biol* **50**, 469-83.

Blair, E. A., Castle, A. M. and Castle, J. D. (1991). Proteoglycan sulfation and storage parallels storage of basic secretory proteins in exocrine cells. *Am J Physiol* **261**, C897-905.

Blazquez, M., Thiele, C., Huttner, W. B., Docherty, K. and Shennan, K. I. (2000). Involvement of the membrane lipid bilayer in sorting prohormone convertase 2 into the regulated secretory pathway. *Biochem J* **349**, 843-52.

Brand, S. H., Laurie, S. M., Mixon, M. B. and Castle, J. D. (1991). Secretory carrier membrane proteins 31-35 define a common protein composition among secretory carrier membranes. *J Biol Chem* **266**, 18949-57.

Braun, M. and Thevenod, F. (2000). Photoaffinity labeling and purification of ZG-16p, a high-affinity dihydropyridine binding protein of rat pancreatic zymogen granule membranes that regulates a K(+)-selective conductance. *Mol Pharmacol* **57**, 308-16.

Brion, C., Miller, S. G. and Moore, H. P. (1992). Regulated and constitutive secretion. Differential effects of protein synthesis arrest on transport of glycosaminoglycan chains to the two secretory pathways. *J Biol Chem* **267**, 1477-83.

Brown, D. A. and Rose, J. K. (1992). Sorting of GPI-anchored proteins to glycolipid-enriched membrane subdomains during transport to the apical cell surface. *Cell* **68**, 533-44.

Cabana, C., Hugon, J. S. and Lamy, F. (1981). Freeze-fracture and deep-etching studies on zymogen-granule membranes of the rat pancreas. *Cell Tissue Res* **214**, 355-67.

Castle, A. M. and Castle, J. D. (1993). Novel secretory proline-rich proteoglycans from rat parotid. Cloning and characterization by expression in AtT-20 cells. *J Biol Chem* **268**, 20490-6.

Castle, A. M. and Castle, J. D. (1998a). Enhanced glycosylation and sulfation of secretory proteoglycans is coupled to the expression of a basic secretory protein. *Mol Biol Cell* **9**, 575-83.

Castle, D. and Castle, A. (1998b). Intracellular transport and secretion of salivary proteins. *Crit Rev Oral Biol Med* **9**, 4-22.

Chamberlain, L. H., Burgoyne, R. D. and Gould, G. W. (2001). SNARE proteins are highly enriched in lipid rafts in PC12 cells: implications for the spatial control of exocytosis. *Proc Natl Acad Sci USA* **98**, 5619-24.

Chanat, E. and Huttner, W. (1991). Milieu-induced, selective aggregation of regulated secretory proteins in the trans-Golgi network. *J Cell Biol* **115**, 1505-19.

Chanat, E., Weiss, U., Huttner, W. and Tooze, S. (1993). Reduction of the disulfide bond of chromogranin B (secretogranin I) in the trans-Golgi network causes its missorting to the constitutive secretory pathways. *Embo J* **12**, 2159-68.

Chang, A. and Jamieson, J. D. (1986). Stimulus-secretion coupling in the developing exocrine pancreas: secretory responsiveness to cholecystokinin. *J Cell Biol* **103**, 2353-65.

Chen, C. Y., Cronshagen, U. and Kern, H. F. (1997). A novel pancreas-specific serpin (ZG-46p) localizes to the soluble and membrane fraction of the Golgi complex and the zymogen granules of acinar cells. *Eur J Cell Biol* **73**, 205-214.

Chuang, J. Z. and Sung, C. H. (1998). The cytoplasmic tail of rhodopsin acts as a novel apical sorting signal in polarized MDCK cells. *J Cell Biol* **142**, 1245-56.

Colomer, V., Kicska, G. A. and Rindler, M. J. (1996). Secretory granule content proteins and the luminal domains of granule membrane proteins aggregate in vitro at mildly acidic pH. *J Biol Chem* **271**, 48-55.

Colomer, V., Lal, K., Hoops, T. C. and Rindler, M. J. (1994). Exocrine granule specific packaging signals are present in the polypeptide moiety of the pancreatic granule membrane protein GP2 and in amylase: implications for protein targeting to secretory granules. *Embo J* **13**, 3711-9.

Cool, D. R., Fenger, M., Snell, C. R. and Loh, Y. P. (1995). Identification of the sorting signal motif within pro-opiomelanocortin for the regulated secretory pathway. *J Biol Chem* **270**, 8723-9.

Cool, D. R., Normant, E., Shen, F. S., Chen, H. C., Pannell, L., Zhang, Y. and Loh, Y. P. (1997). Carboxypeptidase E is a regulated secretory pathway sorting receptor: genetic obliteration leads to endocrine disorders in CPE(fat) mice. *Cell* **88**, 73-83.

Creemers, J. W., Usac, E. F., Bright, N. A., Van de Loo, J. W., Jansen, E., Van de Ven, W. J. and Hutton, J. C. (1996). Identification of a transferable sorting domain for the regulated pathway in the prohormone convertase PC2. *J Biol Chem* **271**, 25284-91.

Cronshagen, U., Voland, P. and Kern, H. F. (1994). cDNA cloning and characterization of a novel 16 kDa protein located in zymogen granules of rat pancreas and goblet cells of the gut. *Eur J Cell Biol* **65**, 366-77.

Dartsch, H., Kleene, R. and Kern, H. F. (1998). In vitro condensation-sorting of enzyme proteins isolated from rat pancreatic acinar cells. *Eur J Cell Biol* **75**, 211-22.

Davis, A. F., Bai, J., Fasshauer, D., Wolowick, M. J., Lewis, J. L. and Chapman, E. R. (1999). Kinetics of synaptotagmin responses to Ca2+ and assembly with the core SNARE complex onto membranes. *Neuron* **24**, 363-76.

De Bie, I., Marcinkiewicz, M., Malide, D., Lazure, C., Nakayama, K., Bendayan, M. and Seidah, N. G. (1996). The isoforms of proprotein convertase PC5 are sorted to different subcellular compartments. *J Cell Biol* **135**, 1261-75.

De Lisle, R. C. (1994). Characterization of the major sulfated protein of mouse pancreatic acinar cells: a high molecular weight peripheral membrane glycoprotein of zymogen granules. *J Cell Biochem* **56**, 385-96.

De Lisle, R. C. (2002). Role of sulfated O-linked glycoproteins in zymogen granule formation. *J Cell Sci* **115**, 2941-52.

De Lisle, R. C. and Ziemer, D. (2000). Processing of pro-Muclin and divergent trafficking of its products to zymogen granules and the apical plasma membrane of pancreatic acinar cells. *Eur J Cell Biol* **79**, 892-904.

Dhanvantari, S., Arnaoutova, I., Snell, C. R., Steinbach, P. J., Hammond, K., Caputo, G. A., London, E. and Loh, Y. P. (2002). Carboxypeptidase E, a prohormone sorting receptor, is anchored to secretory granules via a C-terminal transmembrane insertion. *Biochemistry* **41**, 52-60.

Dhanvantari, S. and Loh, Y. P. (2000). Lipid raft association of carboxypeptidase E is necessary for its function as a regulated secretory pathway sorting receptor. *J Biol Chem* **275**, 29887-93.

Disdier, M., Morrissey, J. H., Fugate, R. D., Bainton, D. F. and McEver, R. P. (1992). Cytoplasmic domain of P-selectin (CD62) contains the signal for sorting into the regulated secretory pathway. *Mol Biol Cell* **3**, 309-21.

Dittie, A. and Kern, H. F. (1992). The major zymogen granule membrane protein GP-2 in the rat pancreas is not involved in granule formation. *Eur J Cell Biol* **58**, 243-58.

Eaton, B. A., Haugwitz, M., Lau, D. and Moore, H. P. (2000). Biogenesis of regulated exocytotic carriers in neuroendocrine cells. *J Neurosci* **20**, 7334-44.

Edwardson, J. M., An, S. and Jahn, R. (1997). The secretory granule protein syncollin binds to syntaxin in a Ca2(+)- sensitive manner. *Cell* **90**, 325-33.

Fiedler, K., Parton, R. G., Kellner, R., Etzold, T. and Simons, K. (1994). VIP36, a novel component of glycolipid rafts and exocytic carrier vesicles in epithelial cells. *Embo J* **13**, 1729-40.

Fiedler, K. and Simons, K. (1995). The role of N-glycans in the secretory pathway. *Cell* **81**, 309-12.

Forsberg, E. and Kjellen, L. (2001). Heparan sulfate: lessons from knockout mice. *J Clin Invest* **108**, 175-80.

Forsberg, E., Pejler, G., Ringvall, M., Lunderius, C., Tomasini-Johansson, B., Kusche-Gullberg, M., Eriksson, I., Ledin, J., Hellman, L. and Kjellen, L. (1999). Abnormal mast cells in mice deficient in a heparin-synthesizing enzyme. *Nature* **400**, 773-6.

Freedman, S. and Scheele, G. (1993a). Reversible pH-induced homophilic binding of GP2, a glycosyl-phosphatidylinositol-anchored protein in pancreatic zymogen granule membranes. *Eur J Cell Biol* **61**, 229-38.

Freedman, S. D., Katz, M. H., Parker, E. M. and Gelrud, A. (1999). Endocytosis at the apical plasma membrane of pancreatic acinar cells is regulated by tyrosine kinases. *Am J Physiol* **276**, C306-11.

Freedman, S. D., Kern, H. F. and Scheele, G. A. (1998a). Acinar lumen pH regulates endocytosis, but not exocytosis, at the apical plasma membrane of pancreatic acinar cells. *Eur J Cell Biol* **75**, 153-62.

Freedman, S. D., Kern, H. F. and Scheele, G. A. (1998b). Cleavage of GPI-anchored proteins from the plasma membrane activates apical endocytosis in pancreatic acinar cells. *Eur J Cell Biol* **75**, 163-73.

Freedman, S. D. and Scheele, G. A. (1993b). Regulated secretory proteins in the exocrine pancreas aggregate under conditions that mimic the trans-Golgi network. *Biochem. Biophys Res Commun* **197**, 992-9.

Freedman, S. D. and Scheele, G. A. (1994). Acid-base interactions during exocrine pancreatic secretion. Primary role for ductal bicarbonate in acinar lumen function. *Ann NY Acad Sci* **713**, 199-206.

Fricker, L. D., Das, B. and Angeletti, R. H. (1990). Identification of the pH-dependent membrane anchor of carboxypeptidase E (EC 3.4.17.10). *J Biol Chem* **265**, 2476-82.

Fukuoka, S., Freedman, S. D. and Scheele, G. A. (1991). A single gene encodes membrane-bound and free forms of GP-2, the major glycoprotein in pancreatic secretory (zymogen) granule membranes. *Proc Natl Acad Sci USA* **88**, 2898-902.

Fukuoka, S., Freedman, S. D., Yu, H., Sukhatme, V. P. and Scheele, G. A. (1992). GP-2/THP gene family encodes self-binding glycosylphosphatidylinositol-anchored proteins in apical secretory compartments of pancreas and kidney. *Proc Natl Acad Sci USA* **89**, 1189-93.

Gaisano, H. Y., Ghai, M., Malkus, P. N., Sheu, L., Bouquillon, A., Bennett, M. K. and Trimble, W. S. (1996a). Distinct cellular locations of the syntaxin family of proteins in rat pancreatic acinar cells. *Mol Biol Cell* **7**, 2019-27.

Gaisano, H. Y., Sheu, L., Foskett, J. K. and Trimble, W. S. (1994). Tetanus toxin light chain cleaves a vesicle-associated membrane protein (VAMP) isoform 2 in rat pancreatic zymogen granules and inhibits enzyme secretion. *J Biol Chem* **269**, 17062-6.

Gaisano, H. Y., Sheu, L., Grondin, G., Ghai, M., Bouquillon, A., Lowe, A., Beaudoin, A. and Trimble, W. S. (1996b). The vesicle-associated membrane protein family of proteins in rat pancreatic and parotid acinar cells. *Gastroenterology* **111**, 1661-9.

Gallagher, J. T. (1989). The extended family of proteoglycans: social residents of the pericellular zone. *Curr Opin Cell Biol* **1**, 1201-18.

Gallagher, J. T., Lyon, M. and Steward, W. P. (1986). Structure and function of heparan sulphate proteoglycans. *Biochem J* **236**, 313-25.

Geisse, N. A., Wasle, B., Saslowsky, D. E., Henderson, R. M. and Edwardson, J. M. (2002). Syncollin homo-oligomers associate with lipid bilayers in the form of doughnut-shaped structures. *J Membr Biol* **189**, 83-92.

Gerona, R. R., Larsen, E. C., Kowalchyk, J. A. and Martin, T. F. (2000). The C terminus of SNAP25 is essential for Ca(2+)-dependent binding of synaptotagmin to SNARE complexes. *J Biol Chem* **275**, 6328-36.

Glombik, M. M. and Gerdes, H. H. (2000). Signal-mediated sorting of neuropeptides and prohormones: secretory granule biogenesis revisited. *Biochimie* **82**, 315-26.

Glombik, M. M., Kromer, A., Salm, T., Huttner, W. B. and Gerdes, H. H. (1999). The disulfide-bonded loop of chromogranin B mediates membrane binding and directs sorting from the trans-Golgi network to secretory granules. *Embo J* **18**, 1059-70.

Goncz, K. K. and Rothman, S. S. (1992). Protein flux across the membrane of single secretion granules. *Biochim Biophys Acta* **1109**, 7-16.

Goncz, K. K. and Rothman, S. S. (1995). A trans-membrane pore can account for protein movement across zymogen granule membranes. *Biochim Biophys Acta* **1238**, 91-3.

Grimes, M. and Kelly, R. B. (1992). Intermediates in the constitutive and regulated secretory pathways released in vitro from semi-intact cells. *J Cell Biol* **117**, 539-49.

Grondin, G., St-Jean, P. and Beaudoin, A. R. (1992). Cytochemical and immunocytochemical characterization of a fibrillar network (GP2) in pancreatic juice: possible role as a sieve in the pancreatic ductal system. *Eur J Cell Biol* **57**, 155-64.

Hansen, L. J., Reddy, M.K., Reddy, J.K. (1983). Comparison of secretory protein and membrane composition of secretory granules isolated from normal and neoplastic pancreatic acinar cells of rats. *Proc Natl Acad Sci USA* **80**, 4379-83.

Hansen, N. J., Antonin, W. and Edwardson, J. M. (1999). Identification of SNAREs involved in regulated exocytosis in the pancreatic acinar cell. *J Biol Chem* **274**, 22871-6.

Harder, T. and Simons, K. (1997). Caveolae, DIGs, and the dynamics of sphingolipid-cholesterol microdomains. *Curr Opin Cell Biol* **9**, 534-42.

Hodel, A., An, S. J., Hansen, N. J., Lawrence, J., Wasle, B., Schrader, M. and Edwardson, J. M. (2001). Cholesterol-dependent interaction of syncollin with the membrane of the pancreatic zymogen granule. *Biochem J* **356**, 843-50.

Hodel, A. and Edwardson, J. M. (2000). Targeting of the zymogen-granule protein syncollin in AR42J and AtT-20 cells. *Biochem J* **350**, 637-43.

Holmskov, U., Lawson, P., Teisner, B., Tornoe, I., Willis, A. C., Morgan, C., Koch, C. and Reid, K. B. (1997). Isolation and characterization of a new member of the scavenger receptor superfamily, glycoprotein-340 (gp-340), as a lung surfactant protein-D binding molecule. *J Biol Chem* **272**, 13743-9.

Hoops, T. C., Ivanov, I., Cui, Z., Colomer-Gould, V. and Rindler, M. J. (1993). Incorporation of the pancreatic membrane protein GP-2 into secretory granules in exocrine but not endocrine cells. *J Biol Chem* **268**, 25694-705.

Hoops, T. C. and Rindler, M. J. (1991). Isolation of the cDNA encoding glycoprotein-2 (GP-2), the major zymogen granule membrane protein. Homology to uromodulin/Tamm-Horsfall protein. *J Biol Chem* **266**, 4257-63.

Huang, C., Sali, A. and Stevens, R. L. (1998). Regulation and function of mast cell proteases in inflammation. *J Clin Immunol* **18**, 169-83.

Humphries, D. E., Wong, G. W., Friend, D. S., Gurish, M. F., Qiu, W. T., Huang, C., Sharpe, A. H. and Stevens, R. L. (1999a). Heparin is essential for the storage of specific granule proteases in mast cells. *Nature* **400**, 769-72.

Humphries, D. E., Wong, G. W., Friend, D. S., Gurish, M. F. and Stevens, R. L. (1999b). 14 Heparin-null Transgenic Mice are Unable to Store Certain Granule Proteases in Their Mast Cells. *J Histochem Cytochem* **47**, 1645D-1646.

Huttner, W. B., Gerdes, H. H. and Rosa, P. (1991). The granin (chromogranin/secretogranin) family. *Trends Biochem Sci* **126**, 27-30.

Huttner, W. B. and Natori, S. (1995). Regulated secretion. Helper proteins for neuroendocrine secretion. *Curr Biol* **5**, 242-5.

Huttner, W. B. and Zimmerberg, J. (2001). Implications of lipid microdomains for membrane curvature, budding and fission. *Curr Opin Cell Biol* **13**, 478-84.

Ikonen, E. (2001). Roles of lipid rafts in membrane transport. *Curr Opin Cell Biol* **13**, 470-7.

Ikonen, E. and Simons, K. (1998). Protein and lipid sorting from the trans-Golgi network to the plasma membrane in polarized cells. *Semin Cell Dev Biol* **9**, 503-9.

Imai, A., Nashida, T. and Shimomura, H. (2001). mRNA expression of membrane-fusion-related proteins in rat parotid gland. *Arch Oral Biol* **46**, 955-62.

Irminger, J. C., Verchere, C. B., Meyer, K. and Halban, P. A. (1997). Proinsulin targeting to the regulated pathway is not impaired in carboxypeptidase E-deficient Cpefat/Cpefat mice. *J Biol Chem* **272**, 27532-4.

Iwanij, V. and Jamieson, J. D. (1982). Biochemical analysis of secretory proteins synthesized by normal rat pancreas and by pancreatic acinar tumor cells. *J Cell Biol* **95**, 734-41.

Jacob, M., Laine, J. and LeBel, D. (1992). Specific interactions of pancreatic amylase at acidic pH. Amylase and the major protein of the zymogen granule membrane (GP-2) bind to immobilized or polymerized amylase. *Biochem Cell Biol* **70**, 1105-14.

Jahn, R. and Sudhof, T. C. (1999). Membrane fusion and exocytosis. *Annu Rev Biochem* **68**, 863-911.

Kalus, I., Hodel, A., Koch, A., Kleene, R., Michael Edwardson, J. and Schrader, M. (2002). Interaction of syncollin with GP-2, the major membrane protein of pancreatic zymogen granules, and association with lipid microdomains. *Biochem J* **362**, 433-42.

Kim, T., Tao-Cheng, J. H., Eiden, L. E. and Loh, Y. P. (2001). Chromogranin A, an “on/off” switch controlling dense-core secretory granule biogenesis. *Cell* **106**, 499-509.

Kleene, R., Classen, B., Zdzieblo, J. and Schrader, M. (2000). SH3 binding sites of ZG29p mediate an interaction with amylase and are involved in condensation-sorting in the exocrine rat pancreas. *Biochemistry* **39**, 9893-900.

Kleene, R., Dartsch, H. and Kern, H. F. (1999a). The secretory lectin ZG16p mediates sorting of enzyme proteins to the zymogen granule membrane in pancreatic acinar cells. *European Journal of Cell Biology* **78**, 79-90.

Kleene, R., Kastner, B., Rosser, R. and Kern, H. (1999b). Complex formation among rat pancreatic secretory proteins under mild alkaline pH conditions. *Digestion* **60**, 305-13.

Kleene, R., Zdzieblo, J., Wege, K. and Kern, H. F. (1999c). A novel zymogen granule protein (ZG29p) and the nuclear protein MTA1p are differentially expressed by alternative transcription initiation in pancreatic acinar cells of the rat. *J Cell Sci* **112**, 2539-48.

Klumperman, J., Kuliawat, R., Griffith, J. M., Geuze, H. J. and Arvan, P. (1998). Mannose 6-phosphate receptors are sorted from immature secretory granules via adaptor protein AP-1, clathrin, and syntaxin 6-positive vesicles. *J Cell Biol* **141**, 359-71.

Koedam, J. A., Cramer, E. M., Briend, E., Furie, B., Furie, B. C. and Wagner, D. D. (1992). P-selectin, a granule membrane protein of platelets and endothelial cells, follows the regulated secretory pathway in AtT-20 cells. *J Cell Biol* **116**, 617-25.

Kolhekar, A. S., Mains, R. E. and Eipper, B. A. (1997). Peptidylglycine alpha-amidating monooxygenase: an ascorbate-requiring enzyme. *Methods Enzymol* **279**, 35-43.

Kolset, S. O. and Gallagher, J. T. (1990). Proteoglycans in haemopoietic cells. *Biochim Biophys Acta* **1032**, 191-211.

Kraemer, J., Schmitz, F. and Drenckhahn, D. (1999). Cytoplasmic dynein and dynactin as likely candidates for microtubule-dependent apical targeting of pancreatic zymogen granules. *Eur J Cell Biol* **78**, 265-77.

Krömer, A., Glombik, M. M., Huttner, W. B. and Gerdes, H. H. (1998). Essential role of the disulfide-bonded loop of chromogranin B for sorting to secretory granules is revealed by expression of a deletion mutant in the absence of endogenous granin synthesis. *J Cell Biol* **140**, 1331-46.

Kundu, A., Avalos, R. T., Sanderson, C. M. and Nayak, D. P. (1996). Transmembrane domain of influenza virus neuraminidase, a type II protein, possesses an apical sorting signal in polarized MDCK cells. *J Virol* **70**, 6508-15.

Laine, J., Pelletier, G., Grondin, G., Peng, M. and LeBel, D. (1996). Development of GP-2 and five zymogens in the fetal and young pig: biochemical and immunocytochemical evidence of an atypical zymogen granule composition in the fetus. *J Histochem Cytochem* **44**, 481-99.

Lang, T., Bruns, D., Wenzel, D., Riedel, D., Holroyd, P., Thiele, C. and Jahn, R. (2001). SNAREs are concentrated in cholesterol-dependent clusters that define docking and fusion sites for exocytosis. *Embo J* **20**, 2202-13.

LeBel, D. and Beattie, M. (1988). The major protein of pancreatic zymogen granule membranes (GP-2) is anchored via covalent bonds to phosphatidylinositol. *Biochem Biophys Res Commun* **154**, 818-23.

Leblond, F. A., Viau, G., Laine, J. and Lebel, D. (1993). Reconstitution in vitro of the pH-dependent aggregation of pancreatic zymogens en route to the secretory granule: implication of GP-2. *Biochem J* **291**, 289-96.

Li, X. J. and Snyder, S. H. (1995). Molecular cloning of Ebnerin, a von Ebner's gland protein associated with taste buds. *J Biol Chem* **270**, 17674-9.

Lin, R. C. and Scheller, R. H. (2000). Mechanisms of synaptic vesicle exocytosis. *Annu Rev Cell Dev Biol* **16**, 19-49.

Lin, S., Naim, H. Y., Rodriguez, A. C. and Roth, M. G. (1998). Mutations in the middle of the transmembrane domain reverse the polarity of transport of the influenza virus hemagglutinin in MDCK epithelial cells. *J Cell Biol* **142**, 51-7.

Logsdon, C. D., Moessner, J., Williams, J. A. and Goldfine, I. D. (1985). Glucocorticoids increase amylase mRNA levels, secretory organelles, and secretion in pancreatic acinar AR42J cells. *J Cell Biol* **100**, 1200-8.

Lutzelschwab, C., Pejler, G., Aveskogh, M. and Hellman, L. (1997). Secretory granule proteases in rat mast cells. Cloning of 10 different serine proteases and a carboxypeptidase A from various rat mast cell populations. *J Exp Med* **185**, 13-29.

Madore, N., Smith, K. L., Graham, C. H., Jen, A., Brady, K., Hall, S. and Morris, R. (1999). Functionally different GPI proteins are organized in different domains on the neuronal surface. *Embo J* **18**, 6917-26.

Martin-Belmonte, F., Alonso, M. A., Zhang, X. and Arvan, P. (2000). Thyroglobulin is selected as luminal protein cargo for apical transport via detergent-resistant membranes in epithelial cells. *J Biol Chem* **275**, 41074-81.

Matsumoto, R., Sali, A., Ghildyal, N., Karplus, M. and Stevens, R. L. (1995). Packaging of proteases and proteoglycans in the granules of mast cells and other hematopoietic cells. A cluster of histidines on mouse mast cell protease 7 regulates its binding to heparin serglycin proteoglycans. *J Biol Chem* **270**, 19524-31.

Meldolesi, J., Jamieson, J. D. and Palade, G. E. (1971). Composition of cellular membranes in the pancreas of the guinea pig. II. Lipids. *J Cell Biol* **49**, 130-49.

Milgram, S. L., Kho, S. T., Martin, G. V., Mains, R. E. and Eipper, B. A. (1997). Localization of integral membrane peptidylglycine alpha-amidating monooxygenase in neuroendocrine cells. *J Cell Sci* **110**, 695-706.

Mitra, A., Song, L. and Fricker, L. D. (1994). The C-terminal region of carboxypeptidase E is involved in membrane binding and intracellular routing in AtT-20 cells. *J Biol Chem* **269**, 19876-81.

Normant, E. and Loh, Y. P. (1998). Depletion of carboxypeptidase E, a regulated secretory pathway sorting receptor, causes misrouting and constitutive secretion of proinsulin and proenkephalin, but not chromogranin A. *Endocrinology* **139**, 2137-45.

Ohnishi, H., Ernst, S. A., Wys, N., McNiven, M. and Williams, J. A. (1996). Rab3D localizes to zymogen granules in rat pancreatic acini and other exocrine glands. *Am J Physiol* 271, G531-8.

Ohnishi, H., Mine, T., Shibata, H., Ueda, N., Tsuchida, T. and Fujita, T. (1999). Involvement of Rab4 in regulated exocytosis of rat pancreatic acini. *Gastroenterology* **116**, 943-52.

Ozawa, H. and Takata, K. (1995). The granin family – its role in sorting and secretory granule formation. *Cell Struct Funct* **20**, 415-20.

Palade, G. E. (1975). Intracellular aspects in the process of protein secretion. *Science* **189**, 347-58.

Palmer, D. J. and Christie, D. L. (1992). Identification of molecular aggregates containing glycoproteins III, J, K (carboxypeptidase H), and H (Kex2-related proteases) in the soluble and membrane fractions of adrenal medullary chromaffin granules. *J Biol Chem* **267**, 19806-12.

Parker, E. M., Zaman, M. M. and Freedman, S. D. (2000). GP2, a GPI-anchored protein in the apical plasma membrane of the pancreatic acinar cell, co-immunoprecipitates with src kinases and caveolin. *Pancreas* **21**, 219-25.

Pimplikar, S. W. and Huttner, W. B. (1992). Chromogranin B (secretogranin I), a secretory protein of the regulated pathway, is also present in a tightly membrane-associated form in PC12 cells. *J Biol Chem* **267**, 4110-8.

Prydz, K. and Dalen, K. T. (2000). Synthesis and sorting of proteoglycans. *J Cell Sci* **113**, 193-205.

Reggio, H. A. and Palade, G. E. (1978). Sulfated compounds in the zymogen granules of the guinea pig pancreas. *J Cell Biol* **77**, 288-314.

Riedel, D., Antonin, W., Fernandez-Chacon, R., Alvarez de Toledo, G., Jo, T., Geppert, M., Valentijn, J. A., Valentijn, K., Jamieson, J. D., Sudhof, T. C. et al. (2002). Rab3D is not required for exocrine exocytosis but for maintenance of normally sized secretory granules. *Mol Cell Biol* **22**, 6487-97.

Rindler, M. J. (1998). Carboxypeptidase E, a peripheral membrane protein implicated in the targeting of hormones to secretory granules, co-aggregates with granule content proteins at acidic pH. *J Biol Chem* **273**, 31180-5.

Rodgers, W., Crise, B. and Rose, J. K. (1994). Signals determining protein tyrosine kinase and glycosyl-phosphatidylinositol-anchored protein targeting to a glycolipid-enriched membrane fraction. *Mol Cell Biol* **14**, 5384-91.

Rodriguez-Boulan, E. and Gonzalez, A. (1999). Glycans in post-Golgi apical targeting: sorting signals or structural props? *Trends Cell Biol* **9**, 291-4.

Ronzio, R. A., Kornquist, K. E., Douglas, S. L., MacDonald, R. J., Mohrlock, S. H. and O'Donell, J. J. (1978). Glycoprotein synthesis in the adult rat pancreas. IV. Subcellular distribution of membrane glycoproteins. *Biochim Biophys Acta* **508**, 65-84.

Röper, K., Corbeil, D. and Huttner, W. B. (2000). Retention of prominin in microvilli reveals distinct cholesterol-based lipid micro-domains in the apical plasma membrane. *Nat Cell Biol* **2**, 582-92.

Rudolf, R., Salm, T., Rustom, A. and Gerdes, H. H. (2001). Dynamics of immature secretory granules: role of cytoskeletal elements during transport, cortical restriction, and F-actin-dependent tethering. *Mol Biol Cell* **12**, 1353-65.

Sargiacomo, M., Sudol, M., Tang, Z. and Lisanti, M. P. (1993). Signal transducing molecules and glycosyl-phosphatidylinositol-linked proteins form a caveolin-rich insoluble complex in MDCK cells. *J Cell Biol* **122**, 789-807.

Scheele, G. and Kern, H. F. (1993). Cellular compartmentation, protein processing, and secretion in the exocrine pancreas. In *The pancreas: biology, pathobiology and disease*, (ed. V. L. W. Go), pp. 121-150. New York: Raven Press.

Scheele, G. A., Fukuoka, S. and Freedman, S. D. (1994). Role of the GP2/THP family of GPI-anchored proteins in membrane trafficking during regulated exocrine secretion. *Pancreas* **9**, 139-49.

Scheele, G. A., Palade, G. E. and Tartakoff, A. M. (1978). Cell fractionation studies on the guinea pig pancreas. Redistribution of exocrine proteins during tissue homogenization. *J Cell Biol* **78**, 110-30.

Schmidt, K., Dartsch, H., Linder, D., Kern, H. and Kleene, R. (2000). A submembranous matrix of proteoglycans on zymogen granule membranes is involved in granule formation in rat pancreatic acinar cells. *J Cell Sci* **113**, 2233-42.

Schmidt, K., Schrader, M., Kern, H. F. and Kleene, R. (2001). Regulated apical secretion of zymogens in rat pancreas: Involvement of the GPI-anchored glycoprotein GP-2, the lectin ZG16p and cholesterol-glycosphingolipid enriched microdomains. *J Biol Chem* **10**, 10.

Scranton, T. W., Iwata, M. and Carlson, S. S. (1993). The SV2 protein of synaptic vesicles is a keratan sulfate proteoglycan. *J Neurochem* **61**, 29-44.

Simons, K. and Ikonen, E. (1997). Functional rafts in cell membranes. *Nature* **387**, 569-72.

Stefanova, I., Horejsi, V., Ansotegui, I. J., Knapp, W. and Stockinger, H. (1991). GPI-anchored cell-surface molecules complexed to protein tyrosine kinases. *Science* **254**, 1016-9.

Stoller, T. J. and Shields, D. (1989). The propeptide of preprosomatostatin mediates intracellular transport and secretion of alpha-globin from mammalian cells. *J Cell Biol* **108**, 1647-55.

Sun, A. Q., Ananthanarayanan, M., Soroka, C. J., Thevananther, S., Shneider, B. L. and Suchy, F. J. (1998). Sorting of rat liver and ileal sodium-dependent bile acid transporters in polarized epithelial cells. *Am J Physiol* **275**, G1045-55.

Tan, S. and Hooi, S. C. (2000). Syncollin is differentially expressed in rat proximal small intestine and regulated by feeding behavior. *Am J Physiol Gastrointest Liver Physiol* **278**, G308-20.

Tartakoff, A. M., Jamieson, J. D., Scheele, G. A. and Palade, G. E. (1975). Studies on the pancreas of the guinea pig. Parallel processing and discharge of exocrine proteins. *J Biol Chem* **250**, 2671-7.

Thevenod, F. (2002). Ion channels in secretory granules of the pancreas and their role in exocytosis and release of secretory proteins. *Am J Physiol Cell Physiol* **283**, C651-72.

Thiele, C., Gerdes, H. H. and Huttner, W. B. (1997). Protein secretion: puzzling receptors. *Curr Biol* **7**, R496-500.

Thiele, C. and Huttner, W. B. (1998a). The disulfide-bonded loop of chromogranins, which is essential for sorting to secretory granules, mediates homodimerization. *J Biol Chem* **273**, 1223-31.

Thiele, C. and Huttner, W. B. (1998b). Protein and lipid sorting from the trans-Golgi network to secretory granules-recent developments. *Semin Cell Dev Biol* **9**, 511-6.

Tooze, J., Kern, H. F., Fuller, S. D. and Howell, K. E. (1989). Condensation-sorting events in the rough endoplasmic reticulum of exocrine pancreatic cells. *J Cell Biol* **109**, 35-50.

Tooze, S. A. (1998). Biogenesis of secretory granules in the trans-Golgi network of neuroendocrine and endocrine cells. *Biochim Biophys Acta* **1404**, 231-44.

Tooze, S. A., Martens, G. J. M. and Huttner, W. B. (2001). Secretory granule biogenesis: rafting to the SNARE. *Trends in Cell Biology* **11**, 116-122.

Ueda, N., Ohnishi, H., Kanamaru, C., Suzuki, J., Tsuchida, T., Mashima, H., Yasuda, H. and Fujita, T. (2000). Kinesin is involved in regulation of rat pancreatic amylase secretion. *Gastroenterology* **119**, 1123-31.

Valentijn, J. A., Valentijn, K., Pastore, L. M. and Jamieson, J. D. (2000). Actin coating of secretory granules during regulated exocytosis correlates with the release of rab3D. *Proc Natl Acad Sci USA* **97**, 1091-5.

Varlamov, O., Fricker, L. D., Furukawa, H., Steiner, D. F., Langley, S. H. and Leiter, E. H. (1997). Beta-cell lines derived from transgenic Cpe(fat)/Cpe(fat) mice are defective in carboxypeptidase E and proinsulin processing. *Endocrinology* **138**, 4883-92.

Wagner, A. C., Strowski, M. Z. and Williams, J. A. (1994). Identification of Rab 5 but not Rab 3A in rat pancreatic zymogen granule membranes. *Biochem Biophys Res Commun* **200**, 542-8.

Wagner, A. C. and Williams, J. A. (1994). Pancreatic zymogen granule membrane proteins: molecular details begin to emerge. *Digestion* **55**, 191-9.

Wang, Y., Thiele, C. and Huttner, W. B. (2000). Cholesterol is required for the formation of regulated and constitutive secretory vesicles from the trans-Golgi network. *Traffic* **1**, 952-62.

Wasmeier, C., Bright, N. A. and Hutton, J. C. (2002). The lumenal domain of the integral membrane protein phogrin mediates targeting to secretory granules. *Traffic* **3**, 654-65.

Wendler, F., Page, L., Urbe, S. and Tooze, S. A. (2001). Homotypic fusion of immature secretory granules during maturation requires syntaxin 6. *Mol Biol Cell* **12**, 1699-709.

Williams, J. A. (2001). Intracellular signaling mechanisms activated by cholecystokinin-regulating synthesis and secretion of digestive enzymes in pancreatic acinar cells. *Annu Rev Physiol* **63**, 77-97.

Yoo, S. H. (1993). pH-dependent binding of chromogranin B and secretory vesicle matrix proteins to the vesicle membrane. *Biochim Biophys Acta* **1179**, 239-46.

Yoo, S. H. (1995). pH- and Ca(2+)-induced conformational change and aggregation of chromogranin B. Comparison with chromogranin A and implication in secretory vesicle biogenesis. *J Biol Chem* **270**, 12578-83.

Yoo, S. H. and Albanesi, J. P. (1990). Ca2(+)-induced conformational change and aggregation of chromogranin A. *J Biol Chem* **265**, 14414-21.

Zhang, C. F., Dhanvantari, S., Lou, H. and Loh, Y. P. (2003). Sorting of carboxypeptidase E to the regulated secretory pathway requires interaction of its transmembrane domain with lipid rafts. *Biochem J* **369**, 453-60.

Zhang, C. F., Snell, C. R. and Loh, Y. P. (1999). Identification of a novel prohormone sorting signal-binding site on carboxypeptidase E, a regulated secretory pathway-sorting receptor. *Mol Endocrinol* **13**, 527-36.

Zhou, A., Webb, G., Zhu, X. and Steiner, D. F. (1999). Proteolytic processing in the secretory pathway. *J Biol Chem* **274**, 20745-8.

Part VI

Domain Dynamics in Disease

Chapter 13

Oxidative Stress, Caveolae and Caveolin-1

Marie-Odile Parat and Paul L. Fox
Departments of Anesthesiology Research and Cell Biology, The Lerner Research Institute, Cleveland Clinic Foundation, 9500 Euclid Ave., Cleveland, OH 44195, USA

Abstract
Oxidative stress underlies a range of pathophysiological conditions. Reactive oxygen species are also generated intracellularly to serve as second messengers and some are linked to caveolae/raft signalling systems. The effect of oxidative stress on caveolin-1 expression, post-translational modifications, membrane trafficking and function are described.

1. CAVEOLAE AND CAVEOLINS

Originally named to describe cave-like invaginations at the cell surface, caveolae are specialized membrane sub-domains enriched in glycosphingolipids, cholesterol and lipid-anchored membrane proteins. Biochemical and morphological experiments have shown the presence of caveolae in most cell types (Simionescu *et al.*, 1982; Lisanti *et al.*, 1994). Caveolae are involved in endothelial transcytosis and potocytosis, cell signaling and cholesterol regulation (Anderson *et al.*, 1992; Schlegel *et al.*, 2000; Mineo and Anderson, 2001). These functions are believed to require caveolin-1, the major protein component of caveolae.

Caveolin-1 is found primarily as two isoforms, caveolin-1α and -1β, differing in their N-termini and derived from alternative translation initiation sites

(Scherer *et al.*, 1995). Caveolin-1 is an integral membrane protein with an unusual hairpin-like conformation (Figure 1A); the N- and C-terminal regions both face the cytosol and are connected by a membrane-embedded, hydrophobic domain containing 33 amino acids (Dupree *et al.*, 1993). Caveolin-1

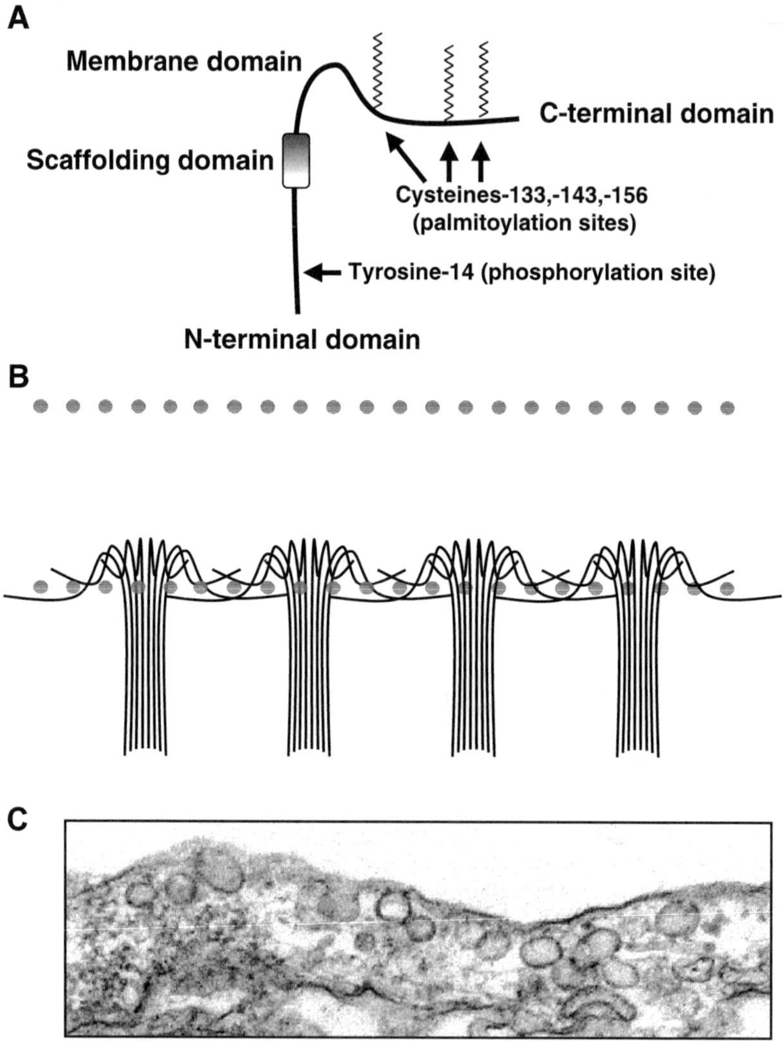

Figure 1. A, schematic of caveolin-1 domains, palmitoylation sites and tyrosine-phosphorylation site. B, schematic of caveolin oligomers interacting to form a scaffold, and their insertion in the membrane. C, electron micrograph of bovine aortic endothelial cell caveolae showing 70 nm invaginations of the plasma membrane.

participates in multiple protein/protein and protein/lipid interactions which are critical for its diverse functions. Interactions of caveolin-1 with adjacent caveolin molecules result in oligomerization and formation of a rigid scaffold (Figure 1B) that maintains the characteristic cave-like morphology (Figure 1C) (Fra *et al.*, 1995; Sargiacomo *et al.*, 1995; Song *et al.*, 1997; Das *et al.*, 1999) and appears as a filamentous coat (Rothberg *et al.*, 1992). In addition to its structural function, caveolin-1 has several important regulatory activities. For example caveolin-1 interacts, via a consensus binding motif (the scaffolding domain, amino acids 82–101), with signaling proteins, including heterotrimeric G-protein α-subunits, protein tyrosine kinases, $p21^{Ha\text{-}ras}$, and EC-specific nitric oxide synthase (eNOS) (Li *et al.*, 1995; Li *et al.*, 1996a; Garcia-Cardena *et al.*, 1997). Interaction with caveolin generally inhibits the signal transducing activity of the protein (Smart *et al.*, 1999). Recent studies suggest that caveolin-1, which is known to bind cholesterol (Murata *et al.*, 1995), interacts with chaperones of the immunophilin family to form a soluble complex (Uittenbogaard *et al.*, 1998) involved in the transport of cholesterol from the endoplasmic reticulum to the caveolae, and subsequently to non-caveolar plasma membranes (Smart *et al.*, 1996).

Caveolin-1 is subject to two types of post-translational modifications that regulate its intracellular localization or activity, namely phosphorylation and palmitoylation. Several stimuli, including epidermal growth factor (EGF) (Kim *et al.*, 2000), induce Src family kinase-dependent phosphorylation of caveolin-1 on tyrosine-14. Caveolin-1 is also phosphorylated on serine-80 in unstimulated cells (Schlegel *et al.*, 2001). Recent evidence suggest that serine phosphorylation of caveolin-1 might also be subject to regulation and inversely correlated to tyrosine 14-phosphorylation (Wang *et al.*, 2002). In addition, like many caveolae-targeted proteins, caveolin-1 is acylated; three Cys residues near the C-terminus are susceptible to palmitoylation (Dietzen *et al.*, 1995). Mutation of the Cys residues impairs caveolin-1 interaction with other acylated proteins (Lee *et al.*, 2001) and its binding and transport of cholesterol (Uittenbogaard and Smart, 2000).

In addition to caveolin-1, the caveolin family contains two other gene products identified to date, namely caveolin-2 and -3. Caveolin-2 is co-expressed with caveolin-1 in most cell types and both proteins tightly interact with each other to form hetero-oligomeric complexes (Scherer *et al.*, 1997). Caveolin-2 is not known to be palmitoylated and does not contain the C-terminal cysteine residues palmitoylated in other members of the family, but phosphorylation on Tyrosine-19 by c-Src has recently been documented (Lee *et al.*, 2002). In contrast, the expression of caveolin-3 is muscle-specific (Tang *et al.*, 1996). Caveolin-3 forms high molecular weight homo-oligomeric complexes and is palmitoylated on the C-terminal Cys residues conserved in caveolin-1 and -3 (Tang *et al.*, 1996; Galbiati *et al.*, 1999).

2. OXIDATIVE STRESS

Reactive oxygen species are normally generated in low amounts during respiration, the process by which molecular oxygen is reduced in the mitochondrial respiratory chain to produce ATP. During this process, a series of single electron transfers leads to the production of partially reduced and highly reactive metabolites of oxygen, namely the superoxide anion free radical, and hydrogen peroxide. In the presence of transition metals, hydrogen peroxide gives rise to the highly reactive hydroxyl radical. Reactive oxygen species are regarded as toxic by-products of metabolism because of their potential to cause damage to nucleic acids, proteins and lipids. To prevent these deleterious effects of reactive oxygen species, cells are equipped with antioxidants and repair enzymes. In addition to protecting themselves against reactive oxygen species, eukaryotic cells synthesize reactive oxygen species in a controlled fashion and use them as essential participants in cell signaling as second messengers (Thannickal and Fanburg, 2000). An excess of reactive oxygen species, also called oxidative stress, can occur in some physio-pathological conditions when antioxidant defenses are overwhelmed by overproduction of reactive oxygen species. Such conditions include inflammation, atherosclerosis, cancer, degenerative diseases, and aging.

Several conditions have been used in an attempt to reproduce in cultured cell systems the physio-pathological effects of oxidative stress, for example direct application of reactive oxygen species by a bolus of hydrogen peroxide added to the cell culture medium. Alternatively, reactive oxygen species-generating systems such as glucose/glucose oxidase or xanthine/xanthine oxidase can continuously deliver hydrogen peroxide or superoxide anions, respectively, over extended periods of time (as long as the enzyme and substrate are available in the medium). Endothelial cells are a privileged cell system to study the effect of oxidative stress on caveolin-1 and caveolae, because they highly express caveolin-1 and exhibit abundant caveolae. Exposure to oxysterols is widely used in studies of caveolae and rafts. Exposure can be achieved through direct introduction of oxysterols, or oxidized LDL, in the cell medium. Production of reactive oxygen species following oxysterol (Lizard *et al.*, 1998) or oxidized LDL (Cominacini *et al.*, 2000) treatment of cultured cells has been reported. Alternatively, caveolae cholesterol can be oxidized by subjecting cells to cholesterol oxidase, a bacterial enzyme which preferentially oxidizes caveolae cholesterol into the non-physiological oxysterol, cholest-4-en-3-one (cholestenone) (Smart and Anderson, 2002). In view of the different effects of hydrogen peroxide and cholesterol oxidase on caveolae, it should be noted that cholesterol oxidase produces one mole of hydrogen peroxide for each mole of cholesterol oxidized (Doukyu and Aono, 1999), whereas cell exposure to hydrogen peroxide (up to 500 μM for 30 min) does not lead to the generation of cholestenone (our unpublished results).

3. CAVEOLAE/RAFTS AS THE SITE OF REACTIVE OXYGEN SPECIES GENERATION

Recent evidence suggests that reactive oxygen species are generated intracellularly to serve as second messengers. This occurs after cell stimulation by cytokines (interleukin-1, interferon-γ, tumor necrosis factor-α), growth factors (platelet-derived growth factor (PDGF), epidermal growth factor (EGF), basic fibroblast growth factor, insulin, granulocyte-macrophage colony-stimulating factor, transforming growth factor–β1) or agonists of G protein-coupled receptors (angiotensin II, serotonin, bradykinin, thrombin, endothelin) (Finkel, 2001; Thannickal and Fanburg, 2000).

Several of the receptors initiating a signal transduction cascade involving reactive oxygen species are concentrated in caveolae or, for cells that do not express caveolin, in lipid rafts. These membrane microdomains, which play a pivotal role in signal transduction (Smart *et al.*, 1999), therefore have been proposed to be a preferred site of reactive oxygen species generation (Gniadecki *et al.*, 2002). In accordance with this hypothesis, the basal level of reactive oxygen species detected using the fluorescent probe carboxy-2′,7′-dichlorodihydroxyfluorescein diacetate in human keratinocytes correlates with the level of raft staining using fluorescently tagged cholera toxin B subunit, and therefore with the amount of lipid rafts. The basal level of reactive oxygen species is dramatically reduced when rafts are disrupted by cholesterol extraction with methyl-β-cyclodextrin, when cholesterol is sequestered by the drug filipin, or when protein compartmentalization into rafts is disturbed by addition of an excess of the raft-enriched ganglioside GM-1. Reactive oxygen species production is restored in cholesterol-repleted cells (Gniadecki *et al.*, 2002). Ultraviolet radiation induces the production of reactive oxygen species in cells, which are involved as mediators of apoptosis, aging and carcinogenesis. In particular, the UVA (320–380 nm) region is less strongly absorbed by nucleic acids and proteins than UVB (290–320 nm) but causes a variety of oxidative events. Raft integrity and protein compartmentalization in rafts are also essential to UVA-induced reactive oxygen species production (Gniadecki *et al.*, 2002).

Caveolae are also involved in the activation/deactivation cycle of endothelial specific nitric oxide synthase (NOS), and the production of nitric oxide (NO), a long-lived free radical with multiple biological actions. NO is relatively unreactive toward most biomolecules, but can give rise to various reactive nitrogen species capable of damaging biological targets. In particular, NO reacts with superoxide to form peroxynitrite, a strong oxidant which is more stable than NO or superoxide (Kojda and Harrison, 1999). NO is synthesized in mammalian cells by three nitric oxide synthases, neuronal NOS, inducible NOS and endothelial NOS (eNOS). Of the three isoforms, eNOS is the only one that is dually acylated by the saturated fatty acids palmitate and myristate (Garcia-Cardena *et al.*, 1996). Reversible palmitoylation of eNOS on two cysteines near the N-terminus is involved in the eNOS activation-deactivation cycle (Michel, 1999). In unstimulated cells, eNOS is maintained in an inactive state in caveolae through its interaction with caveolin-1 scaffolding domain. Upon stimulation (by calcium mobilizing agonists, vascular endothelial growth factor, or shear stress), intracellular calcium increases and/or eNOS undergoes Akt-dependent phosphorylation, eNOS dissociates from caveolin and binds to calmodulin, in a process facilitated by hsp90 (Michel *et al.*, 1997; Garcia-Cardena *et al.*, 1998). At the same time, eNOS is depalmitoylated and translocates to the cytosol, where it generates NO (and sometimes superoxide) (Prabhakar *et al.*, 1998). Deactivation of eNOS involves dissociation from calmodulin, re-association with caveolin, re-palmitoylation and return to plasmalemmal caveolae (Michel, 1999; Govers and Rabelink, 2001; Goligorsky *et al.*, 2002). Thus, caveolae are not the site of nitric oxide production, but the starting point of the eNOS activation cycle.

4. OXIDATIVE STRESS AND CAVEOLIN-1 EXPRESSION

4.1. Effect of exposure to reactive oxygen species

Reports on the effect of oxidative stress on caveolin-1 expression are ambiguous. In an effort to mimic the oxidative stress contributing to atherosclerosis, Peterson *et al.* (1999) subjected bovine aortic endothelial cells (BAECs) for 24 h to a superoxide anion-generating compound (LY83583) and observed no alteration of caveolin-1 expression as detected by immunoblot analysis. In a recent study suggesting that caveolin-1 mediates oxidative stress-induced senescence, Volonte *et al.* (2002) have shown that oxidative stress increased, whereas antioxidants decreased, caveolin-1 expression. Exposure to sub-lethal doses of hydrogen peroxide or UVC induced overexpression of caveolin-1, not caveolin-2, in NIH3T3 cells. Caveolin-1 upregulation was delayed (3–6 days) and sustained (up to 11 days). The hydrogen peroxide-induced up regulation of caveolin-1 was prevented by the antioxidants quercetin and vitamin E. Moreover, the flavonoid quercetin directly down regulated caveolin-1 expression. Regulation of caveolin-1 expression was assessed by immunoblot analysis as well as by transcriptional activity of a luciferase-caveolin-1 promoter construct. Our data suggest that a short exposure of BAECs to hydrogen peroxide (200 μM for 20 min) does not affect the neosynthesis of caveolin-1 as measured by metabolic labeling with [^{35}S]methionine, nor does it have an effect on the amount of caveolar caveolin-1. The apparent contradiction between these results might be explained by the time course of the exposure to oxidative stress (acute versus chronic).

4.2. Effect of oxysterols

Oxysterols are formed by spontaneous or enzymatic oxidation of both free and esterified cholesterol. Fielding *et al.* (1997) have shown that exposure of primary skin fibroblasts to LDL containing cholesterol α-epoxide, cholesterol β-epoxide, 7α-hydroxycholesterol or 7-ketocholesterol caused a decrease of caveolin-1 mRNA levels compared to exposure of LDL containing only free cholesterol (Table 1). This effect was observed after 3 h of exposure to oxysterols. Incubation of MDCK cells for 3 d in the presence of 25-hydroxycholesterol led to a decrease in caveolin-1 mRNA, which the authors attributed to a decrease in cellular cholesterol, and did not affect caveolin-1 protein level (Hailstones *et al.*, 1998). Caveolin-1 has a very slow turnover (Parat and Fox, 2001), and therefore a decrease in caveolin-1 mRNA will only slowly translate into a decrease of the protein. We have assessed the effect of cholestenone on caveolin-1 expression. Cholestenone is not a physiological oxysterol, but results from the oxidation of cholesterol by the bacterial enzyme cholesterol oxidase. Exposure of cultured cells to the enzyme or to its product profoundly disturbs caveolae. In our experiments, caveolin-1 mRNA detected by northern blotting as well as caveolin-1 neosynthesis detected by [^{35}S]methionine labeling are dramatically decreased in bovine aortic endothelial cells subjected to cholestenone treatment for 30 min (unpublished data).

5. OXIDATIVE STRESS AND POST-TRANSLATIONAL MODIFICATIONS OF CAVEOLIN-1

5.1. Tyrosine phosphorylation

Caveolin-1 was first identified as a phospho-protein in Rous sarcoma virus-transformed chicken embryo fibroblasts (Glenney and Zokas, 1989), which led to the hypothesis that caveolin-1 may be a critical target during cellular transformation. It is phosphorylated by oncogenic viral src (v-Src) (Glenney, 1989; Li *et al.*, 1996b; Ko *et al.*, 1998) and by normal cellular Src kinase (c-Src) on tyrosine-14 (Li *et al.*, 1996b; Lee *et al.*, 2000). More recently, caveolin-1 has also been identified as a major phosphoprotein detected in cells overexpressing c-Abl (Sanguinetti and Mastick, 2003).

Table 1. Effect of oxysterols on caveolae and caveolin-1

Oxysterol stress	Concentration	Duration	Cell type	Effect	References
7-ketocholesterol	100 μg/ml	1 h	ECV304	Increased Src activity	(Myers and Stanley, 1999)
7α-OH cholesterol 7β-OH cholesterol Cholesterol α-epoxide	100 μg/ml	1 h	ECV304	Decreased Src activity	(Myers and Stanley, 1999)
7α-OH cholesterol Cholesterol α-epoxide Cholesterol β-epoxide 7-ketocholesterol	10% of free cholesterol (wt/wt)	3 h	Normal skin fibroblasts	Prevents LDL-induced increase of caveolin-1 mRNA	(Fielding *et al.*, 1997)
Cholestenone	900 μM	1 h	NIH3T3	Reduced hyperosmotic shock-induced caveolin-1 phosphorylation and p38 MAPK activation	(Volonte *et al.*, 2001)
Cholestenone	900 μM	1 h	Normal human fibroblasts	Uncouples PDGF activation from neighbouring substrates phosphorylation	(Liu *et al.*, 2000)
Cholesterol oxidase	2 U/ml	2 h	CHO cells	Causes intracellular translocation of caveolin-1 Diverts internalization of endothelin receptor A to chlathrin-coated pits	(Okamoto *et al.*, 2000)
Cholesterol oxidase	0.5 U/ml	1 h	Normal human skin fibroblasts	Displacement of caveolin from the plasma membrane to the Golgi apparatus	(Smart *et al.*, 1994)
Cholesterol oxidase	1.8 U/ml	1 h	Normal human fibroblasts	Displaces caveolin-1, but not other proteins, from caveolae	(Liu *et al.*, 2000)
Cholesterol oxidase	1 U/ml	1 h	Rat adipocytes	Impairs insulin-dependent phosphorylation	(Gustavsson *et al.*, 1999)
Cholesterol oxidase	1.8 U/ml	10–60 min	Normal human fibroblasts	Blocks PDGF-dependent tyrosine phosphorylation	(Liu *et al.*, 2000)
Cholesterol oxidase	1 U/ml	30–60 min	ECV304	Internalization of caveolin and BENE	(de Marco *et al.*, 2001)
Cholesterol oxidase	2 U/ml	2 h	CHO cells	Switched internalization of endothelin receptor A from caveolae to chlathrin coated pits	(Okamoto *et al.*, 2000)
Oxidized LDL	25 μg/ml	24 h	ECV304	Increased Src activity	(Myers and Stanley, 1999)
Oxidized LDL	10 μg/ml	1 h	Porcine pulmonary artery EC HUVEC	Displacement of eNOS and caveolin-1 from caveolae. Inhibition of eNOS activation. Depletion of caveolae Cholesterol	(Blair *et al.*, 1999) (Uittenbogaard *et al.*, 2000)

Phosphocaveolin-1 can be detected *in vivo* in the endothelium of small venules and capillaries, in the basal cells of stratified squamous epithelium (oesophagus and skin), and in intra-hepatic bile duct epithelium of normal adult rats (Aoki *et al.*, 1999). In unstimulated cultured cells, caveolin-1 is phosphorylated at a low (Mastick *et al.*, 1995; Parat *et al.*, 2002) or undetectable level (Li *et al.*, 1996b; Aoki *et al.*, 1999; Nomura and Fujimoto, 1999; Volonte *et al.*, 2001). Caveolin-1 undergoes tyrosine 14-phosphorylation following exposure to various stimuli which include insulin (Mastick *et al.*, 1995), EGF (Kim *et al.*, 2002; Wang *et al.*, 2002), or under stress, such as hyperosmotic shock or oxidative stress (Volonte *et al.*, 2001; Sanguinetti and Mastick, 2003). Bovine pulmonary artery endothelial cells subjected to pervanadate treatment exhibit an increased tyrosine phosphorylation of caveolin-1 as shown by immunoprecipitation with anti-caveolin antibody followed by immunoblotting using anti-phosphotyrosine antibody (Vepa *et al.*, 1997). In human umbilical vein endothelial cells (HUVECs), an increased tyrosine-14 phosphorylation (evidenced by immunoblotting with a specific anti-tyrosine-14-phosphorylated caveolin-1 antibody) was detected after 30 min of treatment with 100 μM hydrogen peroxide. This increased tyrosine phosphorylation was inhibited by herbimycin or genistein, suggesting an involvement of the Src family kinases (Aoki *et al.*, 1999). In their study of cellular stress-induced caveolin-1 phosphorylation, Volonte *et al.* have used hydrogen peroxide at 1 and 5 mM to induce caveolin-1 tyrosine phosphorylation in NIH 3T3 cells within 5–20 min. The highest dose was needed to detect a clear signal on the immunoblot using phosphocaveolin antibody (Volonte *et al.*, 2001). They further showed that p38 MAP kinase activation and c-Src activity were required for hydrogen peroxide-induced caveolin-1 tyrosine phosphorylation. In a recent study, it has been shown that the expression of c-Abl, but not c-Src, was required for hydrogen peroxide-induced caveolin-1 tyrosine phosphorylation, using embryo fibroblasts isolated from kinase-gene disrupted mice (Sanguinetti and Mastick, 2003). Our results confirm in a different cellular model (bovine aortic endothelial cells) the need for a high dose of hydrogen peroxide (at least 1 mM) to induce caveolin-1 tyrosine-14 phosphorylation (Parat *et al.*, 2002).

The only oxysterol whose effect on caveolin-1 phosphorylation has been tested is cholestenone. The introduction of cholestenone in caveolae membranes does not increase caveolin-1 phosphorylation. On the contrary, cholestenone prevents hyperosmotic shock-induced caveolin-1-phosphorylation, most likely by uncoupling the osmotic stress from intracellular signaling (Volonte *et al.*, 2001).

Caveolin-2 has recently been shown to undergo Src-induced phosphorylation on tyrosine 19, following cell exposure to insulin or integrin ligation, but no data is available yet concerning the effect of oxidative stress of caveolin-2 tyrosine 19 phosphorylation.

5.2. Palmitoylation

Palmitoylation is the regulated, post-translational modification of cysteine residues by the saturated fatty acid palmitate, through a reversible, covalent thioester linkage. Three cysteines in the C-terminus of caveolin-1 undergo palmitate modification. Caveolin-1 palmitoylation is highly unusual at least in two aspects. First, palmitoylation is not necessary for caveolin-1 targeting to caveolae, as shown by experiments in which palmitoylation-deficient mutants of caveolin-1 were detected in biochemically isolated caveolae (Dietzen *et al.*, 1995; Uittenbogaard and Smart, 2000). In addition, palmitoylation of caveolin-1 is essentially irreversible and no turnover of the fatty acid can be detected in pulse chase experiments (Parat and Fox, 2001). We have shown that oxidative stress inhibits caveolin-1 palmitoylation by subjecting endothelial cells to hydrogen peroxide, delivered either as a bolus or by continuous delivery with glucose and glucose oxidase (Parat *et al.*, 2002). Hydrogen peroxide inhibits the incorporation of palmitate in newly synthesized caveolin-1, but does not increase the rate of caveolin depalmitoylation. Moreover, the effect of hydrogen peroxide on caveolin-1 palmitoylation occurs at much lower concentrations than those needed to increase caveolin-1 phosphorylation, indicating that the effect on palmitoylation is not a direct consequence of the increased phosphorylation.

6. OXIDATIVE STRESS AND CAVEOLIN-1 TRAFFICKING

An early recognition that caveolin-1 is a sensitive target of oxidative stress came from the observation that oxidation of caveolar cholesterol by cholesterol oxidase causes reversible translocation of caveolin-1 from caveolae to the Golgi apparatus (Smart *et al.*, 1994). Caveolin-1 internalization was later confirmed by immunofluorescence in NIH3T3 fibroblasts exposed to high doses (1 mM) of hydrogen peroxide (Kang *et al.*, 2000), and by biochemical experiments in which treatment of endothelial cells with oxidized LDL, but not native LDL, caused the translocation of caveolin-1 from plasma membrane caveolae to an intracellular membrane fraction (Blair *et al.*, 1999; Uittenbogaard *et al.*, 2000). *In vitro* treatment of bovine aortic endothelial cells caveolae-enriched membranes with superoxide-generating xanthine/xanthine oxidase dissociated caveolin-1 from the membranes (Peterson *et al.*, 1999). In bovine aortic endothelial cells subjected to low doses of hydrogen peroxide (100–200 μM), we did not find a difference in the subcellular localization of caveolin-1 evaluated by immunofluorescence, or a shift from caveolae/rafts to non-raft membranes. However, we detected a decrease in the trafficking of newly synthesized caveolin-1 to caveolae (Parat *et al.*, 2002).

7. OXIDATIVE STRESS AND CAVEOLAE FUNCTION

7.1. Oxidative stress and compartmentalization of signaling molecules in caveolae

Caveolae sterols are oxidized by exposure to cholesterol oxidizing agents, to oxidized LDL or by direct introduction of oxysterols in the cell medium. Oxysterols accumulate in the glycosphingolipid-rich membranes of the cell and perturb caveolae function, possibly by interaction with caveolin-1. The presence of cholestenone in caveolae uncouples PDGF receptor autophosphorylation from tyrosine phosphorylation of neighbouring proteins, thereby interfering with PDGF signal transduction (Liu *et al.*, 2000). The ability of cholestenone to uncouple PDGF receptor from its partners seems to be due to the disruption of the organization of signaling proteins within caveolae rather than to a loss of caveolin-1 from caveolae.

Oxidative stress is also a potent modulator of eNOS function. First, eNOS is released from caveolae-enriched membrane exposed *in vitro* to xanthine/ xanthine oxidase (Peterson *et al.*, 1999). Oxidized LDL also reversibly displaces eNOS from plasmalemmal caveolae, and inhibits acetylcholine-induced activation of eNOS (Blair *et al.*, 1999). Other examples of caveolae-initiated signaling pathways affected by cholesterol oxidation include insulin receptor signaling (Gustavsson *et al.*, 1999). In another study, replacement of a fraction of caveolae cholesterol with 7-ketocholesterol, the major oxysterol found in oxidized LDL, was able to increase the activity of an enzyme found mostly in caveolae, namely Src kinase, by a factor of 2.5, without changing the amounts of Src protein (Myers and Stanley, 1999). By contrast, introduction of 7α-hydroxycholesterol, 7β-hydroxycholesterol or cholesterol epoxide decreased the activity of Src kinase (Myers and Stanley, 1999).

Nitric oxide perturbs signaling initiated by receptors located in caveolae: the $[Ca^{2+}]_i$ increase elicited by bradykinin in human umbilical vascular endothelial cells, or by endothelin-1 in smooth muscle cells, was impaired when cells were pretreated with a nitric oxide donor (Li *et al.*, 2001).

7.2. Oxidative stress and caveolae internalization

Caveolae are endocytosis-competent cell surface invaginations which depend on dynamin for internalization (Henley *et al.*, 1998; Le *et al.*, 2002). Although this process is not very efficient under normal culture conditions (Thomsen *et al.*, 2002), caveolae-mediated endocytosis has been reported for a variety of infectious agents (Norkin, 2001) as well as receptors after ligand binding (Benlimame *et al.*, 1998). Oxidative stress can cause caveolae-mediated endocytosis to be redirected to a different pathway of internalization. This is the case for endothelin receptor type A, which resides in caveolae in the absence of ligand and is internalized by a caveolae-dependent pathway upon occupancy of the receptor by its agonist endothelin-1 (Okamoto *et al.*, 2000). In cholesterol oxidase-treated cells, this pathway is down-regulated and internalization of agonist-occupied receptor instead occurs through clathrin-coated pits.

7.3. Oxidative stress and maintenance of caveolae structure

Smart *et al.* (1994) showed that although caveolin-1 was translocated to the Golgi apparatus upon treatment with cholesterol oxidase, the number or morphology of caveolae was unaffected, raising the possibility that caveolae can exist without caveolin. In contrast, Peterson *et al.* (1999) showed that when bovine aortic endothelial cells were treated with a superoxide-generating naphtoquinolinedione (LY 83583), there was a decrease in caveolae number.

Reactive oxygen species seem to directly affect the organization of caveolae membranes by acting both on the scaffold made of caveolin-1 and on the lipids. Exposure of human umbilical vein endothelial cells to nitric oxide resulted in a partial dissociation of the caveolin-1 scaffold, as shown by an increase of the monomeric form of caveolin-1 (Li *et al.*, 2001). *In vitro* biophysical experiments performed on sphingomyelin/cholesterol rafts with a mixture of the phospholipids dioleoyl-phosphatidylcholine and dioleoyl-phosphatidylethanolamine showed that exposure of one side of a raft to cholesterol oxidase affected lipid clustering in both monolayers, the raft boundary becoming irregular, leading to raft disappearance (Samsonov *et al.*, 2001).

8. CONCLUSIONS

There is increasing available evidence that oxidative stress, which has been implicated in a large number of human diseases, affects caveolin-1 expression, post-translational modifications, subcellular localization and trafficking as well as caveolae functions. This organelle plays key physiological roles, (vasomotor control, cardiopulmonary function, lipid homeostasis, lactation) as evidenced by the increasing list of abnormalities of caveolin-deficient mice. It remains to be shown whether caveolins, and caveolae, are involved in the pathophysiological effects of oxidative stress *in vivo*.

REFERENCES

Anderson, R. G., Kamen, B. A., Rothberg, K. G., and Lacey, S. W. (1992). Potocytosis: Sequestration and transport of small molecules by caveolae. Science *255*, 410-411.

Aoki, T., Nomura, R., and Fujimoto, T. (1999). Tyrosine phosphorylation of caveolin-1 in the endothelium. Exp. Cell Res. *253*, 629-636.

Benlimame, N., Le, P. U., and Nabi, I. R. (1998). Localization of autocrine motility factor receptor to caveolae and clathrin-independent internalization of its ligand to smooth endoplasmic reticulum. Mol. Biol. Cell *9*, 1773-1786.

Blair, A., Shaul, P. W., Yuhanna, I. S., Conrad, P. A., and Smart, E. J. (1999). Oxidized low density lipoprotein displaces endothelial nitric-oxide synthase (eNOS) from plasmalemmal caveolae and impairs eNOS activation. J. Biol. Chem. *274*, 32512-32519.

Cominacini, L., Pasini, A. F., Garbin, U., Davoli, A., Tosetti, M. L., Campagnola, M., Rigoni, A., Pastorino, A. M., Lo Cascio, V., and Sawamura, T. (2000). Oxidized low density lipoprotein (ox-LDL) binding to ox-LDL receptor-1 in endothelial cells induces the activation of NF-κB through an increased production of intracellular reactive oxygen species. J. Biol. Chem. *275*, 12633-12638.

Das, K., Lewis, R. Y., Scherer, P. E., and Lisanti, M. P. (1999). The membrane-spanning domains of caveolins-1 and -2 mediate the formation of caveolin hetero-oligomers. Implications for the assembly of caveolae membranes *in vivo*. J. Biol. Chem. *274*, 18721-18728.

de Marco, M. C., Kremer, L., Albar, J. P., Martinez-Menarguez, J. A., Ballesta, J., Garcia-Lopez, M. A., Marazuela, M., Puertollano, R., and Alonso, M. A. (2001). Bene, a novel raft-associated protein of the mal proteolipid family, interacts with caveolin-1 in human endothelial-like ECV304 cells. J. Biol. Chem. *276*, 23009-23017.

Dietzen, D. J., Hastings, W. R., and Lublin, D. M. (1995). Caveolin is palmitoylated on multiple cysteine residues. Palmitoylation is not necessary for localization of caveolin to caveolae. J. Biol. Chem. *270*, 6838-6842.

Doukyu, N., and Aono, R. (1999). Two moles of O_2 consumption and one mole of H_2O_2 formation during cholesterol peroxidation with cholesterol oxidase from *Pseudomonas* sp. strain ST-200. Biochem. J. *341*, 621-627.

Dupree, P., Parton, R. G., Raposo, G., Kurzchalia, T. V., and Simons, K. (1993). Caveolae and sorting in the trans-golgi network of epithelial cells. EMBO J. *12*, 1597-1605.

Fielding, C. J., Bist, A., and Fielding, P. E. (1997). Caveolin mRNA levels are up-regulated by free cholesterol and down-regulated by oxysterols in fibroblast monolayers. Proc. Natl. Acad. Sci. U.S.A. *94*, 3753-3758.

Finkel, T. (2001). Reactive oxygen species and signal transduction. IUBMB Life *52*, 3-6.

Fra, A. M., Williamson, E., Simons, K., and Parton, R. G. (1995). De novo formation of caveolae in lymphocytes by expression of VIP21-caveolin. Proc. Natl. Acad. Sci. U.S.A. *92*, 8655-8659.

Galbiati, F., Volonte, D., Minetti, C., Chu, J. B., and Lisanti, M. P. (1999). Phenotypic behavior of caveolin-3 mutations that cause autosomal dominant limb girdle muscular dystrophy (LGMD-1C). Retention of LGMD-1C caveolin-3 mutants within the Golgi complex. J. Biol. Chem. *274*, 25632-25641.

Garcia-Cardena, G., Fan, R., Shah, V., Sorrentino, R., Cirino, G., Papapetropoulos, A., and Sessa, W. C. (1998). Dynamic activation of endothelial nitric oxide synthase by hsp90. Nature *392*, 821-824.

Garcia-Cardena, G., Martasek, P., Masters, B. S., Skidd, P. M., Couet, J., Li, S., Lisanti, M. P., and Sessa, W. C. (1997). Dissecting the interaction between nitric oxide synthase (NOS) and caveolin. Functional significance of the NOS caveolin binding domain *in vivo*. J. Biol. Chem. *272*, 25437-25440.

Garcia-Cardena, G., Oh, P., Liu, J., Schnitzer, J. E., and Sessa, W. C. (1996). Targeting of nitric oxide synthase to endothelial cell caveolae via palmitoylation: Implications for nitric oxide signaling. Proc. Natl. Acad. Sci. U.S.A. *93*, 6448-6453.

Glenney, J. R., Jr. (1989). Tyrosine phosphorylation of a 22-kDa protein is correlated with transformation by Rous sarcoma virus. J. Biol. Chem. *264*, 20163-20166.

Glenney, J. R., Jr., and Zokas, L. (1989). Novel tyrosine kinase substrates from Rous sarcoma virus-transformed cells are present in the membrane skeleton. J. Cell Biol. *108*, 2401-2408.

Gniadecki, R., Christoffersen, N., and Wulf, H. C. (2002). Cholesterol-rich plasma membrane domains (lipid rafts) in keratinocytes: Importance in the baseline and UVA-induced generation of reactive oxygen species. J. Invest. Dermatol. *118*, 582-588.

Goligorsky, M. S., Li, H., Brodsky, S., and Chen, J. (2002). Relationships between caveolae and eNOS: Everything in proximity and the proximity of everything. Am. J. Physiol. Renal Physiol. *283*, F1-10.

Govers, R., and Rabelink, T. J. (2001). Cellular regulation of endothelial nitric oxide synthase. Am. J. Physiol. Renal Physiol. *280*, F193-206.

Gustavsson, J., Parpal, S., Karlsson, M., Ramsing, C., Thorn, H., Borg, M., Lindroth, M., Peterson, K. H., Magnusson, K. E., and Stralfors, P. (1999). Localization of the insulin receptor in caveolae of adipocyte plasma membrane. FASEB J. *13*, 1961-1971.

Hailstones, D., Sleer, L. S., Parton, R. G., and Stanley, K. K. (1998). Regulation of caveolin and caveolae by cholesterol in MDCK cells. J. Lipid Res. *39*, 369-379.

Henley, J. R., Krueger, E. W., Oswald, B. J., and McNiven, M. A. (1998). Dynamin-mediated internalization of caveolae. J. Cell Biol. *141*, 85-99.

Kang, Y. S., Ko, Y. G., and Seo, J. S. (2000). Caveolin internalization by heat shock or hyperosmotic shock. Exp. Cell Res. *255*, 221-228.

Kim, Y. N., Dam, P., and Bertics, P. J. (2002). Caveolin-1 phosphorylation in human squamous and epidermoid carcinoma cells: Dependence on erbB1 expression and Src activation. Exp. Cell Res. *280*, 134-147.

Kim, Y. N., Wiepz, G. J., Guadarrama, A. G., and Bertics, P. J. (2000). Epidermal growth factor-stimulated tyrosine phosphorylation of caveolin-1. Enhanced caveolin-1 tyrosine phosphorylation following aberrant epidermal growth factor receptor status. J. Biol. Chem. *275*, 7481-7491.

Ko, Y. G., Liu, P., Pathak, R. K., Craig, L. C., and Anderson, R. G. (1998). Early effects of $PP60^{v\text{-}src}$ kinase activation on caveolae. J. Cell Biochem. *71*, 524-535.

Kojda, G., and Harrison, D. (1999). Interactions between NO and reactive oxygen species: Pathophysiological importance in atherosclerosis, hypertension, diabetes and heart failure. Cardiovasc. Res. *43*, 562-571.

Le, P. U.,Guay, G., Altschuler, Y. and Nabi, I. R. (2002). Caveolin-1 is a negative regulator of caveolae-mediated endocytosis to the endoplasmic reticulum. J. Biol. Chem. *277*, 3371-3379.

Lee, H., Park, D. S., Wang, X. B., Scherer, P. E., Schwartz, P. E., and Lisanti, M. P. (2002). Src-induced phosphorylation of caveolin-2 on tyrosine 19. Phospho-caveolin-2 $(Tyr(P)^{19})$ is localized near focal adhesions, remains associated with lipid rafts/caveolae, but no longer forms a high molecular mass hetero-oligomer with caveolin-1. J. Biol. Chem. *277*, 34556-34567.

Lee, H., Volonte, D., Galbiati, F., Iyengar, P., Lublin, D. M., Bregman, D. B., Wilson, M. T., Campos-Gonzalez, R., Bouzahzah, B., Pestell, R. G., *et al.* (2000). Constitutive and growth factor-regulated phosphorylation of caveolin-1 occurs at the same site (Tyr-14) *in vivo*: Identification of a c-Src/Cav-1/Grb7 signaling cassette. Mol. Endocrinol. *14*, 1750-1775.

Lee, H., Woodman, S. E., Engelman, J. A., Volonte, D., Galbiati, F., Kaufman, H. L., Lublin, D. M., and Lisanti, M. P. (2001). Palmitoylation of caveolin-1 at a single site (Cys-156) controls its coupling to the c-Src tyrosine kinase. Targeting of dually acylated molecules (GPI-linked, transmembrane, or cytoplasmic) to caveolae effectively uncouples c-Src and caveolin-1 (Tyr-14). J. Biol. Chem. *276*, 35150-35158.

Li, H., Brodsky, S., Basco, M., Romanov, V., De Angelis, D. A., and Goligorsky, M. S. (2001). Nitric oxide attenuates signal transduction: Possible role in dissociating caveolin-1 scaffold. Circ. Res. *88*, 229-236.

Li, S., Couet, J., and Lisanti, M. P. (1996a). Src tyrosine kinases, Galpha subunits, and H-Ras share a common membrane-anchored scaffolding protein, caveolin. Caveolin binding negatively regulates the auto-activation of Src tyrosine kinases. J. Biol. Chem. *271*, 29182-29190.

Li, S., Okamoto, T., Chun, M., Sargiacomo, M., Casanova, J. E., Hansen, S. H., Nishimoto, I., and Lisanti, M. P. (1995). Evidence for a regulated interaction between heterotrimeric G proteins and caveolin. J. Biol. Chem. *270*, 15693-15701.

Li, S., Seitz, R., and Lisanti, M. P. (1996b). Phosphorylation of caveolin by Src tyrosine kinases. The alpha-isoform of caveolin is selectively phosphorylated by v-Src *in vivo*. J. Biol. Chem. *271*, 3863-3868.

Lisanti, M. P., Scherer, P. E., Vidugiriene, J., Tang, Z., Hermanowski-Vosatka, A., Tu, Y. H., Cook, R. F., and Sargiacomo, M. (1994). Characterization of caveolin-rich membrane domains isolated from an endothelial-rich source: Implications for human disease. J. Cell Biol. *126*, 111-126.

Liu, P., Wang, P., Michaely, P., Zhu, M., and Anderson, R. G. (2000). Presence of oxidized cholesterol in caveolae uncouples active platelet-derived growth factor receptors from tyrosine kinase substrates. J. Biol. Chem. *275*, 31648-31654.

Lizard, G., Gueldry, S., Sordet, O., Monier, S., Athias, A., Miguet, C., Bessede, G., Lemaire, S., Solary, E., and Gambert, P. (1998). Glutathione is implied in the control of 7-ketocholesterol-induced apoptosis, which is associated with radical oxygen species production. FASEB J. *12*, 1651-1663.

Mastick, C. C., Brady, M. J., and Saltiel, A. R. (1995). Insulin stimulates the tyrosine phosphorylation of caveolin. J. Cell Biol. *129*, 1523-1531.

Michel, J. B., Feron, O., Sacks, D., and Michel, T. (1997). Reciprocal regulation of endothelial nitric-oxide synthase by Ca^{2+}-calmodulin and caveolin. J. Biol. Chem. *272*, 15583-15586.

Michel, T. (1999). Targeting and translocation of endothelial nitric oxide synthase. Braz. J. Med. Biol. Res. *32*, 1361-1366.

Mineo, C., and Anderson, R. G. (2001). Potocytosis. Histochem. Cell Biol. *116*, 109-118.

Murata, M., Peranen, J., Schreiner, R., Wieland, F., Kurzchalia, T. V., and Simons, K. (1995). VIP21/caveolin is a cholesterol-binding protein. Proc. Natl. Acad. Sci. U.S.A. *92*, 10339-10343.

Myers, S. J., and Stanley, K. K. (1999). Src family kinase activation in glycosphingolipid-rich membrane domains of endothelial cells treated with oxidised low density lipoprotein. Atherosclerosis *143*, 389-397.

Nomura, R., and Fujimoto, T. (1999). Tyrosine-phosphorylated caveolin-1: Immunolocalization and molecular characterization. Mol. Biol. Cell *10*, 975-986.

Norkin, L. C. (2001). Caveolae in the uptake and targeting of infectious agents and secreted toxins. Adv. Drug Deliv. Rev. *49*, 301-315.

Okamoto, Y., Ninomiya, H., Miwa, S., and Masaki, T. (2000). Cholesterol oxidation switches the internalization pathway of endothelin receptor type A from caveolae to clathrin-coated pits in chinese hamster ovary cells. J. Biol. Chem. *275*, 6439-6446.

Parat, M. O., and Fox, P. L. (2001). Palmitoylation of caveolin-1 in endothelial cells is post-translational but irreversible. J. Biol. Chem. *276*, 15776-15782.

Parat, M. O., Stachowicz, R. Z., and Fox, P. L. (2002). Oxidative stress inhibits caveolin-1 palmitoylation and trafficking in endothelial cells. Biochem. J. *361*, 681-688.

Peterson, T. E., Poppa, V., Ueba, H., Wu, A., Yan, C., and Berk, B. C. (1999). Opposing effects of reactive oxygen species and cholesterol on endothelial nitric oxide synthase and endothelial cell caveolae. Circ. Res. *85*, 29-37.

Prabhakar, P., Thatte, H. S., Goetz, R. M., Cho, M. R., Golan, D. E., and Michel, T. (1998). Receptor-regulated translocation of endothelial nitric-oxide synthase. J. Biol. Chem. *273*, 27383-27388.

Rothberg, K. G., Heuser, J. E., Donzell, W. C., Ying, Y. S., Glenney, J. R., and Anderson, R. G. (1992). Caveolin, a protein component of caveolae membrane coats. Cell *68*, 673-682.

Samsonov, A. V., Mihalyov, I., and Cohen, F. S. (2001). Characterization of cholesterol-sphingomyelin domains and their dynamics in bilayer membranes. Biophys. J. *81*, 1486-1500.

Sanguinetti, A. R., and Mastick, C. C. (2003). c-Abl is required for oxidative stress-induced phosphorylation of caveolin-1 on tyrosine 14. Cell. Signal. *15*, 289-298.

Sargiacomo, M., Scherer, P. E., Tang, Z., Kubler, E., Song, K. S., Sanders, M. C., and Lisanti, M. P. (1995). Oligomeric structure of caveolin: Implications for caveolae membrane organization. Proc. Natl. Acad. Sci. U.S.A. *92*, 9407-9411.

Scherer, P. E., Lewis, R. Y., Volonte, D., Engelman, J. A., Galbiati, F., Couet, J., Kohtz, D. S., van Donselaar, E., Peters, P., and Lisanti, M. P. (1997). Cell-type and tissue-specific expression of caveolin-2. Caveolins 1 and 2 co-localize and form a stable hetero-oligomeric complex *in vivo*. J. Biol. Chem. *272*, 29337-29346.

Scherer, P. E., Tang, Z., Chun, M., Sargiacomo, M., Lodish, H. F., and Lisanti, M. P. (1995). Caveolin isoforms differ in their N-terminal protein sequence and subcellular distribution. Identification and epitope mapping of an isoform-specific monoclonal antibody probe. J. Biol. Chem. *270*, 16395-16401.

Schlegel, A., Arvan, P., and Lisanti, M. P. (2001). Caveolin-1 binding to endoplasmic reticulum membranes and entry into the regulated secretory pathway are regulated by serine phosphorylation. Protein sorting at the level of the endoplasmic reticulum. J. Biol. Chem. *276*, 4398-4408.

Schlegel, A., Pestell, R. G., and Lisanti, M. P. (2000). Caveolins in cholesterol trafficking and signal transduction: Implications for human diseases. Front. Biosci. *5*, D929-937.

Simionescu, M., Simionescu, N., and Palade, G. E. (1982). Biochemically differentiated microdomains of the cell surface of capillary endothelium. Ann. N.Y. Acad. Sci. *401*, 9-24.

Smart, E. J., and Anderson, R. G. (2002). Alterations in membrane cholesterol that affect structure and function of caveolae. Methods Enzymol. *353*, 131-139.

Smart, E. J., Graf, G. A., McNiven, M. A., Sessa, W. C., Engelman, J. A., Scherer, P. E., Okamoto, T., and Lisanti, M. P. (1999). Caveolins, liquid-ordered domains, and signal transduction. Mol. Cell. Biol. *19*, 7289-7304.

Smart, E. J., Ying Ys, Donzell, W. C., and Anderson, R. G. (1996). A role for caveolin in transport of cholesterol from endoplasmic reticulum to plasma membrane. J. Biol. Chem. *271*, 29427-29435.

Smart, E. J., Ying, Y. S., Conrad, P. A., and Anderson, R. G. (1994). Caveolin moves from caveolae to the Golgi apparatus in response to cholesterol oxidation. J. Cell Biol. *127*, 1185-1197.

Song, K. S., Tang, Z., Li, S., and Lisanti, M. P. (1997). Mutational analysis of the properties of caveolin-1. A novel role for the C-terminal domain in mediating homo-typic caveolin-caveolin interactions. J. Biol. Chem. *272*, 4398-4403.

Tang, Z., Scherer, P. E., Okamoto, T., Song, K., Chu, C., Kohtz, D. S., Nishimoto, I., Lodish, H. F., and Lisanti, M. P. (1996). Molecular cloning of caveolin-3, a novel member of the caveolin gene family expressed predominantly in muscle. J. Biol. Chem. *271*, 2255-2261.

Thannickal, V. J., and Fanburg, B. L. (2000). Reactive oxygen species in cell signaling. Am. J. Physiol. Lung Cell. Mol. Physiol. *279*, L1005-1028.

Thomsen, P., Roepstorff, K., Stahlhut, M., and van Deurs, B. (2002). Caveolae are highly immobile plasma membrane microdomains, which are not involved in constitutive endocytic trafficking. Mol. Biol. Cell *13*, 238-250.

Uittenbogaard, A., Shaul, P. W., Yuhanna, I. S., Blair, A., and Smart, E. J. (2000). High density lipoprotein prevents oxidized low density lipoprotein- induced inhibition of endothelial nitric-oxide synthase localization and activation in caveolae. J. Biol. Chem. *275*, 11278-11283.

Uittenbogaard, A., and Smart, E. J. (2000). Palmitoylation of caveolin-1 is required for cholesterol binding, chaperone complex formation, and rapid transport of cholesterol to caveolae. J. Biol. Chem. *275*, 25595-25599.

Uittenbogaard, A., Ying, Y., and Smart, E. J. (1998). Characterization of a cytosolic heat-shock protein-caveolin chaperone complex. Involvement in cholesterol trafficking. J. Biol. Chem. *273*, 6525-6532.

Vepa, S., Scribner, W. M., and Natarajan, V. (1997). Activation of protein phosphorylation by oxidants in vascular endothelial cells: Identification of tyrosine phosphorylation of caveolin. Free Radic. Biol. Med. *22*, 25-35.

Volonte, D., Galbiati, F., Pestell, R. G., and Lisanti, M. P. (2001). Cellular stress induces the tyrosine phosphorylation of caveolin-1 (Tyr^{14}) via activation of p38 mitogen-activated protein kinase and c-Src kinase. Evidence for caveolae, the actin cytoskeleton, and focal adhesions as mechanical sensors of osmotic stress. J. Biol. Chem. *276*, 8094-8103.

Volonte, D., Zhang, K., Lisanti, M. P., and Galbiati, F. (2002). Expression of caveolin-1 induces premature cellular senescence in primary cultures of murine fibroblasts. Mol. Biol. Cell *13*, 2502-2517.

Wang, X. Q., Sun, P., and Paller, A. S. (2002). Ganglioside induces caveolin-1 redistribution and interaction with the epidermal growth factor receptor. J. Biol. Chem. *277*, 47028-47034.

Chapter 14

The Role of Lipid Microdomains in Virus Biology

Debi P. Nayak and Eric K.-W. Hui
Department of Microbiology, Immunology and Molecular Genetics, UCLA School of Medicine, Los Angeles CA 90095-1747, USA

Abstract
Many of the highly pathogenic viruses including influenza virus, HIV and others of world wide epidemiological importance are enveloped and possess a membrane around the nucleocapsid containing the viral genome. Viral membrane is required to protect the viral genome and provide important functions for attachment, morphogenesis and transmission. Viral membrane is essentially composed of lipids and proteins. While the proteins on the viral envelope are almost exclusively virally encoded, lipids, on the other hand, are all of host origin and recruited from host membrane. However, lipids on the

Abbreviations: AFM, atomic force microscopy; AIDS, acquired immune deficiency syndrome; ALV, nontransforming avian leukosis virus; ASV, avian sarcoma virus; BCR, B cell antigen receptor; BHK cell, baby hamster kidney cell; CD, cyclodextrin; CEC, chick embryo cells; CIV, Chilo Iridescent virus; CLSM, confocal laser scanning microscopy; CMV, cytomegalovirus; CoV, coronavirus; CT, cytoplasmic tail; DIG, detergent insoluble GSL-enriched domain; DRM, detergent resistant membranes; EAV, equine arteritis virus; EBOV, Ebola virus; EBV, Epstein-Barr virus; ER, endoplasmic reticulum; FRT, Fisher rat thyroid cells; FPV, fowl plague virus; FRET, fluorescence energy transfer; Gal Cer, galactosyl ceramide; GEM, GSL-enriched membrane; GlcT-1, glycosyltransferase; GSL, glycosphingolipid; HA, hemagglutinin; HaK, hamster kidney cells; Hat^-, cytoplasmic tail minus HA; HBV, hepatitis B virus; HCV, hepatitis C virus; HIV, human immunodeficiency virus; HSV, herpes simplex virus; IC, intermediate pre-Golgi compartment; I domain, interacting domain; l_c, liquid crystalline phase; l_d, liquid disordered phase; l_o, liquid

Subcellular Biochemistry, Volume 37: Membrane Dynamics and Domains
Edited by Peter J. Quinn. Kluwer Academic / Plenum Publishers, New York, 2004

viral membrane are not incorporated randomly and do not represent average lipid composition of the host membrane. Recent studies support that specific lipid microdomains such as lipid rafts play critical roles in many aspects of the virus infectious cycle including attachment, entry, uncoating, protein transport and sorting as well as viral morphogenesis and budding. Lipid microdomains aid in bringing and concentrating viral components to the budding site. Similarly, specific viral protein plays an important role in organizing lipid microdomains in and around the assembly and budding site of the virus. This review deals with the specific role of lipid microdomains in different aspects of the virus life cycle and the role of specific viral proteins in organizing the lipid microdomains.

1. INTRODUCTION

In a natural setting of viral infection, the human or animal host is infected at a very low moi (multiplicity of infection) with relatively few virus particles. Therefore, cycles of replication leading to release of new progeny viruses and infection of new host cells by the progeny viruses must be repeated many times and are critically required not only for the survival of the virus and spread from one host cell to another but also is fundamentally essential for producing the disease syndrome in the infected host. In most cases viruses must kill, destroy or alter the function of a large number of cells of a specific organ or tissue before the specific functional abnormality in the form of a disease syndrome such as pneumonia, hepatitis, or acquired immune deficiency syndrome (AIDS), etc is manifested. Furthermore, since viruses usually do not release toxins, each virus particle can only infect and kill one cell and in the process of infection of permissive cells, produces many more virus particles. Therefore, virus replication is an obligatory requirement for disease production caused by lytic viruses.

Abbreviations: Continued
ordered phase; LBPA, lysobisphosphatidic acid; L domain, late domain; M1, matrix protein of influenza virus; MAD, membrane attachment domain; MBGV, Marburg virus; MDCK cell, Madin-Darby canine kidney cell; MHC, major histocompatibility complex; MHV, mouse hepatitis virus; M-PMV, Mason-Pfizer monkey virus; MV, Measles virus; NA, neuraminidase; NDV, Newcastle disease virus; PC, phosphatidylcholine; PE, phosphatidylethanolamine; PLD2, phospholipase D2; RSV, respiratory syncytial virus; SFV, Semliki Forest virus; SIN, Sindbis virus; SNOM, scanning near field optical microscopy; SV, Sendai virus; SV5, Sendai virus; SV40, Simian virus 40; TCR, T cell antigen receptor; TGN, *trans* Golgi network; TIM, Triton-insoluble membrane; TM, transmembrane; TMD, transmembrane domain; TR, transfection receptor; TX-100, Triton X-100; VEE, Venezuelan equine encephalitis virus; VLP, virus like particle; VSV, vesicular stomatitis virus; WT, wild type.

Among the living organisms, virus multiplication occurs by a unique process compared to that of either prokaryotes or eukaryotes which multiply as a unit. Unlike any other living forms, viruses do not multiply as a unit and use a unique strategy called the "assembly" process. In this process, viruses after infection of the host cells produce different viral components independently although temporally regulated and often in different amounts and in different location and compartments within the host cell. Finally, the components must be assembled in a specific subcellular location to produce the whole infectious virus. Therefore, the complexity in the assembly process will vary significantly among the viruses depending on the number of components involved and the location and process of assembly. Simple viruses such as polioviruses which have only a few components to assemble, need fewer steps in the assembly process and can be reproduced even in a test tube outside living cells. On the other hand, viruses like herpes- and poxviruses which are highly complex, have many components to assemble and require multiple steps in assembly involving different cellular organelles. The final step, "budding" and release of buds to generate infectious virus, cannot be reproduced outside the living cell.

With respect to the assembly process, the viruses can be classified under two major groups i.e. non-enveloped (naked) and enveloped viruses. In the non-enveloped viruses which contain the nucleocapsid without any envelope, the assembly process consists of two basic events: (i) the assembly of capsid from one or multiple protein components, (ii) incorporation of viral nucleic acid into the capsid. These two basic events which may involve further maturation and processing of proteins during and after assembly will generate infectious virus particles and can often be reproduced *in vitro* outside the living cell. However, with enveloped viruses, the assembly of nucleic acid (DNA or RNA) and capsid is not sufficient for producing infectious virus. The nucleocapsid must be enveloped in a lipid membrane by a process called "budding" for producing infectious virus particles. The envelopment of viral nucleocapsid is a critical event and obligatory requirement for assembly and release of the infectious enveloped viruses. The viral envelope not only protects the nucleocapsid from degradation by nucleases and proteases but also provides the critical functions in the infectious process by providing the exposed transmembrane receptor-binding proteins on the viral envelope for virus attachment to the new host cells and permitting efficient delivery of the viral nucleocapsid into the cytoplasm of the infected cell by a complex process called fusion for initiating the virus replicative cycle. The fusion between the viral membrane and the host membrane can occur either at the plasma membrane at neutral pH or inside the cell with the endosomal membrane at acidic pH. In addition, for some viruses like influenza virus, the envelope possesses a protein component (neuraminidase [NA]) which facilitates the release of virus after budding has occurred and aids in the transmission of virus from cell to cell and host to host.

Since viruses usually do not carry genes for lipid biosynthesis and lipid modification, viral membranes are formed by extension of the host cell membrane while virus budding occurs. During the budding process, all host proteins are essentially excluded and host lipids are selectively included within the viral membrane. Occasionally, a few host proteins which are functionally important for viral infectious cycle are selectively included in some enveloped viruses. How viruses exclude the host proteins is not fully understood. Many viruses suppress or shut off the synthesis and transport of host proteins and thereby prevent the incorporation of newly synthesized host proteins into viral envelope. However, this proposition would not explain exclusion of the preexisting membrane proteins from the viral envelope. It is generally believed that the interaction of the viral membrane proteins with the internal viral components as well as interaction among the internal viral components such as matrix proteins and nucleocapsids and also interaction among viral envelope proteins effectively crowd out the preexisting host proteins from the budding site. In addition, specific lipid composition and state of viral lipids may facilitate exclusion of host proteins from the viral envelope. Finally, although the lipid composition of the viral envelope depends on the host cell, tissues and organelles from which the viruses bud as well as the external environment and factors including the growth medium in which the host cells are grown and viruses are propagated, the viral envelope does not quantitatively represent the lipid composition of organelle or plasma membrane of infected cells. This would suggest that viruses can selectively modify host membranes and acquire different lipids at the budding site. This process in turn would imply that the host cell membrane is not homogenous or uniform with respect to lipid composition but rather is heterogeneous and mosaic in nature containing heterogeneous microdomains with varying lipid composition and that viruses selectively bud from specific microdomains on the cell membrane. In this review we will discuss the nature and heterogeneity of these microdomains and the role these microdomains play in affecting the lipid composition of viral membranes and in budding and envelopment of virus particles.

2. LIPID COMPOSITION OF VIRAL AND HOST MEMBRANES

Although viruses bud from host membranes and all viral lipids are acquired from the host membranes, viral lipid composition does not match quantitatively with that of the host membranes. There are a number of reasons why the lipid composition of the viral membrane is different from that of the host cell membrane and why the lipid composition varies among different viruses.

(i) *Budding site.* Different viruses bud from different subcellular organelles (Figure 1, Table 1). For example, herpes virus, a complex DNA virus, buds from the inner nuclear membrane. However, fully mature infectious herpes virus exits from the basal layer of infected epithelial cells. Hepadna, rota and

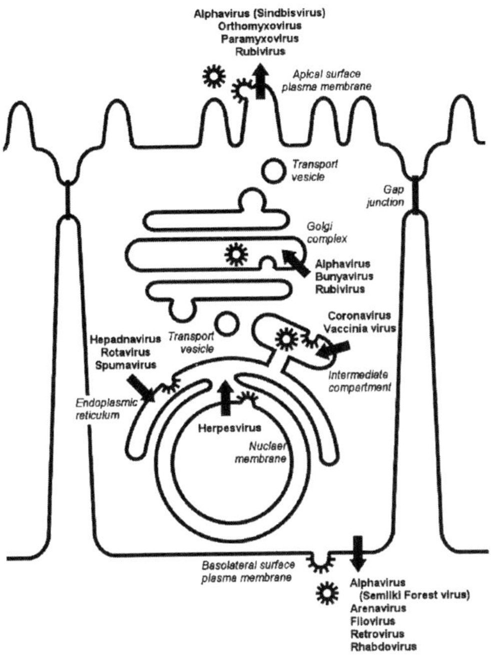

Figure 1. Budding sites of enveloped viruses. The figure depicts a polarized epithelial cell and viruses are budding from the different subcellular compartments and plasma membrane domains are shown.

Table 1. Budding site of enveloped viruses

Host membrane	Budding site of enveloped viruses
Nuclear membrane	Herpesvirus
Endoplasmic reticulum (ER)	Hepadanavirus, Rotavirus, Spumavirus
Intermediate compartment	Coronavirus, Vaccina virus
Golgi complex	Alphavirus, Bunyavirus, Rubivirus
Plasma membrane:	
Apical	Alphavirus (Sindbis virus), Orthomyxovirus, Paramyxovirus, Rubivirus
Basolateral	Alphavirus (Semliki Forest virus), Arenavirus, Filovirus, Retrovirus, Rhabdovirus

spuma viruses bud from the endoplasmic reticulum. It is not clear if they use the specific domain of plasma membrane (apical or basolateral) for their exit and most likely, cell lysis causes release of these viruses from infected cells. Corona and vaccinia viruses acquire their envelope from the intermediate pre-Golgi compartment (IC). Vaccinia virus further undergoes a complex maturation process before being released from the plasma membrane. Both vaccinia and corona viruses bud from the basal plasma membrane. Bunya and toga (alpha- and rubivirus) viruses acquire their envelope from the Golgi complex. However, their budding site at the plasma membrane varies. For example, Sindbis virus (SIN) appears to exit from the apical membrane whereas Semliki Forest virus (SFV) buds from the basolateral membrane although both of them are members of the alpha virus groups. Finally, the assembly and budding of some viruses such as orthomyxo-, paramyxo-, filo-, retro- and rhabdoviruses occur only at the plasma membrane. Therefore complete infectious virions of these virus groups are not found inside the infected cells. Among these viruses, orthomyxo- and paramyxoviruses bud from the apical plasma membrane whereas the filo-, retro- and rhabdoviruses bud from the basolateral plasma membrane. It should be noted that even some non-enveloped viruses such as papilloma, Simian virus 40 (SV40), polio virus, also exit from the apical plasma membrane. Furthermore, the budding or exit site from the apical or basolateral membrane also affects viral pathogensis and the disease syndrome. Usually, viruses budding from the apical plasma membrane such as influenza (orthomyxovirus) viruses do not cause viremia or infect internal organs. On the other hand, viruses budding from basolateral surface cause cell to cell fusion leading to heterokaryon or syncytium formation, vesicle formation as well as viremia and systemic infection of internal organs. However, there are exceptions to this general rule such as measles viruses which although budding predominantly from apical membrane, cause cell to cell fusion, viremia and systemic infection. In addition, other factors such as cleavage of and fusion by F protein at neutral pH play an important role in the phenotype of these viruses.

Lipid composition of viral membrane primarily depends on the budding site. Since different subcellular organelles possess varying lipid composition, viruses budding from membranes of different organelles (Figure 1, Table 1) will have different lipid composition (Table 2). Viruses budding from the endoplasmic reticulum (ER), Golgi, nucleus or plasma membrane will have different lipid composition since the lipid composition of different membranes vary both quantitatively and qualitatively. In addition, viruses budding from the different domains of the same membrane will have different lipid composition. For example, viruses budding from the plasma membrane will have different lipid composition depending on whether the virus buds from the apical or basolateral domain of the plasma membrane in polarized epithelial cells.

Table 2. Phospholipid composition of viral envelope and cellular membrane

Cell/virus	Phospholipid composition (%)										References
	PA	PC	PE	PI	PS	SPH	LPE	LPC	Lec	Other	
Uninfected BHK cells (PM)	-	71.7	13.9	4.3	2.5	7.6	-	-	-	-	van Genderen *et al.*, 1995
EAV from BHK	-	54.9	13.9	7.1	5.1	19.5	-	-	-	-	van Genderen *et al.*, 1995
FPV from BHK	-	38.5	28.6	8.2	6.7	18.1	-	-	-	-	van Meer & Simons, 1986
HSV from BHK	-	55.8	13.0	8.2	7.1	15.9	-	-	-	-	van Genderen *et al.*, 1995
SFV from BHK	5.6	33.0	23.0	1.5	13.0	21.0	-	1.0	-	-	Renkonen *et al.*, 1971
Sindbis virus from BHK	-	26.2	35.4	-	20.3	18.2	-	-	-	-	David, 1971
VSV from BHK	-	41.4	22.2	4.8	10.8	20.8	-	-	-	-	van Genderen *et al.*, 1995
Uninfected calf kidney cells	2.0	-	16.0	5.0	4.0	SPH + LPE: 17.0		-	44.0	4.0	Kates *et al.*, 1961
Influenza virus from calf kidney	12.0	-	8.0	5.0	5.0	SPH + LPE: 14.0		-	36.0	14.0	Kates *et al.*, 1961
Uninfected CE	-	-	18.0	6.0	6.0	SPH + LPE: 20.0		-	48.0	2.0	Kates *et al.*, 1961
Influenza virus from CE	-	-	22.0	6.0	7.0	SPH + LPE: 35.0		-	28.0	2.0	Kates *et al.*, 1961
Uninfected CEC (PM)	-	43.5	23.8	18.8	11.1	3.3	-	-	-	-	David, 1971
Sindbis virus from CEC	-	26.4	32.1	-	20.8	20.5	-	-	-	-	David, 1971
Uninfected CEF (PM)	-	36.0	26.0	-	13.0	22.0	-	-	-	2.0	Quigley *et al.*, 1971
ALV (RAV-2) from CEF	-	20.0	30.0	-	16.0	33.0	-	-	-	1.0	Quigley *et al.*, 1971
ASV (B77) from CEF	-	21.0	29.0	-	17.0	30.0	-	-	-	3.0	Quigley *et al.*, 1971
NDV from CEF	-	26.0	33.0	-	10.0	26.0	-	-	-	4.0	Quigley *et al.*, 1971
RSV from CEF	-	20.0	33.0	-	12.0	29.0	-	-	-	4.0	Quigley *et al.*, 1971
Sendai virus from CEF	-	27.0	31.0	-	14.0	24.0	-	-	-	5.0	Quigley *et al.*, 1971
Sindbis virus from CEF	-	21.0	33.0	-	15.0	29.0	-	-	-	2.0	Quigley *et al.*, 1971
Uninfected CF cells	-	51.2	20.1	7.5	6.6	6.3	-	-	-	8.3	Heydrick *et al.*, 1971
VEE virus from CF	-	35.4	28.1	1.8	16.9	13.1	-	-	-	3.7	Heydrick *et al.*, 1971
Uninfected L cells	-	47.8	16.1	3.9	2.9	9.3	-	-	-	20.0	Heydrick *et al.*, 1971
VEE virus from L cells	-	23.3	29.1	0.5	14.7	26.2	-	-	-	5.1	Heydrick *et al.*, 1971
Cf 124 Ti cells (PM)	-	48.0	25.0	8.5	7.5	0.8	1.4	3.0	-	4	Balange-Orange & Devauche, 1982
CIV from Cf 124 Ti	-	35.0	26.0	27.0	6.0	2.0	4.0	t	-	t	Balange-Orange & Devauche, 1982

Table 2. Continued.

Cell/virus	Phospholipid composition (%)										References
	PA	PC	PE	PI	PS	SPH	LPE	LPC	Lec	Other	
Uninfected H9 cells (PM)	0.6	47.6	23.3	4.9	10.3	9.6	-	-	-	3.4	Aloia *et al.*, 1993
HIV-1_{RF} from H9	1.2	29.9	24.6	0.4	9.0	24.1	-	-	-	8.3	Aloia *et al.*, 1993
HIV-2-L from H9	1.2	27.6	27.3	1.1	15.5	23.0	-	-	-	4.8	Aloia *et al.*, 1993
Uninfected HaK cells (PM)	-	46.8	13.0	11.7	5.0	24.4	-	-	-	-	Klenk & Choppin, 1970
SV5 from HaK	-	43.8	17.1	8.5	5.0	25.8	-	-	-	-	Klenk & Choppin, 1970
Uninfected MDBK cells (PM)	-	44.5	27.2	2.9	2.2	22.8	-	-	-	-	Klenk & Choppin, 1970
SV5 from MDBK	-	23.8	40.9	5.2	2.0	27.3	-	-	-	-	Klenk & Choppin, 1970
Uninfected MDCK cells	-	62.5	20.1	5.2	4.2	7.7	-	-	-	-	van Meer & Simons, 1986
FPV from MDCK (apical)	-	9.6	44.8	2.8	22.4	20.5	-	-	-	-	van Meer & Simons, 1986
VSV from MDCK (basolateral)	-	28.9	33.5	3.2	18.9	15.6	-	-	-	-	van Meer & Simons, 1986
Uninfected Sac(-) cells	-	71.8	16.8	5.9	3.8	1.8	-	-	-	-	van Genderen *et al.*, 1995
MHV from Sac(-)	-	71.1	10.8	10.1	5.7	4.2	-	-	-	-	van Genderen *et al.*, 1995

-, Not determined; t, trace amount.

Compared to the basolateral domain, the apical domain of the plasma membrane of intestinal epithelial cells and Madin-Darby canine kidney (MDCK) cells possesses 2 to 4 fold enrichment of phosphatidylcholine (PC) but similar level of phosphatidylethanolamine (PE). Therefore in polarized epithelial cells like MDCK cells, both influenza virus (fowl plague virus [FPV]) budding from apical surface and vesicular stomatitis virus (VSV) budding from the basolateral surface will have some common lipids representing the plasma membrane but some different lipids representing the domain of their origin. Both FPV and VSV possessed less PC and more SPH and compared to the total cellular lipid representing their origin from the plasma membrane. On the other hand, FPV, an apical virus, has 4 fold less PC compared to VSV, a basolateral virus, when grown in polarized MDCK cells. In polarized cells, tight junctions restrict the lateral diffusion of lipids on the exoplasmic leaflet but permit free diffusion of lipids in the cytoplasmic leaflet between the apical and basolateral domains. This will affect the lipid composition of the apical and basolateral domains and consequently, will affect the lipid composition of viruses budding from these respective plasma membranes. However, when both viruses were grown in baby hamster kidney (BHK) cell, a nonpolarized cell, the PC content of both FPV and VSV were similar (van Meer and Simons, 1986).

(ii) *Other factors.* The lipid composition of the same virus will also depend on the cell type in which the viruses were grown, composition of the growth medium used for cell culture and virus propagation. In addition, the state of the cell monolayer layer such as stationary vs actively growing and association with the neighboring cells, i.e. polarized vs nonpolarized or transformed vs nontransformed, etc. will also affect lipid composition of virus particles.

(iii) *Mosaic nature of the membrane lipids.* Membrane lipids form an amphipathic barrier between the cell and its outside environment and exist as a lipid bilayer. Numerous studies have convincingly shown that lipid bilayers of plasma membrane are not homogenous or uniform but mosaic in nature containing heterogenous microdomains and viruses bud from these microdomains.

(iv) *Viral envelope proteins* are also critical factors and play important roles in determining the lipid composition of the viral membrane. This was evident in early studies describing the requirement and selection of specific lipids in the envelopment of bacteriophage PM2, an enveloped DNA virus (Brewer, 1979). Authors showed that only one structural protein sp6.6, an integral membrane protein, was able to mobilize specific lipids into newly assembled membrane of the desired shape, size and lipid composition. These lipid vesicles containing a single viral protein were composed of phosphatidylglycerol and phosphatidylethanolamine whose ratio was essentially inverse of that found in host membranes but same as in the wild type (WT) virus. With enveloped animal viruses, integral transmembrane viral proteins associate with different lipids present in the liquid-ordered phase (l_o phase) or liquid-disordered phase (l_d phase) during their transport from *trans* Golgi membrane to the plasma membrane and during their residence on the plasma membrane prior to and during the virus budding process. The lipid-protein association is further modified and stabilized by the internal viral proteins which interact with the transmembrane viral proteins and/or with the cytoplasmic leaflet of the lipid bilayer. Accordingly, the lipid composition of virus membrane will vary depending on microdomains from which the virus will bud.

3. LIPID MICRODOMAINS

Membrane lipids are generally viewed to exist as lipid bilayer in two dimensional fluid mosaic state forming an amphipathic barrier between the cell and its external environment. Their main functions are to prevent free leakage of cytoplasmic content into outside environment and unregulated entry of external elements into the cell and also to provide suitable means to respond to the external environmental stimuli by means of signal transduction pathways. It is generally believed that membrane lipids exist in liquid-crystalline (l_c) or l_d phase and function as a solvent for membrane proteins. However, recent studies over the last decade have argued against such simplified view of the physical state of membrane lipids and demonstrated the existence of heterogeneous lipid microdomains. Heterogeneity among the membrane lipids leads to lateral organization and phase separation forming l_o phase and causes asymmetric distribution of different membrane lipids in exoplasmic vs cytoplasmic leaflets forming assembly of various microdomains. These lipid microdomains exhibit a common property of relative insolubility in nonionic detergents such as Triton X-100 (TX-100) at low temperature (Simmons and Ikonen, 1997). However, there appears to be heterogeneity among the microdomains. In general, three types of microdomains with varying lipid composition and function have been identified. These are: (a) caveolae, (b) lysobisphosphatidic acid (LBPA) domain, and (c) lipid rafts enriched in cholesterol and sphingolipids. There are further variations within each of these classes of microdomains.

3.1. Caveolae

Caveolae are flask shaped invaginations (50–100 nm in diameter) formed on the plasma membrane, predominantly on the basolateral surface containing both caveolin 1 and caveolin 2 proteins. Approximately, 14 to 16 caveolin 1 monomers assemble into discrete patch forming homooligomers by the interacting NH_2 termini with NH_2 termini and COOH termini with COOH termini at the cytoplasmic side of the membrane. Membrane attachment of caveolin 1 is mediated by three separate membrane attachment domains (MAD) (N-terminus MAD amino acid [aa] 82–101, transmembrane MAD aa 102–134 and C-terminus MAD aa 135–150) (Galbiati *et al.*, 2001). Caveolin 1 binds to cholesterol, and exhibits TX-100 insolubility supporting the l_o phase separation of membrane lipids. Caveolin containing microdomains are also TX-100 insoluble but can be separated from lipid rafts by antibodies against caveolin 1. Caveolin containing microdomains possess caveolin, glucosylceramide, Ha Ras, and a large quantity of cholesterol but only a small amount of sphingomyelin whereas lipid raft microdomains contain GM3 gangloside, sphingomyelin, cholesterol, c-Src, Rho A but no caveolin. Caveolae have been shown to play important role in endocytosis and in apical transport of proteins in some cells. Antibodies to caveolin 1 inhibit transport of influenza virus hemagglutinin (HA), an apical protein but not of VSV G, a basolateral protein, to the cell surface in MDCK cells (Scheiffele *et al.*, 1998). However, in Fisher rat thyroid (FRT) cells which do not synthesize caveolin 1, both influenza virus HA as well as GPI-anchored PLAP are efficiently transported to apical surface (Martin-Belmonte *et al.*, 2000) suggesting that caveolin may not be an absolute requirement for transport of apical proteins. The role of caveolae in virus biology is not well defined. However, caveolae have been shown to be involved in the entry process of SV40, a nonenveloped double stranded small DNA virus, which replicates in the nucleus although the function of caveolae in SV40 entry and uncoating process remains yet to be elucidated. Specific role of caveolae in virus morphogenesis of enveloped viruses has not been defined. Therefore, properties and function of caveolae will not be discussed further in this review.

3.2. Lysobisphosphatidic acid (LBPA) domain

Recently, the existence of another lipid membrane microdomain which is different from glycosphingolipid (GSL)-rich lipid raft, has been identified (Kobayashi and Hirabayashi, 2001). LBPA microdomain was discovered in GM-95 cells, a mutant melanoma cell line lacking all GSLs due to a deficiency of ceramide glucosyltransferase (GlcT-1), suggesting that TX-100 resistant microdomains can be formed in the absence of GSLs and that GSLs are not essential for the formation of TX-100 resistant raft-like microdomains. LBPA microdomains were shown to be present in endocytic pathway and is enriched in a unique phospholipid, LBPA which is localized primarily in the internal membrane of multivesicular endosome (late endosome), and is involved in lipid and protein sorting within the endosomal system. Like lipid rafts, LPBA microdomains exhibit common properties such as TX-100 resistance supporting the l_o state of the membrane lipids in the LPBA microdomains. However, little is known about the role of this LBPA microdomain in viral membrane biogenesis and in virus budding. Therefore, LPBA microdomains will not be discussed further in this review.

3.3. Lipid rafts

3.3.1. Structure and function of lipid rafts

Lipid rafts are operationally defined as cholesterol dependent microdomains resistant to solublization by nonionic detergents such as TX-100 at low temperature. Lipid rafts consist of sphingolipid-cholesterol clusters, usually varying in size and are present on the plasma membrane, apical transport vesicles as well as on Golgi and *trans* Golgi membranes. Lipid rafts vary in size, ~50 nm in diameter (Pralle *et al.*, 2000) and smaller than the caveolae which also exhibits TX-100 insolubility similar to the lipid rafts. However, as indicated above, caveolae contain caveolin, a cholesterol-binding protein, which is absent in lipid rafts and exhibit a different lipid composition. Furthermore, caveolae are present predominantly in the basolateral membrane whereas lipid rafts are found primarily in the apical domain of the plasma membrane of polarized cells but are also present in non-polarized cells such as fibroblasts.

Lipid raft microdomains are formed by lateral organization and phase separation of lipids between l_o phase and l_d or l_c phases. l_o phase separation also leads to asymmetric distribution of different lipids in the exoplasmic vs cytoplasmic lipid leaflets. These lipid microdomains containing l_o phase have been variously called by different names such as detergent insoluble GSL-enriched domains (DIGs), GSL-enriched membrane (GEMs) or microdomain, detergent resistant membranes (DRMs), Triton-insoluble membranes (TIMs), GSL/or sphingolipid-cholesterol rafts, lipid rafts or simply rafts (Simons and Toomre, 2000). We will refer these microdomains as lipid rafts or rafts in this review. As mentioned above, although these rafts are found in the plasma membrane primarily on the apical domain, they are also present in the basolateral domain of plasma membrane as well as in the Golgi and *trans* Golgi network (TGN) membranes.

Detailed structural and functional analysis of lipid rafts will not be presented here as they have been described in a number of recent reviews (Simons and Ikonen, 1997; Brown and London, 1998a, 1998b; Nayak and Barman, 2002). Our emphasis in this review will be to discuss the functional and structural properties of the membrane lipids that affect virus membrane assembly and virus budding and in turn, aid in the incorporation of specific lipids into the viral membrane and thereby affect the lipid composition of virus particles. The lipid composition of viral membrane and host cell domain are not only important in the final steps of the infection cycle such as virus assembly and budding but for many viruses, play critical roles in the initial steps such as virus entry, infectivity including virus attachment, fusion between viral and cell membrane and delivery of viral genome into the host cell for initiating the infectious cycle (Samsonov *et al.*, 2002; Chatterjee *et al.*, 2002). However, the major emphasis of this review will be to discuss the physical state of membrane lipids at the budding site and their role in the final steps of the infectious cycle, namely the virus assembly, envelopment and release of virus particles.

Specific classes of lipids are critical for the formation of functional microdomains including lipid rafts. It has been shown that sphingolipids and cholesterol when mixed *in vitro* can form detergent resistant raft-like artificial membranes in the absence of any protein. This would argue against the absolute requirement of proteins in the formation of lipid rafts and support that proteins may be present but are not absolutely essential for maintaining the physical state of lipids such as lateral packing of lipids leading to the formation of lipid rafts. To explain the behavior of certain lipids particularly sphingolipids (glycosphingolipids, GSL) and sphingomyelin and cholesterols as well as GPI-anchored proteins to non-ionic detergent treatments such as Triton X-100, NP40, Brij at low temperature (4°C), Simons and colleagues (1997) proposed the "lipid raft" hypothesis. They proposed that lipid rafts are formed as microdomains or specific clusters of lipids by lipid–lipid interactions depending on the biophysical characteristic of the lipid components. GPI anchored proteins which also exhibited TX-100 insolubility was also considered as part of the lipid raft microdomains. However, since demonstration of these complexes was dependent on extraction by TX-100 detergents at low temperature, questions were raised whether these microdomains were present in the living cell membrane or formed during TX-100 treatment. Later, however, lipid rafts were demonstrated in living cells and studied by a number of independent techniques such as fluorescence energy transfer (FRET), electron microscopy (EM) with immunogold beads, atomic force microscopy (AFM), scanning near field optical microscopy (SNOM), confocal laser scanning microscopy (CLSM) (Damjanovich *et al.*, 2002). Therefore it is now an accepted view that lipid rafts are present in living cells and provide critical role in a number of important cellular functions such as endocytosis (Sharma *et al.*, 2002; Nabi and Le, 2003), protein transport (Alonso and Millán, 2001), protein sorting (Ikonen, 2001; Aït Slimane and Hoekstra, 2002), signal transduction (Muller, 2002), immunological responses to specific antigens (Langlet *et al.*, 2000; Katagiri *et al.*, 2001; Miceli *et al.*, 2001; Alonso and Millán, J. 2001; Vogt *et al.*, 2002) as well as in virus assembly and budding (Barman *et al.*, 2001; Campbell *et al.*, 2001; Ono and Freed, 2001; van der Goot and Harder,

2001; Nayak and Barman, 2002; Suomalainen, 2002; Briggs *et al.*, 2003; Chazal and Gerlier, 2003; Harder, 2003).

3.3.2. Requirement of specific lipids in lipid rafts

Lipid rafts are dynamic in nature. They can coalesce or dissociate to form larger or smaller rafts with individual lipids moving in and out of the raft complex (Kusumi and Sako, 1996). Since cholesterol-sphingolipid complexes are particularly insoluble in TX-100 extraction at low temperature and float to the top of the gradient because of low density and high lipid content, this procedure has been used extensively to isolate and study the property of the lipid rafts. Such analysis demonstrated varying degrees of insolubility to TX-100 treatment of different membrane functions and therefore raised the possibility of existence of different classes of lipid rafts. Bulk membranes are enriched in glycophospholipids (frequently containing unsaturated fatty acids) and are highly soluble to TX-100 extraction at low temperature. Lipids in these membranes are loosely packed in l_d state and display a state of high fluidity. Lipid raft microdomains, on the other hand, contain glycerophospholipids, (glyco)sphingolipids, GPI lipids bearing predominantly saturated fatty acids and cholesterol. These lipids can form tight packing and cholesterol in the region contributes to tight packing by filling the interstitial space between the long saturated acyl chains and sphingolipids resulting in the formation of l_o state of lipids in lipid raft microdomains. Most of the membrane proteins including the transmembrane (TM) proteins except the GPI-anchored proteins are excluded from these microdomains. These microdomains also contain gangliosides such as GMI and GM2 (Alonso and Millán, 2001). The tight lateral packing of sphingolipids and cholesterol leads to TX-100 insolubility at low temperature. However, these microdomains are not homogenous and vary in TX-100 insolubility and lipid composition (Kobayashi and Hirabayashi, 2001). Resistance of these microdomains to cold nonionic detergents and alkaline sodium bicarbonate are sensitive to cholesterol depletion indicating the role of cholesterol in the formation and stability of lipids and lipid packing in lipid rafts (Müller, 2002). In addition to cholesterol, sphingolipids are also essential components of lipid rafts. There are two major sphingolipid complexes- GSLs and sphingomyelin in mammalian cells. Studies using inhibitors for GSL biosynthesis and mutant cells showed that the presence of GSLs is not required for raft formation. Therefore, the presence of either GSLs or sphingomyelin on the outer leaflet along with cholesterol may be sufficient for the formation of raft-like microdomains.

Organization of lipids in membranes can lead to different physical state of the membranes: (i) quasi solid gel phase, (ii) l_c, also called l_d phase and (iii) l_o phase. However, the quasi solid gel phase is usually formed in test tube *in vitro* and unlikely to be found in living cells. Living cells contain lipids in l_c and l_o phases depending on the physical state of the lipids. The majority of lipids in the membrane of living cells are in an intermediate (l_c or l_d) phase exhibiting high fluidity and low melting point. These lipids are less packed and have a high rotational and lateral mobility. On the other hand, lipid rafts are specialized lipid microdomains containing lipids in l_o phase. The organization of lipids in the lipid raft depends on the physical states of lipids, which differ in packing, degree of the order and mobility of the constitutent lipids. In l_o phase, hydrocarbon chains are in extended conformation and lipids are highly ordered, densely packed and have reduced mobility in the presence of cholesterol. In the test tube, they exhibit a relatively high melting point, resist solubilization by TX-100 and possess raft-like properties. Cholesterol tends to intercalate its rigid planar hydrophobic ring between the hydrocarbon chains of lipid with its hydroxyl group oriented towards the water phase. Cholesterol aligns with phospholipids and sphingolipids and supports the extended *trans*conformation and increases the order but decreases the lateral and rotational mobility. Glycoglycerophospholipids with kinked structure interferes with packing and have lower affinity for l_o phase. On the other hand, ceramide backbone of sphingolipids contains saturated hydrocarbon chains and favors packing and l_o phase formation. Cholesterol aligns with saturated hydrocarbon leading to the formation of sphingomyelin-cholesterol coupling and increases resistance to TX-100 and Brij 58. Glycosphingolipids as well as sphingomyelins are present in lipid rafts. Phospholipids that are present in the lipid raft are enriched with saturated and monounsaturated but depleted in polyunsaturated acyl chains. Although lipid components including ceramides as well as cholesterol are synthesized ER and may be assembled as preraft-like microdomains in ER, formation of detergent resistant raft domains does not occur until in the Golgi where ceramides are further modified by glycerol transferases. Sphingomyelin/cholesterol ratio as well as total cholesterol concentration in the raft are critical for TX-100 insolubility (Slimane and Hoekstra, 2002). Also, glycosphingolipids may distribute differently in rafts causing variation in concentration in individual rafts and result in varying detergent resistance within lipid rafts.

3.3.3. Requirement of proteins in lipid rafts

As indicated above, although lipid bilayer representing lipid rafts can be formed using only different lipid molecules which can assemble into microdomains *in vitro*, lipid rafts present in cellular membrane invariably contain protein molecules. However, it is not clear whether proteins assemble microdomains around them by attracting particular lipids or partition into already formed microdomains. These proteins include GPI anchored proteins exposed outside and linked to outer plasma membrane leaflet via phosphatidylinositol moiety. The saturated acyl and alkyl chains of GPI increase the affinity for ordered lipids in the membrane phase. On the other hand, the majority of the transmembrane (TM) proteins are excluded from the lipid rafts. However, some TM receptors are selectively present in raft microdomains. For example CD44 (type I), CD26 (type II) are found in raft microdomains. Also, multispanning proteins such as MAL (tetraspan) are present lipid rafts and in apical transport system. Similarly, CD20 (tetraspan), LMP1 (hexaspan) in Epstein-Barr virus (EBV)-infected cells as well as influenza virus NA without any lipid modification are also present in lipid rafts (Hoessli *et al.*, 2001). Lipid rafts are present in Golgi and subapical compartments (SAC) along with MAL and annexin IIIB. Protein– lipid and protein–protein interactions may contribute to coalescence, growth and stability of some lipid rafts. Sphingolipid-cholesterol packing may coexist with more fluid membrane lipid and actin-based cytoskeleton may aid in creating diffusion-restricted membrane compartments. Cytoskeleton-associated TM proteins like CD44, CD43 may affect domain organization at basolateral surface. GM3, GM1 enriched rafts segregate at the leading edge of uropod. Identification of proteins as a component of raft requires the detergent insolubility of the protein and in turn, depends on the presence of cholesterol in membrane. GPI-anchored proteins are present on the outer leaflet whereas proteins with palmitoylation, prenylation, acylation, myristoylation are usually present in the inner leaflet of lipid bilayer.

Although rafts have been shown to exclude most membrane proteins, particularly TM proteins, they are responsible for concentrating GPI-linked and myristylated proteins such as doubly acylated src kinases as well as G proteins for receptor signaling and receptor trafficking in polarized epithelial and endothelial cells. Rafts function in immune cell activation upon ligand binding by (i) concentrating key receptors such as T-cell antigen receptor (TCR) (Janes *et al.*, 1999; Langlet *et al.*, 2000; Drevot *et al.*, 2002), B cell antigen receptor (BCR) (Guo *et al.*, 2000) as well as certain Fc receptors such as FcεRI (Holowka *et al.*, 2000) and FcγRIIB1 (Aman *et al.*, 2001), (ii) recruiting additional components such as Src family kinase Lyn and others for receptor signaling, (iii) providing platform for trafficking for BCR. In addition, entry of receptors into rafts may be controlled to regulate cell activation. Finally, pathogens may use a strategy to target rafts for blocking immune cell function. For example, LMP2 of EBV is present in the rafts, blocks BCR entry into rafts and BCR trafficking and thus shuts down BCR function (Cherukuri *et al.*, 2001).

On the cytoplasmic sides, Src kinases and some Gα subunit of heterotrimeric GTPases interact via their acyl chains (e.g. saturated palmitic, and also N-terminal myristic acids) with the cytoplasmic leaflet of lipid rafts. Although most of the TM proteins are excluded from the raft, some TM proteins like influenza virus HA, human immunodeficiency virus (HIV) gp160 use acyl modification of cysteine as well as specific amino acid residues of in the TM for interaction with lipid rafts. However, now it is well established that some TM protein like influenza virus NA (type II protein) which does not have any acyl modification of transmembrane domain (TMD) also associates with lipid raft (Barman and Nayak, 2000; Zhang *et al.*, 2000). Furthermore, cytoplasmic tail can also affect lipid raft-association of NA (van der Goot and Harder, 2001).

Proteins in the raft may provide stability in raft domains since rafts can vary in size (Sheets *et al.*, 1997; Pralle *et al.*, 2000; Schütz *et al.*, 2000). Raft-like patches as large as 350–700 nm can be assembled from multiple small elementary raft-ordered membrane domains through protein-based lattices or anchored to cytoskeletal elements such as actin (Harder and Simons, 1999).

Furthermore, plasma membrane is in a dynamic equilibrium between l_o raft and l_d non-raft phases. Proteins such as Src related tyrosine kinase Lck remain separated from CD45-phosphatase which regulates Lck's kinase activity. They remain segregated in raft and nonraft regions of the plasma membrane. Also, raft-associated proteins could be separated in different raft domains. Ligand binding causes linking outside of the membrane that brings different raft or non-raft-associated proteins together by coalescing. Their contents are mixed causing lateral cross-linking of raft-associated proteins. This type of clustering will affect functional activation and initiate chain reaction as in signal induction.

Presence of specific peptides in a specific conformation may increase the order of the lipids in the lipid raft. The influenza virus HA TM peptide is present in an α-helical conformation in detergent micelles and in phospholipid bilayers. The helicity of the peptide increased in lipid bilayers composed of acidic lipids and in turn, the presence of the peptide also increased the acyl chain order of the lipid bilayer. The mutual interactions of the TM peptide and the surrounding lipids may be reciprocal. Ordered lipids attract TM domains and TM domains in turn increase the order of the lipids surrounding them. This process may aid in targeting HA and NA TM proteins to the ordered lipid rafts and organizing ordered lipid rafts around them (Tatulian and Tamm, 2000). However, incorporation of HA alone is not sufficient to organize the ordered lipid environment since HA incorporated in VSV envelope is TX-100 soluble (Scheiffele *et al.*, 1999).

In addition, raft-dependent protein–protein interactions could bring proteins that are present in less-ordered membrane to lipid rafts by interaction with raft-associated proteins. Interaction between influenza virus matrix (M1) and HA brings M1, a nonraft-associated protein, into lipid raft by its interaction with HA (Ali *et al.*, 2000a). Also raft-ordered membrane domains may be formed *de novo* around TM proteins on the plasma membrane such as the engaged immune receptors. Affinity of the l_o domains can be increased by organization, acylation, coupling to raft-associated molecules or by conformational changes (Harder *et al.*, 1998).

Despite extensive studies of the nature and the function of lipid rafts, less is known about the heterogeneity among the lipid rafts. It is not clear if there are specific classes of distinct lipid rafts. The common criteria of TX-100 insolubility used for assaying raft-association of membrane proteins show wide variation among different proteins or among the various mutants of the same protein. These results could be due to either varying affinity of different proteins for the same class of lipid rafts or similar affinity for different classes of lipid rafts exhibiting different TX-100 insolubility (Röper *et al.*, 2000).

3.3.4. Function of lipid rafts

Lipid rafts have been shown to be critically involved in numerous cellular functions (Simons and Ikonen, 1997). These include clathrin-independent endocytosis, protein transport, protein sorting, protein targeting, membrane-trafficking, membrane-signaling, antigen-mediated immune response, as well as in the assembly and budding of some enveloped viruses.

In polarized epithelial cells such as MDCK cells, lipid rafts have been implicated in cell surface transport as well as in sorting and targeting apical proteins. Although the majority of these apical vesicles contain lipid rafts as evidenced from their lipid composition and TX-100 insolubility, some apical proteins such as bovine enteropeptidase (Zheng *et al.*, 1999), intestinal brush border enzymes (Danielsen, 1995), transferrin family proteins (Danielsen and van Deurs, 1995) as well as influenza C virus HEF and influenza A virus M2 protein (Hughey *et al.*, 1992), both of which are acylated (Viet *et al.*, 1990; 1991), do not use lipid rafts (Zhang *et al.*, 2000) for apical delivery.

Using influenza virus HA, a model transmembrane apical protein as a marker, apical carrier vesicles which are involved both in the transport of HA to cell surface in nonpolarized cells such as fibroblasts as well as to the apical surface in polarized epithelial cells such as MDCK were shown to contain lipid raft lipids as well as proteins such as caveolin 1, annexin XIIIb, MAL/VP17, phospholipase D2 (PLD2). Therefore these proteins are presumed to function in apical targeting. Antibody to caveolin-1 inhibited the transport of influenza virus HA to the cell surface and to the apical but not of VSV G, basolateral protein, implicating the presence of caveolin-1 in apical vesicle and its function in apical transport (Scheiffele *et al.*, 1998). However, caveolin-1 is not an essential element for apical transport of HA since in FRT cells which do not express caveolin 1/VIP21 (Zurzolo *et al.*, 1994), HA as well as GPI-anchored placental alkaline phosphatase (PLAP) are efficiently transported to the apical plasma membrane (Martin-Belmote *et al.*, 2000). Annexin XIIIb, a member of the annexin XIII sub-family, in its myristoylated form is present in sphingolipid-cholesterol rafts and stimulates apical transport (Lafont *et al.*, 1998). MAL/VIP17, a tetraspanning membrane proteolipid, is a nonglycosylated integral membrane protein, present in lipid rafts and functions as an element in apical transport pathway since depletion of MAL inhibits both the cell surface transport and apical transport of influenza virus HA (Puertollano *et al.*, 1999). PLD2 is present in TX-100 insoluble lipid rafts and is upregulated by caveolin-1 and therefore may function as a component of apical vesicles (Czarny *et al.*, 1999). Others such as Thy-1 CD59 as well as ganglioside GM1 are preferentially partitioned into lipid rafts, although their function in apical transport pathway is not well defined.

Lipid rafts are involved not only in protein sorting and protein targeting to apical membrane in polarized cells but also in post Golgi transport of apical proteins to cell surface in nonpolarized cells. Depletion of cholesterol which leads to disorganization of lipid rafts causes missorting of apical proteins like HA, also reduces the overall transport of HA to the cell surface in both polarized and nonpolarized cells (Keller and Simons, 1998). Lipid rafts are also involved in endocytosis of proteins from apical surface via endocytic pathways independent of clathrin-coated vesicles or caveolae (Danielsen and van Deurs, 1995), as well as in endocytosis and transcytosis of proteins from basolateral surface via caveolae (Schnitzer *et al.*, 1994). Lipid rafts are also involved in membrane signaling. Several signaling molecules such as trimeric G proteins, and Ras are also associated with lipid rafts (Li *et al.*, 1995; Song *et al.*, 1996). Src family kinases, Lck ($p56^{lck}$) and Fyn ($p59^{fyn}$) implicated in antigen-mediated B and T-cell activation, are also associated with lipid rafts. Raft-association may concentrate several members of proteins involved in signal transduction and thus facilitate cross-talk among the protein molecules involved.

4. ROLE OF LIPID RAFTS IN VIRUS BIOLOGY

Lipid rafts have been shown to be involved in many aspects of virus life cycle. They have been implicated in virus attachment, entry, uncoating, RNA replication, protein transport, and most significantly, assembly as well as budding (Nayak and Barman, 2002; Chazal and Gerlier, 2003). Although lipid rafts clearly play the most important role in budding, they may also play critical roles in other processes of virus life cycle.

4.1. Role of lipid rafts in virus–host interactions, virus entry and virus replication

Rafts have been shown to be the preferential site for interaction of many pathogenic organisms (bacteria, parasite, viruses) with target cells (Campbell *et al.*, 2001). With respect to viruses, lipid rafts may provide platforms for

Table 3. Role of lipid rafts in viral entry

Virus	Mechanism of use	References
Semliki Forest virus (SFV)	Fusion of viral envelope requires the presence of lipid rafts.	Kielian & Helenius, 1984 Phalen & Kielian, 1991 Waarts *et al.*, 2002
Sindbis virus (SIN)	Highly dependent on cholesterol for fusion. Fusion of viral envelope involves lipid rafts.	Lu *et al.*, 1999 Waarts *et al.*, 2002
HIV-1	HIV-1 enters epithelial cells via raft-mediated endocytosis and transcytosis.	Nguyen & Hildreth, 2000 Alfsen *et al.*, 2001 Narayan *et al.*, 2003
Simian virus 40 (SV40)	Virus enters the cell by binding MHC class 1 molecules that are located in caveolae.	Stang *et al.*, 1997
Ebola virus	Viral entry requires functional rafts.	Bavari *et al.*, 2002
Marburg virus	Viral entry requires functional rafts.	Bavari *et al.*, 2002

attachment and entry by providing means to concentrate receptors, thereby increasing the binding affinity of viruses. Certain viruses may preferentially interact with raft as the site for attachment and internalization (Table 3). For example, SV40, a DNA containing non-enveloped virus was shown to interact with major histocompatibility complex (MHC) class I molecules leading to recruitment of caveolin around the attached virus and generating caveolae like invagination (caveosome) with virus attached and thereby facilitating internalization of the attached virus (Pelkmans and Helenius, 2003). However, virus in the invaginated raft somehow escapes proteolytic degradation and finds its way into the endoplasmic reticulum and then into the nucleus, the site of virus replication. During HIV infection, the HIV-host cell fusion was shown to require reorganization of membrane lipids. It has been proposed that initial interaction of gp120 with CD4 promotes clustering of raft domains and bringing coreceptors at the infection site. It has been shown that HIV receptor CD4, coreceptor CXCR4 as well as other associated molecules favoring HIV infection such as glycosphingolipids, CD44, all reside in lipid rafts. Recently, SFV was shown to require lipid raft for promoting fusion of viral membranes with the cell membrane (van der Goot and Harder, 2001). The role of lipid rafts in the fusion process between the viral and plasma membrane has been demonstrated for a number of viruses. For SFV, the presence of sphingolipids was critical for the fusion of viral membrane to the target membrane (Nieva *et al.*, 1994). Also, fusion among measles virus-infected Vero cells was affected by drugs that inhibit cholesterol biosynthesis (Malvoisin and Wild, 1990).

In the case of HIV, which causes AIDS by complex pathogenic process involving both virus and host, rafts are involved in multiple steps of the HIV replication process involving entry, signal transduction, trafficking and assembly of HIV proteins as well as budding of HIV particles. Lipid raft-association of HIV promotes viral entry. Cholesterol depletion by the methyl-β-cyclodextrin which disrupts rafts, inhibits HIV infection (Dimitrov, 2000; Hug *et al.*, 2000). Inhibitors of GSL biosynthesis affected HIV infection and syncytia formation and could be reversed by addition of specific GSL. Also, GSL-depleted cells lost their ability to induce conformational changes in Env-expressing cells although gp120 could interact with CD4 and coreceptors (CCR5 and CXCR4). Some receptor protein such as ATRC1 for ecotropic murine leukemia virus is raft-associated. SFV requires the presence of cholesterol and sphingolipids in the target membrane. E1 protein of SFV in low pH underwent conformation at changes, interacted with target membrane via insertion of its fusion peptide leading to association of E1 with lipid rafts. However, the fusion peptide of influenza virus HA did not associate with lipid raft. Cleaved HA requires its membrane domain for complete fusion since GPI-linked HA lacking TMD, although more strongly raft-associated than the transmembrane HA, leads to hemifusion but not complete fusion of membrane.

Lipid rafts not only facilitate HIV entry during fusion and uncoating of infecting virus but also play an important role in pathogenesis by aiding the virus in penetrating the intact barriers of epithelial cells present at the surface of gastrointestinal and genitounrinary tracts. It is believed that HIV uses transcytosis to gain entry into the human host but the infection process does not involve infection of surface epithelial cells. According to this model, HIV binds to glycosphingolipid galactosyl ceramide (Gal Cer) which has been shown to function as an alternative receptor of HIV glycoproteins (Alfsen *et al.*, 2001). Gal Cer is a component of the lipid raft, enriched in the apical membrane of epithelial cells and HIV becomes associated with lipid rafts and the intact virus then traverses the epithelial cell from the apical to the basal side by transcytosis and subsequently, the transcytosed virion infects the mononuclear cells containing CD4 and chemokine co-receptors which are also raft-associated leading to entry into host and productive infection process. It has also been shown that raft-association of HIV protein may play a critical role in altering signal transduction and disease progression. Nef (negative factor) protein which becomes myristoylated and raft-associated (Wang *et al.*, 2000), has been implicated with high viral load and increased pathogenesis. Nef is a raft-associated protein. Association of Nef with lipid raft may affect clustering of signaling molecules such as Lck, Hck, Lyn, Fyn, and may interfere with T cell activation (Cheng *et al.*, 1999). Nef contains a highly conserved proline rich motif (PXXP) which interacts with SH3 domains of Hck and Fyn (Collette *et al.*, 2000; Hiipakka and Saksela, 2002).

Also, host proteins such as HLA class II which are raft-associated and become incorporated in the HIV envelope, may influence host signaling to facilitate HIV entry into the target cells (Bastiani *et al.*, 1997). HIV envelope protein (gp160 or gp120/gp41) is a type I transmembrane basolateral protein and HIV also buds from the basolateral surface of polarized epithelial cells. However, unlike most basolateral proteins and basolateral viruses which are not raft-associated, HIV as well as HIV env are raft-associated and raft-association of HIV is critical for virus infectivity. Ebola virus (EBOV) and Marburg virus (MBGV) are among the two most pathogenic viruses and belong to the filoviruses. It has been shown that transient treatment of cells with filipin and nystatin, raft disrupting agents, interfered with virus infection suggesting that lipid raft may be important as a gateway of entry of filoviruses. However, it was noted that virus entry and not virus attachment involved the folate receptor which can function as a cellular cofactor for filoviruses and was also shown to reside in lipid raft (Bavari *et al.*, 2002).

In addition to attachment and entry, lipid raft may provide important function in RNA replication. For many positive strand RNA viruses, whether non-enveloped (picornavirus) or enveloped (hepatitis C virus [HCV]), RNA replication occurs in membrane-associated compartments. For HCV, the membrane-associated replication complex seems to be associated with lipid raft (Shi *et al.*, 2003). HCV nonstructural proteins including NS5A, NS4B as well as the newly synthesized HCV RNA were present in a detergent-resistant membrane complex and cofractionated with caveolin 2, a lipid raft-associated intracellular membrane protein. HCV RNA synthesis was protected from RNase in the membrane complex but became sensitive after detergent treatment. Depletion of cellular cholesterol also reduced HCV RNA synthesis. These results suggest that HCV RNA synthesis occurred in raft-associated membrane complex.

4.2. Role of lipid rafts in assembly and budding processes

In the virus life cycle, budding and release of the virus particles into the outside environment are last events. For budding to occur, three steps are obligatorily required. However, successful completion of these three steps does not guarantee that the budding and release of infectious virus particles will occur since additional steps may be required for the completion of the bud and release of the particles from the host cell as well as maturation of virus particles. These three obligatory steps are: (i) Following or concomitant to the synthesis of different viral components, they must be sorted, targeted and delivered to the assembly site. (ii) During this transport process or upon arrival to the assembly site, different viral components must interact with each other in an orderly fashion for bud formation to occur. (iii) Physical process of bud formation including the curvature of the membrane and eventual closure of the bud by membrane fusion and fission must occur to separate the virus particle from the host membrane. Lipid rafts appear to play important roles in each of these three steps.

4.2.1. Transporting and targeting viral proteins to the assembly site

As indicated earlier, lipid rafts actively participate in transporting and targeting many of the integral membrane proteins to the apical domains of the plasma membrane. Protein components of enveloped viruses are synthesized either in membrane-associated polyribosomes in ER or in free cytoplasmic polyribosomes. The proteins that are translated on the membrane-associated polyribosomes are either the integral membrane proteins present in the virus envelope or the secretory proteins lacking the membrane anchor region, usually not encoded by the viral genome. The other viral proteins which constitute the internal core viral components including both capsid and matrix proteins, are synthesized in the free polyribosomes as soluble cytoplasmic proteins. As indicated earlier, lipid components of lipid rafts are also synthesized in the ER and Golgi compartments and use the same exocytic pathway and vesicular transport system used by the integral membrane proteins to reach the plasma membrane. Close proximity of the integral membrane proteins with the lipid components in the exocytic pathways and their affinity for each other may facilitate their interaction and sorting in the same vesicle during their exocytic transport from the *trans* Golgi compartment to the plasma membrane. In polarized epithelial cells, integral membrane proteins are destined for either the apical or basolateral domains of the plasma membranes. It has been shown that both the basolateral (eg. VSV G protein) and apical (eg. influenza virus NA and HA) are transported together by the same vesicles to the TGN. However, the transport of apical and basolateral proteins from the TGN to the plasma membrane occurs via separate transport vesicles. Therefore, it is postulated that assembly and budding of vesicles containing apical and basolateral proteins occur at the separate domains of the TGN membrane and that the presence of lipid rafts in TGN and the affinity of the specific transmembrane proteins for lipid rafts provide platforms for segregation of apical proteins from the basolateral proteins in TGN. Influenza viral proteins have been extensively used as model for studying raft-association of proteins during transport and sorting. Influenza viral envelope possesses three integral transmembrane proteins namely HA, NA and M2. Among these both HA (type I) and NA (type II) proteins use lipid raft as a platform for apical transport (Nayak and Barman, 2002; Nayak and Hui, 2002) but M2, although an integral membrane protein, does not use lipid raft for apical transport (Zhang *et al.*, 2000). Furthermore, in the envelope of released virions, both HA and NA remain raft-associated but M2 does not associate with the lipid raft, indicating that the influenza viral envelope also exhibits a mosaic mixture of both raft and non-raft lipid microdomains even though the majority of lipids present in the lipid bilayer of the viral envelope are in l_o phase. Influenza HA associates with lipid raft and becomes TX-100 insoluble after acquiring the complex sugars in the Golgi complex. However, association of HA with lipid rafts is not dependent on oligosaccharide modification or its association with other viral proteins or its assembly into virus particles. Neither, the association of HA with lipid rafts depends on the polarity of cells (Skibbens *et al.*, 1989). TMD of HA has critical determinants for interacting with lipid rafts since chimeric HGH or HCH proteins containing TMD of VSV G or HSV C proteins and the ectodomain and cytoplasmic tail of HA did not associate with the lipid rafts. Furthermore, the exoplasmic half of the HA TMD and the amino acids contained in this region were critical for lipid raft-association (Scheiffele *et al.*, 1997; Lin *et al.*, 1998). The interaction of HA TMD with lipid raft also depended on a number of factors such as palmitoylation of three cysteine residues present in the HA TMD and cytoplasmic tail as well as the structural features such as α-helical conformation of HA TMD peptide. Furthermore, as indicated earlier, the HA TMD peptide was also shown to increase the acyl chain order of the lipid bilayer and thereby may have actively participated in stabilizing lipid rafts. In addition to TMD, the HA cytoplasmic tail may have an effect on the preferential association of HA with lipid raft since cytoplasmic tail minus HA (Hat^-) exhibited markedly reduced TX-100 resistance both in released virus particles and in cDNA transfected cells (Zhang *et al.*, 2000). This could be partly due to the loss of two cysteine residues in the cytoplasmic tail.

NA, a type II integral influenza virus protein, associates also with lipid raft via its TMD during intracellular transport. Chimeric analysis of using human transfection receptor (TR), also a type II protein, demonstrated that the NA TMD was responsible for interacting with lipid raft and apical transport signal (Kundu *et al.*, 1996; Barman and Nayak, 2000). However, unlike HA, interaction of NA TMD with lipid raft was not dependent on acylation of cysteine residues. Also like HA, the cytoplasmic tail of NA affected the interaction of NA in the lipid raft since removal of the conserved cytoplasmic tail reduced raft-association and increased TX-100 solubility of NA. Furthermore, NA has been implicated in facilitating the raft-association of HA (Zhang *et al.*, 2000). Like influenza virus, Sendai virus (SV) F and HN protein also associate with lipid rafts. However, both F and HN appear to associate with different classes of lipid rafts since TX-100 resistance of mature F and HN varied and was also different from that of influenza virus HA or NA (Ali *et al.*, 2000a). Similarly, SV5 as well as respiratory syncytial virus (RSV), members of paramyxovirus group, also bud from apical domain of plasma membrane are likely to contain lipid rafts in their viral envelope. However, detailed information regarding the presence of ordered lipids in the viral envelope or the raft-association of these glycoproteins is not available. Measles virus (MV), another number of paramyxovirus, also bud from the apical surface in polarized epithelial cells (Blau and Compans, 1995; Vincent *et al.*, 2000) and become raft-associated. However, both the process of raft-association as well as apical budding of MV appear to be the result of complex interactions among the viral proteins during intracellular transport, sorting and budding on the plasma membrane. In the MV-infected cells, both surface glycoproteins (F and H proteins) as well as internal M and N proteins became lipid raft-associated. However, in cDNA transfected cells, measles virus F but not H protein when expressed alone, became raft-associated. But upon coexpression of H protein with F protein, both F and H proteins became lipid raft-associated suggesting that F protein somehow facilitated raft-association of H protein (Vincent *et al.*, 2000). On the other hand, M and N proteins, the core viral proteins, did not depend on H and F proteins for lipid raft-association (Manié *et al.*, 2000). However, it is unclear how M and N proteins which are synthesized by free polyribosomes and not transported by exocytic pathways, became raft-associated. Furthermore, relative ratios of raft-associated cell surface MV glycoproteins such as H protein was lower than that of intracellular proteins suggesting that H protein become dissociated from lipid raft after it reached cell surface. Analysis of virus envelope from released viruses showed that only about 30% of H and 10% of F glycoproteins present in the viral envelope were raft-associated (Manié *et al.*, 2000). This would suggest that MV budding occurred not from the typical lipid raft as determined by TX-100 resistance or that H and F proteins became dissociated from lipid rafts during or after budding and/or release of MV particles. Since specific domains of glycoproteins were involved in raft-association, it is expected that both the TMD of F and as well as palmitoylation of the membrane proximal cysteine residue of F protein are likely to be involved in lipid raft-association.

It has been shown that palmitoylation of HIV env protein in the Golgi is critical for raft-association and for packaging in viral envelope (Rousso *et al.*, 2000). Association of HIV env protein with cholesterol sphingolipid rafts in the Golgi is also critical for intracellular trafficking of env protein to reach the plasma membrane microdomain for assembly and budding of HIV to occur. In addition, another HIV protein, the Gag protein present in the core complex of HIV, is translated as a soluble cytoplasmic protein using free polyribosome, but becomes raft-associated because of its myristoylation and interaction with the inner leaflet of lipid raft. It is likely that raft-association of both HIV env and Gag proteins are required for targeting HIV protein complex to the proper assembly site and that Gag-env interaction is required for conformational change and proteolytic cleavage of Gag-pol for budding and release of infectious virions. As mentioned earlier, Nef another critical HIV protein in virus budding is also raft-associated (Wang *et al.*, 2000; Zheng *et al.*, 2001). It was further shown that mature HIV released in the medium contains lipid raft in its envelope suggesting that the budding process may also be facilitated by the presence of lipid rafts at the assembly site (Aloia *et al.*, 1993). It appears that lipid rafts are not only involved in targeting HIV protein complexes to the assembly site and in budding and release of the virion, but is also critically required for maintaining the structure and conformation of matured HIV virion particles. Treatment of mature HIV with 2-hydroxy-propyl-β-cyclodextrin or methyl-β-cyclodextrin which removes cholesterol or with cholestenone (cholesterol analogue), destabilizes HIV structure and reduces infectivity. Furthermore, virion particles become denser, smaller and noninfectious suggesting that cholesterol (or lipid raft) is present on HIV envelope and is required for maintaining the integrity of HIV structure (Campbell *et al.*, 2001; Graham *et al.*, 2003). SFV treated with cyclodextrin (CD) also showed morphological defect (Scheiffele *et al.*, 1999).

Budding of the Rubella virus, a member of the Togaviridae family, genus Rubivirus, occurs from the Golgi membrane and virus particles are released from the apical side of cultured polarized epithelial cells (Garbutt *et al.*, 1999). However, little is known about the interaction of E1 and E2 proteins with lipid rafts or the role of lipid raft in the transport of envelope protein and in virus release. Two other Toga virus, SIN and SFV, bud apically in FRT cells but basolaterally from CaCo-2 cells and the distribution of p62/E2 envelope proteins is also apical in FRT cells but basolateral in CaCo-2 cells (Zurzolo *et al.*, 1992). Again, the role of lipid raft in neither of the transport of p62/E2 protein nor budding of these viruses is known. SFV and SIN are members of the alphavirus of Togavirdae family and possess similar structure as the rubella virus but in addition, they have a 6K TM protein. The 6K protein undergoes palmitoylation and is likely to interact with lipid rafts and thereby may facilitate virus budding in BHK cells. However, 6K protein appears not to be required for budding in virus-infected insect cells which possess more fluid membrane and compared to BHK cells, contain less sphingomyelin (9% vs 25%), and PC (14% vs 27%) (Cadd *et al.*, 1997). Lipid rafts may not actively participate in the assembly and budding of SFV and possibly, other members of alpha viruses since SFV budding requires precise interaction of trimeric envelope proteins forming spike with hexagonal capsid. On the other hand, for majority of enveloped viruses, such precise interaction of envelope proteins with capsids is exception rather than the rule and the incorporation of other proteins in viral envelope is often permitted (Briggs *et al.*, 2003).

Filovirus assembly and budding is also raft-associated (Bavari *et al.*, 2002). Filovirions incorporated raft-associated lipid molecules such as GMI. Ebola and Marburg virus glycoprotein (GP) were found to reside in lipid and the lipid raft-association of GPs was dependent on the palmitoylation of cysteine residues. Lipid raft-associated patches on the plasma membrane were involved in the assembly and budding of filovirus particles. In summery, lipid rafts often interact with the envelope as well as core proteins of many viruses and thereby facilitate their sorting and transport to the assembly site.

4.2.2. Selection of assembly site

As noted earlier, some enveloped viruses assemble and bud from the intracellular organelles (Hobman, 1993), while others bud from the plasma membrane (Figure 1, Table 1) (Garoff *et al.*, 1998). Among the viruses that assemble and bud at the plasma membrane, some bud predominantly from the apical while others fiom the basolateral but usually not from both domains of the plasma membrane of polarized epithelial cells (Compans, 1995). However, what makes the virus to assemble and bud from a specific site in a cell or a specific domain of plasma membrane remains unclear.

It is generally accepted that viral glycoproteins determine the site of virus assembly and budding. This notion comes from the fact that viral glycoproteins accumulate at the site of the virus budding even when expressed alone. For example, glycoproteins of viruses such as coronaviruses (CoV), hepatitis B virus (HBV), bunyaviruses and others which bud from the internal sub-cellular organelles, possess intrinsic determinants for the same sub-cellular localization as the site of virus budding (Hobman, 1993). On the other hand, glycoproteins of viruses that bud from the plasma membrane are not retained in the internal organelles but are transported to the plasma membrane. Moreover, for the latter group of viruses, viral glycoproteins possess either apical or basolateral sorting signals and are directed to the apical or basolateral surface, the specific site where virus assembly and virus budding occur in polarized epithelial cells. Furthermore, in different cells and tissues where some viruses bud from the opposite membrane, their glycoproteins are distributed accordingly. For example, SFV bud apically from FRT cells but basolaterally from CaCo-2 cells; similarly, p62/E2, the envelope glycoproteins of SFV, are targeted apically in FRT cells but basolaterally in CaCo-2 cells (Zurzolo *et al.*, 1992). However, since apical and basolateral distribution of p62/E2 glycoproteins were determined in virus-infected cells, it can not be ascertained from these experiments if other viral components affected the distribution of envelope glycoproteins of these viruses.

For retroviruses, particularly HIV, which buds from basolateral surface in polarized epithelial cells, it appears that HIV env protein plays a critical role in determining the basolateral budding site. It was observed that HIV capsid proteins when expressed alone released particles randomly from both apical and basolateral surfaces, whereas upon co-expression with the envelope protein gp160, virus particles were released predominantly from basolateral surface (Owens *et al.*, 1991). Authors concluded from these studies that the HIV envelope protein, which is targeted to the basolateral surface in polarized epithelial cells, determines the site of virus budding. Further role of the envelope protein in polarized budding of HIV was supported by a number of experiments where mutation in the cytoplasmic tail of gp41 caused non-polar distribution of the envelope protein resulting in the non-polar budding of HIV (Lodge *et al.*, 1994; 1997) and affecting cell to cell transmission (Deschambeault *et al.*, 1999). However, attempts to redirect virus budding into ER and Golgi by introducing specific retention signals into the HIV envelope protein did not succeed (Salzwedel *et al.*, 1998). Although, in this case, one can argue that the Golgi and ER environments are much different from that of the plasma membrane and that the glycoproteins may not attain the proper structural conformation for interacting with HIV capsids to cause budding. On the other hand, mutations in the matrix domain of the Gag protein caused redistribution of both Gag and env protein on the plasma membrane (Hermida-Matsumoto and Resh, 2000) suggesting that the envelope protein may not be the only determinant for selecting the site for HIV budding.

Experiments with measles virus have shown that budding site in polarized epithelial cells was not determined by glycoproteins (Maisner *et al.*, 1998). Although viral glycoproteins H and F were transported in a random fashion and to basolateral membrane respectively, virus budding occurred predominantly from the apical surface of polarized MDCK cells. Similarly, although Marburg virus buds predominantly from the basolateral surface, its glycoprotein is transported to the apical surface (Sanger *et al.*, 2001). With human cytomegalovirus (CMV), p150, a nonenvelope tegument protein, was critical for the site of virus assembly (Sanchez *et al.*, 2000). Influenza virus, which assembles and buds from the apical plasma membrane in polarized epithelial cells, has been used extensively as a model for studying protein targeting. As indicated earlier, it possesses three transmembrane proteins HA, NA and M2, all of which are apical. Both HA and NA interact with lipid raft by the sequences present in their transmembrane domains which also possess signals for apical transport (Kundu *et al.*, 1996; Lin *et al.*, 1998; Barman and Nayak, 2000; Barman *et al.*, 2001). M2, which is also an apical protein (Hughey *et al.*, 1992), does not interact with lipid raft (Zhang *et al.*, 2000) and the apical signal of M2 is unknown at present. Of these three envelope proteins, HA, the major glycoprotein, constitutes over 80% of the envelope proteins present in virus particle. In transfected cells, a single amino acid change (Cys543→Tyr543) in HA (H2) was shown to direct HAtyr to the basolateral side and did not affect the intracellular transport and cell surface expression of the mutant protein (Brewer and Roth, 1991). However, a mutant virus containing the basolateral HA (HAtyr), budded almost exclusively from the apical side (>99%) i.e. from the same site as the wild type virus, even though majority of HAtyr was directed to the basolateral side (Mora *et al.*, 2002; Barman *et al.*, 2003). These results demonstrated that HA, the major envelope protein, was not the determinant for the site of assembly and budding of influenza virus particles. More recently, it was also shown that VSV G protein (targeted basolaterally) does not determine the basolateral budding of that virus (Zimmer *et al.*, 2002). These studies suggest that viral glycoproteins may not be the only or major determinant for selecting the budding site.

Since the major glycoprotein was not responsible for selecting the site for virus assembly and budding, other factors of both host and virus origin are likely to be involved in selecting the budding site. Among the viral components, the matrix protein (M1) may play critical role in site selection. We have recently shown that influenza virus M1 possesses a "late" (L) domain (Hui *et al.*, 2003) which may interact with host components for virus budding. Such host components may be restricted at the apical domain for influenza virus budding and at the basolateral side for VSV budding. With measles virus, which buds predominantly from the apical surface, although its F and H are expressed predominantly on the basolateral surface, it has been suggested that polarized expression of M protein or interaction of M protein with glycoproteins at the apical plasma membrane domain may favor apical budding (Naim *et al.*, 2000). In addition to M1 protein, NP or vRNP or M1/vRNP complex of influenza A virus could have polarized distribution and may facilitate apical budding. However, how such cytoplasmic proteins which are not acylated could have a targeted polarized distribution remains unknown.

It is also likely that influenza as well as other viruses require lipid rafts as budding platform (Table 4), which are predominantly present on the apical membrane to produce virus particles with ordered lipid membrane (Brown and London, 1998a; 1998b) and that the fraction of HAtyr transported to the apical surface (but not to the basolateral surface) interacted with lipid raft. Furthermore, it has been shown that lipid rafts play an important role in influenza virus budding. With HAtail$^-$/NAtail$^-$ viruses where HA was more TX-100 soluble and less lipid raft-associated, virus production was low and virus particles were deformed and bizarre in shape (Zhang *et al.*, 2000). Lipid rafts in addition to transporting, targeting and concentrating viral and host components to the assembly site may have some intrinsic properties for budding to occur and thereby facilitate budding from a specific site of a membrane for a given virus.

Although viruses can bud and form particles in the absence of glycoproteins and although the Gag protein of HIV and matrix proteins of many

Table 4. Role of lipid rafts in viral exit

Virus	Mechanism of use	References
Sindbis virus (SIN)	Virus buds from cholesterol-rich domains.	Lu *et al.*, 1999
Vesicular stomatitis virus (VSV)	VSV-M and -G protein redistribute lipids during virus formation.	Luan *et al.*, 1995
HIV-1	Virus exits by budding selectively from lipid rafts. Gag specifically associates with raft like membrane.	Nguyen & Hildreth, 2000 Alfsen *et al.*, 2001 Ono & Freed, 2001
Influenza virus	Virus exits by assembling viral glycoproteins in lipid rafts and trafficking to the apical cell membrane.	Scheiffele *et al.*, 1997; Keller & Simons, 1998; Scheiffele *et al.*, 1999; Barman *et al.*, 2001
Measles virus (MV)	Virus assembly occurs in lipid rafts.	Manié *et al.*, 2000
Sendai virus (SV)	Structural glycoproteins assemble in lipid rafts for virus exit.	Ali & Nayak, 2000
Respiratory syncytial virus (RSV)	Matures at regions of the cell surface that are enriched in the protein caveolin-1 and lipid raft ganglioside GM1.	Brown *et al.*, 2002a Brown *et al.*, 2002b
Ebola virus (EBOV)	Viral proteins within lipid rafts during viral assembly and budding.	Bavari *et al.*, 2002
Marburg virus (MBGV)	Viral proteins within lipid rafts during viral assembly and budding.	Bavari *et al.*, 2002
Murine leukemia virus (MuLV)	Envelope protein is lipid raft associated through palmitoylation.	Li *et al.*, 2002

negative strand viruses can bud and acquire envelope, the lipid composition of such virus like particles (VLPs) are not known. Whether such particles contain lipid raft in their envelope or glycoproteins are required for acquiring lipid rafts in their envelope remains to be determined. If such VLPs do not possess lipid rafts, it is likely that virus budding can occur from plasma membrane outside the lipid raft microdomain. If, on the other hand, such VLPs do possess lipid rafts as found in infectious virus particles, it would imply that the core components including matrix proteins can interact with lipid raft in the absence of glycoproteins and that budding requires the presence of lipid raft microdomain.

4.2.3. Interaction among viral components

Lipid rafts play important roles in different steps of the budding process. One important step involves the interaction among the viral components required for producing the infectious virus. Lipid rafts not only promote transporting and targeting viral components to the assembly site but aids in the interaction among the viral components in a number of ways.

(a) Firstly, it can bring the viral components in the proximity of each other by partitioning and concentrating the proteins in specific microdomains. This is evident for proteins which can independently associate with lipid rafts and therefore, can interact with each other because of their proximity. This appears to be the case in HIV-infected cells. In HIV-infected cells, coalition of lipid rafts will concentrate HIV env and MA proteins both of which are acylated, and already present in the similar lipid compartments and thereby will facilitate their interaction with each other and the budding process. Some host proteins such as CD55, CD59, inhibitors of complement pathways, as well as other GPI-linked proteins that are present in lipid rafts are also included in HIV particles although they are not known to interact with any specific viral proteins (Nguyen and Hildreth, 2000).

In MV-infected cells, F and HN, the TM proteins, become raft-associated. However, it has been reported that M and N the intracellular cytoplasmic proteins synthesized on free polysomes, also become raft-associated (Manié *et al.*, 2000), although the mechanism by which these two proteins which are not known to be acylated, become raft-associated is unclear. Raft-association of N, M, F and HN will facilitate interactions among these viral components.

(b) Secondly, lipid rafts can also bring nonraft-associated proteins in the raft microdomain because of their interaction with raft-associated components. This appears to the case with a number of viruses. For example, Ortho- and Paramyxovirus glycoproteins are lipid raft-associated whereas their matrix proteins (M and M1) are nonraft membrane-associated. Matrix proteins after their interaction with glycoproteins become raft-associated (Ali *et al.*, 2000a, 2000b). Interactions of matrix proteins with glycoproteins as well as raft-association of matrix proteins will facilitate interaction among the matrix proteins by increasing their concentration at the budding site and this interaction among the matrix proteins is critical for virus budding.

Also, raft-ordered membrane domains may be formed de novo around TM proteins on the plasma membrane such as the engaged immune receptors. Affinity of the l_o domains can be increased by organization, acylation, coupling to raft-associated molecules or by conformational changes (Harder *et al.*, 1998). All of these processes may aid in the M1/M1 and M1/glycoprotein interactions.

In studies involving VSV G and M proteins, it was shown that both proteins were needed for inducing sph- and cholesterol-enriched domains and either of these proteins alone were not effective (Luan *et al.*, 1995), again implying the role of lipid domains in bringing the interacting proteins together. Therefore, lipid rafts play an important role in promoting interaction among viral components required for assembly and budding.

(c) Thirdly, pseudotyping is a common phenomenon observed in cells mixedly infected with two or more enveloped viruses where progeny viruses containing the genome (and capsid) of one virus and the envelope proteins of a second virus are formed. This type of mixing of core (capsid) components with envelope components has been observed with many DNA and RNA viruses and provides an important tool for gene delivery and is required for either broadening or selecting host range and tissue tropism. This well-documented phenomenon of pseudotyping is at odds with the common notion that specific interaction between the core component and envelope proteins governs the assembly and budding of most viruses. It appears that lipid forms a common platform for mixing the envelope proteins of different viruses and cellular membrane proteins. Even viruses belonging to diverse groups such herpes virus and VSV can produce pseudotype viruses. A common property among all these diverse viral and cellular proteins is that they are raft-associated and often myristoylated or palmitoylated. Even basolateral proteins like VSV G are palmitoylated and raft-associated. (Harder *et al.*, 1998). It was observed that in mixed infection, envelope proteins of different viruses as well as some of the core proteins such as HIV Gag are often raft-associated and non-ionic detergent insoluble (Pickl *et al.*, 2001). Therefore, detergent-resistant lipid rafts on the plasma membrane are common meeting grounds for core and transmembrane envelope proteins of different viruses. Also, these lipid rafts function as a platform for envelopment and budding resulting in the production of pseudotype viruses. Some of the host components such as CD4, CXCR4, as well as env proteins of other retroviruses and GPI-anchored proteins can be incorporated in viral envelope. Lipid rafts therefore provide the basis for promiscuity of incorporation of foreign proteins into number of virus particles such as VSV, HIV and support the passive incorporation of integral membrane proteins into virus particles. However, envelope protein and core interactions also play a critical role in selecting the incorporation of specific viral proteins and excluding most membrane-associated host proteins from the budding site and virus particles. Therefore, lipid microdomains such as lipid rafts facilitate mixing and interaction among the viral components required for assembly and budding of infectious viruses as well as in pseudotype formation.

4.2.4. Budding process and release of virus buds

4.2.4.1. Budding site: Role of viral proteins and lipid rafts

As indicated earlier, viruses do not bud randomly from host membrane but selectively from specific organelles or from the specific domains (apical or basolateral) of plasma membrane. It also appears that within the specific membrane domain, virus budding occurs from a few select "hot spots" as if the same site in being used for multiple rounds of virus budding. Observation that multiple incomplete virus buds are joined together in HIV-infected cells defective in Tsg101 function supports this hypothesis of hot spots of virus budding (Garrus *et al.*, 2001; Demirov *et al.*, 2002; Goila-Gaur *et al.*, 2003). Release of influenza virus particles in HeLa cells after cytoskeletal disrupting agents from a few spots on the membrane (Gujuluva *et al.*, 1994) as well as incomplete virus budding of virus particles joined together budding defective influenza virus mutants also support virus budding from selective spots on the membrane (Roberts and Compans, 1998; Simpson-Holley *et al.*, 2002; Barman *et al.*, 2003; Hui *et al.*, 2003). The biophysical and biochemical properties of these hotspots of budding site are yet unknown. We speculate that at least lipid rafts are an important contributing factor for the budding site since many viruses use lipid rafts for budding and incorporation into the membrane of mature virus particles. It then becomes important to determine how the membrane is modified for budding site. Viral membranes are not only enriched with lipids assembled as lipid rafts but also contain virally encoded proteins. Two types of proteins are associated with viral membrane: (i) TM proteins that traverse the entire lipid bilayer and may have acyl modifications in TMD and cytoplasmic tails but some proteins such as influenza virus NA, a type II TM protein, are purely TM proteins without any acyl modification. TM proteins with or without acyl chains primarily interact with exoplasmic leaflet but also may interact with the cytoplasmic leaflet of the lipid bilayer by the acyl chains. (ii) Proteins present in the cytoplasmic side interact with the inner leaflet of the lipid bilayer. These cytoplasmic proteins are either modified with acyl groups and thus interact with the lipid bilayer or interact with the cytoplasmic tail of TM proteins and thereby become indirectly associated with lipid raft. As indicated earlier, some TM proteins, particularly the apical proteins also use lipid raft for sorting and transporting to the virus assembly site.

Since viral envelope proteins are present in the lipid rafts and since lipid rafts essentially constitute viral membranes, it becomes an important question whether the viral proteins present in the lipid rafts are passively incorporated into virions or they play an active role in the assembly and budding process. One possibility is that the lipid rafts are assembled into microdomains due to the physical and structural properties of lipid molecules as has been shown for the assembly of the lipid rafts in artificial membranes and viral proteins or proteins in general are just passive participants and become incorporated into lipid rafts because they are present in the vicinity and have affinity for the lipid raft. These lipid rafts containing viral TM proteins become the assembly and budding site of enveloped viruses. The second possibility is that the viral TM proteins actively participate in selecting and assembling lipids and packing the lipids into lipid rafts for functioning as viral membranes. Current evidence suggests that proteins and lipids can interact with each other at multiple levels of organization of rafts and their incorporation into viral envelope may vary with different enveloped virus groups. For example, as indicated earlier, the influenza HA TMD peptide assumes a predominantly alpha helical conformation in detergent micelles and phospholipid bilayers. Furthermore, depending on the composition of lipid bilayer such as lipid bilayer composed of pure phosphatidylcholines, the helicity of HA peptide is increased. On the other hand, the HA TMD peptide also increased the acyl chain order and packing of lipid bilayer supporting the role of protein component in assembling raft and preferential association with lipid rafts. Possible interaction of the specific amino acids of HA and NA TMD peptides with the outer bilayer of lipid rafts has been supported by mutational analysis (Scheiffele *et al.*, 1997; Lin *et al.*, 1998; Barman *et al.*, 2001). Furthermore, cytoplasmic tail of HA may also affect interaction of HA TMD with lipid raft as the tail minus HA becomes more TX-100 soluble (Zhang *et al.*, 2000). In addition to the amino acid sequence of HA TMD, the palmitoylation of cysteine residues and their interaction with the inner leaflet of lipid bilayer also affect HA interaction with lipid rafts. Mutational analysis also showed that for NA, a type II protein, which does not undergo any acyl modification, the amino acid sequences at COOH terminus of TMD interact with the exoplasmic lipid leaflet of the raft containing glycosphingolipids and cholesterol. Also like HA, removal of the NA cytoplasmic tail affected its interaction with lipid rafts. It is not clear how the cytoplasmic tail outside the lipid bilayer would affect its association with the lipid raft. It is possible that the removal of cytoplasmic tail may affect structure of the TMD such as alpha helicity which is known to have affinity of lipid raft. In addition, oligomerization among the NA TMD as well as interaction of the cytoplasmic influenza virus proteins such as M1 may help in cluster formation and stabilization of lipid rafts.

Even VSV G, the model basolateral protein, becomes palmitoylated (Brown *et al.*, 1989) and associates with raft-like structures present in basolateral membrane. In fact VSV G protein facilitates the formation of cholesterol- and sphingomyelin-rich lipid domain in vesicles and in VSV particles (Harder *et al.*, 1998). Therefore, initially viral components are concentrated at the localized lipid raft like regions, the assembly site. Subsequently, lateral interaction among the viral components amplify the changes in local lipid composition, which in turn facilitate further concentration of viral components leading to bud formation.

For retroviruses, particularly HIV, interaction among env protein (gp160, gp120/gp41) and MA protein play critical roles for association with lipid rafts and budding. As indicated earlier, the env protein although a basolateral protein, can independently associate with lipid rafts in the absence of MA protein. This interaction of env protein to the lipid raft is due to the palmitoylation of the two cysteine residue in the cytoplasmic tail (CT) of env protein and occurs during the transport of the env protein to the plasma membrane. Thus with HIV env protein, unlike the TMD influenza virus HA or NA, the palmitoylation of the cytoplasmic tail is critical for its interaction of with lipid rafts. Therefore, it is possible that env protein may be passively incorporated into the virus particles because it is present in lipid rafts, the site of virus budding and that interaction of env proteins with MA is not required for virus budding. HIV gag protein (cytoplasmic protein) become myristoylated, independently interacts with lipid raft and bud efficiently in the absence of HIV env protein producing VLPs. However, whether these VLPs containing Gag only have envelope with similar lipid raft composition as the virus env protein remains to be determined. It is known that budding of virus particles in polarized epithelial cells in the absence of env protein is random compared to basolateral budding of HIV in the presence of env protein. This would suggest that the presence of env proteins in lipid rafts and its interaction with Gag protein further facilitate the accumulation of Gag proteins in the lipid rafts leading Gag-Gag lateral clustering and multimerization. Furthermore, interaction of env with Gag protein may prevent internalization and promote budding. Unlike the assembly and budding of C-type viruses which occur only at the plasma membrane, the B- and D-type viruses, preassemble the viral core in the cytoplasm. A single mutation in the MA protein of Mason-Pfizer monkey virus (M-PMV; type-D retrovirus), abolishes cytoplasmic assembly and renders the mutant virus to assemble its core at the plasma membrane like the C type virus.

Although envelope proteins and the interaction of envelope proteins with matrix proteins facilitate budding of virus particles, matrix proteins of most viruses alone can become membrane-bound and bud efficiently. This is particularly true for VSV, a basolateral virus, which does not associate with typical lipid rafts. However, VSV virions possess a lipid composition which is different from that of the host plasma membrane. Therefore, it is likely that VSV G and M protein interaction as well as interaction among M proteins would lead to clustering of lipids binding to G proteins and formation of patches in the plasma membrane. These patches although different from typical lipid rafts as demonstrated by high TX-100 solubility have a high phosphatidic acid, phosphatidylserine, cholesterol, sphingomyelin and phosphatidyl ethanol amine (Luan *et al.*, 1995). G and M protein interaction may alter the lateral distribution of lipids in the plasma membrane. Furthermore, since VSV M may exist in phosphorylated and unphosphorylated forms, unphosphorylated forms are likely to favor lipid patch formation whereas phosphorylation of M leads to loss of lipid patch and dispersion of M into the membrane lipids.

As indicated above, plasma membrane proteins associate with lipid raft or affect the lipid order by interacting with the outer leaflet of the lipid bilayer. These protein-lipid interactions are mediated by GPI in integral membrane proteins without TMD or acyl modification of the amino acids in the TMD and cytoplasmic tail or by direct amino acid/lipid interaction with TMD. However, lipid rafts possess asymmetric distribution of lipids in both outer and inner leaflets of the lipid bilayer. It is known that in the host cell, a number of proteins in the cytoplasmic side interact with the inner leaflet prominently by the acyl groups such as acylated Src kinases. Viral proteins, particularly the matrix proteins comprising the protein component of the inner core can interact with the inner leaflet of the lipid bilayer. Some of these proteins like M1 of the influenza virus or the M protein of VSV and other negative strand viruses appear to directly interact with the inner leaflet of the lipid bilayer by protein/lipid interaction. For some viral matrix proteins, the lipid/proteins interactions are believed to occur by the hydrophobic domains or the exposed hydrophobic surface of the amphipathic helix or by the electrostatic interactions with the positive charges present in the matrix proteins. The lipid/protein interactions are however not lipid raft specific as determined by TX-100 resistance. Only when these matrix proteins interact with integral viral envelope proteins present in the lipid rafts, these matrix proteins become indirectly raft-associated (Ali *et al.*, 2000a). In many cases, these membrane-associated viral matrix proteins interact with each other, form complexes and bud as VLP. The lipid composition of the envelope of VLPs is not known and may aid in understanding the role of lipid raft in the budding process. Matrix proteins of some viruses like HIV and other retroviruses undergo acyl group modification, therefore interact with the inner leaflet of the lipid bilayer with the acyl groups. These matrix proteins like Src kinases are likely to interact with the inner leaflet of the lipid raft-like domain.

Gag protein (p55) of HIV has been shown to bud efficiently from the plasma membrane. p55 after proteolytic cleavage generates a number of proteins: MA (p17) – CA (p24) – p2 – NC (p7) – p6. Several reports have shown that Gag is targeted to lipid raft microdomains (Nguyen and Hildreth, 2000; Rousso *et al.*, 2000; Lindwasser and Resh, 2001; Ono and Freed, 2001; Zheng *et al.*, 2001; Holm *et al.*, 2003). MA is a homologue of the matrix proteins of other virus and possesses a M domain which undergoes myristoylation. However, the interaction of different protein components of p55 with inner leaflet of the lipid bilayer is complex. It has been reported that the M domain possesses determinants within its N terminal region which mediates Gag/membrane interaction. It has also been shown that the interaction domain (I domain) present in NC, functions as an assembly domain for Gag multiimer formation and is also involved in plasma membrane targeting of the Gag proteins (Sandefur *et al.*, 2000). Furthermore, the C terminal position of Gag is also required from lipid raft-association (Lindwasser and Resh, 2001; Ono and Freed, 2001) suggesting that multimerization as well as myristoylation of Gag proteins are required for the assembly and association of Gag with lipid raft-like domains. However, it has been recently reported that although HIV Gag protein interacts with plasma membrane and forms raft-like detergent resistant barges, these microdomains are different from typical lipid rafts. These detergent resistant lipid/Gag protein complexes were much denser than lipid rafts and did not incorporate classical lipid raft proteins such as GPI-anchored PLAP and were not disrupted by cholesterol extraction. It appeared that Gag protein complexes did not function as a raft-patching agents from the cytoplasmic leaflet of the membrane since typical raft proteins were not associated with Gag associated raft-like membranes. However, it is possible that Gag binds to a different class of lipid rafts since the existence of the different classes of lipid rafts with varying protein and lipid composition has been reported (Röper *et al.*, 2000).

4.2.4.2. Bud formation and release

Bud formation and release require bending of membrane, incorporation of the cargo into the bud, growth of the bud and eventually, pinching off of the bud from the parent membrane. Bending of the membrane requires the pulling and pushing forces at the site of bud. Release of bud requires fusion of apposing membrane of the bud and the donor cell with each other causing fission of the bud from the donor cell and releasing of the particles into the environment. Although this overall mechanism and the physical forces for budding may be same, the contributing factors that govern these processes vary. Clearly, the requirements for budding intracellular transport vesicles are different from that for virus budding and virus release. Clathrin, COPI or COPII required for budding of transport vesicles are not involved in virus budding. In addition, the mechanism of fission of virus buds are also clearly different from that of intracellular transport vesicles. Although budding and vesicle formation go hand in hand, we will analyze these two processes separately.

4.2.4.2.1. Formation of virus buds. Bending of membrane is an absolute requirement for bud formation (Farsad and De Camilli, 2003). This will in general involve a transition from more planar structure to a curved structure. Although the physical forces that cause the curvature to occur at the budding site remains unclear, both lipids and proteins can contribute to causing the bending of membrane. Asymmetry in lipid bilayers can cause intrinsic curvature of one monolayer relative to other monolayer leading membrane bending. Selective transfer of lipids between the lipid bilayers, interaction of cholesterol into the budding leaflet as well as hydrolytic cleavage of phosphocholine head group of sphingomyelin by sphingomyelinase generating ceramides with smaller head group have been proposed to cause membrane deformation and budding (Holopaineu *et al.*, 2000). Specific lipid microdomains such as lipid rafts have been shown to be a contributing factor in virus budding. Depletion of cholesterol, a major component for lipid raft formation, often suppresses virus budding indicating the role of lipid raft in bud formation.

In addition, to specific lipid microdomain, virus bud formation requires specific viral proteins. However, interaction TM viral proteins which partition with and stabilizes lipid rafts may not be sufficient for bud formation although their absence may affect budding partially. It has been observed that the expression of the TM viral proteins such as influenza HA, NA, VSV G or HIV env protein on the cell surface does not lead to either bud formation or extracellular release of bud containing these proteins. Similarly, other observations clearly suggest that lipid rafts with or without viral membrane proteins as well as the pulling force of glycoproteins are not sufficient for initiating bud formation. On the other hand, matrix proteins of many viruses often expressed alone are sufficient for bud formation and cause efficient release of buds such as VLPs. Matrix proteins which binds to the cytoplasmic leaflet of lipid bilayers, can cause bending of membrane and lead to generating the pushing force to initiate virus buds. Interaction of matrix proteins with each other at the cytoplasmic side will also mimick asymmetric lipid distribution and lipid to lipid interaction causing membrane bending. However, it is not clear if in these budding processes involving matrix protein alone, lipid microdomains play any role. Since it has been shown that matrix proteins of influenza and SV unlike that of HIV expressed alone do not interact with lipid rafts, it would suggest that lipid rafts are not required for virus budding to occur. However, lipid component of released particles containing matrix protein alone has not been analyzed to show whether budding of these VLP particles occurred from lipid microdomains or randomly from plasma membrane. In addition, cytoskeletal components like actin filaments, membrane tethered myosin motors also cause membrane deformation in the formation of cell surface tubular microvilli. However, their role in virus budding remains unknown at present.

4.2.4.2.2. Release of virus particles. Subsequent to bud formation, buds are released by a mechanism of fusion of the apposing membrane and fission of the bud from the cell. These processes determine the size and shape of the particles. It appears that a number of factors both virus and host may affect this process. Among the viral factors both glycoproteins as well as nucleocapsids and matrix proteins are critical components. Virus particles released in the medium can be either spherical or filamentous of varying sizes. Icosahedral nucleocapsids such as found with Semliki Forest virus clearly determines the spherical shape of the released virus particles. Similarly, helical VSV nucleocapsid is critical in determining the bullet shape and the length of the virus particles. Defective interfering VSV particles contain smaller nucleocapsids which is responsible for producing small virus particles. These observations indicate that at least with these viruses, separation of virus buds from the host membranes depends on the cargo nucleocapsid and occurs immediately after the enclosure of the nucleocapsid. However, many viruses such as influenza viruses are flexible and pleomorphic and can produce spherical or filamentous particles. With these viruses, a number of factors may play critical roles in causing the fusion and fission processes and determining the size and shape of released virus particles.

Among the viral factors, matrix proteins as well as glycoproteins have been shown to affect virus shape and size. Deletion of cytoplasmic tail of HA and NA was shown to generate bizarre filamentous shape of virus particles. Reduced lipid raft-association of tail minus HA and NA was proposed to be the cause of such abnormal virus particles suggesting the role of lipid raft in both budding and fission of virus particles (Zhang *et al.*, 2000). Secondly, some influenza virus strains exhibited filamentous morphology and matrix proteins contributed to this strains specific filamentous shape (Roberts *et al.*, 1998; Bourmakina and Garcia-Sastre, 2003; Hui *et al.*, 2003). We have recently shown that influenza virus matrix protein like many other viruses possesses "late" domain (L domain) which affects fission of virus buds. It remain to be seen if this mechanism i.e. lack of fission caused by M1 L domain, was also responsible for filamentous form of A/Udorn virus (Roberts *et al.*, 1998).

In addition to viral factors, a number of host components have been shown to be involved in the budding process. A number of host proteins including ubiquitin, Tsg101, Vps4, Nedd4, and other members of vacuolar protein sorting pathway, which are involved in the formation multivesicular bodies in the endosome have been shown to be involved in the budding process of a number of virus (Luban, 2001; Perez and Nolan, 2001; Cimarelli and Darlix, 2002; Freed 2002). They possess WW domain and interact with the L domain of virus matrix protein. All of these host proteins in some way facilitate the fusion and fission process so that any defect in the interaction of these virus and host components will lead to defective or incomplete virus release, often forming multiple particles joined together. However, how the host proteins or host protein/M interactions will help in the process of fusion and fission is not clear. Furthermore, these particles were not completely filamentous or tubular but exhibited clover leaf-like structure suggesting incomplete fusion and fission. It will be interesting to determine if these particles represented a state similar to hemifusion in which only the inner leaflet undergoes fusion and therefore can not undergo complete fission and release from the host membrane and separation from each other.

In addition, cytoskeletal components particularly actin microfilaments have been shown to contribute to filamentous form of influenza virus (Roberts and Compans, 1998). It has been proposed that microfilaments which bind to vRNP may provide outward pushing force in bud formation. However, if actin is involved the budding process, the fusion of membrane in the bud and fission of particles will require disassembly of actin filaments at the last stage of the budding process. Enhanced release of virus particles from influenza virus-infected HeLa cells with actin disruptive agents (Gujuluva *et al.*, 1994) as well as increase in production of spherical over filamentous particles in influenza- and parainfluenza virus infected polarized MDCK cells (Roberts and Compans, 1998) support the role of actin in the budding process. Bud formation and virus release have been shown to be an energy dependent process requiring ATP (Hui and Nayak, 2001; 2002). However, specific role of ATP in the budding process remains to be elucidated.

Finally, lipid rafts can affect both bud formation and fusion and fission process at multiple steps. As indicated earlier, asymmetry in the lipid bilayer can cause membrane curvature leading to the formation of buds. Assembly of lipid rafts at the budding site will affect physical properties of the membrane including lipid heterogeneity, lipid/protein interaction, increased viscosity and rigidity, slow diffusion. Presence of lipid heterogeneity could cause increased budding, increased fission and release of buds. For example, if one leaflet of the lipid microdomain is in raft and the other in nonraft state, one leaflet of the lipid microdomain will be higher in density and viscosity. This would lead to the formation of curvature provided the asymmetry in lipid leaflets are stably maintained by interaction with proteins in and around the raft. Heterogeneity in viscosity in the budding site due to the presence of both raft and nonraft lipids will facilitate fusion and fission. However, although we know that many viruses bud from lipid raft microdomain and possess lipid raft microdomain in their envelope, we do not know about the topology of microheterogeneity in lipid composition, viscosity in budding site for assessing the role of lipid raft, specifically in the budding process itself such as membrane bending, bud growth, fusion and fission leading to bud release. However, it is evident that lipid raft plays an important role in the budding process since destabilizing lipid raft by depleting cholesterol affects virus budding.

5. CONCLUSION

Lipid microdomains, particularly lipid rafts, play important roles in many aspects of virus biology including receptor binding, entry, uncoating including fusion, RNA replication, protein transport, protein sorting, protein targeting, budding, fusion, fission and release of buds as well as maintaining the integrity and stability of virus envelope and virus structure. Furthermore, lipid rafts provide the critical elements for signal transduction by many viral components. However, specific aspects and functions of lipid rafts in virus biology are poorly understood. One of the problems in defining the association and involvement of lipid rafts in virus biology is the heterogeneity among the membrane lipid rafts as determined by varying TX-100 insolubility. It will be therefore important to define the classes of lipid rafts and determine the association of a specific class of lipid rafts with a specific viral membrane and viral function. Secondly, lipid rafts, on the one hand, increase selective inclusion of viral proteins and exclusion of nonviral host components from the budding site whereas, on the other hand, they provide platforms for pseudotype formation by bringing the foreign proteins to the budding site and thereby facilitating their incorporation in viral envelope. The latter property of lipid rafts aids in pseudotype formation and will be an important area of research for targeting enveloped virus vectors to specific cells and tissues either by broadening or narrowing their host specificity. Finally, involvement of lipid rafts in virus replication may provide a novel antiviral approach. Topical application of β-CD, a lipid raft destabilizer, shows the promise of antiviral effect in HIV transmission (Khanna *et al.*, 2002) possibly by preventing virus entry and affecting virus budding and virus structure as well.

ACKNOWLEDGMENTS

This work was supported by USPHS grants (AI 16348, AI 41681).

REFERENCES

Aït Slimane, T. and Hoekstra, D. (2002) Sphingolipid trafficking and protein sorting in epithelial cells. FEBS Lett. 529:54-59.

Alfsen, A., Iniguez, P., Bouguyon, E. and Bomsel, M. (2001) Secretory IgA specific for a conserved epitope on gp41 envelope glycoprotein inhibits epithelial transcytosis of HIV-1. J. Immunol. 166:6257-6265.

Ali, A., Avalos, R.T., Ponimaskin, E. and Nayak, D.P. (2000) Influenza virus assembly: effect of influenza virus glycoproteins on the membrane association of M1 protein. J. Virol. 74:8709-8719.

Ali, A. and Nayak, D.P. (2000) Assembly of Sendai virus: M protein interacts with F and HN proteins and with the cytoplasmic tail and transmembrane domain of F protein. Virology 276:289-303.

Aloia, R.C., Tian, H. and Jensen, F.C. (1993) Lipid composition and fluidity of the human immunodeficiency virus envelope and host cell plasma membranes. Proc. Natl. Acad. Sci. USA 90:5181-5185.

Alonso, M.A. and Millán, J. (2001) The role of lipid rafts in signalling and membrane trafficking in T lymphocytes. J. Cell Sci. 114:3957-3965.

Aman, M.J., Tosello-Trampont, A.-C. and Ravichandran, K. (2001) FcγRIIB1/SHIP-mediated inhibitory signaling in B cells involves lipid rafts. J. Biol. Chem. 276:46371-46378.

Balange-Orange, N. and Devauchelle, G. (1982) Lipid composition of an Iridescent virus type 6 (CIV). Arch. Virol. 73:363-367.

Barman, S. and Nayak, D.P. (2000) Analysis of the transmembrane domain of influenza virus neuraminidase, a type II transmembrane glycoprotein, for apical sorting and raft association. J. Virol. 74:6538-6545.

Barman, S., Ali, A., Hui, E.K.-W., Adhikary, L. and Nayak, D.P. (2001) Transport of viral proteins to the apical membranes and interaction of matrix protein with glycoproteins in the assembly of influenza viruses. Virus. Res. 77:61-69.

Barman, S., Adhikary, L., Kawaoka, Y. and Nayak, D.P. (2003) Influenza A virus hemagglutinin containing basolateral localization signal does not alter the apical budding of a recombinant influenza A virus in polarized MDCK cells. Virology 305:138-152.

Bastiani, L., Laal, S., Kim, M. and Zolla-Pazner, S. (1997) Host cell-dependent alterations in envelope components of human immunodeficiency virus type 1 virions. J. Virol. 71: 3444-3450.

Bavari, S., Bosio, C.M., Wiegand, E., Ruthel, G., Will, A.B., Geisbert, T.W., Hevey, M., Schmaljohn, C., Schmaljohn, A. and Aman, M.J. (2002) Lipid raft microdomains: a gateway for compartmentalized trafficking of Ebola and Marburg viruses. J. Exp. Med. 195:593-602.

Blau, D.M. and Compans, R.W. (1995) Entry and release of measles virus are polarized in epithelial cells. Virology 210:91-99.

Bourmakina, S.V. and Garcia-Sastre, A. (2003) Reverse genetics studies on the filamentous morphology of influenza A virus. J. Gen. Virol. 84:517-527.

Brewer, G.J. (1979) In vivo assembly of a biological membrane of defined size, shape, and lipid composition. J. Virol. 30:875-882.

Brewer, C.B. and Roth, M.G. (1991) A single amino acid change in the cytoplasmic domain alters the polarized delivery of influenza virus hemagglutinin. J. Cell Biol. 114:413-421.

Briggs, J.A.G., Wilk, T. and Fuller, S.D. (2003) Do lipid rafts mediate virus assembly and pseudotyping? J. Gen. Virol. 84:757-768.

Brown, D.A. and London, E. (1998a) Functions of lipid rafts in biological membranes. Annu. Rev. Cell Dev. Biol. 14, 111-136.

Brown, D.A. and London, E. (1998b) Structure and origin of ordered lipid domains in biological membranes. J. Membr. Biol. 164, 103-114.

Brown, D.A., Crise, B. and Rose, J.K. (1989) Mechanism of membrane anchoring affects polarized expression of two proteins in MDCK cells. Science 245:1499-1501.

Brown, G., Aitken, J., Rixon, H.W. and Sugrue, R.J. (2000a) Caveolin-1 is incorporated into mature respiratory syncytial virus particles during virus assembly on the surface of virus-infected cells. J. Gen. Virol. 83:611-621.

Brown, G., Rixon, H.W. and Sugrue, R.J. (2002b) Respiratory syncytial virus assembly occurs in GM1-rich regions of the host-cell membrane and alters the cellular distribution of tyrosine phosphorylated caveolin-1. J. Gen. Virol. 83:1841-1850.

Cadd, T.L., Skoging, U., and Liljestrom, P. (1997) Budding of enveloped viruses from the plasma membrane. Bioessays 19:993-1000.

Campbell, S.M., Crowe, S.M. and Mak, J. (2001) Lipid rafts and HIV-1: from viral entry to assembly of progeny virions. J. Clin. Virol. 22:217-227.

Chatterjee, P.K., Eng, C.H. and Kielian, M. (2002) Novel mutations that control the sphingolipid and cholesterol dependence of the Semliki Forest virus fusion protein. J. Virol. 76:12712-12722.

Chazal, N. and Gerlier, D. (2003) Virus entry, assembly, budding, and membrane rafts. Microbiol. Mol. Biol. Rev. 67:226-237.

Cheng, H., Hoxie, J. and Parks, W.P. (1999) The conserved core of human immunodeficiency virus type 1 Nef is essential for association with Lck and for enhanced viral replication in T-lymphocytes. Virology 264:5-15.

Cherukuri, A., Dykstra, M. and Pierce, S.K. (2001) Floating the raft hypothesis: lipid rafts play a role in immune cell activation. Immunity 14:657-660.

Cimarelli, A. and Darlix, J.-L. (2002) Assembling the human immunodeficiency virus type 1. Cell. Mol. Life Sci. 59:1166-1184.

Collette, Y., Arold, S., Picard, C., Janvier, K., Benichou, S., Benarous, R., Olive, D. and Dumas, C. (2000) HIV-2 and SIV Nef proteins target different Src family SH3 domains than does HIV-1 Nef because of a triple amino acid substitution. J. Biol. Chem. 275:4171-4176.

Compans, R.W. (1995) Virus entry and release in polarized epithelial cells. Curr. Top. Microbio. Immunol. 202:209-219.

Czarny, M., Lavie, Y., Fiucci, G. and Liscovitch, M. (1999) Localization of phospholipase D in detergent-insoluble, caveolin-rich membrane domains: modulation by caveolin-1 expression and caveolin-182-101. J. Biol. Chem. 274:2717-2724.

Damjanovich, S., Mátyus, L., Damjanovich, L., Bene, L., Jenei, A., Matkó, J., Gáspár, Jr., R. and Szöllösi, J. (2002) Does mosaicism of the plasma membrane at molecular and high hierarchical levels in human lymphocytes carry information on the immediate history of cells? Immunol. Lett. 82:93-99.

Danielsen, E.M. (1995) Involvement of detergent-insoluble complexes in the intracellular transport of intestinal brush border enzymes. Biochemistry 34:1596-1605.

Danielsen, E.M. and van Deurs, B. (1995) A transferrin-like GPI-linked iron-binding protein in detergent-insoluble noncaveolar microdomains at the apical surface of fetal intestinal epithelial cells. J. Cell Biol. 131:939-950.

David A.E. (1971) Lipid composition of Sindbis virus. Virology 46:711-720.

Demirov, D.G., Ono, A., Orenstein, J.M. and Freed, E.O. (2002) Overexpression of the N-terminal domain of TSG101 inhibits HIV-1 budding by blocking late domain function. Proc. Natl. Acad. Sci. USA 99:955-960.

Deschambeault, J., Lalonde, J.P., Cervantes-Acosta, G., Lodge, R., Cohen, E.A. and Lemay, G. (1999) Polarized human immunodeficiency virus budding in lymphocytes involves a tyrosine-based signal and favors cell-to-cell viral transmission. J. Virol. 73, 5010-5017.

Dimitrov, D.S. (2000) Cell biology of virus entry. Cell 101:697-702.

Drevot, P., Langlet, C., Guo, X.J., Bernard, A.M., Colard, O., Chauvin, J.P., Lasserre, R. and He, H.T. (2002) TCR signal initiation machinery is pre-assembled and activated in a subset of membrane rafts. EMBO J. 21:1899-1908.

Farsad, K. and De Camilli, P. (2003) Lipids in endocytic membrane transport and sorting. Curr. Opin. Cell Biol. 15:372-

Freed, E.O. (2002) Viral late domain. J. Virol. 76:4679-4687.

Galbiati, F., Razani, B. and Lisanti, M.P. (2001) Emerging themes in lipid rafts and caveolae. Cell 106:403-411.

Garbutt, M., Chan, H. and Hobman, T.C. (1999) Secretion of rubella virions and virus-like particles in cultured epithelial cells. Virology 261:340-346.

Garoff, H., Hewson, R. and Opstelten, D.-J.E. (1998) Virus maturation by budding. Microbiol. Mol. Biol. Rev. 62:1171-1190.

Garrus, J.E., von Schwedler, U.K., Pornillos, O.W., Morham, S.G., Zavitz, K.H., Wang, H.E., Wettstein, D.A., Stray, K.M., Côte, M., Rich, R.L., Mysxka, D.G. and Sundquist, W.I. (2001) Tsg101 and the vacuolar protein sorting pathway are essential for HIV-1 budding. Cell 107:55-65.

Goila-Gaur, R., Demirov, D.G., Orenstein, J.M., Ono, A. and Freed, E.O. (2003) Defects in human immunodeficiency virus budding and endosomal sorting induced by TSG101 overexpression. J. Virol. 77:6507-6519.

Graham, D.R.M., Chertova, E., Hilburn, J.M., Arthur, L.O. and Hildreth, J.E.K. (2003) Cholesterol depletion of human immunodeficiency virus type 1 and simian immunodeficiency virus with b-cyclodextrin inactivates and permeabilizes the virions: evidence for virion-associated lipid rafts. J. Virol. 77:8237-8248.

Gujuluva, C.N., Kundu, A., Murti, K.G. and Nayak, D.P. (1994) Abortive replication of influenza virus A/WSN/33 in HeLa cells: defective viral entry and budding processes. Virology 204:491-505.

Guo, B., Kato, R.M., Garcia-Lloret, M., Wahl, M.I. and Rawlings, D.J. (2000) Engagement of the human pre-B cell receptor generates a lipid raft-dependent calcium signaling complex. Immunity 13:243-253.

Harder, T. (2003) Formation of functional cell membrane domains: the interplay of lipid- and protein-mediated interactions. Philos. Trans. R. Soc. Lond. B Biol. Sci. 358:863-868.

Harder, T. and Simons, K. (1999) Clusters of glycolipid and glycosylphosphatidylinositol-anchored proteins in lymphoid cells: accumulation of actin regulated by local tyrosine phosphorylation. Eur. J. Immunol. 29:556-562.

Harder, T., Scheiffele, P., Verkade, P. and Simons, K. (1998) Lipid domain structure of the plasma membrane revealed by patching of membrane components. J. Cell Biol. 141:929-942.

Hermida-Matsumoto, L. and Resh, M.D. (2000) Localization of human immunodeficiency virus type 1 Gag and Env at the plasma membrane by confocal imaging. J. Viral. 74, 8670-8679.

Heydrick, F.P., Comer, J.F. and Wachter, R.F. (1971) Phospholipid composition of Venezuelan equine encephalitis virus. J. Virol. 7:642-645.

Hiipakka, M. and Saksela, K. (2002) Capacity of simian immunodeficiency virus strain mac Nef for high-affinity Src homology 3 (SH3) binding revealed by ligand-tailored SH3 domains. J. Gen. Virol. 83:3147-3152.

Hobman, T.C. (1993) Targeting of viral glycoproteins to the Golgi complex. Trends Microbiol. 1:124-130.

Hoessli, D.C., Ilangumaran, S, Soltermann, A., Robinson, P.J., Borisch, B. and Nasir-Ud-Din. (2001) Signaling through sphingolipid microdomains of the plasma membrane: the concept of signaling platform. Glycoconjugate J. 17:191-197.

Holm, K., Weclewicz, K., Hewson, R. and Suomalainen, M. (2003) Human immunodeficiency virus type 1 assembly and lipid rafts: $Pr55^{gag}$ associates with membrane domains that are largely resistant to Brij98 but sensitive to Triton X-100. J. Virol. 77:4805-4817.

Holopainen, J.M., Angelora, M.I. and Kinnunen, P.K. (2000) Vectorial budding of vesicles by asymmetrical enzymatic formation of ceramide in giant liposomes. Biophys. J. 78:830-838.

Holowka, D., Sheets, E.D. and Baird, B. (2000) Interactions between FcϵRI and lipid raft components are regulated by the actin cytoskeleton. J. Cell Sci. 113:1009-1019.

Hug, P., Lin, H.-M.J., Korte, T., Xiaodong, X., Dimitrov, D.S., Wang, J.M., Puri, A. and Blumenthal, R. (2000) Glycosphingolipids promote entry of a broad range of human immunodeficiency virus type 1 isolates into cell lines expressing CD4, CXCR4, and/or CCR5. J. Virol. 74:6377-6385.

Hughey, P.G., Compans, R.W., Zebedee, S.L. and Lamb, R.A. (1992) Expression of the influenza A virus M2 protein is restricted to apical surfaces of polarized epithelial cells. J. Virol. 66:5542-5552.

Hui, E.K.-W. and Nayak, D.P. (2001) Role of ATP in influenza virus budding. Virology 290:329-341.

Hui, E.K.-W and Nayak, D.P. (2002) Role of G protein and protein kinase signalling in influenza virus budding in MDCK cells. J. Gen. Virol. 83:3055-3066.

Hui, E.K.-W., Barman, S., Yang, T.Y. and Nayak, D.P. (2003) Basic residues of the helix six domain of influenza virus M1 involved in nuclear translocation of M1 can be replaced by PTAP and YPDL late assembly domain motifs. J. Virol. 77:7078-7092.

Ikonen, E. (2001) Roles of lipid rafts in membrane transport. Curr. Opin. Cell Biol. 13:470-477.

Janes, P.W., Ley, S.C. and Magee, A.I. (1999) Aggregation of lipid rafts accompanies signaling via the T cell antigen receptor. J. Cell Biol. 147:447-461.

Katagiri, Y.U., Liyokawa, N. and Fujimoto, J. (2001) A role for lipid rafts in immune cell signaling. Microbiol. Immunol. 45:1-8.

Kate, M., Allison, A.C., Tyrrell, D.A.J. and James, A.T. (1961) Lipids of influenza virus and their relation to those of the host cell. Biochim. Biophys. Acta 52:455-466.

Keller, P. and Simons, K. (1998) Cholesterol is required for surface transport of influenza virus hemagglutinin. J. Cell Biol. 140:1357-1367.

Khanna, K.V., Whaley, K.J., Zeitlin, L., Moench, T.R., Mehrazar, K., Cone, R.A., Liao, Z., Hildreth, J.E.K., Hoen, T.E., Shultz, L. and Markham, R.B. (2002) Vaginal transmission of cell-associated HIV-1 in the mouse is blocked by a topical, membrane-modifying agent. J. Clin. Invest. 109:205-211.

Kielian, M.C. and Helenius, A. (1984) Role of cholesterol in fusion of Semliki Forest virus with membranes. J. Virol. 52:281-283.

Klenk, H.-D. and Choppin, P.W. (1970) Plasma membrane lipids and parainfluenza virus assembly. Virology 40:939-947.

Kobayashi, T. and Hirabayashi, Y. (2001) Lipid membrane domains in cell surface and vacuolar systems. Glycoconjugate J. 17:163-171.

Kundu, A., Avalos, R.T., Sanderson, C.M. and Nayak, D.P. (1996) Transmembrane domain of influenza virus neuraminidase, a type II protein, possesses an apical sorting signal in polarized MDCK cells. J. Virol. 70, 6508-6515.

Kusumi, A. and Sako, Y. (1996) Cell surface organization by the membrane skeleton. Curr. Opin. Cell Biol. 8:566-574.

Lafont, F., Lecat, S., Verkade, P. and Simons, K. (1998) Annexin XIIb associates with lipid microdomains to function in apical delivery. J. Cell Biol. 142:1413-1427.

Langlet, C., Bernard, A.M., Drevot, P. and He, H.T. (2000) Membrane rafts and signaling by the multichain immune recognition receptors. Curr. Opin. Immunol. 12:250-255.

Li, M., Yang, C., Tong, S., Weidmann, A. and Compans, R.W. (2002) Palmitoylation of the murine leukemia virus envelope protein is critical for lipid raft association and surface expression. J. Virol. 76:11845-11852.

Li, S., Okamoto, T., Chun, M., Sargiacomo, M., Casanova, J.E., Hansen, S.H., Nishimoto, I. And Lisanti, M.P. (1995) Evidence for a regulated interaction between heterotrimeric G proteins and caveolin. J. Biol. Chem. 270:15693-15701.

Lin, S., Naim, H.Y., Rodriguez, A.C. and Roth, M. G. (1998) Mutations in the middle of the transmembrane domain reverse the polarity of transport of the influenza virus hemagglutinin in MDCK epithelial cells. J. Cell Biol. 142:51-57.

Lindwasser, O.W. and Resh, M.D. (2001) Multimerization of human immunodeficiency virus type 1 Gag promotes its localization to barges, raft-like membrane microdomains. J. Virol. 75:7913-7924.

Lodge, R., Gottlinger, H., Gabuzda, D., Cohen, E.A. and Lemay, G. (1994). The intracytoplasmic domain of gp41 mediates polarized budding of human immunodeficiency virus type 1 in MDCK cells. J. Virol. 68:4857-4861.

Lodge, R., Lalonde, J.P., Lemay, G. and Cohen, E. A. (1997) The membrane-proximal intracytoplasmic tyrosine residue of HIV-1 envelope glycoprotein is critical for basolateral targeting of viral budding in MDCK cells. EMBO J. 16, 695-705.

Lu, Y.E., Cassese, T. and Kielian, M. (1999) The cholesterol requirement for Sindbis virus entry and exit and characterization of a spike protein region involved in cholesterol dependence. J. Virol. 73:4272-4278.

Luan, P., Yang, L. and Glaser, M. (1995) Formation of membrane domains created during the budding of vesicular stomatitis virus: a model for selective lipid and protein sorting in biological membranes. Biochemistry 34:9874-9883.

Luban, J. (2001) HIV-1 and Ebola virus: the getaway driver nabbed. Nat. Med. 7:1278-1280.

Maisner, A., Klenk, H., and Herrler, G. (1998) Polarized budding of measles virus is not determined by viral surface glycoproteins. J. Virol. 72:5276-5278.

Malvoisin, E. and Wild, F. (1990) Effect of drugs which inhibit cholesterol synthesis on syncytia formation in vero cells infected with measles virus. Biochim. Biophys. Acta 1042:359-364.

Manié, S.N., Debreyne, S., Vincent, S. and Gerlier, D. (2000) Measles virus structural components are enriched into lipid raft microdomains: a potential cellular location for virus assembly. J. Virol. 74:305-311.

Martin-Belmonte, F., Puertollano, R., Millan, J. and Alonso, M.A. (2000) The MAL proteolipid is necessary for the overall apical delivery of membrane proteins in the polarized epithelial Madin-Darby canine kidney and fischer rat thyroid cell lines. Mol. Biol. Cell 11:2033-2045.

Miceli, M.C., Moran, M., Chung, C.D., Patel, V.P., Low, T. and Zinnznti, W. (2001) Co-stimulation and counter-stimulation: lipid raft clustering controls TCR signaling and functional outcomes. Semin. Immunol. 13:115-128.

Mora, R., Rodriguez-Boulan, E., Palese, P. and Garcia-Sastre, A. (2002) Apical budding of a recombinant influenza A virus expressing a hemagglutinin protein with a basolateral localization signal. J. Virol. 76, 3544-3553.

Müller, G. (2002) Dynamics of plasma membrane microdomains and cross-talk to the insulin signalling cascade. FEBS Lett. 531:81-87.

Nabi, I.R. and Le, P.U. (2003) Caveolae/raft-dependent endocytosis. J. Cell Biol. 161:673-677.

Naim, H. Y., Ehler, E., and Billeter, M.A. (2000) Measles virus matrix protein specifies apical virus release and glycoprotein sorting in epithelial cells. EMBO J. 19, 3576-3585.

Narayan, S., Barnard, R.J.O. and Young, J.A.T. (2003) Two retroviral entry pathways distinguished by lipid raft association of the viral receptor and differences in viral infectivity. J. Virol. 77:1977-1983.

Nayak, D.P. and Barman, S. (2002) Role of lipid rafts in virus assembly and budding. Adv. Virus Res. 58:1-28.

Nayak, D.P. and Hui, E.K.-W. (2002) Assembly and morphogenesis of influenza virus. Recent Res. Dev. Virol. 4:35-54.

Nguyen, D.H. and Hildreth, J.E.K. (2000) Evidence for budding of human immunodeficiency virus type 1 selectively from glycolipid-enriched membrane lipid rafts. J. Virol. 74:3264-3272.

Nieva, J.L., Bron, R., Corver, J. and Wilschut, J. (1994) Membrane fusion of Semliki Forest virus requires sphingolipids in the target membrane. EMBO J. 13:2797-2804.

Ono, A. and Freed, E.O. (2001) Plasma membrane rafts play a critical role in HIV-1 assembly and release. Proc. Natl. Acad. Sci. USA 98:13925-13990.

Owens, R.J., Dubay, J.W., Hunter, E. and Compans, R.W. (1991) Human immunodeficiency virus envelope protein determines the site of virus release in polarized epithelial cells. Proc. Natl. Acad. Sci. USA 88:3987-3991.

Pelkmans, L. and Helenius, A. (2003) Insider information: what viruses tell us about endocytosis. Curr. Opin. Cell Biol. 15:414-422.

Perez, O.D. and Nolan, G.P. (2001) Resistance is futile: assimilation of cellular machinery by HIV-1. Immunity 15:687-690.
Phalen, T. and Kielian, M. (1991) Cholesterol is required for infection by Semliki Forest virus. J. Cell Biol. 112:615-623.
Pickl, W.F., Pimentel-Muiños, F.X. and Seed, B. (2001) Lipid rafts and pseudotyping. J. Virol. 75:7175-7183.
Pralle, A., Keller, P., Florin, E.-L., Simons, K. and Hörber, J.K.H. (2000) Sphingolipid-cholesterol rafts diffuse as small entities in the plasma membrane of mammalian cells. J. Cell Biol. 148:997-1008.
Puertollano, R., Martin-Belmonte, F., Millan, J., de Marco, M.C., Albar, J.P., Kremer, L. and Alonso, M.A. (1999) The MAL proteolipid is necessary for normal apical transport and accurate sorting of the influenza virus hemagglutinin in Madin-Darby canine kidney cells. J. Cell Biol. 145:141-151.
Quigley, J.P., Rifkin, D.B. and Reich, E. (1971) Phospholipid composition of Rous sarcoma virus, host cell membranes and other enveloped RNA viruses. Virology 46:106-116.
Renkonen, O., Kääräinen, L., Simons, K. and Gahmberg, C.G. (1971) The lipid class composition of Semliki Forest virus and of plasma membranes of the host cells. Virology 46:318-326.
Roberts, P.C. and Compans, R.W. (1998) Host cell dependence of viral morphology. Proc. Natl. Acad. Sci. USA 95:5746-5751.
Roberts, P.C., Lamb, R.A. and Compans, R.W. (1998) The M1 and M2 proteins of influenza A virus are important determinants in filamentous particle formation. Virology 240:127-137.
Röper, K., Corbeil, D. and Huttner, W.B. (2000) Retention of prominin in microvilli reveals distinct cholesterol-based lipid micro-domains in the apical plasma membrane. Nat. Cell Biol. 2:582-592.
Rousso, I., Mixon, M.B., Chen, B.K. and Kim, P.S. (2000) Palmitoylation of the HIV-1 envelope glycoprotein is critical for viral infectivity. Proc. Natl. Acad. Sci. USA 97:13523-13525.
Salzwedel, K., West Jr., J.T., Mulligan, M.J. and Hunter, E. (1998) Retention of the human immunodeficiency virus type 1 envelope glycoprotein in the endoplasmic reticulum does not redirect virus assembly from the plasma membrane. J. Virol. 72:7523-7531.
Samsonov, A.V., Chatterjee, P.K., Razinkov, V.I., Eng, C.H., Kielian, M. and Cohen, F.S. (2002) Effects of membrane potential and sphingolipid structures on fusion of Semliki Forest virus. J. Virol. 76:12691-12702.
Sanchez, V., Greis, K.D., Sztul, E. and Britt, W.J. (2000) Accumulation of virion tegument and envelope proteins in a stable cytoplasmic compartment during human cytomegalovirus replication: characterization of a potential site of virus assembly. J. Virol. 74, 975-986.
Sandefur, S., Smith, R.M., Varthakavi, V. and Spearman, P. (2000) Mapping and characterization of the N-terminal I domain of human immunodeficiency virus type 1 $Pr55^{Gag}$. J. Virol. 74:7238-7249.
Sanger, C., Muhlberger, E., Ryabchikova, E., Kolesnikova, L., Klenk, H.D. and Becker, S. (2001) Sorting of marburg virus surface protein and virus release take place at opposite surfaces of infected polarized epithelial cells. J. Virol. 75:1274-1283.
Scheiffele, P., Roth, M.G. and Simon, K. (1997) Interaction of influenza virus haemagglutinin with sphingolipid-cholesterol membrane domains via its transmembrane domain. EMBO J. 16:5501-5508.
Scheiffele, P., Verkade, P., Fra, A.M., Virta, H., Simons, K. and Ikonen, E. (1998) Caveolin-1 and -2 in the exocytic pathway of MDCK cells. J. Cell Biol. 140:795-806.
Scheiffele, P., Rietveld, A., Wilk, T. and Simons, K. (1999) Influenza viruses select ordered lipid domains during budding from the plasma membrane. J. Biol. Chem. 274:2038-2044.

Schnitzer, J.E., Oh, P., Pinney, E. and Allard, J. (1994) Filipin-sensitive caveolae-mediated transport in endothelium: reduced transcytosis, scavenger endocytosis, and capillary permeability of select macromolecules. J. Cell Biol. 127:1217-1232.

Schütz, G.J., Kada, G., Pastushenko, V.P. and Schindler, H. (2000) Properties of lipid microdomains in a muscle cell membrane visualized by single molecule microscopy. EMBO J. 19:892-901.

Sharma, P., Sabharanjak, S. and Mayor, S. (2002) Endocytosis of lipid rafts: an identity crisis. Semin. Cell Dev. Biol. 13:205-214.

Sheets, E.D., Lee, G.M., Simson, R. and Jacobson, K. (1997) Transient confinement of a glycosylphosphatidylinositol-anchored protein in the plasma membrane. Biochemistry 36:12449-12458.

Shi, S.T., Lee, K.J., Aizaki, H., Hwang, S.B. and Lai, M.M. (2003) Hepatitis C virus RNA replication occurs on a detergent-resistant membrane that cofractionate with caveolin-2. J. Virol. 77:4160-4168.

Simons, K. and Ikonen, E. (1997) Functional rafts in cell membranes. Nature 387:569-572.

Simons, K. and Toomre, D. (2000) Lipid rafts and signal transduction. Nat. Rev. Mol. Cell Biol. 1:31-39

Simpson-Holley, M., Ellis, D., Fisher, D., Elton, D., McCauley, J. and Digard, P. (2002) A functional link between the actin cytoskeleton and lipid rafts during budding of filamentous influenza virions. Virology 301:212-225.

Skibbens, J.E., Roth, M.G. and Matlin, K.S. (1989) Differential extractability of influenza virus hemagglutinin during intracellular transport in polarized epithelial cells and nonpolar fibroblasts. J. Cell Biol. 108:821-832.

Slimane, T.A. and Hoekstra, D. (2002) Sphingolipid trafficking and protein sorting in epithelial cells. FEBS Lett. 529:54-59.

Song, K.S., Shengwen, L., Okamoto, T., Quilliam, L.A., Sargiacomo, M. and Lisanti, M.P. (1996) Co-purification and direct interaction of Ras with caveolin, an integral membrane protein of caveolae microdomains: detergent-free purification of caveolae microdomains. J. Biol. Chem. 271:9690-9697.

Stang, E., Kartenbeck, J. and Parton, R.G. (1997) Major histocompatibility complex class I molecules mediate association of SV40 with caveolae. Mol. Biol. Cell 8:47-57.

Suomalainen, M. (2002) Lipid rafts and assembly of enveloped viruses. Traffic 3:705-709.

Tatulian, S.A. and Tamm, L.K. (2000) Secondary structure, orientation, oligomerization, and lipid interactions of the transmembrane domain of influenza hemagglutinin. Biochemistry 39:496-507.

van der Goot, F.G. and Harder, T. (2001) Raft membrane domains: from a liquid-ordered membrane phase to a site of pathogen attack. Seminars Immunol. 13:89-97.

van Genderen, I.L., Godeke, G.-J., Rottier, P.J.M. and van Meer, G. (1995) The phospholipid composition of enveloped viruses depends on the intracellular membrane through which they bud. Biochem. Soc. Trans. 23:523-526.

van Meer, G. and Simons, K. (1986) The function of tight junctions in maintaining differences in lipid composition between the apical and the basolateral cell surface domains of MDCK cells. EMBO J. 5:1455-1464.

Veit, M., Herrler, G., Schmidt, M.F., Rott, R. and Klenk, H.D. (1990) The hemagglutinating glycoproteins of influenza B and C viruses are acylated with different fatty acids. Virology 177:807-811.

Veit, M., Klenk, H.D., Kendal, A. and Rott, R. (1991) The M2 protein of influenza A virus is acylated. J. Gen. Virol. 72:1461-1465.

Vincent, S., Gerlier, D. and Manié, S.N. (2000) Measles virus assembly within membrane rafts. J. Virol. 74:9911-9915.

Vogt, A.B., Spindeldreher, S. and Kropshofer, H. (2002) Clustering of MHC-peptide complexes prior to their engagement in the immunological synapse: lipid raft and tetraspan microdomains. Immunol. Rev. 189:136-151.
Waarts, B.-L., Bittman, R. and Wilschut, J. (2002) Sphingolipid and cholesterol dependence of alphavirus membrane fusion. J. Biol. Chem. 277:38141-38147.
Wang, J.K., Kiyokawa, E., Verdin, E. and Trono, D. (2000) The Nef protein of HIV-1 associates with rafts and primes T cells for activation. Proc. Natl. Acad. Sci. USA 97:394-399.
Zhang, J., Pekosz, A. and Lamb, R.A. (2000) Influenza virus assembly and lipid raft microdomains: a role for the cytoplasmic tails of the spike glycoproteins. J. Virol. 74:4634-4644.
Zheng, X., Lu, D. and Sadler, J.E. (1999) Apical sorting of bovine enteropeptidase does not involve detergent-resistant association with sphingolipid-cholesterol rafts. J. Biol. Chem. 274:1596-1605.
Zheng, Y.-H., Plemenitas, A., Linnemann, T., Fackler, O.T. and Peterlin, B.M. (2001) Nef increases infectivity of HIV via lipid rafts. Curr. Biol. 11:875-879.
Zimmer, G., Zimmer K.-P., Trotz, I. and Herrler, G. (2002) Vesicular somatitis virus glycoprotein does not determine the site of virus release in polarized epithelial cells. J. Virol. 76:4103- 4107.
Zurzolo, C., Polistina, C., Saini, M., Gentile, R., Aloj, L., Migliaccio, G., Bonatti, S. and Nitsch, L. (1992) Opposite polarity of virus budding and of viral envelope glycoprotein distribution in epithelial cells derived from different tissues. J. Cell Biol. 117:551-564.
Zurzolo, C., van't Hof, W., van Meer, G. and Rodriguez-Boulan, E. (1994) VIP21/caveolin, glycosphingolipid clusters and the sorting of glycosylphosphatidylinositol-anchored proteins in epithelial cells. EMBO J. 13:42-53.

Index